PRINCIPLES OF FEEDBACK CONTROL

VOLUME 2: ADVANCED CONTROL TOPICS

PRINCIPLES OF FEEDBACK CONTROL

VOLUME 2: ADVANCED CONTROL TOPICS

George Biernson
Communication Systems Division
GTE Government Systems

WILEY

A Wiley-Interscience Publication

JOHN WILEY & SONS

New York · **Chichester** · **Brisbane** · **Toronto** · **Singapore**

Library of Congress Cataloging in Publication Data:

Biernson, George.
    Principles of feedback control/George Biernson.
        p.   cm.
    "A Wiley-Interscience publication."
    Bibliography: p.
    Includes indexes.
        Contents: v. 1. Feedback system design—v. 2. Advanced
control topics.
        ISBN 0-471-82167-5 (v. 1)
        ISBN 0-471-50120-4 (v. 2)
        1. Feedback control systems. I. Title.

TJ216.B45   1988
629.8'312—dc 19   87-30539

Printed in the United States of America

10 9 8 7 6 5 4 3 2 1

To my loving wife, Trudy, and to my children,
Jane, Cindy, Tom and Nancy.

Preface

EMPHASIS OF BOOK

This is an advanced textbook on feedback-control principles contained in two volumes. It is assumed that the reader has a basic background in Laplace-transform and frequency-response methods of feedback-control analysis. The volumes build on these concepts to provide a broad foundation of theoretical and practical tools for feedback-control design. Included are studies of sampled-data (digital) systems, digital-computer simulation, AC control systems, mechanical structural dynamics, statistical analysis of signals, radar tracking systems, and nonlinear control.

Frequency-response analysis is extended to develop approximation methods, which allow complex multiloop control systems to be analyzed in a simple manner. Concepts of frequency response, transient response, poles and zeros, and the general response to an arbitrary input are tied together coherently.

The approximation analysis provides a basic frame of reference for explaining control-system dynamics. More accurate information can be obtained by supplementing this with digital-computer simulation. The book presents the author's "serial" method of digital simulation, which is implemented in the same convenient step-by-step manner as analog-computer simulation. This serial-simulation technique can be easily applied using BASIC computer language on a personal computer, to model very complicated dynamic systems.

ORGANIZATION OF THE TWO VOLUMES

Principles of Feedback Control is separated into two closely related but independent volumes, which are entitled:

Volume 1: Feedback System Design
Volume 2: Advanced Control Topics

Volume 1 presents theoretical and practical principles for feedback control system design, which are applied to realistic multi-loop control systems. Volume 2 augments this with advanced control topics, which include:

Digital computer simulation
Sampled-data (digital) control systems
AC control systems
Mechanical structural dynamics
Statistical analysis of random signals
Radar tracking
Nonlinear control
Advanced feedback analysis techniques
Design of a servo with ultra-high accuracy

The combination of these two volumes gives a broad engineering foundation for developing feedback control systems.

The Introduction in Chapter 1 contains an overview of the feedback system concepts and terminology developed in Volume 1. Chapter 1 is also included in Volume 2 to provide the reader with sufficient background to understand nearly all of the material without requiring that he read Volume 1. Thus, the two volumes are independent. Nevertheless, they are closely interrelated by multiple cross-references, and strongly supplement one another.

Volume 1 contains Chapters 1 to 7. Volume 2 repeats Chapter 1 and contains Chapters 8 to 15. Problems relating to Chapters 2 to 7 are in Volume 1, and those for later chapters are in Volume 2.

In order to help tie the two volumes together, each volume contains the Table of Contents for both volumes, the References for both volumes, and a common Index, which refers to material in both volumes.

AUTHOR'S FEEDBACK-CONTROL EXPERIENCE

This book presents feedback-control theory that is directed to the solution of practical engineering problems. Therefore, it seems desirable to summarize my own experience in this field.

I worked at the Servomechanisms Laboratory of the Massachusetts Institute of Technology from 1950 to 1956. This included the design of hydraulic control valves, AC positional servos, a pressure control system for an iron lung, and several years of analysis and measurement of a complex multiloop fire-control system for tail defense of an aircraft. In the tests on this fire-control system, I came to appreciate the complexity of mechanical structural resonance, and its fundamental importance in the design of control systems.

Inspired by these practical feedback-control problems, I investigated many theoretical issues of feedback control dynamics, which resulted in several papers in AIEE, IRE, and IEEE journals. The major ones are given in Refs [1.11, 6.1, 8.4, 11.6, 12.1, 13.4, 14.1]. There was also an appreciable amount of work that subsequently became buried in the classified literature, particularly the analysis of multiloop control systems.

In 1956 I joined the Applied Research Laboratory of Sylvania Electric Products, which later was absorbed into General Telephone and Electronics (GTE). First, I managed a feasibility program for a pulse-doppler radar missile seeker. Then, I led the control systems design for two high-performance antenna systems: the hydraulically actuated MPQ-32 radar antenna for tracking artillery shells, and the 60-ft Advent antenna for satellite communications. The Advent antenna is described in Chapter 10, Section 10.5.2.2. The MPQ-32 radar failed to achieve adequate clutter rejection, and so did not reach production. However, its servos satisfied their ultrahigh accuracy requirements.

In the 1960s, I strayed from the servo field into the exciting area of bionics research, which evolved into techniques for enhancement of photographic and television imagery. A consequence of this work was a 1966 *IEEE Proceedings* paper: "A Feedback Control Model of Human Vision" [1.15]. What does the eye have to do with feedback control? one may ask. The human retina operates over a range of light intensities of 3 billion to one, and under good lighting conditions can clearly distinguish among at least 10 million different color samples. It is obvious that very effective adaptation (or feedback control) processes must be implemented within the retina to achieve this amazing performance.

After the Applied Research Laboratory was dissolved, I left GTE and joined ITEK in 1971, where I returned to the servo field. At ITEK I became very familiar with techniques for high-accuracy control of optical systems.

In 1974–76 I worked at GCA, where I designed the control system for a new step-and-repeat camera for fabricating integrated circuits. This servo, which is described in Chapter 15, positions an optical stage to a precision of $\frac{1}{15}$ wavelength of light over a 6-inch travel. It was the foundation for wafer-stepping photolithography, which revolutionized integrated-circuit fabrication. As explained in Chapter 15, this process allowed the density of integrated circuitry to be increased by 25 : 1 in the late 1970s.

In 1978 I rejoined GTE. There I redesigned the servo drives for the 150-ft Altair radar antenna (located in the Marshall Islands), which tracks ballistic-missile warheads. I also worked on a software upgrade of the Altair tracking system, which included a study of its Kalman filter tracking algorithms. This is the basis for my book *Optimal Radar Tracking Systems*, given in Ref [1.14], which presents a practical explanation of Kalman optimal-filter theory. Another area of investigation was the effect of wind loads on large satellite communications antennas, which is described in Chapter 11, Section 11.4.

In addition to these major control systems, I have designed many other feedback control circuits and systems, including phase-lock loops, wide-band video feedback amplifiers, automatic gain-control circuits, tape transport drives, and linear sweep circuits for cathode-ray tubes.

About 1960, I started to write a book on feedback control, to incorporate the concepts of my technical papers. However, I found it difficult to present them in a clear pedagogical manner. In 1973, as a result of prodding by an engineering associate, Arthur J. Bellemore, I began to teach evening classes at the University of Lowell, Massachusetts. This has involved basic and advanced courses in feedback control, and a course on operational amplifiers. This teaching has provided the discipline to organize and develop the concepts coherently, which has resulted in this textbook.

GENERALITY OF FEEDBACK CONTROL MATERIAL

The discussion of my experience in feedback control does not imply that this is a book on practical control-system design techniques. Every control system has its own unique engineering problems, and it is beyond the scope of this work to document these engineering details. Rather, this is a presentation of general feedback-control theory that has broad applicability in practical system design.

An effective means of teaching engineering principles is to apply them to specific cases, and this book emphasizes that approach. Chapter 7 presents a design study of a DC-motor servo for positioning an optical instrument stage. The signal-flow diagram of this servo is used as a model for computer simulation in Chapter 9. In Chapter 10, the simulation is extended to include the effects of mechanical structural resonance. Although this deals with a particular system, the principles involved are readily extrapolated to a wide variety of servo applications. For example, the analysis of this servo, with its 100-watt servo motor, is very similar to that for the 150-foot Altair antenna, which uses 100-kilowatt servo motors.

Although the primary examples are servomechanisms (i.e., systems that control mechanical elements), the concepts presented are general, and apply to all systems where the feedback control principle is implemented. This includes such areas as process control, opamp circuits, digital phase-lock loops, and optimal digital-controlled tracking systems.

Great care has been taken to present control principles in as simple a manner as possible. The reason for this is to provide tools that can be applied to complicated problems. A control system can be very confusing for a number of reasons, which include: (1) the dynamic complexity of the controlled process and the multiple feedback loops of the control system, (2) nonlinearity in the process and controller, and (3) uncertainty of system characteristics, particularly those of the process and the disturbances it experiences. Unless an analytical concept is clearly understood, it is very difficult to apply it effectively in this environment.

ACKNOWLEDGMENT

The author is grateful for the support provided by GTE Government Systems in preparing this book.

GCA Corporation and Hewlett-Packard were very helpful in approving the publication of the material in Chapter 15, which describes the control system design of an ultra-accurate step-and-repeat camera for fabricating integrated circuits, which was designed by the author. Analog Devices graciously provided the illustrations for describing noise signals given in Figs 11.1-7 and 11.1-8.

I want to thank Denny D. Pidhayny for the material he provided relative to antenna structural dynamics, given in Chapter 10, and for his very helpful review of the manuscript.

GEORGE BIERNSON

Concord, Massachusetts

Contents

14   RESPONSE TO AN ARBITRARY INPUT   509

Contents

Chapter 1

Introduction

This is an advanced textbook on feedback-control principles, giving a broad background in theoretical and practical tools for developing feedback-control systems. It is assumed that the reader has had a basic course in feedback control, which includes the following:

1. Use the Laplace transform in feedback-control analysis.
2. Preparation of magnitude and phase plots (Bode plots) of the frequency response of the open-loop transfer function of a feedback loop.
3. Use of the polar locus plot of the open-loop transfer function, and the Nichols chart, to obtain magnitude and phase plots of the closed-loop transfer function.

Volume 1 builds on these principles to develop general practical methods for designing complex multiloop control systems. Volume 2 extends this with studies of AC control systems, sampled-data (digital) systems, digital-computer simulation, mechanical structural dynamics, statistical analysis of signals, radar tracking, and nonlinear control.

1.1 PHILOSOPHY OF THE BOOK

1.1.1 Historical Development of Feedback-Control Theory

During World War II, the requirements for high-accuracy feedback-control systems in fire control and other military systems prompted research to obtain more effective methods of control-system analysis. A conceptual breakthrough occurred at that time when it was discovered that the frequency-response techniques that had been developed for designing feedback amplifiers, primarily at Bell Telephone Laboratories, could also be applied to the nuts-and-bolts problems of a servomechanism (i.e., a feedback-control system that controls a mechanical element). The complex dynamic effects in a servomechanism become greatly simplified when they were analyzed in terms of

1

frequency-response characteristics, and the insight arising from this simplification led to considerable improvements in servomechanism design.

Most of this work was classified during the war, except for the monograph by A. C. Hall [1.1] published in 1943, based on his Ph.D. thesis research. This presented the basic principle of feedback-control system design using the polar locus. Immediately after the war, the new feedback-control theory was set forth in a number of textbooks. Volume 25 of the Radiation Laboratory Series, by James, Nichols, and Phillips [1.2], which was published in 1947, provided a detailed engineering treatment of servomechanism design. A much more effective teaching text was available in 1948 with the book by Brown and Campbell [1.3]. In 1950, Chestnut and Mayer [1.4] presented a much longer and more thorough textbook for teaching feedback-control principles.

The early textbooks on feedback control reached a summit with the 1956 publication in French of an extensive treatise by Gille, Pelegrin, and Decaulne. An English version of this was published in 1959 [1.5], and it was also translated into German, Russian, and Polish. This book provides a very thorough treatment of feedback control that is both theoretical and practical. Another textbook of special merit (which will be discussed subsequently) was presented in 1958 by Bower and Schultheiss [1.6].

Since that time, a large array of textbooks in control theory have been published. The material has been expanded greatly in mathematical complexity, to the point where the frequency-response techniques that evolved in the 1940s are often regarded as "old hat" in comparison with the new "modern control theory".

One might assume, therefore, that by now feedback-control theory should have matured to the point where at least the basic concepts would be well understood. A teacher should have at his disposal textbooks that can provide the student with a solid theoretical foundation for understanding and developing feedback-control systems. However, this is not so. In many respects, the textbooks of today give the student an inferior base for designing feedback control systems than did the early works written just after World War II. The new material of practical value that is included is usually offset by superficiality in presenting earlier principles, and is often confused by mathematical techniques that have practical usefulness only in specialized areas.

This does not mean that the early textbooks on feedback control had thoroughly covered the theoretical issues. On the contrary, as shown in Section 1.1.4, they contain fundamental conceptual gaps. However, as the control field has evolved, it has often become preoccupied with theory for the sake of theory, with the result that basic questions have been sidetracked, and understanding has been sacrificed in favor of mathematical sophistication.

1.1.2   Computer Simulation in Feedback-Control Design

One of the reasons that early feedback-control analysis techniques have been downplayed in recent years is that they are often considered to be obsolete in today's age of the computer. It is certainly true that many calculations performed in those early days would not have been done if computer simulation had been easier. Nevertheless, the basic frequency-response analysis methods developed at that time are still essential today to provide an understanding of feedback-control principles.

This book uses frequency response as the foundation for explaining feedback-control dynamics. By means of approximations, this is extended to yield simple methods for designing complicated multiloop control systems. The effect of inaccuracy due to the approximations can be determined by supplementing the analysis with computer simulation.

Almost every engineer today has at his disposal at least a personal computer, which is capable of powerful control-system simulation. A modern textbook on feedback control should train the engineer to take advantage of today's computation capability, and this book emphasizes that philosophy.

Chapter 9 presents a digital programming procedure, developed by the author, called "serial simulation", which allows one to simulate any control system, regardless of its complexity, by using no more than a personal computer programmed in BASIC computer language. The serial simulation procedure allows one to use digital simulation in the same convenient step-by-step manner as analog computer simulation. Consequently, it is a very user-friendly computer technique for the design engineer.

As explained in Chapter 9, a dynamic transfer function is implemented in a digital computer by means of a difference equation, which is a weighted sum of present and past values of the input and output signals for a computation step. The Laplace transform of a difference equation can be taken by expressing each time delay of one computation cycle as the transfer function $\exp[-sT]$, where T is the sample period. The equivalent signal-frequency transfer function that is implemented by a difference equation can be obtained by replacing each $\exp[-sT]$ factor with the expression

$$\exp[-sT] = \frac{1 - pT/2}{1 + pT/2} \tag{1.1-1}$$

The variable p is a pseudo complex frequency, which is approximately equal to the true complex frequency s for frequencies up to $\frac{1}{4}$ of the sampling frequency. (Note that a sampled signal can only convey information at frequencies less than $\frac{1}{2}$ of the sampling frequency.) The relation between the pseudo complex frequency p and the true complex frequency s is explained in Chapter 9, to provide a rigorous theory for designing a sampled-data difference equation to satisfy a desired signal transfer function.

The serial simulation procedure is based on the following principles:

1. The relation of Eq 1.1-1 is applied to obtain the computer algorithm that corresponds to each transfer function being simulated.

2. A feedback loop in a computation process must include a time delay of one computation cycle, because the feedback data are derived from data computed in the previous cycle. The phase lag caused by this time delay is compensated for by inserting the lead factor $(1 + pT)$ into the computation loop.

The theory presented in Chapter 9 yields simple practical constraints for establishing the proper computations and sampling frequency, to assure that the serial simulation provides a stable and accurate solution. This serial simulation technique is applied in Chapter 10 to simulate a control system with complicated structural dynamics, and in Chapter 12 to simulate a highly nonlinear system for controlling pressure in a vacuum chamber.

An important advantage of serial simulation is that it allows the engineer to determine the effect of inserting a dynamic element within his control system, simply by adding corresponding steps to his simulation program. With the conventional state-variable method of digital computer simulation, which uses Runge-Kutta integration, a new state-equation matrix must be constructed for any such change.

The author has found serial simulation to be a very effective pedagogical tool in teaching an advanced course on feedback control. The students learn the approach very quickly, and apply it on whatever computer is most convenient, using a variety of computer languages, including BASIC, FORTRAN, and PASCAL. The technique ties their simulation closely to the dynamics of the system, and so gives them a good physical feel for the control problem. Once a model has been formulated, it is very easy to extend it to include nonlinear constraints and additional dynamic elements.

One may ask: with an effective means of computer simulation, why does one need analysis? Why not rely completely on computer simulation for designing a control system? The answer is that simulation without analysis is an ineffective approach to system design.

Brute-force simulation of a control system that is at all complex can become hopelessly bogged down in a mass of parameters. When an engineer changes a resistor, or a potentiometer setting, or a capacitor, it is very cumbersome to factor that change into the computer simulation program, to determine its effect. Approximation theory, which allows one to calculate directly the approximate effect of such a change, enormously simplifies the design task. Without it, the engineer is often overwhelmed by the details of his job, and is reduced to a trial-and-error technician.

A serious problem often encountered in simulation studies is that the simulated model may bear little relation to the actual system. Dynamic characteristics of a system, particularly structural dynamics, are often simulated poorly. A theoretical understanding is essential to guide measurements on the system, to assure that the system and the model are consistent. A key issue in such tests is measurement of the "control bandwidths" of the various feedback loops, which is discussed in Sections 1.1.4 and 1.2.3 of this Introduction, and in Chapter 5, Section 5.6.

Approximations provide basic understanding, to greatly simplify the problem; while simulation allows one to investigate the more subtle aspects of the system, and to obtain more accurate results than can be achieved by approximations. The combination of approximation analysis and simulation is very much more effective than simulation alone.

In short, simulation is not a substitute for understanding. The engineer needs both analysis and simulation, and the purpose of this book is to give him that background.

1.1.3 Optimal Control

Control-system textbooks in recent years have often placed strong emphasis on mathematical concepts of optimal control. However, much of this material is so abstract it is difficult to apply in practical situations.

This book provides a background in principles that are essential for an effective engineering approach to optimal control, which include:

1. Methods for analysis of complex multiloop control systems.
2. Sampled-data theory, including methods for digital-computer simulation of dynamic systems.
3. An explanation of the state variable, the state equation, and the transition matrix.
4. The statistical analysis of signals, and its application to radar tracking.
5. Nonlinear methods for optimizing saturated transients of a control system.

A direct study of optimal control is beyond the scope of this book. However, the author has addressed this issue in the book *Optimal Radar Tracking Systems* [1.14]. That reference presents a study of the Altair radar, which applies Kalman filtering to achieve optimal tracking of ballistic missile warheads. This sophisticated computer-controlled radar system has a huge mechanical steerable antenna, 150 ft in diameter, and is located on Kwajalein Atoll in the Marshall Islands. The book explains Kalman filter theory, and from its basic equations derives the tracking-filter algorithms of the Altair system.

This examination of the Altair radar system provides a means of explaining Kalman optimal-filter theory in a simple practical manner. It allows one to separate those aspects of system performance that are due to the advantages of the Kalman theory from those that are the result of conventional tracking concepts.

1.1.4   The Meaning of "Control-System Bandwidth"

As was stated previously, early textbooks on feedback control contain fundamental conceptual gaps, which have been sidetracked as the control textbooks
have moved in the direction of mathematical sophistication. To illustrate this
point, let us consider a misconception that has persisted in this field ever since
the control engineers during World War II borrowed their frequency-response
concepts from the communication field. Let us address the question: what do
we mean by the bandwidth of a feedback-control system?

The closed-loop frequency response of a feedback-control system usually
has a magnitude and phase curve of the general form shown in Fig 1.1-1.
Alternatively, these curves could represent the normalized frequency response
of an amplifier used in a communication system. The communication engineer
generally takes as his bandwidth criterion the half-power frequency, which is
indicated in the figure as ω_{hp}. This is the frequency where the magnitude has
decreased to $1/\sqrt{2}$ or 0.707 times the value at zero frequency.

This bandwidth criterion is very meaningful in the communication field,
even when one is considering transient response, because the half-power
frequency ω_{hp} directly constrains the rise time of the step response. As
illustrated in Fig 1.1-2, the rise time T_R is defined as the time for the step
response to rise from 10% to 90% of the final value. For a low-pass transfer
function having good stability, the rise time T_R of the step response is related
approximately as follows to the half-power frequency ω_{hp} (expressed in
radians per second):

$$\text{Rise time: } T_R \doteq 2/\omega_{hp} \qquad (1.1\text{-}2)$$

The greater the half-power frequency ω_{hp}, the steeper is the rise in the step
response, and so (if the response has good stability) the more accurately the
step response conforms to the shape of the waveform of the input step. This
relationship applies also to an input signal of complex arbitrary shape. The
greater the half-power "bandwidth" of the filter, the more closely the output
waveform duplicates the shape of the input waveform.

On the other hand, the fact that the input and output waveforms have
nearly the same shape does not necessarily mean that they are nearly equal to
one another, because there can be an appreciable time delay between the two
waveforms. The time delay between a step input and the response is characterized by the *delay time* T_d. As shown in Fig 1.1-2, the delay time T_d is
defined as the time, after the step input, for the step response to reach 50% of
the final value.

In terms of frequency response, time delay between input and output
sinusoids represents phase lag. Therefore, the delay time in the step response is
directly related to the phase curve of the frequency response. The delay time
T_d is approximately equal to the reciprocal of the frequency, designated ω_ϕ,

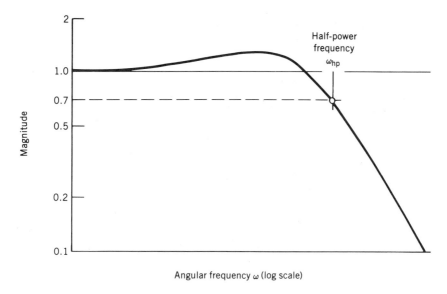

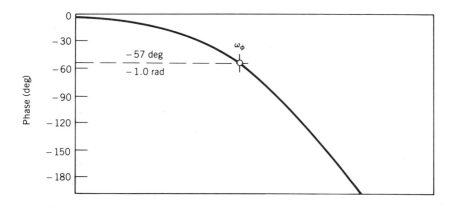

Figure 1.1-1 Typical closed-loop frequency response of feedback control system.

where the phase lag is -1 radian (or $-57°$):

$$\text{Delay Time: } T_d \doteq 1/\omega_\phi \qquad (1.1\text{-}3)$$

This frequency parameter ω_ϕ is shown in Fig 1.1-1.

Time delay between the input and output is generally of little concern to the communication engineer, as long as the output waveform has essentially

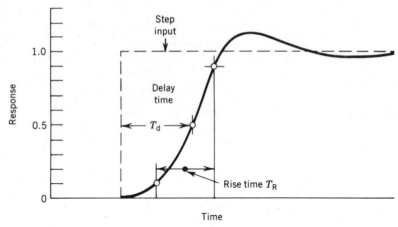

Figure 1.1-2 Typical transient response of feedback-control system, which defines rise time T_R and delay time T_d.

the same shape as the input waveform. However, to the control engineer, time delay is crucially important. He is generally concerned with minimizing the error between the input and output signals, and time delay between these signals represents dynamic error. Waveform fidelity is usually not important in a control application. In fact, control loops often have low-frequency compensation networks (integral networks) that purposely distort the waveform of the output signal in order to minimize the low-frequency components of the error.

Thus, the goals of the communication and control engineers are quite different: the communication engineer is generally concerned with duplicating waveform shape, while the control engineer is generally concerned with minimizing dynamic error. Consequently, the communication engineer should measure bandwidth from the magnitude of the frequency response, because this curve characterizes the ability of the amplifier to duplicate waveform shape. In contrast, the control engineer should measure bandwidth from the phase of the frequency response, because this curve characterizes the time delay between the input and the output, which is the cause of dynamic error. Thus ω_ϕ is a good measure of control-system bandwidth; but ω_{hp} is not, even though it is a good measure of communication-system bandwidth.

It should be noted in this regard that bandwidth measurements derived from magnitude and phase plots cannot be determined from one another, as is often assumed. For example, Fig. 1.1-3 shows two step responses that have the same delay time, but one has five times the rise time as the other. The corresponding frequency responses would have essentially the same values for ω_ϕ, but the values for the half power frequency ω_{hp} would differ by about a factor of 5.

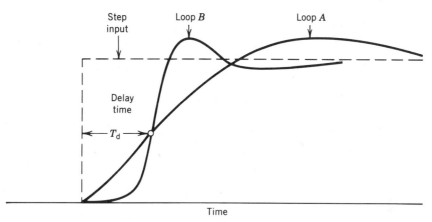

Figure 1.1-3 Typical step responses of two feedback control loops having same delay time, but 5 : 1 difference in rise time.

The issues discussed above were not understood by the control engineers when they borrowed their frequency-response techniques from the communication field. Therefore, it was natural that they, like the communication engineers, should also measure bandwidth from the magnitude of the closed-loop frequency response, rather than from the phase curve. Despite the extensive control-system research that has occurred over the subsequent half century, this very important basic concept is still largely misunderstood. Textbooks on feedback control almost universally follow the erroneous approach of specifying control-system "bandwidth" in terms of the magnitude of the closed-loop frequency response. Also, when a frequency response is taken, it is common practice to omit the phase measurement, because it is much easier to measure only the magnitude. As a result, control-system engineers have often found frequency-response specifications to be ambiguous, and so have turned to much more complicated, and much less effective, approaches for systems design. These include the root-locus technique, and various automatic optimization procedures.

The 1958 control system textbook by Bower and Schultheiss [1.6] is a notable exception to this confusion concerning the meaning of control system bandwidth. Their book develops general principles relating the step-response rise time to the magnitude of the frequency response, and the delay time to the phase of the frequency response. The analysis of "bandwidth" in Chapter 5, Section 5.5 of this book is an extension of their approach. A summary of bandwidth parameters is given in Section 1.2.3.

The general failure of control systems textbooks to explain what is meant by control-system "bandwidth" is only one of a number of areas where the basic concepts of feedback control are poorly understood. The purpose of this book is to set forth these principles, and thereby to provide the student with a far better background for control-system design.

1.1.5   Design in Terms of Closed-Loop Poles

In recent years, feedback control textbooks have generally used the root-locus method by Evans [1.13] as a primary engineering design tool. However, this book presents the root-locus method (in Chapter 13) as a supplementary analysis tool, along with other techniques for calculating closed-loop poles. None of these methods, including those developed by the author, are considered by the author to be effective tools for direct design.

The reason that the root-locus method has become so popular is that frequency response has not been properly developed and explained in earlier textbooks. Since an engineer could not interpret frequency-response information reliably, he looked to the root locus to provide more meaningful answers. As will be shown in this book, when frequency-response information is properly used, it yields much more meaningful information concerning transient response than do the values of the dominant closed-loop poles.

A serious limitation of the root-locus approach is that it assumes that the values of the dominant closed-loop poles characterize the transient response of the system with reasonable accuracy. However, this assumption is unjustified, as will be demonstrated in Section 2.4.2. For example, in practical feedback control systems having 25% peak overshoot of the step response, the damping ratio of the dominant closed-loop poles can vary all the way from 0.4 to unity. On the other hand, in such systems, the maximum magnitude of the closed-loop frequency response can vary only from 1.25 to 1.35.

1.2   SUMMARY OF FEEDBACK-CONTROL DESIGN APPROACHES

1.2.1   The Feedback-Control Principle

The principle on which feedback control is based is illustrated in the signal-flow diagram of Fig 1.2-1a. The controlled variable X_c is measured to form the feedback signal X_b. The feedback signal X_b is subtracted from the input signal X_i to produce the error signal X_e. The error signal X_e is amplified and used to change the controlled variable X_c in the direction that reduces the error. When the error is close to zero, the feedback signal X_b is approximately equal to the input X_i:

$$X_b \doteq X_i \qquad (1.2\text{-}1)$$

The sensitivity of the circuit that measures the controlled variable X_c to form the feedback signal X_b is defined as the feedback gain K_f. Thus, the feedback

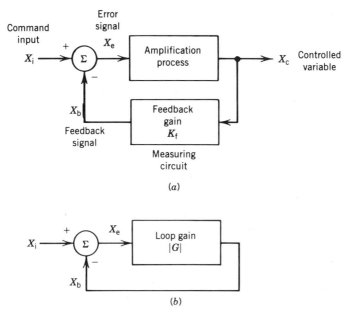

Figure 1.2-1   Signal-flow diagrams: (a) feedback loop to provide feedback control, (b) basic form of feedback loop.

signal X_b is equal to

$$X_b = K_f X_c \qquad (1.2\text{-}2)$$

Solve Eq 1.2-2 for X_c, and combine this with Eq 1.2-1. This gives the following approximation relating the controlled variable X_c to the input signal X_i:

$$X_c = \left(\frac{1}{K_f}\right)X_b \doteq \left(\frac{1}{K_f}\right)X_i \qquad (1.2\text{-}3)$$

Thus, the feedback loop works to keep the controlled variable X_c proportional to the input X_i.

The feedback loop can be expressed in a more general form by combining the amplification process and the measuring circuit into a single block, called the "loop gain", as shown in Fig. 1.2-1b. The controlled variable X_c is ignored, and the basic feedback loop is concerned only with the three general loop variables X_i, X_e, X_b. If one knows how accurately the feedback signal X_b follows the input X_i, one can readily determine the response of the controlled variable X_c by setting X_c equal to $(1/K_f)X_b$, in accordance with Eq 1.2-3. Thus, the controlled variable X_c need not be considered in the basic dynamic analysis of the feedback loop.

The relationships among the three basic loop variables X_i, X_b, X_c can be expressed in a simple fashion by assuming that the input X_i is a sinusoidal signal of varying frequency. At any frequency, these three signals can be

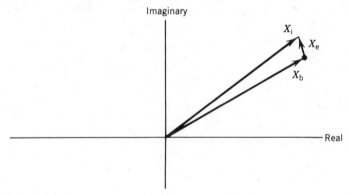

Figure 1.2-2 Typical vectors (or phasors) representing input (X_i), error (X_e), and feedback (X_b) signals for sinusoidal input X_i.

represented by complex vectors (or "phasors") as shown in Fig 1.2-2. Figure 1.2-1 shows that the loop error X_e is equal to the loop input X_i minus the loop feedback signal X_b:

$$X_e = X_i - X_b \qquad (1.2\text{-}4)$$

Therefore, as shown in Fig 1.2-2, the error vector X_e is the difference between the input signal vector X_i and the feedback signal vector X_b.

The loop gain, which is designated $|G|$, is the ratio of the magnitude of the feedback vector X_b divided by the magnitude of the error vector X_e. Thus,

$$\text{loop gain} = |G| = \frac{|X_b|}{|X_e|} \qquad (1.2\text{-}5)$$

where $|X_e|$ is the length of the vector X_e and $|X_b|$ is the length of the vector X_b. If the loop gain $|G|$ is much greater than unity, the feedback vector X_b in Fig 1.2-2 is much longer than the error vector X_e. For this condition, the feedback vector X_b must nearly coincide with the input vector X_i. Thus, at a frequency where the loop gain $|G|$ is much greater than unity, the feedback signal X_b is nearly equal to the input X_i. The exact value of the loop gain is not important, provided that the loop gain is much greater than unity, and the loop has good stability.

The loop gain varies with frequency in the general manner indicated in Fig 1.2-3. Since the loop gain is zero at infinite frequency, it must decrease with increasing frequency. The frequency at which the loop gain drops below unity is called the gain crossover frequency, which is designated ω_{gc}. The feedback loop has effective feedback-control action only at frequencies below ω_{gc}, where the loop gain is greater than unity. Hence, the gain crossover frequency ω_{gc} is

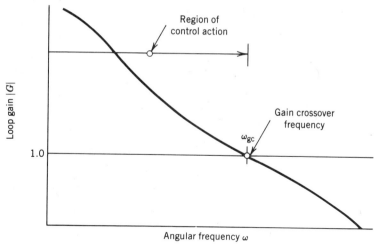

Figure 1.2-3   Typical plot of loop gain versus frequency.

the upper limit to the region of feedback control action, and so is a good criterion for the "control bandwidth" of the feedback loop.

The discussion in Section 1.1.4 showed that ω_ϕ, the frequency of one radian of phase lag of the closed-loop response, is also a good criterion for feedback-control "bandwidth". These two criteria are consistent, because the parameters ω_{gc} and ω_ϕ are approximately equal, for any feedback control loop that has good stability and reasonably high loop gain at low frequency. Thus

$$\omega_{gc} \doteq \omega_\phi \qquad (1.2\text{-}6)$$

Note that ω_{gc} is derived from the magnitude of the open-loop frequency response, while ω_ϕ is derived from the phase of the closed-loop frequency response.

The transfer function relating the loop error X_e to the loop feedback signal X_b is called the loop transfer function, and is designated G:

$$G = X_b/X_e = \text{loop transfer function} \qquad (1.2\text{-}7)$$

The loop transfer function G is characterized by frequency response plots of its magnitude and phase. The magnitude of G, designated $|G|$, is called the *loop gain*, and Ang[G], the phase of G, is called the *loop phase*. Note that the phase angle of a complex quantity F is designated Ang[F] in this book.

Figure 1.2-4a shows a general signal-flow diagram of a feedback loop. The transfer functions relating the loop input X_i to the feedback signal X_b and to the error signal X_e are called the feedback and error transfer functions of the

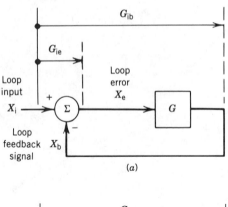

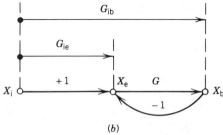

Figure 1.2-4 Basic feedback loop, illustrating the loop, feedback, and error transfer functions G, G_{ib}, and G_{ie}; (a) Signal-flow diagram; (b) Signal-flow graph.

loop. These are designated as follows:

$$G_{ib} = X_b/X_i = \text{feedback transfer function} \qquad (1.2\text{-}8)$$

$$G_{ie} = X_e/X_i = \text{error transfer function} \qquad (1.2\text{-}9)$$

The loop transfer function G is commonly called the "open-loop" transfer function, and the feedback transfer function G_{ib} is commonly called the "closed-loop" transfer function. However, G_{ie} might also be called a "closed-loop" transfer function. The "closed-loop" frequency-response plots that were given in Fig 1.1-1 are the magnitude and phase of G_{ib}.

The transfer functions G_{ib} and G_{ie} can be related to the loop transfer function G by combining Eqs 1.2-4, -7, -8, and -9, to obtain

$$G_{ib} = \frac{G}{1 + G} \qquad (1.2\text{-}10)$$

$$G_{ie} = \frac{1}{1 + G} \qquad (1.2\text{-}11)$$

A fundamental problem with the use of feedback for achieving control is that feedback can also produce an oscillator. To illustrate this point, suppose that G were equal to -1 at a particular frequency. The expression $(1 + G)$ would be zero, and so G_{ib} and G_{ie} in Eqs 1.2-10, -11 would be infinite. This indicates that the loop would oscillate at that frequency. This oscillatory condition, $G = -1$, corresponds to the following values of magnitude and phase of G:

$$|G| = 1, \qquad \text{Ang}[G] = -180° \qquad (1.2\text{-}12)$$

This shows that a feedback loop exhibits a stable oscillation if the loop phase lag is 180° at the gain crossover frequency. It can also be proven that a feedback loop exhibits a growing oscillation if the phase lag is greater than 180° at this frequency.

Thus, a stable feedback loop must have less than 180° of phase lag at gain crossover. A well-designed control system must not only be absolutely stable; it must have *good* stability. It can be shown that for good stability a feedback loop should have no more than 135° of phase lag at gain crossover.

This requirement is a necessary but not sufficient condition for good stability. There are certain conditions, based on the Nyquist stability theorem, that must be satisfied, to assure that the loop is stable in an absolute sense. If the loop has absolute stability, the primary criterion for good stability is that the maximum value of $|G_{ib}|$ should not exceed 1.3. When $\text{Max}|G_{ib}| = 1.3$, the peak overshoot of the step response is about 25%.

This book uses the term "signal-flow diagram" to represent what is commonly called a "block diagram" in the feedback control literature. The term "signal-flow diagram" is an extension of the "signal-flow graph" concept developed by S. J. Mason [1.7, 1.8, 1.9]. Figure 1.2-4b shows the signal-flow graph of the basic feedback loop, which is equivalent to the signal-flow diagram of Fig 1.2-4a. In a signal-flow graph, the transforms of the signals are represented by nodes, and the signal-path lines between those nodes specify the transfer functions. The signal transform at a particular node is equal to the sum of the contributions from paths leading to that node. The contribution from an incoming path is equal to the transfer function of the path, multiplied by the signal transform of the node from which it emerges.

The signal-flow diagram and the signal-flow graph are mathematically equivalent. However, the signal-flow diagram is much easier to use when the transfer functions are complicated. The signal-flow graph has particular advantages in the design of feedback circuits.

A signal-flow diagram describes the dynamic equations of a system in terms of the transforms of the signals and the transfer functions that relate those transforms to one another. Hence the signal-flow diagram only applies to a linear system. When the system is nonlinear, the signal-flow diagram can only represent a linear approximation of the system. System signal diagrams that actually include nonlinear functions are called "block diagrams" or "mathematical diagrams" in this book.

1.2.2 Approximations to Simplify Mathematics

The preceeding discussion has examined feedback control in terms of the action of a single feedback loop. However, a feedback control system often has many feedback loops, one inside another, where each loop provides feedback control of a different controlled variable. Consequently, the dynamic equations of a feedback-control system can be extremely complicated. To optimize performance, one must adjust a multitude of parameters, which set the loop gains and other dynamic elements of the many feedback loops.

On the other hand, despite the confusing complexity of a multiloop control system, well-designed feedback-control systems are relatively insensitive to parameter variations. It is as if the control system itself does not realize how complicated it really is. Although the various control loops are interrelated, they perform their functions largely independently of one another. In other words, underlying the apparent complexity there is inherent simplicity. The engineer should approach the control-system design in a manner that takes advantage of this simplicity.

The key to understanding the inherent simplicity of feedback-control dynamics is approximation. In a well-designed control system, all feedback-control loops should have good stability. When this condition is satisfied, the G_{ib} and G_{ie} transfer functions of each loop can be approximated quite simply.

An important step in making such approximations is to express the loop transfer function of each feedback-control loop in terms of the asymptotic

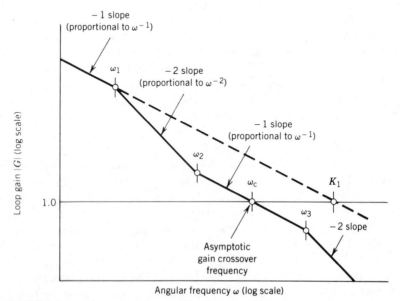

Figure 1.2-5 Approximate (asymptote) log-log plot of loop gain versus frequency for typical feedback loop.

gain crossover frequency of the loop, which is designated ω_c. The asymptotic gain crossover frequency ω_c is always approximately equal to the actual gain crossover frequency ω_{gc}, which characterizes the control "bandwidth" of the feedback loop.

The asymptotic gain crossover frequency ω_c is based on the asymptote plot of loop gain $|G|$ versus frequency. A typical plot is shown in Fig 1.2-5. The magnitude of G is plotted on a logarithmic scale, versus angular frequency ω on a logarithmic scale. The log–log magnitude plot of $|G|$ can be approximated by a series of straight-line segments, called asymptotes. The angular frequencies $\omega_1, \omega_2, \omega_3$ at the intersections of these asymptotes are called the break frequencies.

The asymptote between the break frequencies ω_1, ω_2 has a logarithmic slope of -2, which indicates that this segment is proportional to ω^{-2} (or $1/\omega^2$). We are particularly interested in the asymptote between frequencies ω_2, ω_3, because this asymptote passes through unity gain. It has a slope of -1, and so is proportional to ω^{-1} (or $1/\omega$). The frequency where this asymptote of -1 slope passes through unity gain is called the asymptotic gain crossover frequency and is designated ω_c. Since this asymptote segment is proportional to $1/\omega$, and is equal to unity at the frequency $\omega = \omega_c$, the equation for this asymptote segment is ω_c/ω.

The loop transfer function, corresponding to the magnitude asymptote plot of Fig 1.2-5, it normally expressed in the following form:

$$G = \frac{K_1(1 + j\omega/\omega_2)}{j\omega(1 + j\omega/\omega_1)(1 + j\omega/\omega_3)} \tag{1.2-13}$$

The parameter K_1 is called the velocity constant. As the frequency ω approaches zero, all of the factors in parentheses approximate unity. Hence, at very low frequencies, G approximates $K_1/j\omega$, and the loop gain $|G|$ approximates K_1/ω. This indicates that the equation for the low-frequency magnitude asymptote in Fig 1.2-5 (at frequencies below ω_1) is K_1/ω. The expression K_1/ω is equal to unity at the frequency $\omega = K_1$. Therefore, when the low-frequency asymptote is extended to high frequencies (as shown by the dashed line), it crosses unity gain at the frequency $\omega = K_1$.

This book expresses the equation for a loop transfer function in a different manner. The loop transfer function corresponding to the plot of $|G|$ in Fig 1.2-5 is written in the following form:

$$G = \frac{\omega_c(1 + \omega_2/j\omega)}{j\omega(1 + \omega_1/j\omega)(1 + j\omega/\omega_3)} \tag{1.2-14}$$

Note that the terms in the factors that correspond to the low-frequency breaks ω_1, ω_2 are inverted relative to those in Eq 1.2-13. These factors approximate unity at high frequencies, rather than at low frequencies. With this method of writing the loop transfer function, the gain constant of the equation is the asymptotic gain crossover frequency ω_c. The principles of writing the loop transfer function in this form are explained in Chapter 3, Section 3.2.

Writing the loop transfer function in the alternate form of Eq 1.2-14 yields a gain constant ω_c that has fundamental significance, because ω_c is always approximately equal to the actual gain crossover frequency ω_{gc}. In contrast, the conventional form of Eq 1.2-13 yields a gain constant K_1 that merely characterizes the low-frequency asymptote, which is generally not very important. If the velocity constant K_1 and the break frequency ω_1 are varied together in such a manner as to keep ω_c constant, this variation has little effect on the dynamic performance of the loop.

Because the gain parameter ω_c of the loop transfer function has fundamental significance, writing the loop transfer function in terms of ω_c generalizes and greatly simplifies the analysis. The next step in this approximation process is to derive approximate plots of $|G_{ib}|$ and $|G_{ie}|$ from the plot of the loop gain $|G|$, using the following relations:

For $|G| > 1$ (for $\omega < \omega_c$),

$$G_{ib} \doteq 1 \tag{1.2-15}$$

$$G_{ie} \doteq 1/G \tag{1.2-16}$$

For $|G| < 1$ (for $\omega > \omega_c$),

$$G_{ib} \doteq G \tag{1.2-17}$$

$$G_{ie} \doteq 1 \tag{1.2-18}$$

These approximations are applied in Fig 1.2-6. In diagram a, the solid curve is the plot of $|G|$ obtained from Fig 1.2-5. The dashed curve is the approximate plot of $|G_{ib}|$. By Eqs 1.2-15, -17, the approximate plot of $|G_{ib}|$ is unity at frequencies below ω_c, and coincides with the plot of $|G|$ at frequencies above ω_c.

In diagram b of Fig 1.2-6, the solid curve is a plot of $|1/G|$, which is obtained by inverting the plot of $|G|$ in Fig 1.2-5. (Since the magnitude scale is logarithmic, inverting the plot of $|G|$ forms the negative of the logarithm of $|G|$, which is the logarithm of $|1/G|$.) The dashed curve is the approximate plot of $|G_{ie}|$. By Eqs 1.2-16,-18, the approximate plot of $|G_{ie}|$ follows the plot of $|1/G|$ at frequencies below ω_c, and is unity at frequencies above ω_c.

The following approximations for the G_{ib} and G_{ie} transfer functions can be derived by inspection from these approximate plots of $|G_{ib}|$ and $|G_{ie}|$:

$$G_{ib} \doteq \frac{1}{(1 + j\omega/\omega_c)(1 + j\omega/\omega_3)} \tag{1.2-19}$$

$$G_{ie} \doteq \frac{1 + \omega_1/j\omega}{(1 + \omega_2/j\omega)(1 + \omega_c/j\omega)} \tag{1.2-20}$$

This principle can be applied to any control system, and yields relatively simple approximate transfer functions even for a complicated multiloop control system.

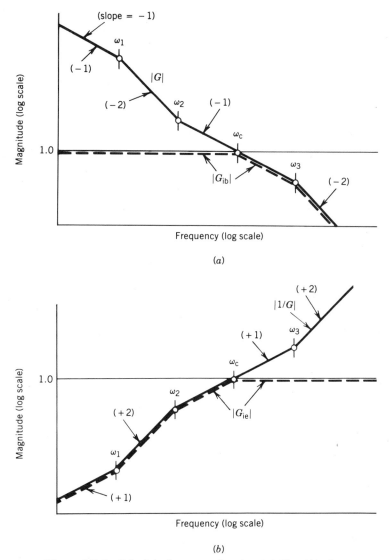

Figure 1.2-6   Principle for approximating (*a*) G_{ib}, (*b*) G_{ie}.

There are many subtle issues associated with this simple approximation principle, which are explained in Chapter 3 and are extended in Chapter 13. A key requirement for making such approximations is that each feedback loop must have good stability. Methods are presented in Section 3.4 that provide simple means for assuring that this stability requirement is satisfied.

In approximating the closed-loop transfer functions G_{ib} and G_{ie}, one is in effect approximating the closed-loop poles. In Chapter 13, methods for calculating the closed-loop poles are presented, which show that these poles are approximately given by

1. The low-frequency zeros of G (those lower than ω_c in frequency).
2. The high-frequency poles of G (those higher than ω_c in frequency).
3. The approximate closed-loop pole: $s = -\omega_c$.

The approximation of G_{ib} in Eq 1.2-19 contains the approximate high-frequency closed-loop poles, while the approximation of G_{ie} in Eq 1.2-20 contains the approximate low-frequency closed-loop poles. Both the G_{ib} and G_{ie} approximations have the closed-loop pole $s = -\omega_c$. The G_{ib} approximation does not have the low-frequency closed-loop poles, because these are approximately canceled by zeros of G_{ib}. Similarly, the G_{ie} approximation does not have the high-frequency closed-loop poles, because these are approximately canceled by zeros of G_{ie}.

The error transfer function G_{ie} is often ignored in feedback-control textbooks. However, this transfer function is essential for describing the response of a feedback control system to a low-frequency input, which is the type of input that can be followed accurately by the control loop. The approximation for G_{ie} is the basis for the discussion in Chapter 5, which explains the general time behavior of feedback-control loops.

In fact, this G_{ie} approximation is probably the most important concept for removing the mystery from feedback control. It was first described by Harris, Kirby, and Von Arx [1.10]. An excellent summary of this principle is presented in the following comment by Maurice J. Kirby and D. C. Beaumariage, which was given in a discussion of a paper by the author [1.11]:

> When it is necessary to study the behavior of a closed-loop transfer function over the entire complex-frequency plane, Mr. Biernson's method will be found to give very valuable assistance. It is not necessary, however, to cover the entire complex-frequency plane in order to obtain the simple but valuable approximations which relate the properties of the open-loop frequency response to the properties of the transient response of a system. These relations have been pointed out in various earlier papers [1.10, 1.12].
>
> It is well known that the upper break frequencies above the crossover frequency in the open-loop response do not give rise to significant terms in the transient response, if the system is well stabilized. They do, however, set the limit of stability beyond which the performance of the system cannot be pushed. Furthermore, the transient response of a well stabilized system can be approximated by a sum of exponential terms having real exponents [1.10]. In this sum, one term corresponds to the crossover frequency and one to each break frequency in the asymptotic open-loop transfer function [where the plot] is concave upward. The exponents of these terms are equal to the corresponding break frequencies.
>
> Physically this means that the transient response of many common systems consists of a rise having a time constant inversely proportional to the crossover frequency, and subsequent decays or "tails" having time constants inversely proportional to the breaks below the crossover frequency [that are concave upward, such as ω_2 of

Fig 1.2-5]. The amplitudes of these terms, and hence the degree of overshoot in the response to a step of velocity, can be approximated rather simply. From this it is obvious that integral networks or others which add low-frequency breaks [that are concave upward] to the open-loop transfer function add corresponding tails to the transient response; these decay with time constants [equal to the reciprocals of the breaks, such as $1/\omega_2$ for the example of Fig 1.2-5].

These principles are applied in Chapter 5 to explain the general time behavior of a feedback control loop.

When considering the use of approximations, one must clearly distinguish between reliable and unreliable approximations. Many approximations (or "rules of thumb") have been applied to feedback control systems that only work some of the time. A common example is the assumption that G_{ib} can be approximated by a second-order transfer function—an approximation that often does not hold. The approximations presented here can be used all of the time, and so provide reliable though approximate answers.

Usually, the parameters derived from these approximations are adequate for designing the control equipment, and are sufficiently close to the final settings to allow the system to be optimized by trial-and-error testing. Final parameter settings can often be established from tests performed on the actual control system. To obtain more accurate parameters than are obtainable with the approximations, simulation is often a more effective tool than more accurate analysis. The use of simulation was discussed in Section 1.1.2.

1.2.3   Summary of Bandwidth Criteria

A critical parameter of any feedback-control loop is the control "bandwidth" of the loop. The bandwidth criteria that have been discussed are summarized as follows:

ω_{gc} = gain crossover frequency: the frequency where the magnitude of the *open-loop* transfer function G drops below unity;

ω_c = asymptotic gain crossover frequency: an approximation of ω_{gc} which directly establishes the amplifier gain settings of a feedback loop;

ω_{hp} = half-power frequency; the frequency where the magnitude of the *closed-loop* transfer function G_{ib} drops below $1/\sqrt{2} = 0.707$;

ω_ϕ = frequency of one radian (57°) of phase lag of the *closed-loop* transfer function G_{ib}.

The gain crossover frequency ω_{gc}, and the frequency ω_ϕ of one radian of phase lag of G_{ib}, are both good criteria for feedback-control bandwidth. These criteria are consistent, because ω_{gc} and ω_ϕ are approximately equal. The gain crossover frequency ω_{gc} is derived from the magnitude of the *open-loop* transfer function G, while the frequency ω_ϕ is derived from the phase of the *closed-loop* transfer function G_{ib}.

In determining whether an actual control system is consistent with a design based on simulation or analysis, the most critical parameters to measure are the gain crossover frequencies of the various feedback loops. It is usually much easier to measure ω_ϕ than ω_{gc}. Hence, when frequency-response tests are performed on a feedback-control system, the primary bandwidth parameter to measure for each feedback loop is the frequency ω_ϕ of one radian of phase lag of G_{ib}. When only transient response tests are taken, one can measure the step-response delay time T_d, which is approximately equal to $1/\omega_\phi$ or $1/\omega_{gc}$. The delay time T_d is the time after the input step for the step response to reach 50% of its final value. The principles for measuring feedback-control dynamic response are discussed in Chapter 5, Section 5.6.

The asymptotic gain crossover frequency ω_c is a convenient bandwidth parameter that can be derived directly from the equation of the loop transfer function. The asymptotic gain crossover frequency ω_c is approximately equal to the actual gain crossover frequency ω_{gc}. Hence, when loop transfer functions are expressed in terms of ω_c, the gain settings of amplifiers in a control system can be related directly to the actual gain crossover frequencies of the various feedback loops.

Although the half-power frequency ω_{hp} is a good bandwidth criterion in the design of amplifiers and filters in communication applications, it has limited meaning in the design of feedback-control systems. The half-power frequency ω_{hp} should *never* be used as an indirect measure of gain crossover frequency ω_{gc}, because the ratio ω_{hp}/ω_{gc} can vary widely for practical control systems.

In a feedback control system, the half-power frequency is useful for describing noise transmission. As is explained in Chapter 11, this is important in radar tracking loops, which act as filters of receiver noise in the radar signal. The noise transmission of a tracking feedback loop is characterized by the noise bandwidth of the loop, which usually is roughly equal to the half-power frequency ω_{hp}.

In comparing communication and control systems, note that many amplifiers used for communications use feedback, and so are themselves feedback control systems. Feedback is used in an amplifier to stabilize gain and to reduce the effect of nonlinearity. The gain crossover frequency ω_{gc} of a feedback amplifier is the upper frequency at which the feedback loop has effective feedback-control action. However, in terms of signal transmission, the full half-power bandwidth ω_{hp} of the amplifier is important in characterizing the ability of the amplifier to duplicate the waveform shape.

1.3 SUMMARY OF CONTENTS OF THE BOOK

Volume 1 contains Chapters 1 to 7, which present theoretical and practical principles for designing feedback control systems. Volume 2 repeats this Introduction (Chapter 1), and contains Chapters 8 to 15, which present advanced control system topics. This includes digital computer simulation, and studies of sampled-data (digital) control systems, AC control systems, mechanical structural dynamics, statistical analysis of random signals, radar tracking, nonlinear control, and advanced feedback analysis techniques. The following summarizes the contents of the two volumes.

1.3.1 Summary of Volume 1

Although it is assumed that the reader has had a course in the Laplace transform, those aspects of the Laplace transform that are needed in the book are summarized in Chapter 2. The Fourier transform is also summarized, to provide a broad background in transform theory. Also in Chapter 2, the feedback-control principle is introduced, along with terminology for characterizing feedback-control systems. Based on the transient responses and frequency responses of a number of basic types of feedback-loop transfer functions, approximations are developed for relating the characteristics of transient and frequency response.

It is assumed that the reader has a basic background in feedback control theory, including the use of the polar locus of G and the Nichols chart to calculate the magnitude and phase plots of G_{ib}. Section 3.1 uses the polar locus of G to explain principles for assuring that a feedback loop is stable in an absolute sense, and has good stability. A consequence of these stability constraints is that essentially all practical feedback control loops have a rather broad frequency region, at or near gain crossover, where the magnitude asymptote is proportional to $1/\omega$. The frequency where this magnitude asymptote (extended if necessary) crosses unity gain is called the asymptotic gain crossover frequency ω_c. Hence, the equation of this asymptote is ω_c/ω. Section 3.2 shows how loop transfer functions are written in terms of this ω_c/ω asymptote.

Section 3.3 develops approximate expressions for the feedback transfer function G_{ib} and the error transfer function G_{ie}, which are derived by inspection from the loop transfer function G.

Section 3.4 presents methods for optimizing the parameters of a loop transfer function, consistent with a given maximum value of the magnitude of G_{ib} (the stability specification). Except for transfer functions with strongly underdamped poles and zeroes, these methods allow optimum system parameters to be selected without graphical construction.

Chapter 4 applies the techniques developed in Chapter 3 to the design of three types of "servomechanisms". A servomechanism (or "servo") is defined as a feedback control system that controls a mechanical device. The three servos are: (1) a simple uncompensated servo using a DC electric motor, (2) an extension of the simple servo which has tachometer feedback and integral-network compensation, and (3) an extension of the simple servo which has lead-network compensation and integral-network compensation. The error due to a friction load torque is calculated.

Chapter 5 describes the general time behavior of feedback-control loops. Graphical methods for approximating the ramp and step responses are developed, which allow one to relate open-loop frequency response characteristics directly to transient response. By expanding the approximate G_{ie} transfer function, a technique is derived for calculating the response of a feedback loop to a low-frequency arbitrary input, using error coefficients. Also, analyses are presented that justify the approximations relating the rise time T_R and delay time T_d of the step response to the frequency-response parameters ω_{gc}, ω_ϕ, and ω_{hp}. The chapter concludes with a discussion of dynamic feedback control measurements, which emphasizes the necessity of measuring phase as well as magnitude in frequency-response tests.

Chapter 6 analyzes the effects of load disturbances. General approximations are developed for computing the error due to a torque disturbance, particularly static friction, from the gain crossover frequencies of the control loops.

Chapter 7 presents a detailed study of the design of a control system for positioning an optical instrument stage. This control system includes a position loop (which provides feedback control of linear stage displacement), a tachometer velocity loop (which provides feedback control of motor velocity), and a current loop (which provides feedback control of motor current and motor torque). Many practical issues are considered, including the effects of motor inertia and stage mass, friction forces and torques, and limits on the maximum values of velocity and torque that are set by the control loops. General principles are developed for analyzing a complex multiloop control system by using the G_{ie} and G_{ib} approximations. This control example is used as a model for subsequent computer simulation studies in Chapters 9 and 10.

1.3.2 Summary of Volume 2

Chapter 8 analyzes AC-modulated signals in feedback-control systems. It shows how AC (or carrier-frequency) networks affect the signal-frequency response of a feedback loop. The chapter discusses AC transducers, such as synchros and resolvers, and the conversion of the AC signals from these transducers to digital data. The principles presented in this chapter can also be applied in a communication or radar amplifier, to determine the effect of a bandpass IF filter on the envelope of the IF waveform.

Chapter 9 analyzes sampled-data systems and their use in digital data processing. The basic principles for generating and demodulating sampled data are presented. A method is given for designing a sampled-data algorithm to satisfy a specified frequency response. A general procedure, called "serial simulation", is developed for simulating a feedback-control system on a digital computer. This is used to simulate the stage-positioning servo described in Chapter 7. The procedure can be readily implemented on a personal computer, using BASIC computer language, to simulate any control system, regardless of its complexity. The general discussion of sampled-data principles provides theoretical constraints on the sampling frequency, which assure that the simulation is stable and accurate.

A general discussion of digital simulation techniques is presented in Section 9.4, which includes the Runge-Kutta method of digital integration. The state-variable concept is described, and is extended to derive the transition matrix. State variables and the transition matrix are important elements in the theory of the Kalman filter. A practical explanation of Kalman filtering is given by the author in Ref. [1.14] (Chapter 8).

Feedback-control loops that include digital and analog data are discussed in Section 9.5. The z-transform is described, and combined with frequency-response analysis to characterize, in a general manner, the dynamic performance of a feedback-control system that uses sampled data.

Chapter 10 introduces the issue of structural dynamics. Mechanical structural resonance is generally the primary limitation on the dynamic performance of a system that controls a mechanical element. A computer model is developed to simulate the effect of compliance in the gear train between a servo motor and an inertial load. This is applied to the system studied in Chapter 7, to provide a very realistic simulation of a practical servomechanism.

Chapter 11 analyzes the response of a control system to a random input that is characterized statistically. It describes thermal noise and shot noise, the concept of noise bandwidth, and the Gaussian distribution. Autocorrelation and cross-correlation functions are presented, along with a general method of computing noise bandwidth.

Section 11.3 examines the processes of range and angle tracking in a radar system, showing how statistical signal concepts are used to calculate the errors due to receiver noise. Section 11.4 shows how wind forces can be characterized statistically and is used to calculate the tracking error of a large satellite communication antenna.

Chapter 12 investigates the effects of nonlinearities in feedback-control systems. Saturated lock-on transients are studied with piecewise analysis and the phase plane. The describing-function method of nonlinear analysis is used to determine the effects of backlash in a servomechanism gear train. The principle of linearizing a nonlinearity about an operating point is applied to the design of a control system that controls pressure in a vacuum processing chamber. This approximate analysis is combined with an exact nonlinear simulation study, which uses the simulation methods developed in Chapter 9.

Chapter 13 presents advanced techniques for the analysis of stability and transient response. Section 13.1 describes the Nyquist criterion and the Routh criterion for determining the absolute stability of a feedback loop. Section 13.2 presents two methods for computing the closed-loop poles of a feedback loop. The first method is a generalized frequency-response approach that was developed by Kusters and Moore and extended by the author. The second method is an iterative procedure developed by the author. The methods of Section 13.2 are extended in Section 13.3 to yield a general approach for approximating transient response from open-loop frequency response. This provides a mathematical justification for the simple approximations developed in Chapter 3.

Section 13.4 summarizes the root-locus method of Evans [1.13] for calculating closed-loop poles. It also shows how a root-locus plot can be derived by applying the generalized frequency-response method described in Section 13.2.

Chapter 14 presents a method for analyzing the response of a system to an arbitrary input signal. This provides background for the concepts developed in Chapter 5 for characterizing the general time behavior of feedback-control loops.

Chapter 15 describes a unique control system that was developed by the author. This was the basis for the GCA wafer-stepping photolithography equipment for fabricating integrated circuits, which exposes the image directly on the silicon wafer. An optical stage is positioned to a precision of $\frac{1}{15}$ of a wavelength of light over a 6-inch travel. This servo (along with the wafer-stepping approach) provided greatly improved registration accuracy, which allowed the line widths of integrated circuits to be reduced from 5 microns (micrometres) to less than 1 micron. This equipment has resulted in a tremendous increase in the density of integrated circuitry since 1977.

1.4 NEED FOR FEEDBACK CONTROL THEORY

The material in this book is concerned with general concepts of feedback control and related theory. The primary examples used in presenting this material pertain to systems controlling mechanical elements (servomechanisms). However, the concepts are much more general, and apply to all systems where the feedback-control principle is implemented. Servos are particularly good examples, because they are quite complex and are physically easy to comprehend.

Two of the areas where feedback control is most often experienced by electrical engineers are feedback circuitry using operational amplifiers (opamps) and phase-lock loops. Although simple opamp circuits are applied in the examples, the book is not able to explore opamps and phase-lock loops in detail. Nevertheless, the general approaches that are presented should provide the principles needed to achieve a clear understanding of these applications.

Process control is an important area requiring feedback-control theory. In certain respects process-control problems are usually simpler than servo problems, but nonlinearities are generally much more severe. In Chapter 12, Section 12.4 analyzes and simulates a nonlinear pressure control system, which should provide a good introduction to nonlinear process control.

Generally, the most important constraint in a control system that moves a mechanical device is resonance in the mechanical structure. Hence, control theory should be just as important to the mechanical engineer as to the electrical engineer. NASTRAN is a family of sophisticated computer subroutines, developed under NASA sponsorship, which allows the dynamic responses of a complicated structure to be computed accurately. The structure is broken down into a large array of elements (or "nodes"), each of which is characterized by its mass and moments of inertia, and by the compliances (including damping) between the element and adjacent elements. However, for the mechanical engineer to apply these powerful NASTRAN tools effectively in control systems, he should understand the control issues himself. The introduction to structural dynamics presented in Chapter 10 should be particularly useful in relating mechanical structural analysis to the resultant control-system performance.

Many engineers experience control applications only in terms of algorithms they program on a digital computer. The general discussions of sampled-data theory given in Chapter 9 should serve as a good introduction to that field. In the book *Optimal Radar Tracking Systems* [1.14] (Chapter 8), the author has extended these principles, to show how digital-computer algorithms are applied in a sophisticated computer-controlled radar tracking system.

Feedback control is an important aspect of radar system engineering. Chapter 11 presents an analysis of a radar tracking system, and the effects of wind gusts on a microwave antenna.

Thus, a knowledge of feedback control concepts is very important in many engineering disciplines. The purpose of this book is to present these concepts in a simple yet thorough manner, to provide adequate background to use them in any application.

1.5 NOTES ON SYMBOLISM AND TERMINOLOGY

In this book, functional relationships are expressed exclusively with square brackets [ ]. These brackets are used to represent multiplication only in cases where they cannot be confused with functional relationships. For example, $F[x + y]$ always indicates that F is a function of $(x + y)$, and cannot represent F times $(x + y)$; whereas $F(x + y)$ always means F times $(x + y)$.

Although angular frequencies, designated by the symbol ω, have the units of radian/second, the term "radian" is generally omitted in this book, in order to simplify the calculation of units. For example, the expression $\omega_1 = 4$ rad/sec is usually written $\omega_1 = 4 \text{ sec}^{-1}$.

The sampled-data literature uses the symbol z to represent $\exp[sT]$, the Laplace transform for a time shift in the future of one sample period. In contrast, this book deals with $\exp[-sT]$, the Laplace transform for a one-period delay, which is represented as $\bar{z}$. This variable $\bar{z}$ is therefore equal to $1/z$. The conventional symbolism has the disadvantage that z normally occurs only in the form z^{-n}, because time shifts in the future are not realizable.

The logarithm to base 10 of x is expressed as $\log[x]$, while the logarithm to base e (the natural logarithm) is expressed as $\ln[x]$. To represent a logarithmic signal-amplitude ratio, the book generally uses the decilog (abbreviated dg) instead of the decibel (abbreviated dB). The decilog, which is 10 times the logarithm of the signal amplitude ratio, is discussed in Section 2.3.4. An important advantage of the decilog is that a signal gain in decilogs can be easily interpreted in terms of the corresponding signal amplitude ratio; whereas a signal gain in decibels cannot.

Chapter 8

AC Modulated Signals

This chapter analyzes AC modulated signals in feedback-control systems. It shows the manner in which AC (or carrier frequency) networks affect the signal-frequency response of a feedback loop. Included in this chapter is a discussion of AC transducers, such as synchros and resolvers, and the conversion of the AC signals from these transducers to digital data. The principles presented in this chapter can also be applied in communication systems to explain the effect of IF filtering on the envelope of the IF waveform.

Many of the position transducers used in control systems, for measuring both angular and linear displacement, generate AC modulated signals. The amplitude of the AC signal is proportional to the magnitude of the displacement from null, and the sign of the displacement is determined by the phase of the AC signal, relative to an AC reference. The AC signal is in phase with the reference for positive displacements, and it is 180° out of phase for negative displacements.

AC modulated signals are also encountered in the control of two-phase AC servomotors, which have been widely used in small servomechanisms. One winding of the two-phase servomotor is excited by the AC control signal, while the other winding is excited by a fixed AC reference, which is 90° out of phase relative to the control signal.

8.1 MODULATION AND DEMODULATION

8.1.1 Analysis of Ideal Modulation and Demodulation Processes

Many position transducers, measuring linear displacement or angular rotation, provide an AC modulated voltage e_x of the form

$$e_x[t] = Kx[t]\cos[\omega_r t] = Kx[t]\cos[2\pi f_r t] \qquad (8.1-1)$$

where $x[t]$ is the linear or angular displacement from the center (null) position, K is a constant, ω_r is the frequency of the modulating carrier (or reference) in rad/sec, and f_r is that carrier frequency in hertz. The most common carrier frequencies are 60 and 400 Hz, but other frequencies are often used.

Figure 8.1-1 illustrates the basic modulation process. Diagram *a* shows a DC position signal $x[t]$, and diagram *b* shows an AC carrier (or reference) signal $R[t]$. The signals of diagrams *a* and *b* are multiplied together to form the AC modulated signal $e_x[t]$ shown in diagram *c*. Note that at point (A), where the input $x[t]$ is positive, the AC modulated signal $e_x[t]$ of diagram *c* is

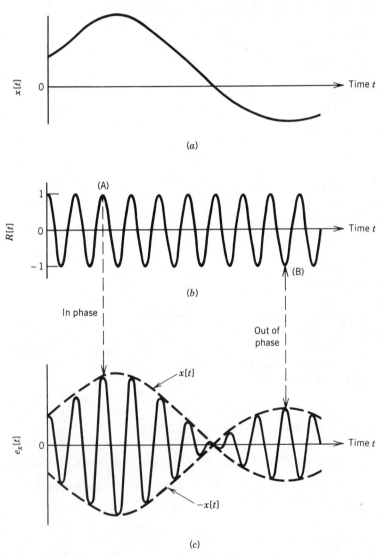

Figure 8.1-1   Waveforms illustrating AC modulation process: (*a*) position signal $x[t]$; (*b*) sinusoidal reference (or carrier) signal $R[t]$; (*c*) modulated signal $e_x[t]$.

in phase with the reference $R[t]$ of diagram b. At point (B), where $x[t]$ is negative, the signal $e_x[t]$ of diagram c is 180° out of phase with respect to the reference of diagram b.

In analyzing a control system using such a transducer, it is convenient to assume that the position displacement $x[t]$ is modulated sinusoidally at a modulation frequency ω_m:

$$x[t] = X_m\cos[\omega_m t] \tag{8.1-2}$$

where X_m is the maximum displacement of the sinusoidal modulation. For simplicity, the constant K in Eq 8.1-1 is normalized to unity. Combining Eqs 8.1-1 and 8.1-2 gives for the modulated voltage $e_x[t]$

$$e_x[t] = X_m\cos[\omega_m t]\cos[\omega_r t] \tag{8.1-3}$$

This can be expanded by applying the formula

$$\cos[a]\cos[b] = \tfrac{1}{2}(\cos[a + b] + \cos[a - b]) \tag{8.1-4}$$

Equation 8.1-3 becomes

$$e_x[t] = \frac{X_m}{2}\{\cos[(\omega_r + \omega_m)t] + \cos[(\omega_r - \omega_m)t]\} \tag{8.1-5}$$

This shows that the modulated signal $e_x[t]$ consists of two sidebands, at the frequencies $(\omega_r + \omega_m)$ and $(\omega_r - \omega_m)$. These sidebands are displaced above and below the carrier reference frequency ω_r by the modulation frequency ω_m. The spectra of the position signal $x[t]$ and the modulated signal $e_x[t]$ are shown in Fig 8.1-2.

In an amplitude-modulated (AM) radio signal, the modulating signal $x[t]$ in Eq 8.1-2 is the sum of a DC level plus the audio signal. Because of this DC level, the modulated signal $e_x[t]$ of Eq 8.1-5 also has a component at the carrier frequency, which is proportional to $\cos[\omega_r]$. On the other hand, in a control application, the modulating signal $x[t]$ does not normally have a fixed DC component, and so the modulated signal $e_x[t]$ does not normally have a component at the carrier frequency. Such an amplitude-modulation process is sometimes called "suppressed-carrier amplitude modulation."

To obtain the position variable $x[t]$, the signal $e_x[t]$ can be demodulated by feeding it through a phase-sensitive detector, which uses the carrier as a reference signal. This reference can be expressed as

$$R[t] = \cos[\omega_r t] \tag{8.1-6}$$

A mathematically ideal phase-sensitive detector is one that multiplies the modulated signal by the reference and lowpass-filters the result. Multiplying

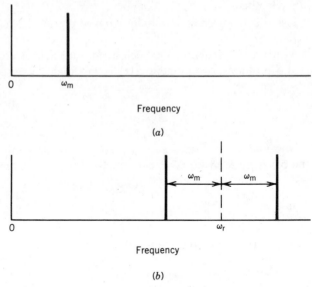

Frequency

(a)

Frequency

(b)

Figure 8.1-2 Effect of AC modulation on signal spectrum: (a) spectrum of position signal $x[t]$; (b) spectrum of modulated signal $e_x[t]$.

the signal $e_x[t]$ in Eq 8.1-5 by twice the reference $R[t]$ in Eq 8.1-6 gives

$$e[t] = 2e_x[t]R[t]$$
$$= X_m\{\cos[(\omega_r + \omega_m)t] + \cos[(\omega_r - \omega_m)t]\}\cos[\omega_r t]$$
$$= X_m\cos[(\omega_r + \omega_m)t]\cos[\omega_r t] + X_m\cos[(\omega_r - \omega_m)t]\cos[\omega_r t] \quad (8.1\text{-}7)$$

Applying the trigonometric identity of Eq 8.1-4 to this gives

$$e[t] = \frac{X_m}{2}\{\cos[(2\omega_r + \omega_m)t] + \cos[\omega_m t]\}$$
$$+ \frac{X_m}{2}\{\cos[(2\omega_r - \omega_m)t] + \cos[\omega_m t]\}$$
$$= X_m\cos[\omega_m t] + \frac{X_m}{2}\{\cos[(2\omega_r + \omega_m)t] + \cos[(2\omega_r - \omega_m)t]\} \quad (8.1\text{-}8)$$

This signal $e[t]$ is fed through a lowpass filter, which passes the signal at the frequency ω_m essentially unchanged, and blocks the double harmonic signals at the frequencies $(2\omega_r + \omega_m)$ and $(2\omega_r - \omega_m)$. The filter output, which is designated $\tilde{x}[t]$, is approximately

$$\tilde{x}[t] \doteq X_m\cos[\omega_m t] \tag{8.1-9}$$

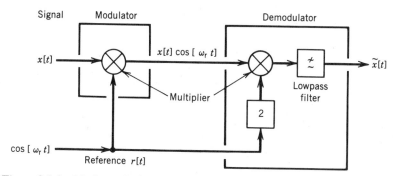

Figure 8.1-3 Mathematically ideal processes for modulation and demodulation.

Thus the modulation and demodulation processes can be represented as shown in Fig 8.1-3. The signal is modulated by multiplying the signal by the reference $\cos[\omega_r t]$. It is demodulated by multiplying the modulated signal by the same reference, multiplied by 2, and feeding the result through a lowpass filter. The output is represented by the signal $\tilde{x}[t]$, where the ($\tilde{}$) indicates that the $x[t]$ information has been filtered by the lowpass filter.

Note that multiplication is represented by the symbol $\otimes$. Some writers use this same symbol to represent summation in feedback-control signal-flow diagrams, but this is poor practice. This book uses the symbol Σ to represent summation and reserves $\otimes$ for multiplication.

8.1.2 Practical AC Modulators

Linear Variable Differential Transformer (LVDT). The LVDT was applied in Chapter 7 to measure linear displacement of the stage. The basic transducer consists of three coils wound about a hollow tube, which contains a movable iron slug. As shown in Fig 8.1-4, the reference winding A is excited by a constant AC reference voltage E_r. This signal couples magnetically through the iron slug to the two output windings $B1$ and $B2$. The windings $B1$, $B2$ are connected in series opposition, so that the AC voltages generated in these windings tend to cancel.

The iron slug is allowed to move axially. When the slug is in the central (null) position, there is equal magnetic coupling from winding A to windings $B1$, $B2$. Hence voltages E_1, E_2 are equal, and the winding output voltage E_x is zero. When the slug is displaced in the direction x from the null position, the magnetic coupling to winding $B1$ increases and that to winding $B2$ decreases. Voltage E_1 is now greater than E_2, and the difference between them, which is E_x, is in phase with the reference voltage E_r applied to winding A. The amplitude of E_x is proportional to the displacement of the slug from null. When the slug is displaced from null in the opposite direction, voltage E_2 becomes greater than E_1, and so E_x is 180° out of phase with the reference E_r.

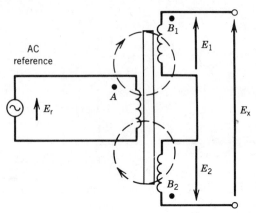

Figure 8.1-4   Circuit using linear variable differential transformer (LVDT).

Figure 8.1-5 shows a plot of the in-phase AC voltage E_x versus the displacement x of the slug from the null position. A positive value indicates that the AC voltage is in phase with the reference, whereas a negative value indicates that it is 180° out of phase with the reference. There is a large central region where the device is very linear, which is labeled the linear range. In this range, the response of the LVDT can be described by

$$e_x = E_m \cos[\omega_r t] \frac{x}{X_m} \tag{8.1-10}$$

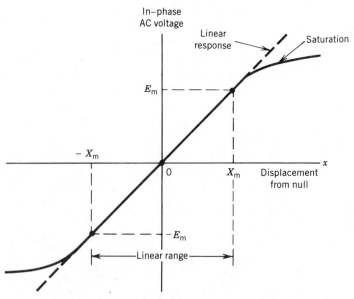

Figure 8.1-5   Characteristic curve of LVDT.

The variable x is the displacement from null, $\pm X_m$ is the extent of the linear range, and E_m is the peak value of the AC signal, when the LVDT is at either limit of this range.

Phase shift in the circuit generates a small quadrature output voltage that is proportional to $\sin[\omega_r t]$. The demodulation process, which converts the AC signal e_x to a DC signal, discriminates against this quadrature signal, and ideally rejects it.

As was explained in Chapter 7, modern LVDT devices often contain electronic circuitry, which generates the AC reference signal and demodulates the signal e_x, to provide a DC signal proportional to displacement. An excitation of several kilohertz is generally used. The high carrier frequency improves the magnetic efficiency of the device, and minimizes the dynamic effects associated with modulation and demodulation.

Rotary Variable Differential Transformer (RVDT). The rotary variable differential transformer (RVDT) is a transducer for measuring rotary motion which is related to the LVDT. It has a characteristic similar to that of Fig 8.1-5, except that x is replaced by angular desplacement Θ. Its response is described by

$$e_\Theta = E_m \cos[\omega_r t] \frac{\Theta}{\Theta_m} \qquad (8.1\text{-}11)$$

The angle Θ is the angular displacement from null, the linear range is $\pm\Theta_m$, and E_m is the peak output AC voltage, which occurs at the extremes $\pm\Theta_m$ of the linear range. This device can be used for angular control over a peak-to-peak range of about $120°$.

AC Resolver. The AC resolver is a magnetic device for sensing angular position, and has the circuit shown in Fig 8.1-6. The rotor generally has two

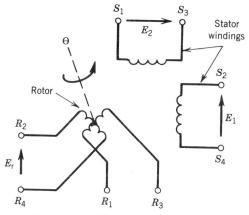

Figure 8.1-6 Circuit for AC resolver.

windings in space quadrature. One of the rotor windings is excited by an AC reference voltage E_r, expressed in the time domain as

$$e_r[t] = E_m\cos[\omega_r t] = E_m\cos[2\pi f_r t] \tag{8.1-12}$$

The reference frequency f_r is usually 60 or 400 Hz. The stator has two windings in space quadrature. These deliver the voltages E_1, E_2, expressed as follows in the time domain:

$$e_1[t] = E_m\cos[\Theta]\cos[\omega_r t] \tag{8.1-13}$$
$$e_2[t] = E_m\cos[\Theta - 90°]\cos[\omega_r t] = E_m\sin[\Theta]\cos[\omega_r t] \tag{8.1-14}$$

where Θ is the angle of the resolver shaft. These voltages provide accurate measurement of angle over a complete cycle, or over many cycles.

The AC synchro is closely related to the AC resolver, but has a three-phase winding on the stator, rather than a two-phase winding. The use of the synchro and the resolver for position sensing is discussed in Section 8.4.

AC Modulation of DC Electrical Signal. A chopping device is generally employed to convert a DC signal to an AC modulated signal. Vibrating mechanical switches were once commonly used for this purpose, but the most common modulating device used today is the integrated-circuit FET switch. Figure 8.1-7 shows an AC modulator employing FET switches. This could be

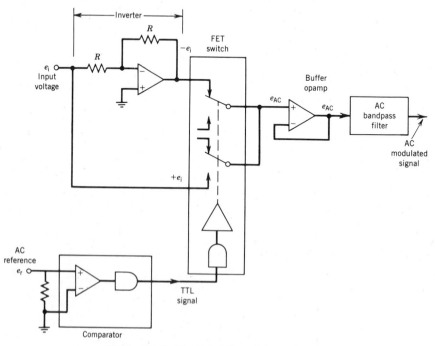

Figure 8.1-7   Typical AC modulator circuit.

implemented with a Siliconix DG191 integrated circuit, which has two SPDT (single-pole double-throw) FET switches in a 16-pin dual-inline package (DIP). Each switch is controlled by a TTL digital signal, and has one switch state for a TTL HIGH (2–5 V) and the other state for a TTL LOW (0–0.8 V).

The TTL command for controlling the FET switch can be derived by feeding the AC reference signal into a comparator with a TTL output (for example, a Harris HA4900, which has four comparators in a 16-pin DIP). Diagram a of Fig 8.1-8 shows the AC reference signal input to the compara-

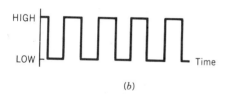

(a)

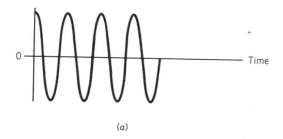

(b)

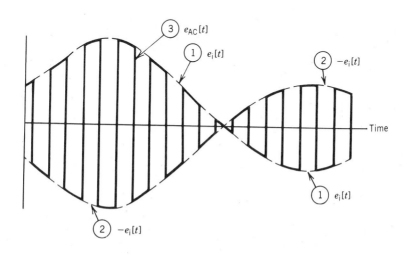

(c)

Figure 8.1-8   Modulated DC signal generated by circuit of Fig 8.1-7: (a) AC reference; (b) TTL signal; (c) modulated signal.

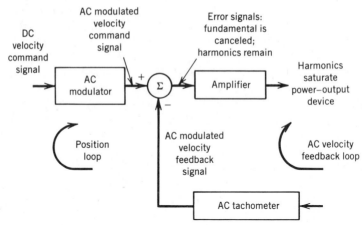

Figure 8.1-9   Saturation effect of harmonics of AC modulated signal in feedback-control system.

tor. The output, shown in diagram b, is a TTL-compatible square wave in synchronism with the reference. In diagram c, the dashed curve ① is the input DC signal $e_i[t]$ that is being modulated, which is the same as the signal shown in Fig 8.1-1a. The dashed curve ② is the output $-e_i[t]$ from the inverter opamp, which is the negative of the input $e_i[t]$. The solid curve ③ is the square-wave modulated voltage $e_{AC}[t]$ at the outputs of the FET switches. This voltage $e_{AC}[t]$ is fed through a buffer opamp to an AC bandpass filter tuned to the reference frequency. The bandpass filter attenuates the harmonics of the square-wave modulated signal (curve ③) to produce a smooth waveform similar to that shown in Fig 8.1-1c.

In feedback-control applications, strong filtering of harmonics is generally desirable in the AC modulator. The reason is illustrated in Fig 8.1-9. It is assumed that an AC modulator converts a DC command to an AC modulated signal, which acts as the velocity command signal for an AC velocity feedback loop. An AC tachometer generates an AC feedback signal for the tachometer loop. This has an AC waveform with a modulation proportional to the tachometer velocity.

The signal-frequency components of the AC command and feedback signals nearly cancel one another, but the harmonics remain. When these harmonics are amplified, they may saturate the power amplifier that drives the AC motor. If this occurs, dynamic performance is severely degraded. Hence, harmonic and quadrature components in the input modulated signal must be kept small.

As will be explained in Section 8.2, the bandpass filter that attenuates harmonics in the AC modulated waveform adds phase lag to the feedback loop at the signal frequency, which limits the loop performance. Better dynamic response could be achieved if the modulator were a true analog multiplier (as illustrated in Fig 8.1-1), rather than a switch. Since the harmonics generated by the modulator would be much lower, less filtering would be needed.

In communication systems, signal modulation is commonly provided by a balanced mixer, which acts like an ideal multiplier. This approach minimizes unwanted sidebands in the modulated signal. However, the balanced mixer operates at low signal levels, and generally does not allow accurate control of DC offset. A switching modulator can operate at high signal levels, while providing a very accurate zero. Hence the switching modulator is generally used in feedback-control applications, despite its dynamic limitations.

8.1.3 Practical AC Demodulators

The same circuit used to modulate the AC signal can be used to demodulate it, except that the filtering is different. Figure 8.1-10 shows a typical demodulating circuit, which is called a phase-sensitive detector. Two demodulator approaches are included in the figure: one uses a lowpass filter to attenuate the AC harmonics, and the other uses a sample-and-hold circuit.

Figure 8.1-11 shows the waveforms of the demodulation process. Diagram a shows the input AC modulated waveform, which is the same as the waveform shown earlier in Fig 8.1-1c. Diagram b shows the AC reference signal, and diagram c shows the TTL-switch control waveform that is derived from the AC reference. Diagram d is the demodulated waveform at the output of the FET switches. This is equal to the waveform a when the TTL command c is HIGH, and it is equal to the negative of the waveform a when the TTL switch command c is LOW.

The signal of diagram d is fed through the lowpass filter, which attenuates the harmonics of the waveform. The resultant filtered signal is approximately the same as the original DC signal, which was shown in Fig 8.1-1a.

Another way to eliminate harmonics is to employ a sample-and-hold circuit. The sampling pulse train is shown in diagram e of Fig 8.1-11. These are short pulses that are timed to occur at the peaks of the AC waveform. The sampling pulses can be derived using digital circuitry from the TTL switch-control pulses of diagram c. These sampling pulses (e) sample the peak values of the demodulated waveform (d), which are indicated by the points ○. Each sampled value is stored on the capacitor C in Fig 8.1-10, which holds the output voltage fixed between sampling instants. Diagram f in Fig 8.1-11 shows the waveform at the output of the sample-and-hold circuit. A small amount of additional lowpass filtering is generally needed after the sample-and-hold circuit to attenuate the remaining harmonics.

A sample-and-hold circuit is a very effective means of reducing AC harmonics in a demodulator. To achieve equivalent harmonic reduction with a lowpass filter results in large phase lag in the demodulated signal, which may severely degrade feedback-control performance.

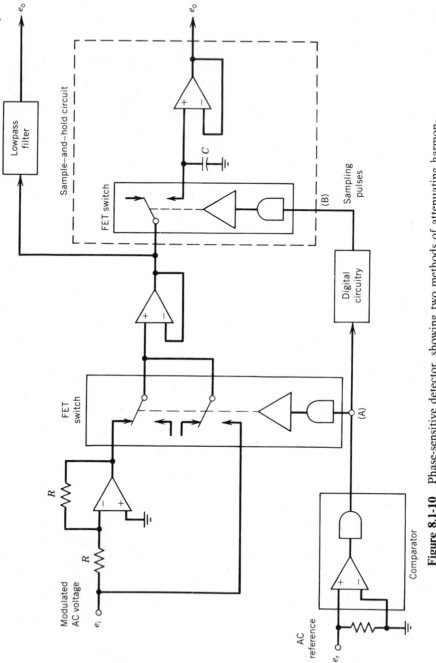

Figure 8.1-10 Phase-sensitive detector, showing two methods of attenuating harmonics.

40

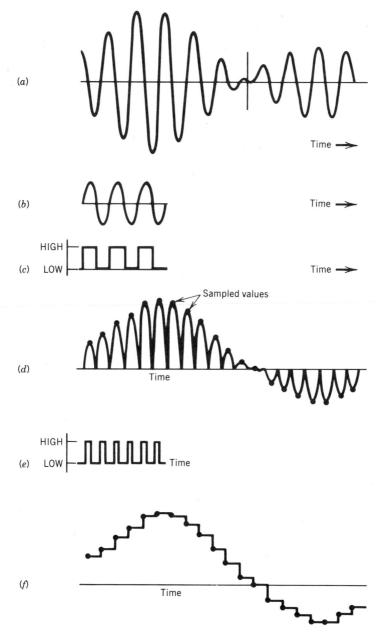

(a)

Time →

(b)

Time →

(c) HIGH / LOW

Time →

Sampled values

(d)

Time

(e) HIGH / LOW

Time

(f)

Time

Figure 8.1-11 Waveforms of phase-sensitive detector of Fig 8.1-9: (a) input AC modulated waveform; (b) AC reference; (c) TTL pulses controlling FET switches; (d) demodulated signal before lowpass filtering; (e) TTL sampling pulses; (f) output from sample-and-hold circuit.

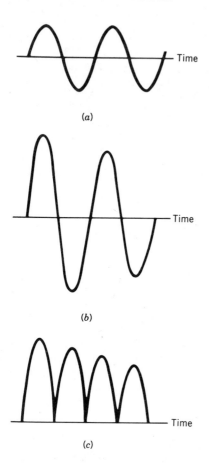

(a)

(b)

(c)

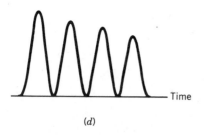

Figure 8.1-12 Comparison of waveforms of mathematically ideal and practical phase-sensitive detectors: (a) AC reference; (b) AC modulated signal; (c) output from practical demodulator; (d) output from mathematically ideal demodulator.

(d)

Figure 8.1-12 compares the waveform of the practical phase-sensitive demodulator with that assumed in the ideal mathematical analysis of Section 8.1.1. Diagram a shows the AC reference, and diagram b shows the assumed AC modulator input. Diagram c is the switch output from the practical demodulator shown in Fig 8.1-9. The mathematical analysis assumes that the demodulator operates by multiplying the reference in diagram a by the AC modulated signal in diagram b. The resultant product (appropriately scaled in amplitude) is shown in diagram d.

Figure 8.1-12 shows that the practical demodulator generates somewhat lower content at the second harmonic than the mathematically ideal demodulator, but generates strong components at the fourth and higher harmonics, which are not produced by the ideal demodulator.

Two-Phase Servo Motor. The two-phase AC induction servo motor was used extensively during World War II and the following decade. However since then it has been largely replaced by the DC servo motor, which has higher efficiency and better dynamic performance. (A discussion of servo-motor efficiency is given in Chapter 12, Section 12.3.) The two-phase servo motor has two stator windings that are excited by AC voltages 90° out of phase. It has an induction rotor, usually of squirrel-cage design.

One of the stator windings (the reference winding) is excited with a fixed AC voltage, while the other (the control winding) is excited with a varying AC voltage. At zero speed, the torque generated by the motor is proportional to the amplitude of the control-winding voltage. The direction of the torque depends on the phase of the voltage on the control winding. Torque is exerted in one direction when the control voltage leads the reference voltage by 90°, and in the other when the control voltage lags the reference voltage by 90°.

Thus, the two-phase servo motor acts like a phase-sensitive detector. Its response can be approximated as follows. The reference winding voltage $e_r[t]$ and the control winding voltage $e_c[t]$ are assumed to be

$$e_r[t] = E_r \sin[\omega_r t] \tag{8.1-15}$$

$$e_c[t] = E_c[t] \cos[\omega_r] t \tag{8.1-16}$$

The amplitude of the reference AC voltage is fixed at E_r. The control voltage $e_c[t]$ has an amplitude $E_c[t]$ that varies over the range $-E_r < E_c[t] < E_r$. The resultant torque is approximately given by

$$T[t] = e_c[t] \frac{2T_m}{E_r} \cos[\omega_r t] \tag{8.1-17}$$

where T_m is the maximum torque, which is generated in the motor at maximum control voltage. Let us assume that the amplitude of the control voltage $E_c[t]$ is held constant at the value E_c. Substituting Eq 8.1-16 into Eq 8.1-17 gives for the torque

$$T[t] = 2T_m (E_c/E_r) \cos^2[\omega_r t] = T_m \frac{E_c}{E_r}(1 + \cos[2\omega_r t]) \tag{8.1-18}$$

Hence, there is a torque proportional to the amplitude E_c of the control voltage, plus an oscillating torque component at the second harmonic of the

AC reference frequency ω_r. The second-harmonic torque is of such high frequency it is strongly attenuated by the motor inertia, and so produces only small perturbations in the resultant angular-velocity and angular-position signals.

AC Induction Tachometer. When one stator winding of a two-phase induction servo motor is excited with a constant reference AC voltage, the other winding develops an AC voltage with an amplitude proportional to speed. This voltage is in phase with the reference voltage for one direction of rotation, and is 180° out of phase for the other direction. Thus, an AC servo motor acts as an AC tachometer. To optimize the performance as a tachometer, a drag-cup rotor (rather than a squirrel-cage rotor) is used. This minimizes inertia and reduces cogging perturbations of the tachometer signal.

When the voltage on the reference winding has the form given in Eq 8.1-15, the tachometer voltage on the other winding is

$$E = KE_r \frac{\Omega}{\Omega_m} \cos[\omega_r t] \qquad (8.1\text{-}19)$$

where Ω is the tachometer angular velocity, and Ω_m is the maximum angular velocity for linear operation (typically 5500 RPM for a 400-Hz tachometer). A typical value for the constant K is 0.15. AC tachometers have achieved accuracies as high as one part in 2000.

According to Eq 8.1-17, the AC two-phase induction servo motor acts like a mathematically ideal AC demodulator, and Eq 8.1-19 indicates that the AC induction tachometer acts like a mathematically ideal AC modulator. However, there appear to be dynamic lags in the signal-frequency responses of both of these devices that are not included in the theoretical analyses. In particular, Gibson and Tuteur [8.1] (p. 300) report that there is a time lag in the AC tachometer response with a time constant "of the order of" one period of the excited frequency. For a 60-Hz carrier frequency, this would correspond to a lowpass filter in the signal frequency response with a break frequency of $60/2\pi$ Hz, which is 9.5 Hz. There is probably also a comparable dynamic lag in the signal frequency response of an AC servo motor.

Detailed discussions of the AC two-phase servo motor and the AC induction tachometer are given by Gibson and Tuteur [8.1] (Chapter 7) and by Davis and Ledgerwood [8.2] (Chapters 5, 6).

8.2 APPROXIMATE EFFECT OF AC NETWORKS ON THE SIGNAL-FREQUENCY RESPONSE

8.2.1 Signal-Frequency to Carrier-Frequency Transformation

Let us now consider the dynamic effect on the signal information of a network placed within the AC path. The transfer function of an element within the AC path is designated $F[\omega]$, while the transfer function of an element within the

DC signal path is designated $H[\omega]$. To simplify the analysis, the input is represented as the real part of a rotating vector. The input is assumed to have unity magnitude and zero phase. Hence the vector input signal is

$$X_i = e^{j\omega_m t} \tag{8.2-1}$$

The time-domain input signal is the real part of this:

$$x_i[t] = \text{Re}[X_i] = \text{Re}[e^{j\omega_m t}] = \cos[\omega_m t] \tag{8.2-2}$$

This is multiplied by the following reference signal:

$$r[t] = 2\cos[\omega_r t] = (e^{j\omega_r t} + e^{-j\omega_r t}) \tag{8.2-3}$$

The factor of 2 is included to achieve a net gain of unity in the AC section. The modulated vector signal is

$$X_{i(AC)} = X_i r[t] = (e^{j\omega_m t})(e^{j\omega_r t} + e^{-j\omega_r t})$$
$$= e^{j(\omega_m + \omega_r)t} + e^{j(\omega_m - \omega_r)t} \tag{8.2-4}$$

This has two frequency components at the frequencies $(\omega_m + \omega_r)$ and $(\omega_m - \omega_r)$. This AC signal is fed through a filter of transfer function $F[\omega]$. Each component is modified by the complex value of the transfer function $F[\omega]$ at its frequency $(\omega_m + \omega_r)$ or $(\omega_m - \omega_r)$. Hence, the output from the filter is

$$X_{o(AC)} = e^{j(\omega_m + \omega_r)t} F[\omega_m + \omega_r] + e^{j(\omega_m - \omega_r)t} F[\omega_m - \omega_r] \tag{8.2-5}$$

This signal is demodulated by multiplying it by the reference $\cos[\omega_r t]$. This gives the DC output signal X_o, which is

$$X_o = \cos[\omega_r t] X_{o(AC)} = \tfrac{1}{2}(e^{j\omega_r t} + e^{-j\omega_r t}) X_{o(AC)}$$
$$= \tfrac{1}{2}(e^{j\omega_r t} + e^{-j\omega_r t})$$
$$\times \{e^{j(\omega_m + \omega_r)t} F[\omega_m + \omega_r] + e^{j(\omega_m - \omega_r)t} F[\omega_m - \omega_r]\}$$
$$= \tfrac{1}{2}(F[\omega_m + \omega_r] + F[\omega_m - \omega_r])e^{j\omega_m t}$$
$$+ \tfrac{1}{2}F[\omega_m + \omega_r]e^{j(\omega_m + 2\omega_r)t} + \tfrac{1}{2}F[\omega_m - \omega_r]e^{j(\omega_m - 2\omega_r)t} \tag{8.2-6}$$

For the moment, we shall consider only the frequency components of X_o at the modulation frequency ω_m, and assume that the components at the frequencies $(\omega_m + 2\omega_r)$ and $(\omega_m - 2\omega_r)$ are eliminated by filtering. By Eq 8.2-1, the vector input signal X_i is $e^{j\omega_m t}$. Hence the transfer function X_o/X_i for the $e^{j\omega_m t}$ component is

$$H[\omega] = \frac{X_o}{X_i} = \tfrac{1}{2}(F[\omega_m + \omega_r] + F[\omega_m - \omega_r]) \tag{8.2-7}$$

This transfer function X_o/X_i, which is designated $H[\omega]$, is the signal-frequency equivalent of the AC transfer function $F[\omega]$.

When $\omega_m < \omega_r$, the frequency $(\omega_m - \omega_r)$ is negative. Let us consider what is meant by a negative frequency. If the frequency parameter ω in any transfer function is replaced by $-\omega$, each factor $j\omega$ is replaced by $-j\omega$. The resultant transfer function has the same magnitude, but the sign of the phase is reversed. Thus, for a transfer function $F[\omega]$, the magnitude and phase of $F[-\omega]$ are

$$|F[-\omega]| = |F[\omega]| \qquad (8.2\text{-}8)$$

$$\text{Ang}[F[-\omega]] = -\text{Ang}[F[\omega]] \qquad (8.2\text{-}9)$$

Hence $F[-\omega]$ is the conjugate of $F[\omega]$:

$$F[-\omega] = F[\omega]^* \qquad (8.2\text{-}10)$$

where the asterisk (*) denotes the conjugate. If this principle is applied to Eq 8.2-7, the transfer function $F[\omega_m - \omega_r]$ can be replaced by $F[\omega_r - \omega_m]^*$ when the quantity $(\omega_m - \omega_r)$ is negative. Hence, Eq 8.2-7 can be expressed in the following forms, depending on whether ω_m is greater than or less than ω_r:

$$H[\omega_m] = \tfrac{1}{2}(F[\omega_m + \omega_r] + F[\omega_r - \omega_m]^*) \qquad \text{for} \quad \omega_m < \omega_r \quad (8.2\text{-}11)$$

$$H[\omega_m] = \tfrac{1}{2}(F[\omega_m + \omega_r] + F[\omega_m - \omega_r]) \qquad \text{for} \quad \omega_m > \omega_r \quad (8.2\text{-}12)$$

This representation eliminates negative frequencies. On the other hand, during analysis it is convenient to ignore the issue of whether the frequency is positive or negative. Hence the form $F[\omega_m - \omega_r]$ is used even when ω_m is less than ω_r. Thus for analysis purposes it is convenient to deal with a fictitious negative frequency.

Let us find the ideal filter response $F[\omega]$ that operates on the AC modulated signal to produce the same effect as a specified signal-frequency response $H[\omega]$. This could theoretically be achieved by replacing every frequency ω in $H[\omega]$ by $(\omega - \omega_r)$. In Fig 8.2-1, curves ① and ② are the magnitude and phase plots of the desired signal-frequency transfer function $H[\omega]$. The dashed portions show the frequency response at negative frequencies. If ω is replaced by $(\omega - \omega_r)$ in the transfer function $H[\omega]$, the frequency-response plots ① and ② are shifted upward in frequency to form curves ③ and ④, which are the plots of the ideal AC frequency response $F[\omega]$. Hence

$$F[\omega_r + \omega_m] = H[\omega_m] \qquad (8.2\text{-}13)$$

$$F[\omega_r - \omega_m]^* = H[-\omega_m]^* = H[\omega_m] \qquad (8.2\text{-}14)$$

Comparing Eqs 8.2-13, -14 with Eq 8.2-7 shows that $F[\omega]$ satisfies the requirements of Eq 8.2-7.

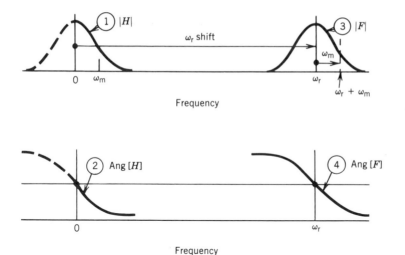

Figure 8.2-1 Shift of signal-frequency transfer function (lowpass filter) to reference (or carrier) frequency, to produce equivalent carrier-frequency transfer function (bandpass filter).

This ideal frequency response is not physically realizable, because $j\omega$ would be replaced by $j(\omega - \omega_r)$. This is equivalent to $(s - j\omega_r)$, which is not real for real values of s. On the other hand, the ideal frequency response can be closely approximated at frequencies close to ω_r by making the following substitution in the signal-frequency transfer function $H[\omega]$:

$$\omega \rightarrow \frac{(\omega - \omega_r)(\omega + \omega_r)}{2\omega} = \frac{\omega^2 - \omega_r^2}{2\omega} = \frac{\omega}{2} - \frac{\omega_r^2}{2\omega} \qquad (8.2\text{-}15)$$

For frequencies near the reference frequency ω_r, the factor $(\omega + \omega_r)/(2\omega)$ is approximately unity, and so this transformation approximates the ideal: $\omega \rightarrow (\omega - \omega_r)$. In terms of the complex frequency $j\omega$, the transformation of Eq 8.2-15 is

$$j\omega \rightarrow j\frac{\omega}{2} - j\frac{\omega_r^2}{2\omega} = \frac{j\omega}{2} + \frac{\omega_r^2}{2\,j\omega} \qquad (8.2\text{-}16)$$

Expressing this in terms of s gives

$$s \rightarrow \frac{s}{2} + \frac{\omega_r^2}{2s} = \frac{s^2 + \omega_r^2}{2s} \qquad (8.2\text{-}17)$$

This substitution converts a signal-frequency (or DC) transfer function into an equivalent carrier-frequency (or AC) transfer function. The transfer function can be implemented directly in terms of circuit components in the following manner. Let us define the impedance of an inductor L_1 in the signal-frequency (DC) transfer function $H[\omega]$ as Z_{DC}:

$$Z_{\mathrm{DC}} = sL_1 \qquad (8.2\text{-}18)$$

To form the equivalent AC filter, s is replaced by the expression of Eq 8.2-17, and so the signal-frequency impedance Z_{DC} is replaced by the following AC impedance Z_{AC}:

$$Z_{\mathrm{AC}} = \left(\frac{s}{2} + \frac{\omega_r^2}{2s} \right) L_1 = s\frac{L_1}{2} + \frac{\omega_r^2 L_1}{2s} \qquad (8.2\text{-}19)$$

As shown in Fig 8.2-2a, this equivalent impedance Z_{AC} in the AC filter is an

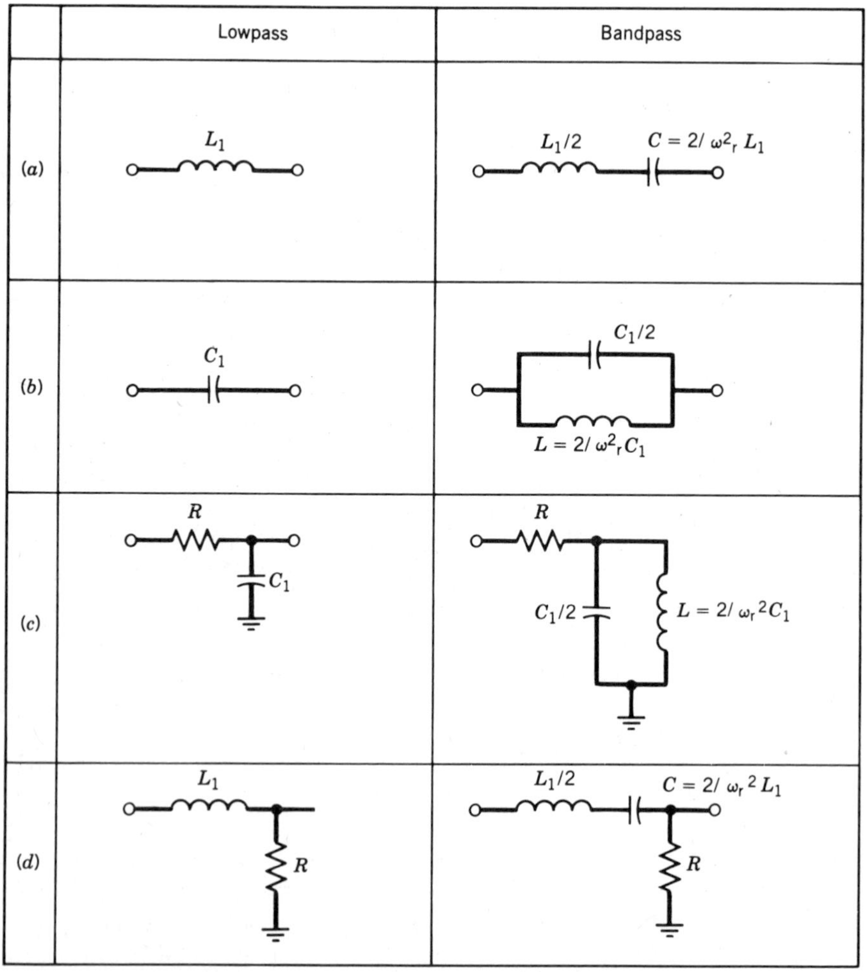

Figure 8.2-2 Equivalent-circuit elements to achieve signal-frequency to carrier-frequency transformation.

inductor of inductance $L_{AC} = L_1/2$ in series with a capacitor of capacitance $C_{AC} = 2/\omega_r^2 L_1$. The impedance Z_{AC} experiences series resonance at the frequency $1/\sqrt{L_{AC}C_{AC}}$, which is equal to

$$\frac{1}{\sqrt{L_{AC}C_{AC}}} = \frac{1}{(L_1/2)(2/\omega_r^2 L_1)} = \omega_r \qquad (8.2\text{-}20)$$

Thus, the series resonant frequency of the AC impedance Z_{AC} is equal to the reference (or carrier) frequency ω_r. In similar fashion, let us define the admittance of a capacitor C_1 of the signal-frequency (DC) filter $H[\omega]$ as Y_{DC}:

$$Y_{DC} = sC_1 \qquad (8.2\text{-}21)$$

In the equivalent AC filter, the capacitor C_1 is replaced by the following admittance Y_{AC}:

$$Y_{AC} = \left(\frac{s}{2} + \frac{\omega_r^2}{2s}\right)C_1 = s\frac{C_1}{2} + \frac{\omega_r^2 C_1}{2s} \qquad (8.2\text{-}22)$$

As shown in Fig 8.2-2b, this equivalent admittance Y_{AC} is a capacitor of capacitance $C_{AC} = C_1/2$ in parallel with an inductor of inductance $L_{AC} = 2/\omega_r^2 C_1$. This admittance experiences parallel resonance at the frequency ω_r.

In Fig 8.2.2, diagrams c and d show RL and RC signal-frequency lowpass filters, along with the corresponding AC (carrier-frequency) bandpass filters. These are constructed in accordance with the transformations of diagrams a and b.

Let us consider this signal-frequency to carrier-frequency transformation in terms of the transfer function. For every single-order factor of the form $(s + \omega_x)$ in the signal-frequency transfer function, the carrier-frequency transfer function has a quadratic factor of the form

$$(s + \omega_x) \rightarrow \frac{s^2 + 2\omega_x s + \omega_r^2}{2s} \qquad (8.2\text{-}23)$$

Thus, the carrier-frequency transfer function for a single-order low-pass filter of break frequency ω_x is obtained as follows:

$$\frac{\omega_x}{s + \omega_x} \rightarrow \frac{2\omega_x s}{s^2 + 2\omega_x s + \omega_r^2} \qquad (8.2\text{-}24)$$

Setting $s = j\omega$ gives the following frequency response of the equivalent carrier-frequency filter:

$$F[j\omega] = \frac{j2\omega\omega_x}{(\omega_r^2 - \omega^2) + j2\omega\omega_x} \qquad (8.2\text{-}25)$$

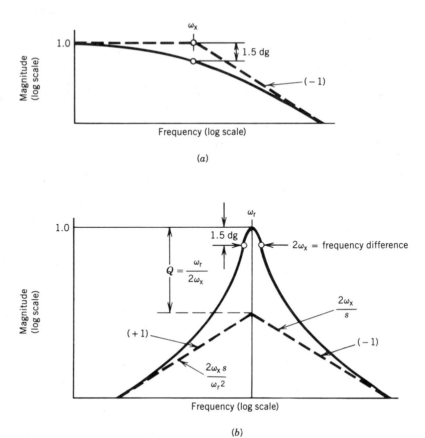

Figure 8.2-3   Frequency-response magnitude plots for (*a*) a single-order lowpass filter and (*b*) the equivalent bandpass filter.

Diagram *a* of Fig 8.2-3 shows the magnitude of the frequency response of the signal-frequency filter. The magnitude of this low-pass response is -1.5 dg at the break frequency ω_x. Diagram *b* shows the magnitude response of the equivalent carrier-frequency filter. The bandwidth of this bandpass response between the two -1.5-dg points is $2\omega_x$.

It can be shown from Eq 8.2-25 that the carrier-frequency filter has a low-frequency magnitude asymptote equal to $2\omega\omega_x/\omega_r^2$ and a high-frequency magnitude asymptote equal to $2\omega_x/\omega$. These asymptotes are indicated as dashed lines in Fig 8.2-3*b*. At the reference frequency ω_r, the magnitude asymptote value is $2\omega_x/\omega_r$, while the actual magnitude value is unity. The Q of the circuit is equal to the ratio of the peak magnitude value divided by the asymptote value at the resonant frequency ω_r, which is equal to

$$Q = \omega_r/2\omega_x \qquad (8.2\text{-}26)$$

Hence the width of the bandpass frequency response, between the half-power points (the points of -1.5-dg magnitude), is equal to the resonant frequency ω_r divided by Q:

$$2\omega_x = \omega_r/Q \tag{8.2-27}$$

When the signal-frequency filter has an underdamped quadratic factor, the corresponding carrier-frequency filter has a pair of quadratic factors stagger-tuned about the reference frequency. Consider for example the following lowpass signal-frequency transfer function:

$$H[s] = \frac{\omega_n^2}{s^2 + 2\zeta\omega_n s + \omega_n^2} \tag{8.2-28}$$

It can be shown, by using the signal-frequency to carrier-frequency transformation of Eq 8.2-17, that the equivalent AC transfer function is as follows:

$$F[s] = \frac{2(\omega_n/K)s}{s^2 + (1/Q)(\omega_r/K)s + (\omega_r/K)^2} \frac{2(K\omega_n)s}{s^2 + (1/Q)(K\omega_r)s + (K\omega_r)^2} \tag{8.2-29}$$

This transfer function consists of two tuned factors of the same Q, one tuned to the frequency ω_r/K and the other tuned to the frequency $K\omega_r$. The parameters of the transfer functions of Eqs 8.2-28, -29 are related by

$$4\zeta\omega_n = \frac{\omega_r(K + 1/K)}{Q} \tag{8.2-30}$$

$$(2\omega_n)^2 = \left(\frac{\omega_r}{Q}\right)^2 + \left(K\omega_r - \frac{\omega_r}{K}\right)^2 \tag{8.2-31}$$

Solve Eq 8.2-30 for ω_r/Q and substitute this into Eq 8.2-31. This gives the following for the difference between the two tuned frequencies ω_r/K and $K\omega_r$:

$$K\omega_r - \omega_r/K = 2\omega_n\sqrt{1 - \frac{4\zeta^2}{(K + 1/K)^2}} \tag{8.2-32}$$

Equation 8.2-30 gives the following for the amplification Q, which is the same for the two factors:

$$Q = \frac{\omega_r(K + 1/K)}{4\zeta\omega_n} \tag{8.2-33}$$

If the bandwidth of the stagger-tuned filter is reasonably narrow relative to the reference frequency ω_r, the factor $(K + 1/K)$ can be closely approximated by

$$K + \frac{1}{K} \doteq 2 \tag{8.2-34}$$

Substituting this approximation into Eqs 8.2-32, -33 gives the approximations

$$K\omega_r - \frac{\omega_r}{K} \doteq 2\omega_n\sqrt{1 - \zeta^2} \tag{8.2-35}$$

$$Q \doteq \omega_r/2\zeta\omega_n \tag{8.2-36}$$

In Fig 8.2-4, diagram *a* shows the magnitude of a quadratic lowpass signal-frequency filter of damping ratio $\zeta = 0.5$, and diagram *b* shows the

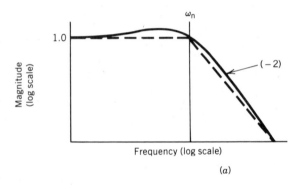

(a)

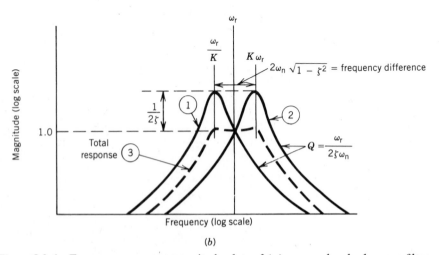

(b)

Figure 8.2-4 Frequency-response magnitude plots of (*a*) a second-order lowpass filter and (*b*) the equivalent bandpass filter.

dashed curve ③ the magnitude response of the resultant bandpass carrier-frequency filter. This bandpass response is the cascade effect of two responses ① and ②, tuned below and above the reference frequency ω_r at the frequencies ω_r/K and $K\omega_r$. These responses have the same Q, which is approximately equal to $\omega_r/2\zeta\omega_n$. The difference in frequency between $K\omega_r$ and ω_r/K is approximately equal to $2\omega_n\sqrt{1-\zeta^2}$. (Remember that on a logarithmic scale the spacing between frequencies ω_r/K and ω_r is equal to the ratio of the frequencies ω_r and $K\omega_r$, not their difference.) The gains of the two factors at the reference frequency ω_r are the same and are equal to

$$\text{Gain} = \frac{\omega_r}{2\omega_n Q} = \frac{2\zeta}{K + 1/K} \doteq \zeta \qquad (8.2\text{-}37)$$

8.2.2 AC Lead Networks

A network commonly implemented in the AC portion of a feedback control loop is the AC lead network. It is the carrier-frequency equivalent of the following signal-frequency transfer function:

$$H[s] = \frac{s + \omega_L}{s + \alpha\omega_L} \qquad (8.2\text{-}38)$$

The magnitude asymptote plot of this signal-frequency lead transfer function is shown in Fig 8.2-5b. Applying the transformation of Eq 8.2-23 gives the following equivalent AC transfer function:

$$F[s] = \frac{s^2 + 2\omega_L s + \omega_r^2}{s^2 + 2\alpha\omega_L s + \omega_r^2} \qquad (8.2\text{-}39)$$

Figure 8.2-5a shows an LC network for achieving this AC transfer function. The transfer function of this network is

$$F[s] = \frac{sL + R_2 + 1/sC}{sL + (R_1 + R_2) + 1/sC} = \frac{s^2 + s(R_2/L) + 1/LC}{s^2 + s(R_1 + R_2)/L + 1/LC}$$

Comparing Eqs 8.2-39, -40 gives

$$\omega_r = \frac{1}{\sqrt{LC}} \qquad (8.2\text{-}41)$$

$$\omega_L = \frac{R_2}{2L} \qquad (8.2\text{-}42)$$

$$\alpha = \frac{R_1 + R_2}{R_2} = 1 + \frac{R_1}{R_2} \qquad (8.2\text{-}43)$$

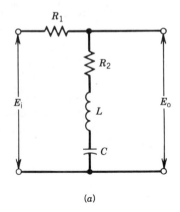

(a)

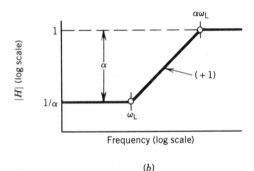

Frequency (log scale)

(b)

Figure 8.2-5 AC lead network using inductor: (a) network; (b) signal-frequency response.

This circuit requires a high-Q inductor, which is rather expensive and bulky. Therefore, AC lead networks are generally implemented with circuits using only resistance and capacitance components, such as the parallel Tee. An early study of the use of parallel-Tee networks in AC servos was given by Sobczyk [8.3], and was summarized in James, Nichols, and Phillips [1.2] (Section 3.14).

Figure 8.2-6a shows a symmetric form of the parallel-Tee circuit, which is often used as an AC lead network. Its transfer function is

$$\frac{E_o}{E_i} = \frac{s^2 + (2/\gamma - 1)\omega_r s + \omega_r^2}{s^2 + (1 + \gamma + 2/\gamma)\omega_r s + \omega_r^2} \qquad (8.2\text{-}44)$$

where

$$\omega_r = 1/RC \qquad (8.2\text{-}45)$$

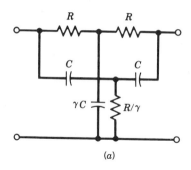

(a)

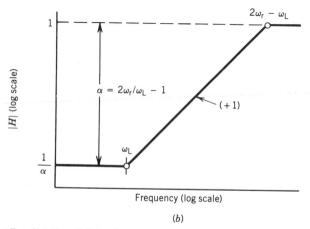

(b)

Figure 8.2-6 Parallel-Tee AC lead network: (a) network; (b) signal-frequency response.

Comparing Eq 8.2-40 to -44 gives

$$2\omega_L = \left(\frac{2}{\gamma} - 1\right)\omega_r \tag{8.2-46}$$

$$2\alpha\omega_L = \left(1 + \gamma + \frac{2}{\gamma}\right)\omega_r \tag{8.2-47}$$

Solve Eq 8.2-46 for γ and substitute this into Eq 8.2-47. This gives

$$\alpha\omega_L = \left(1 + \frac{\omega_L}{\omega_r} + \frac{1}{1 + 2\omega_L/\omega_r}\right)\omega_r \tag{8.2-48}$$

Since $2\omega_L/\omega_r \ll 1$, the following approximation holds to high accuracy:

$$\frac{1}{1 + 2\omega_L/\omega_r} \doteq 1 - \frac{2\omega_L}{\omega_r} \qquad (8.2\text{-}49)$$

This gives

$$\alpha\omega_L \doteq \left(2 - \frac{\omega_L}{\omega_r}\right)\omega_r = 2\omega_r - \omega_L \qquad (8.2\text{-}50)$$

Hence the parallel-Tee transfer function of Eq 8.2-44 can be approximated accurately by

$$F[s] = \frac{E_o}{E_i} = \frac{s^2 + 2\omega_L s + \omega_r^2}{s^2 + 2(2\omega_r - \omega_L)s + \omega_r^2} \qquad (8.2\text{-}51)$$

In accordance with the transformation of Eq 8.2-23, the equivalent signal-frequency transfer function is

$$H[s] = \frac{s + \omega_L}{s + 2\omega_r - \omega_L} \qquad (8.2\text{-}52)$$

A magnitude asymptote plot of this is shown in Fig 8.2-6*b*.

This network has the disadvantage that the upper break frequency $(2\omega_r - \omega_L)$ of the signal-frequency transfer function is very high and cannot be varied. To remedy this, the network can be placed in a feedback loop as shown in Fig 8.2-7. To simplify the analysis, the factor $(2\omega_r - \omega_L)$ in the denominator of Eq 8.2-51 is approximated by $2\omega_r$. The feedback loop has an amplifier gain K in cascade with the transfer function of the parallel Tee, giving a total loop transfer function of

$$G = \frac{K\left(s^2 + 2\omega_L s + \omega_r^2\right)}{s^2 + 4\omega_r s + \omega_r^2} \qquad (8.2\text{-}53)$$

The function $(1 + G)$ is equal to

$$1 + G = \frac{(1 + K)s^2 + (4\omega_r + 2K\omega_L)s + (1 + K)\omega_r^2}{s^2 + 4\omega_r s + \omega_r^2} \qquad (8.2\text{-}54)$$

There are two output signals in Fig 8.2-7: X_{o1}, which provides an AC lead network response; and X_{o2}, which provides the AC equivalent of a lowpass filter. The transfer functions for these two signals are

$$F_1 = \frac{X_{o(1)}}{X_i} = G_{ib} = \frac{G}{1 + G} \qquad (8.2\text{-}55)$$

$$F_2 = \frac{X_{o(2)}}{X_i} = KG_{ie} = \frac{K}{1 + G} \qquad (8.2\text{-}56)$$

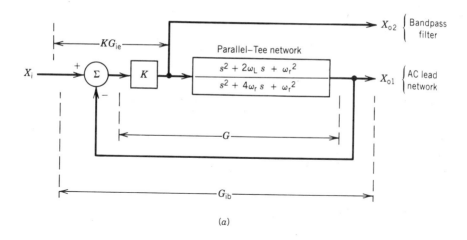

(a)

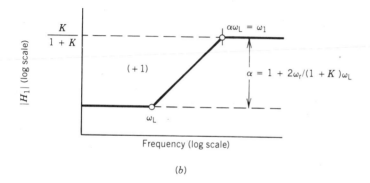

(b)

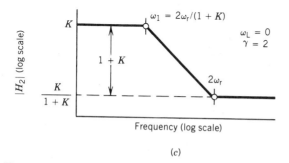

(c)

Figure 8.2-7 Feedback circuit using parallel-Tee network to achieve carrier-frequency lead network and bandpass filter: (a) circuit; (b) signal-frequency response of lead network, obtained from X_{o1}; (c) signal-frequency response of bandpass filter, obtained from X_{o2}.

Substituting Eqs 8.2-53, -54 into Eqs 8.2-55, -56 gives

$$F_1 = \frac{X_{o(1)}}{X_i} = \frac{1}{1 + K} \frac{s^2 + 2\omega_L s + \omega_r^2}{s^2 + 2\omega_1 s + \omega_r^2} \qquad (8.2\text{-}57)$$

$$F_2 = \frac{X_{o(2)}}{X_i} = \frac{K}{1 + K} \frac{s^2 + 4\omega_r s + \omega_r^2}{s^2 + 2\omega_1 s + \omega_r^2} \qquad (8.2\text{-}58)$$

where ω_1 is

$$\omega_1 = \frac{2\omega_r + K\omega_L}{1 + K} = \frac{2\omega_r}{1 + K} + \frac{K}{1 + K}\omega_L \qquad (8.2\text{-}59)$$

This can be approximated by

$$\omega_1 \doteq \frac{2\omega_r}{1 + K} + \omega_L \qquad (8.2\text{-}60)$$

In accordance with the transformation of Eq 8.2-23, the signal-frequency responses equivalent to the AC responses of Eqs 8.2-56, -57 are

$$H_1 = \frac{K}{1 + K} \frac{s + \omega_L}{s + \omega_1} \qquad \text{(for signal } X_{o1}) \qquad (8.2\text{-}61)$$

$$H_2 = \frac{K}{1 + K} \frac{s + 2\omega_r}{s + \omega_1} \qquad \text{(for signal } X_{o2}) \qquad (8.2\text{-}62)$$

Magnitude asymptote plots of these signal-frequency responses are shown in diagrams b and c of Fig 8.2-7. When the signal $X_{o(2)}$ is used to obtain the equivalent of a lowpass filter, the value of γ for the network is generally set equal to 2, so that ω_L is zero. Hence, the response of diagram c is shown for $\omega_L = 0$.

Figure 8.2-8 shows a practical opamp AC lead network circuit using a parallel Tee. The parallel-Tee network is adjusted so that $\gamma = 2$, which makes $\omega_L = 0$ for the network. A parallel signal path is provided by resistor R_i and potentiometer $\mathscr{E}R_i$, which produces the appropriate value for ω_L. If $R_i \ll R$, the transfer function of the combined network shown as F_n in the figure is

$$F_n = \mathscr{E} + \frac{s^2 + \omega_r^2}{s^2 + 4\omega_r s + \omega_r^2}$$

$$= \frac{s^2(1 + \mathscr{E}) + 4\mathscr{E}\omega_r s + (1 + \mathscr{E})\omega_r^2}{s^2 + 4\omega_r s + \omega_r^2}$$

$$= \frac{(1 + \mathscr{E})\left(s^2 + 4\omega_r s/(1 + \mathscr{E}) + \omega_r^2\right)}{s^2 + 4\omega_r s + \omega_r^2} \qquad (8.2\text{-}63)$$

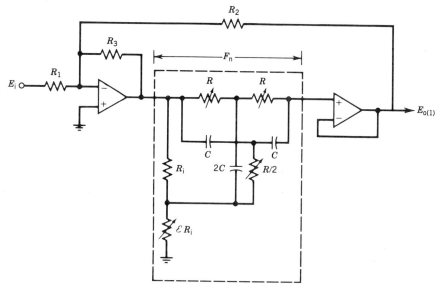

Figure 8.2-8 Convenient parallel-Tee circuit for implementing AC lead network.

Hence the potentiometer $\mathscr{E}R_i$ provides the following value for ω_L:

$$\omega_L = \frac{2\mathscr{E}\omega_r}{1 + \mathscr{E}} \doteq 2\mathscr{E}\omega_r \qquad (8.2\text{-}64)$$

The loop gain K is

$$K = \frac{R_3}{R_2}(1 + \mathscr{E}) \doteq \frac{R_3}{R_2} \qquad (8.2\text{-}65)$$

Using the expression for ω_1 in Eq 8.2-60 gives the following for the attenuation factor α for the lead network:

$$\alpha = \frac{\omega_1}{\omega_L} = 1 + \frac{2\omega_r}{(1 + K)\omega_L} = 1 + \frac{R_2}{\mathscr{E}(R_3 + R_2)} \qquad (8.2\text{-}66)$$

Potentiometers are placed in the resistive arms of the parallel Tee, which are adjusted to compensate for tolerance errors in the values of the capacitors. To perform this adjustment, the bypass potentiometer $\mathscr{E}R_i$ is set to zero, and the resistors are adjusted to obtain a deep null at the reference frequency ω_r. Then potentiometer $\mathscr{E}R_i$ is increased to achieve the desired value of ω_L, in accordance with Eq 8.2-64. Increasing $\mathscr{E}R_i$ shifts the null frequency of the filter downward by an amount approximately given by

$$\Delta\omega_r \doteq \frac{R_i}{R}\omega_L \qquad (8.2\text{-}67)$$

This shift can be compensated for by varying the parallel-Tee potentiometers slightly until the phase shift through the network at the frequency ω_r is zero.

8.3 EXACT ANALYSIS OF THE RESPONSE OF AN AC FEEDBACK LOOP

8.3.1 Effect of Nonideal Characteristics of the AC Transfer Function

Let us now consider how the signal frequency response is affected by the nonideal characteristics of the AC transfer function. In Fig 8.3-1a, the dashed curves show the magnitude and phase plots of the ideal (or desired) AC transfer function $F_d[\omega]$ that would be formed by replacing ω by $(\omega - \omega_m)$ in the desired signal-frequency transfer function $H_d[\omega]$. The solid curves show the actual magnitude and phase plots formed by replacing ω by the expression $(\omega - \omega_r)(\omega + \omega_r)/2\omega$. The region of the frequency response for frequencies less than the reference frequency ω_r is called the *lower sideband*, and the region for frequencies greater than ω_r is called the *upper sideband*.

As shown by the author [8.4], the actual plots of magnitude and phase are shifted upward in frequency relative the ideal plots by an amount $\Delta\omega$ given by

$$\Delta\omega = \sqrt{\omega_m^2 + \omega_r^2} - \omega_r \qquad (8.3\text{-}1)$$

As shown, ω_m is the frequency difference between the reference frequency ω_r and the frequency of a particular point on the ideal $F_d[\omega]$ plot.

The frequency-response curves in diagram b of Fig 8.3-1 are formed from those of diagram a by shifting the zero-frequency axis up to the carrier frequency ω_r, and then rotating the lower sideband about the new zero-frequency axis, while reversing the sign of the phase. The dashed upper and lower sidebands of the desired (ideal) transfer function $F_d[\omega]$ fold onto one another in diagram b to produce single dashed plots of the magnitude and phase, shown as curves ①, ④. These are the magnitude and phase plots of the desired signal-frequency response $H_d[\omega]$. The solid curves in diagram b are derived from the solid lower and upper sideband plots of diagram a. The magnitude and phase curves ③ and ⑥, which are labeled $H_{(+)}$, are derived from the upper sidebands of $F[\omega]$ in diagram a. The magnitude and phase curves ② and ⑤, where are represented as $H_{(-)}^*$, are derived from the lower sidebands of $F[\omega]$ in diagram a.

As shown in Fig 8.3-1b, the upper sideband plots of $H_{(+)}$, represented by curves ③, ⑥, can be formed from the desired signal-frequency response

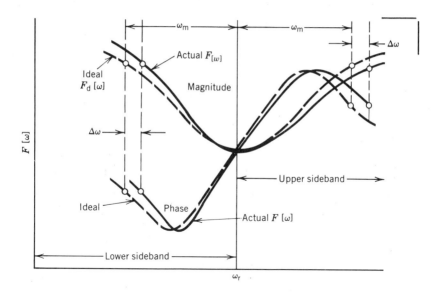

(a)

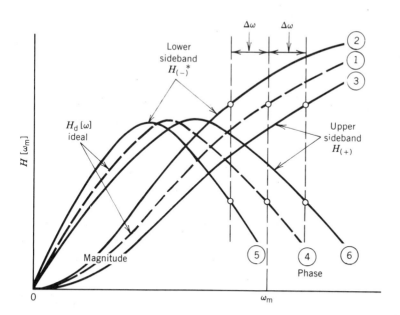

Frequency (linear scale)

(b)

Figure 8.3-1 Graphical approach for calculating equivalent signal-frequency response of a carrier-frequency filter: (*a*) magnitude and phase plots for carrier-frequency filter: (*b*) translation of carrier-frequency magnitude and phase plots to signal frequency, to compute equivalent signal-frequency response.

61

$H_d[\omega_m]$ by shifting the points of curves ①, ④ upward in frequency (relative to the ω_m frequency scale) by the amount $\Delta\omega$ given by Eq 8.3-1. Similarly, the lower sideband plots of $H^*_{(-)}$, represented by curves ②, ⑤, can be formed from the desired signal-frequency response $H_d[\omega_m]$ by shifting the points of curves ①, ④ downward in frequency (relative to the ω_m frequency scale) by this same amount $\Delta\omega$.

Equation 8.2-11 gave an expression for the equivalent signal-frequency transfer function $H[\omega_m]$ in terms of the carrier-frequency transfer function $F[\omega]$. In terms of the sideband plots of $H_{(+)}$ and $H^*_{(-)}$ shown in Fig 8.3-1*b*, Eq 8.2-11 becomes

$$H[\omega_m] = \tfrac{1}{2}\big(H_{(+)}[\omega_m] + H^*_{(-)}[\omega_m]\big) \qquad (8.3\text{-}2)$$

Let us apply this to a practical example. Consider the following signal-frequency transfer function, which is a single-order lowpass filter of half-power frequency $0.25\omega_r$:

$$H_d[\omega_m] = \frac{1}{1 + s/(0.25\omega_r)} \qquad (8.3\text{-}3)$$

In Fig 8.3-2, curve (1) shows the magnitude plot of $H_d[\omega_m]$ for frequencies up to $10\omega_r$. Curve ③ is the magnitude of $H_{(+)}[\omega_m]$, the upper sideband, which is derived by shifting the points on curve ① upward in frequency by the

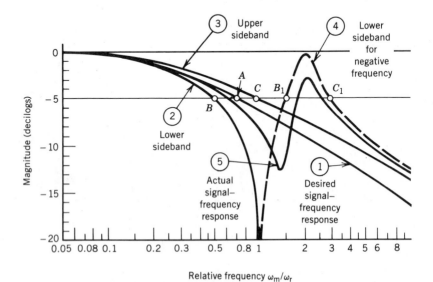

Figure 8.3-2 Calculation of actual signal-frequency magnitude response for a desired response of $1/(1 + s/\omega_x)$, where ω_x is $0.25\omega_r$.

amount $\Delta\omega$ given by Eq 8.3-1. Curve ② is the magnitude of $H^*_{(-)}[\omega_m]$, the lower sideband, which is derived by shifting the points on curve ① downward in frequency by $\Delta\omega$. Consider, for example, point A at the frequency $\omega_m = 0.72\omega_r$, where the magnitude of the ideal transfer function $H_d[\omega_m]$ is -5 dg. The frequency shift for this point A is

$$\Delta\omega = \sqrt{\omega_m^2 + \omega_r^2} - \omega_r = \sqrt{(0.72\omega_r)^2 + \omega_r^2} - \omega_r$$

$$= 0.23\omega_r \qquad\qquad (8.3\text{-}4)$$

Thus, the frequencies of -5-dg magnitude are

Lower sideband $H^*_{(-)}[\omega_m]$ (point B):

$$\omega_m = \omega_{m(B)} = 0.72\omega_r - 0.23\omega_r = 0.49\omega_r \qquad (8.3\text{-}5)$$

Upper sideband $H_{(+)}[\omega_m]$ (point C):

$$\omega_m = \omega_{m(C)} = 0.72\omega_r + 0.23\omega_r = 0.95\omega_r \qquad (8.3\text{-}6)$$

For the lower sideband $H^*_{(-)}[\omega_m]$ (curve ②), the point $\omega_m = \omega_r$ corresponds to $\omega = 0$ for the AC frequency response $F[\omega]$. For higher ω_m frequencies, the lower sideband is described by the dashed curve ④, which is derived from the negative-frequency portion of the response $F[\omega]$. Curve ④ can be explained with the help of Fig 8.3-3. Figure 8.3-3a shows, on linear scales, a magnitude plot of the AC frequency response $F[\omega]$ of the bandpass filter. The dashed portion shows the plot of $F[\omega]$ for negative frequencies. Figure 8.3-3b shows the magnitude plots of the corresponding upper and lower sidebands $H_{(+)}[\omega_m]$ and $H^*_{(-)}[\omega_m]$. These sideband plots are obtained by folding the plot of $F[\omega]$ about the frequency $\omega = \omega_r$.

Note that the lower-sideband magnitude plot of $H^*_{(-)}[\omega_m]$ in Fig 8.3-3b has a peak at $\omega_m = 2\omega_r$. This peak is derived from the value of $F[\omega]$ at $\omega = -\omega_r$, which has the same magnitude as at $\omega = +\omega_r$. In like fashion, the dashed curve ④ in Fig 8.3-2 has a peak at $\omega_m = 2\omega_r$, which is equal to the magnitude of $F[\omega]$ at $\omega = \omega_r$.

In Fig 8.3-4, curve ① is the phase plot of $H_d[\omega_m]$, the desired signal-frequency transfer function. Curve ② is the phase of the lower sideband $H^*_{(-)}$ (for $\omega_m < \omega_r$), and curve ③ is the phase of the upper sideband $H_{(+)}$. These sideband plots are obtained in the same manner as the magnitude plots of Fig 8.3-2. Consider, for example, point A on the $H_d[\omega_m]$ plot at a phase of $-60°$, which occurs at a frequency of $0.433\omega_r$. By Eq 8.3-1, $\Delta\omega = 0.090\omega_r$. The frequencies of $-60°$ phase on the lower and upper sidebands (which are $\omega_m \pm \Delta\omega$) are: $\omega_{m(B)} = 0.343\omega_r$, $\omega_{m(C)} = 0.523\omega_r$.

The dashed curve ④ in Fig 8.3-4 shows the phase of the lower sideband $H^*_{(-)}[\omega_m]$ for $\omega_m > \omega_r$. Just as for curve ④ in Fig 8.3-2, this is derived from

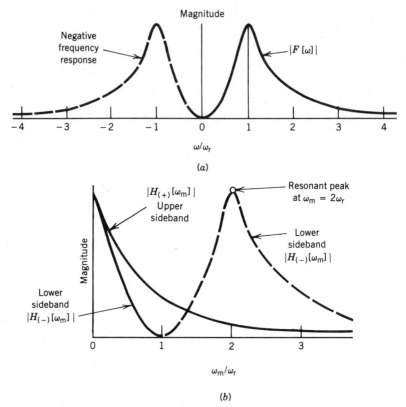

Figure 8.3-3 Magnitude plots on linear scales for bandpass filter tuned to reference frequency ω_r: (*a*) frequency response of bandpass filter about frequency ω_r: (*b*) upper and lower sidebands.

the negative-frequency portion of $F[\omega]$. However, the phase plot is somewhat more complicated because the phase of $H^*_{(-)}[\omega_m]$ is the negative of the corresponding phase of $F[\omega]$, and the phase of $F[-\omega]$ is the negative of the phase of $F[\omega]$.

By applying Eq 8.3-2, the equivalent signal-frequency transfer function $H[\omega_m]$ can be calculated from the magnitude and phase plots for the lower and upper sidebands given as curves ②, ③, ④ in Figs 8.3-2, -4. The resultant plots of magnitude and phase are shown as curve ⑤ in Figs 8.3-2, -4. The following illustrates the calculation of the values at the frequency $\omega_m/\omega_r = 0.7$. At this frequency, the sideband magnitude and phase values in Figs 8.3-2, -4 are:

Upper sideband at $\omega_m/\omega_r = 0.7$:

$$\left|H_{(+)}[\omega_m]\right| = -3.87 \text{ dg} = 0.410 \qquad (8.3\text{-}7)$$

$$\text{Ang}\left[H_{(+)}[\omega_m]\right] = -65.8° \qquad (8.3\text{-}8)$$

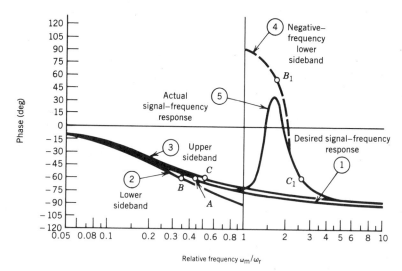

Figure 8.3-4 Calculations of the actual signal-frequency phase response for a desired response of $1/(1 + s/\omega_x)$, where ω_x is $0.25\omega_r$.

Lower sideband at $\omega_m/\omega_r = 0.7$:

$$\left|H^*_{(-)}[\omega_m]\right| = -7.89 \text{ dg} = 0.163 \qquad (8.3\text{-}9)$$

$$\text{Ang}\left[H^*_{(-)}[\omega_m]\right] = -80.6° \qquad (8.3\text{-}10)$$

By Eq 8.3-2, $H[\omega_m]$ at this frequency is the complex mean value of these upper and lower sideband values, which is

$$H[\omega_m] = \tfrac{1}{2}\left(0.410\underline{/-65.8°} + 0.163\underline{/-80.6°}\right)$$

$$= \tfrac{1}{2}\left[(0.0168 - j0.374) + (0.027 - j0.161)\right]$$

$$= 0.0975 - j0.2675 = 0.285\underline{/-70.0°} \qquad (8.3\text{-}11)$$

Since 0.285 corresponds to -5.45 dg, the function $H[\omega_m]$ at the frequency $\omega_m/\omega_r = 0.7$ has a magnitude of -5.45 dg and phase of $-70.0°$. These values are shown on curve ⑤ in Figs 8.3-2, -4.

This graphical approach provides insight into the characteristics of the signal-frequency transfer function $H[\omega_m]$. However, to obtain an accurate plot of $H[\omega]$, it is more convenient to use Eq 8.2-7 along with digital computation. Equation 8.2-7 is repeated as follows:

$$H[\omega_m] = \tfrac{1}{2}\left(F[\omega_m + \omega_r] + F[\omega_m - \omega_r]\right) \qquad (8.3\text{-}12)$$

In accordance with Eq 8.2-24, the following is the carrier-frequency transfer function for achieving the desired signal-frequency transfer function, which was given in Eq 8.3-3:

$$F[\omega] = \frac{j\omega(\omega_r/2)}{(\omega_r^2 - \omega^2) + j\omega(\omega_r/2)} \qquad (8.3\text{-}13)$$

Substitute this into Eq 8.3-12. This gives the following for the signal-frequency transfer function $H[\omega_m]$:

$$H[u] = \frac{u^2 - 1 + j2u(u^2 - 2)}{17u^2 - 1 - 4u^4 + j4u(u^2 - 2)} \qquad (8.3\text{-}14)$$

where u is the normalized ω_m frequency, equal to

$$u = \omega_m/\omega_r \qquad (8.3\text{-}15)$$

The magnitude and phase of this transfer function are plotted as curve ⑤ in Figs 8.4-2, -4.

A comparison of curves ① and ⑤ in Figs 8.3-2, -4 shows that the actual signal-frequency transfer function $H[\omega_m]$ (curve ⑤) follows the desired signal-frequency transfer function (curve ①) quite accurately for ω_m frequencies up to the reference frequency ω_r. The reason for this is that the two sidebands $H_{(+)}$, $H_{(-)}^*$ shift in opposite directions relative to $H_d[\omega_m]$, and so the mean of their complex values, which is $H[\omega_m]$, is quite close to $H_d[\omega_m]$. On the other hand for frequencies greater than ω_r, $H[\omega_m]$ (curve ⑤) departs strongly from the desired signal-frequency transfer function (curve ①), and in particular has a resonance peak at $\omega_m = 2\omega_r$.

Thus, the nonideal characteristics of the AC transfer functions are unimportant for modulation frequencies below the carrier frequency, because the frequency shifts of the two sidebands tend to cancel one another for $\omega_m < \omega_r$. The actual response is very close to the ideal.

At frequencies greater than $\omega_m = \omega_r$, the equivalent DC frequency response $H[\omega_m]$ departs strongly from the ideal. However, at such high frequencies, the theory becomes much more complicated in any case, because the effect of harmonics must be included in the analysis.

8.3.2 Effect on Signal-Frequency Response of Harmonics Produced by Modulation and Demodulation

In Eq 8.2-6, the components at the frequencies $(\omega_m + 2\omega_r)$ and $(\omega_m - 2\omega_r)$ were ignored. In a feedback loop, these harmonics are translated back to the signal-frequency band in the modulation and demodulation processes of the next pass around the loop. Hence they can have important effect on the overall

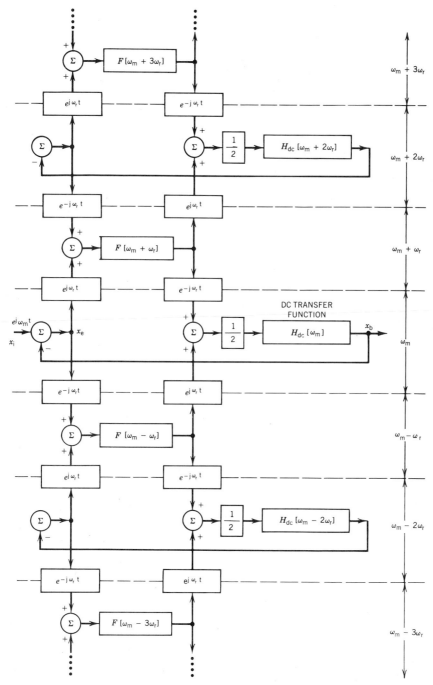

Figure 8.3-5 Signal-flow diagram for exact calculation of response of AC feedback-control system, assuming ideal modulation and demodulation.

frequency response of the loop. A whole series of sidebands are generated in the modulation and demodulation processes, which makes the analysis very complicated.

Figure 8.3-5 shows the signal-flow diagram of an analytical model that the author developed for calculating the effects of the various harmonics. It assumes that modulation and demodulation are mathematically ideal multiplication processes, as described in Section 8.1.1. On the other hand, as was shown in Fig 8.1-11, practical demodulators usually generate quite different harmonics than is assumed in the mathematical model. Hence, the model is at best only an approximation of the actual situation. The following are the results of the analysis, which ignores the effects of harmonics above the fourth.

The transfer functions of the DC and AC portions of the feedback loop are designated $H_{DC}[\omega]$ and $F[\omega]$. The following transfer functions are defined:

$$H_{AC}[\omega_x] = \tfrac{1}{2}(F[\omega_x + \omega_r] + F[\omega_x - \omega_r]) \qquad (8.3\text{-}16)$$

$$G[\omega_x] = H_{DC}[\omega_x]H_{AC}[\omega_x] \qquad (8.3\text{-}17)$$

$$G_{ib}[\omega_x] = G[\omega_x]/(1 + G[\omega_x]) \qquad (8.3\text{-}18)$$

From these the transfer functions $A_{(+)}$ and $A_{(-)}$ are defined as

$$A_{(+)} = \frac{F[\omega_m + 3\omega_r]^2 G_{ib}[\omega_m + 2\omega_r]G_{ib}[\omega_m + 4\omega_r]}{4H_{AC}[\omega_m + 2\omega_r]H_{AC}[\omega_m + 4\omega_r]} \qquad (8.3\text{-}19)$$

$$A_{(-)} = \frac{F[\omega_m - 3\omega_r]^2 G_{ib}[\omega_m - 2\omega_r]G_{ib}[\omega_m - 4\omega_r]}{4H_{AC}[\omega_m - 2\omega_r]H_{AC}[\omega_m - 4\omega_r]} \qquad (8.3\text{-}20)$$

The resultant signal-frequency transfer function is designated $G^{(AC)}[\omega_m]$, and is equal to

$$\frac{G^{(AC)}[\omega_m]}{G[\omega_m]} = 1 - \frac{F[\omega_m + \omega_r]^2 G_{ib}[\omega_m + 2\omega_r](1 + A_{(+)})}{4H_{AC}[\omega_m + 2\omega_r]H_{AC}[\omega_m]}$$

$$- \frac{F[\omega_m - \omega_r]^2 G_{ib}[\omega_m - 2\omega_r](1 + A_{(-)})}{4H_{AC}[\omega_m - 2\omega_r]H_{AC}[\omega_m]} \qquad (8.3\text{-}21)$$

8.3.3   Sampled-Data Model of an AC Feedback Loop

A much simpler analytical model results when a sample-and-hold circuit is used in the demodulation process. This converts the loop to a sampled-data feedback loop, which can be analyzed much more simply. The AC modulated system is equivalent to a sampled-data feedback system with a sampling frequency equal to twice the AC carrier frequency ω_r.

Chapter 9 shows that the maximum gain crossover frequency that can be sustained with good stability in a sampled-data feedback loop is $\frac{1}{8}$ of the sampling frequency. Hence, the maximum allowable gain crossover frequency for the AC carrier system is $\frac{1}{4}$ of the reference frequency. With this criteria, the maximum achievable gain crossover frequency should be 15 Hz, for a 60-Hz carrier frequency, and 100 Hz for a 400-Hz carrier frequency. Experience suggests that these limits are, if anything, greater than can be achieved with an AC servo. Hence, the sampled-data model provides a reasonable estimate of the dynamic effect of AC modulation in a feedback-control system, even when a sample-and-hold circuit is not used in the demodulation process.

8.4 DISCUSSION OF PRACTICAL AC CONTROL SYSTEMS AND AC POSITION TRANSDUCERS

8.4.1 Analog AC Control Systems

Figure 8.4-1 shows a common type of AC servo used in the days of vacuum-tube amplifiers. The position sensor is a synchro, which is equivalent to a resolver with a three-phase stator winding. The shaft of one synchro is connected to the command angle Θ_k, and the other is connected to the controlled angle Θ_c. The rotor for the command synchro (which is called a synchro transmitter) is excited by an AC reference voltage $e_r[t]$ given by

$$e_r[t] = E_r \cos[\omega_r t] \tag{8.4-1}$$

The second synchro has a higher impedance than the synchro transmitter, and is called a synchro control transformer. The rotor voltage from the second synchro provides an error voltage $e_e[t]$ equal to

$$e_e[t] = E_r \sin[\Theta_k - \Theta_c]\cos[\omega_r t] \doteq E_r(\Theta_k - \Theta_c)\cos[\omega_r t] \tag{8.4-2}$$

The amplitude of this voltage is proportional to the sine of the angular difference $(\Theta_k - \Theta_c)$ between the command and control angles. For small angles, this amplitude is approximately proportional to the angular error $(\Theta_k - \Theta_c)$.

This AC modulated error signal is fed through an AC amplifier, an AC lead network, a 90° phase-shift network, and an AC power amplifier. (The AC lead network may be replaced by tachometer feedback, provided by an AC tachometer, to achieve greater accuracy in the presence of static friction.) The AC power amplifier drives the control winding of a two-phase AC servo motor. The reference winding of the servo motor is excited by the same voltage that excites the command synchro. Because of the 90° phase-shift network, the AC excitations on the two servo-motor windings are in quadrature. The AC servo motor drives the controlled member through the gear train, and changes the

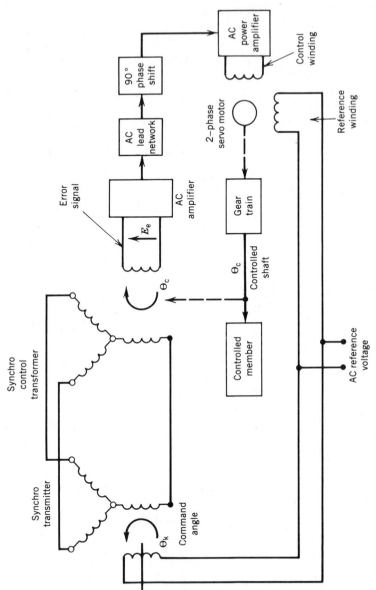

Figure 8.4-1 Block diagram of AC servo with synchro data.

Control winding

Reference winding

AC reference voltage

2-phase servo motor

Gear train

Controlled member

Controlled shaft

Θ_c

Θ_c

AC power amplifier

90° phase shift

AC lead network

AC amplifier

Error signal

E_e

Synchro control transformer

Synchro transmitter

Command angle

Θ_k

controlled angle Θ_c until the AC error signal is reduced to zero. This occurs when $\Theta_c = \Theta_k$.

The 90° phase-shift network degrades the dynamic performance of the AC servo, because the phase-shift network adds phase lag to the signal-frequency response of the servo loops. This network can be eliminated by phase-shifting the winding of the AC motor relative to the reference by 90°. However, the latter is more difficult, because it requires rather large AC capacitors.

With the development of the transistor, DC amplification became much easier. As a result, AC servo motors have been largely replaced with DC servo motors, which have much better efficiency and better dynamic performance. (This point is discussed in Chapter 12, Section 12.3.) With a DC servo motor, the same synchro data system can be used simply by feeding the AC error signal from the synchro to a phase-sensitive demodulator, as was illustrated in Fig 8.1-10.

Very often servos with synchro data systems use two sets of synchros, which provide fine–coarse synchro data. The coarse synchro is directly coupled to the controlled shaft, or geared 1 : 1. The fine synchro is accurately geared up relative to the controlled shaft, so that it rotates many revolutions for one revolution of the controlled angle. This approach allows the synchros to provide much greater angular accuracy.

The coarse synchro signal is used only to bring the controlled shaft to the approximate angle, to within an error equivalent to about 90° of rotation of the fine synchro. At smaller errors, the coarse synchro signal is switched OFF, and the fine synchro signal provides accurate positioning. Various techniques have been developed over the years to allow the servo amplifier to switch between the fine and coarse synchro signals.

Another type of synchro is the synchro differential, which has a three-phase winding on the rotor as well as on the stator. A synchro transmitter, synchro differential, and synchro control transformer are commonly connected as shown in Fig 8.4-2. The rotor of the synchro transmitter is excited with the reference voltage $E_r\cos[\omega_r t]$. The voltage sensed on the winding of the control transformer is

$$E = E_r\cos[\omega_r t]\sin[\Theta_1 + \Theta_2 + \Theta_3] \qquad (8.4\text{-}3)$$

where $\Theta_1, \Theta_2, \Theta_3$ are the shaft angles of the three synchros.

8.4.2 AC Transducers in Digital Control Systems

Modern position-control systems generally operate with a digital measure of position. However, the synchro and resolver are still very popular, because of the synchro-to-digital and resolver-to-digital converters, which are now supplied by many electronics firms.

A resolver is very similar to a sychro, but has a two-phase stator winding rather than a three-phase winding. As was shown in Eqs 8.1-12 to -14, the

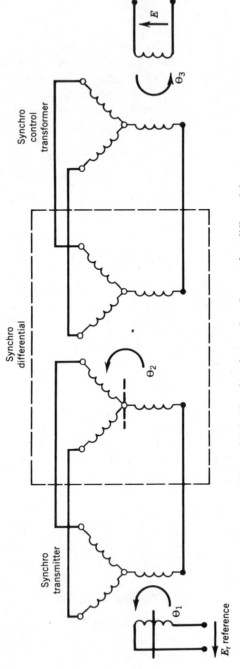

Figure 8.4-2 Synchro-data train using synchro differential.

72

rotor winding of the resolver is excited with a reference voltage $E_r\cos[\omega_r t]$, and the two stator windings deliver the following voltages:

$$e_1[t] = E_r\cos[\Theta]\cos[\omega_r t] \tag{8.4-4}$$

$$e_2[t] = E_r\sin[\Theta]\cos[\omega_r t] \tag{8.4-5}$$

where Θ is the angle of the resolver shaft. The two resolver voltages and the reference voltage are fed to a resolver-to-digital converter, which derives from these AC voltages a digital measure of angle. The resolution is generally 12 to 14 bits. The accuracy for a high-quality 14-bit converter is typically ± 4 arc minute, exclusive of resolver error. This corresponds to ± 3 quanta at 14 bits.

In a synchro-to-digital converter, a Scott-Tee transformer is used to convert the three-phase stator voltages of the synchro to two-phase voltages, equivalent to those from a resolver. Figure 8.4-3a shows the standard representation of the three phase wndings of a synchro, and a Scott-Tee circuit for converting the synchro signals to equivalent resolver signals. The time-domain expressions for the voltages E_1, E_2, E_3 on the synchro windings (relative to neutral) are

$$E_1 = E_{0-s1} = E_r\sin[\Theta]\cos[\omega_r t] \tag{8.4-6}$$

$$E_2 = E_{0-s2} = E_r\sin[\Theta - 120°]\cos[\omega_r t] \tag{8.4-7}$$

$$E_3 = E_{0-s3} = E_r\sin[\Theta + 120°]\cos[\omega_r t] \tag{8.4-8}$$

where Θ is the shaft angle. These voltages E_1, E_2, E_3 are represented in vector form in diagrams b and c. The projection of a vector along the vertical (imaginary) axis is the amplitude of the vector. Positive amplitude represents an AC voltage in phase with the AC reference that is applied to the synchro rotor winding; negative amplitude represents an AC voltage 180° out of phase with that reference. Diagram b shows the vectors for a shaft angle Θ of 30°; diagram c, for 90°.

The voltage difference between winding terminals $S1$ and $S3$ is designated E_{13}, and is represented as a vector in diagrams b and c. The voltage E_A, measured across the output winding A of the Scott-Tee, is proportional to this voltage E_{13}. Point X lies on the winding midway between terminals $S1$ and $S2$. The voltage E_X at this point X, measured relative to neutral, is represented as a vector in diagrams b, c. The voltage difference between point X and winding terminal $S2$ is designated E_{X2}, and is shown as a vector in diagrams b, c. The voltage E_B, measured across output winding B of the Scott-Tee transformer, is proportional to the voltage E_{X2}. Vectors E_{13} and E_{X2} are in quadrature with one another, and therefore the voltages E_A and E_B at the output windings of the Scott-Tee transformer are in quadrature, relative to the angle Θ. (Remember, however, that the actual AC voltages on these two windings are in phase.)

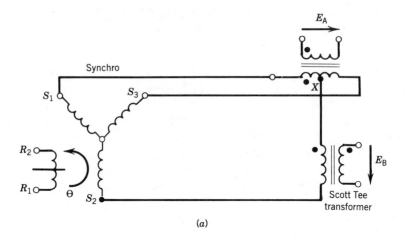

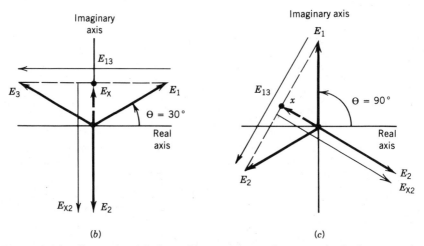

Figure 8.4-3   Conversion of three-phase synchro voltages to equivalent two-phase resolver voltages with Scott-Tee transformer network: (a) circuit using Scott-Tee transformer; (b, c) vector diagrams of synchro and Scott-Tee voltages.

A Scott-Tee transformer is required to transform synchro signals to equivalent resolver signals before converting the information to digital data. However, an isolation transformer is also required when the signals are derived directly from a resolver. Therefore the hardware is essentially equivalent in resolver-to-digital and synchro-to-digital converter modules.

An accurate size-11 synchro (1.1 in. diameter) typically has an accuracy of about ± 6 arc minutes, while an accurate size-11 resolver typically has about half this error. For a size-25 resolver (2.5 in. diameter), an accuracy of ± 0.5 arc minute is typical.

There are also fine–coarse resolver-to-digital and synchro-to-digital converters, which convert to digital data the two sets of signals from fine–coarse resolvers and synchros. The resolution of such converters is typically 16 bits. The accuracy of the converter alone (not including resolver or synchro error) is typically ± 20 to ± 30 arc second.

A fine–coarse resolver (or synchro) system requires an accurate and bulky gear train to gear up the fine resolver relative to the controlled shaft. To avoid this gearing, transducers are available that provide equivalent acuracy when coupled directly to the controlled shaft. One such device is the multispeed resolver, which has many poles, and generates many electrical resolver cycles for one mechanical revolution. A separate one-speed winding is included to provide the coarse resolver signal. This device can be coupled directly to the controlled shaft. Its signals are fed to a fine–coarse resolver-to-digital converter to achieve a resolution of 16 bits or greater.

Another AC angular sensing device is the Inductosyn (the trademark of a sensor made by Farrand). This consists of a rotating plate, which contains radial strips of magnetic material. The instrument provides output signals equivalent to those from a multispeed resolver. However, it does not include a coarse data signal. The generated signals are much lower than those of a resolver, and so modern versions of this instrument usually contain integrated electronics to raise the signal level. Typically the Inductosyn provides 256 cycles of fine data per mechanical cycle. A resolution of 20 bits has been achieved with such equipment with an Inductosyn unit of 5-in. diameter.

The Inductosyn is also commonly used in machine tools for linear measurement. In such instruments the magnetic strips are arranged as a linear array.

The preceeding devices all use AC modulation in the basic instrument. However, in modern applications the final output is generally digital. The same requirements can also be satisfied by digital optical encoders.

There are two broad classes of optical encoders: incremental and absolute. An incremental encoder provides direction-sensed pulses, which are counted to determine the change of angular position of the shaft. The encoder counts up and down relative to a reference to obtain the position measurement. An additional synchronizing pulse from another sensor is required to establish the zero reference. In contrast, the absolute digital encoder provides an absolute angular measurement with each readout.

An incremental optical encoder is much less expensive than an absolute encoder, and so the device is used very widely. However, in critical applications, it may not be practical to synchronize the instrument with the zero reference when power is lost. Besides, the accuracy with which the synchronizing pulse can be measured and integrated with the optical-encoder data is limited. Hence, the absolute encoder can generally provide greater accuracy.

Simple optical encoders usually operate by shining a beam of light through a single slit. However, an accurate optical encoder shines the light through a

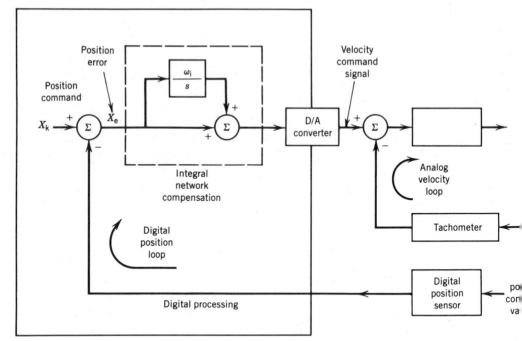

Figure 8.4-4 Digital computation to form position-loop error signal, and to provide integral-network compensation in position loop.

great many slits, and senses an interference pattern between the slits of the rotating disk and the slits on a fixed reticle. The sensing process is analog, and provides a resolution that is a fraction of a slit width. Optical encoders have been designed to very high resolution and accuracy. For example, Itek Measurement Systems has an optical encoder, of 15-in. diameter with an 8-in. central hole, which has a resolution of 0.3 arc second (22 bits).

When an angular transducer of very high accuracy is coupled to a controlled shaft, the coupling mechanism is critical and can be rather large. The design of such coupling devices requires considerable mechanical expertise.

With a conventional synchro-data system, the synchro itself performs the process of subtracting the control angle from the command angle to obtain the angular error. This subtraction is performed to high accuracy when fine–coarse synchro data are used.

With digital position sensing, the subtraction process to obtain the position error signal is implemented digitally, as illustrated in Fig 8.4-4. The position loop is digital, but the subsequent tachometer velocity feedback loop is usually analog. A digital-to-analog (D/A) converter placed after the digital subtraction process converts the position error signal to analog form.

Often the digital circuitry includes dynamic processing to achieve integral compensation, as illustrated by the dashed block in Fig 8.4-4. Digital algorithms for providing dynamic transfer functions are explained in Chapter 9.

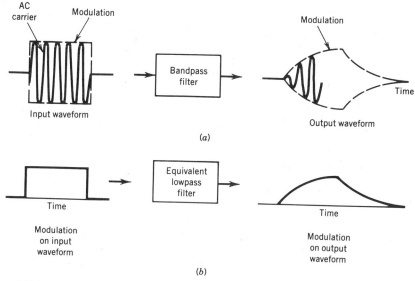

Figure 8.5-1 Application of signal-frequency to carrier-frequency transformation to radar and communication systems: (*a*) pulse-modulated carrier signal fed through bandpass filter; (*b*) equivalent effect at signal frequency.

When the integral network is included in the digital portion, the output from the D/A converter is a velocity command signal, which is fed directly as a command to the velocity feedback loop. This configuration has the advantage that the operation of the system can be conveniently controlled optimally with digital computation. Section 12.2 discusses optimal control of a servomechanism in response to large saturated commands.

8.5 APPLICATION OF AC-MODULATION PRINCIPLES TO SIGNAL PROCESSING IN COMMUNICATION SYSTEMS

The principles of modulation, AC filtering, and demodulation in AC servos also apply to signal processing at much higher frequencies, in communication and radar systems. As illustrated in Fig 8.5-1, the signal-frequency to carrier-frequency (lowpass to bandpass) filter transformation is a powerful concept for explaining signal processing. In diagram (a), a pulsed waveform is fed through a bandpass filter. The waveform is a pulse-modulated carrier, and the bandpass filter is tuned to the frequency of the carrier. One would like to know the envelope of the waveform at the output of the bandpass filter.

This can be found as shown in diagram *b*. Using the approach of Section 8.2, one calculates the signal-frequency lowpass filter that is equivalent to the carrier-frequency bandpass filter. The envelope of the input pulse waveform is fed into this equivalent signal-frequency filter, and the output is determined. This output from the signal-frequency lowpass filter is the same as the envelope of the output from the carrier-frequency bandpass filter.

Chapter 9

Sampled Data and Computer Simulation

When a signal is carried in the form of digital data, each digital quantity represents the value of the signal at a particular instant of time. Between those instants, the signal is not specified. Such a signal, which is defined only at discrete points in time, is called a sampled signal.

The primary application of sampled-data theory is digital signal processing. Nevertheless, not all sampled-data signals are digital. For example, in the phase-sensitive detector described in Chapter 8, Section 8.1.3, a sampling circuit converts an AC modulated signal to analog sampled data (a series of narrow pulses), which are subsequently changed to a continuous signal in a hold circuit.

This chapter analyzes sampled-data systems and their use in digital data processing. The basic principles for generating and demodulating sampled data are presented. A method is given for designing a sampled-data algorithm to satisfy a specified frequency response. Using this, a general procedure (called "serial simulation") is developed for simulating a feedback-control system, which is applied to the stage-positioning servo described in Chapter 7. This procedure can be readily implemented on a personal computer, using BASIC computer language, to simulate any control system, regardless of its complexity. A general analysis of sampled-data principles is given, which provides theoretical constraints on the sampling frequency needed to assure that the simulation is stable and accurate.

A general discussion of digital simulation techniques is presented in Section 9.4, which includes the Runge–Kutta method of digital integration. The state variable concept is described, and is extended to derive the transition matrix. State variables and the transition matrix are important elements in the theory of the Kalman filter. A practical explanation of Kalman filtering is given by the author in Ref [1.14] (Chapter 8).

Feedback-control loops that include digital and analog data are discussed in Section 9.5. That section describes the z-transform, and combines the concepts of the z-transform and frequency response to develop general constraints on the dynamic performance of feedback-control systems that use sampled data.

9.1 SAMPLED-DATA MODULATION AND DEMODULATION

9.1.1 Spectrum of Sampled Signal

A fundamental limitation of a sampled signal is that it cannot convey information at frequencies greater than half the sampling frequency. This can be demonstrated by examining the spectrum of the sampled signal. Chapter 8 described the spectrum of an AC modulated signal, and showed that the modulation process generates two sidebands displaced above and below the carrier frequency by the modulation frequency ω_m. The spectrum of a sampled-data signal has an infinite series of sidebands, which are displaced above and below the sampling frequency, and all harmonics of the sampling frequency, by the modulation frequency ω_m.

Sampling a signal is equivalent to modulating (or multiplying) the signal by a train of very narrow pulses having unit amplitude, as shown in Fig 9.1-1. Diagram *a* shows the input signal, and diagram *b* shows the modulating train of unit-amplitude pulses, which is called the reference signal. The sampling pulse period is designated T and the pulse width is designated τ. The pulses are made so narrow there is negligible variation of the continuous signal within the duration of the pulse, and so the signal is effectively sampled at discrete instants of time.

The continuous signal $s[t]$ of diagram *a* is multiplied by the reference-signal pulse train $r[t]$ of diagram *b*, to obtain the sampled signal $s^s[t]$, shown in diagram *c*:

$$s^s[t] = s[t]r[t] \tag{9.1-1}$$

The symbol $s^*[t]$ is commonly used to represent a sampled signal. This book uses the subscript s instead of the asterisk, so that the latter can be reserved to represent the conjugate of a complex quantity.

The reference pulse train can be represented as follows by a Fourier series. The general expression for a Fourier series expansion is

$$r[t] = \frac{a_0}{2} + \sum_{n=1}^{\infty} \left\{ a_n \cos \frac{2\pi nt}{T} + b_n \sin \left[\frac{2\pi nt}{T} \right] \right\} \tag{9.1-2}$$

The Fourier coefficients of Eq 9.1-2 are

$$a_n = \frac{2}{T} \int_{-T/2}^{+T/2} r[t] \cos \left[\frac{2\pi nt}{T} \right] dt \tag{9.1-3}$$

$$b_n = \frac{2}{T} \int_{-T/2}^{+T/2} r[t] \sin \left[\frac{2\pi nt}{T} \right] dt \tag{9.1-4}$$

The $t = 0$ point is set at the center of a pulse, which makes the sine coefficients

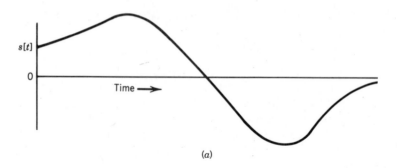

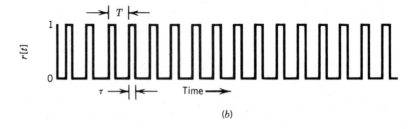

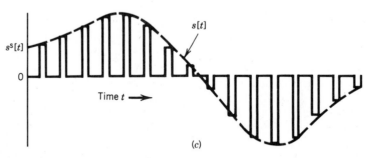

Figure 9.1-1 Waveforms illustrating the sampling process: (a) continuous signal $s[t]$; (b) sampling reference signal $r[t]$; (c) sampled signal $s^s[t]$.

b_n zero. The reference $r[t]$ is unity for $-\tau/2 < t < +\tau/2$, and is zero elsewhere within the $\pm T/2$ range of the integration. Hence Eq 9.1-3 gives for the coefficients a_n

$$a_n = \frac{2}{T} \int_{-\tau/2}^{+\tau/2} \cos\left[\frac{2\pi nt}{T}\right] dt \qquad (9.1\text{-}5)$$

Solving this integral gives

$$a_0 = \frac{2\tau}{T} \tag{9.1-6}$$

$$a_n = \frac{2}{\pi n} \sin\left[\frac{\pi n \tau}{T}\right] \quad (\text{for} \quad n > 0) \tag{9.1-7}$$

If the pulse width τ is made vanishingly small, $\sin[\pi n\tau/T]$ can be replaced by $\pi n\tau/T$. For $n > 0$, this gives for a_n

$$a_n = \frac{2\tau}{T} \quad \text{for} \quad n > 0 \tag{9.1-8}$$

This is the same as the expression for a_0 in Eq 9.1-6. Hence, all of the coefficients a_n are equal to a_0, and so the Fourier series of Eq 9.1-2 becomes

$$r[t] = \frac{\tau}{T}\left\{1 + 2\sum_{n=1}^{\infty}\cos\left[\frac{2\pi n t}{T}\right]\right\} \tag{9.1-9}$$

In this infinite series, each component is proportional to τ and so is vanishingly small.

Assume that the continuous signal $s[t]$ is a sinusoid of unit amplitude, given by

$$s[t] = \cos[\omega_m t] \tag{9.1-10}$$

By Eq 9.1-1, the sampled signal is the product of Eqs 9.1-9, -10, which is

$$s^s[t] = s[t]r[t] = \frac{\tau}{T}\left\{\cos[\omega_m t] + \sum_{n=1}^{\infty} 2\cos[n\omega_s t]\cos[\omega_m t]\right\} \tag{9.1-11}$$

where ω_s is the sampling frequency in rad/sec. The following sampling parameters are defined:

T = sampling period
$F = 1/T$ = sampling frequency in hertz
$\omega_s = 2\pi F$ = sampling frequency in rad/sec

Apply the following trigonometric identity to Eq 9.1-11:

$$\cos[a]\cos[b] = \tfrac{1}{2}(\cos[a + b] + \cos[a - b]) \tag{9.1-12}$$

Equation 9.1-11 becomes

$$s^s[t] = \frac{\tau}{T}\cos[\omega_m t]$$

$$+ \frac{\tau}{T}\sum_{n=1}^{\infty}\left\{\cos[(n\omega_s - \omega_m)t] + \cos[(n\omega_s + \omega_m)t]\right\} \tag{9.1-13}$$

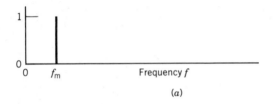

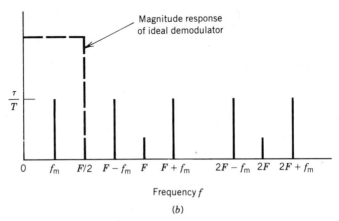

Figure 9.1-2 Spectra of continuous and sampled signals: (a) continuous signal $s[t]$; (b) sampled signal $s^s[t]$.

This has frequency components at the modulation frequency ω_m and at the harmonic sideband frequencies ($n\omega_s \pm \omega_m$), where n is an integer varying from 1 to infinity. Figure 9.1-2 shows the spectra of the continuous and sampled signals expressed in hertz. Diagram a shows the spectrum of the continuous signal $s[t]$, which is a unit-amplitude line at the frequency $f_m = \omega_m/2\pi$. Diagram b shows the spectrum of the sampled signal $s^s[t]$. This has an infinite series of lines of amplitude τ/T, at the frequencies f_m, ($F - f_m$), ($F + f_m$), ($2F - f_m$), ($2F + f_m$), etc.

9.1.2   Ideal Demodulation of Sampled Signal

To recreate the continuous signal $s[t]$, the sampled signal $s^s[t]$ can be demodulated by filtering it to pass the signal at the frequency f_m, while eliminating all the other lines. As the signal frequency f_m is increased relative to the sampling frequency F, the two sidebands at the frequencies f_m and ($F - f_m$) move toward one another. When the signal frequency f_m is equal to $F/2$, these two sidebands coincide and so cannot be separated. Hence information in the continuous signal can only be recovered if the signal frequency is less than half the sampling frequency.

If all of the frequency components of the continuous signal are at less than half the sampling frequency, the signal can theoretically be recovered exactly by passing the sampled data through a filter with no phase lag, having a flat magnitude response out to half the sample frequency and zero response at higher frequencies. The required filter response $H[f]$ is shown by the dashed plot in Fig 9.1-2b, and is defined by

$$H(f) = T/\tau \qquad \text{for} \quad 0 < f < F/2 \qquad (9.1\text{-}14)$$

$$H(f) = 0 \qquad \text{for} \quad f > F/2 \qquad (9.1\text{-}15)$$

Since this filter has magnitude attenuation but no phase shift, it cannot be realized in real time. However, it can be implemented by non-real-time processing if the complete set of sampled-data values is available.

The inverse Fourier transform of this filter frequency-response is the unit-impulse response, which is

$$h[t] = \mathscr{F}^{-1}[H[f]] = \frac{1}{\tau} \frac{\sin[\pi t/T]}{\pi t/T} \qquad (9.1\text{-}16)$$

Each sampling pulse has unit amplitude and a pulse width τ, and so has an area equal to τ. As the pulse width τ is made vanishingly small, the pulse becomes an impulse of area τ. Hence the response of the filter $H[f]$ to a sampling pulse is equal to the unit-impulse response of Eq 9.1-16 multiplied by the impulse area τ. This response, which is designated $p[t]$, is

$$p[t] = \tau h[t] = \frac{\sin[\pi t/T]}{\pi t/T} \qquad (9.1\text{-}17)$$

A plot of $p[t]$ is shown in Fig 9.1-3a. This is the ideal demodulated response for a single sampled value equal to unity.

The ideal continuous output signal corresponding to a series of sampled values can be calculated as shown in Fig 9.1-3b. Each of the sampled values is multiplied by the response $p[t]$ of diagram a, with the time axis of each $p[t]$ shifted to coincide with the sampling instant. The contributions for all of the sampled values are added to obtain the total plot. The points (a), (b), (c) in diagram b are the sampled values. The curves Ⓐ, Ⓑ, Ⓒ are the plots of $p[t]$ multiplied by these sampled values. These curves, along with the curves for all of the other points, are added to produce the ideally demodulated signal shown by the dashed curve.

The figure shows only the contributions from three sampled values, over a limited time interval. The contributions from a great many more sampled values must be included in the summation before an accurate representation of the continuous signal over this time interval can be obtained. Such a summation would be very difficult to perform with hand calculation, but is easy with a digital computer.

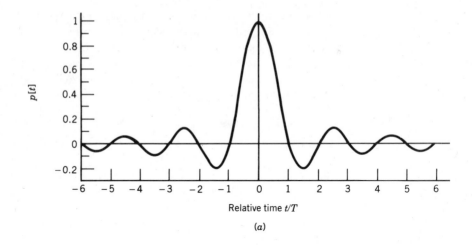

(a)

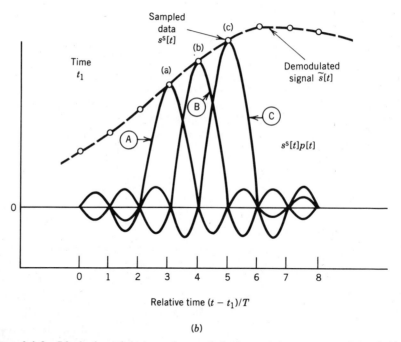

(b)

Figure 9.1-3 Ideal demodulation of sampled data: (a) response $p[t]$ of ideal sampled-data demodulator to a sample value equal to unity; (b) calculation of the ideal continuous signal that corresponds to a set of sampled values.

This ideal demodulation process exactly duplicates the original continuous signal $s[t]$ if the signal has no components at frequencies greater than half the sampling frequency. When this condition is not satisfied, an effect called "foldover" or "aliasing" occurs in the sampling process. This distorts the information, and prohibits the original continuous data from being recovered exactly.

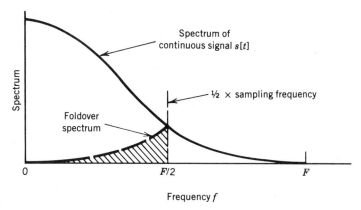

Figure 9.1-4   Spectrum foldover (or aliasing) caused by sampling, when signal has components at frequencies greater than half the sampling frequency.

Foldover (or aliasing) is illustrated in Fig 9.1-4. The solid curve shows the spectrum of a continuous signal. When that signal is sampled, the portion of this spectrum at frequencies greater than $F/2$ is folded over into lower frequencies, to produce the contribution shown by the cross-hatched area. The components represented by the cross-hatched foldover spectrum distort the signal. The sampling process loses all information at frequencies greater than $F/2$. Any components at frequencies greater than $F/2$ that are present in the continuous signal add distortion to the resultant sampled data.

The preceding discussion can be summarized as follows:

1. The sampled signal can only convey information at frequencies that are less than $\frac{1}{2}$ of the sampling frequency.
2. Before a signal is sampled, it should be filtered to remove components at frequencies greater than $\frac{1}{2}$ of the sampling frequency, because any such components add distortion to the resultant data.
3. An ideal sampled-data demodulation process (which cannot be implemented in real time) can be achieved by transmitting, without attenuation or phase shift, all frequency components out to $\frac{1}{2}$ of the sampling frequency, and eliminating all components at higher frequencies. This ideal sampled-data demodulator would exactly recover the continuous signal that is sampled, provided that the signal has no components at frequencies greater than $\frac{1}{2}$ of the sampling frequency. When that condition is not satisfied, this demodulation process still gives the best possible *a priori* approximation of the continuous signal.

The ideal demodulated signal is a smooth curve that passes through the sampled points. One can obtain a reasonable approximation of this ideal response by using a French-curve template to draw a smooth plot through the data points.

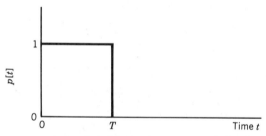

Figure 9.1-5   Response of simple-hold circuit to a sampled value of unity at time $t = 0$.

9.1.3   Simple-Hold Circuit for Demodulation of Sampled Data

A digital sampled signal is usually converted to analog form by a digital-to-analog (D/A) converter, which has a circuit that holds the output fixed between sampling instants. We call this a *simple-hold* circuit, but it is commonly called a "zero-order hold" circuit in the sampled-data literature. Figure 9.1-5 shows the response $p[t]$ of a simple-hold circuit to a sampled value of unity occurring at time $t = 0$. This response $p[t]$ can be represented as a positive unit step occurring at time $t = 0$, followed by a negative unit step occurring at $t = T$. Hence $p[t]$ can be expressed as

$$p[t] = u[t] - u[t - T] \qquad (9.1\text{-}18)$$

where $u[t]$ is a unit step occurring at time $t = 0$. The Laplace transform of this is

$$P[s] = \mathscr{L}[p[t]] = \frac{1}{s} - \frac{1}{s}e^{-sT} = \frac{1}{s}(1 - e^{-sT}) \qquad (9.1\text{-}19)$$

The sampled value $s^s[t]$ is considered to be a pulse of unit amplitude and infinitesimal pulse width τ occurring at time $t = 0$. This can be represented as an impulse of amplitude τ occurring at time $t = 0$:

$$s^s[t] = \tau\delta[t] \qquad (9.1\text{-}20)$$

where $\delta[t]$ is a unit impulse occurring at $t = 0$. The Laplace transform of this is

$$S^s[s] = \mathscr{L}[s^s[t]] = \tau \qquad (9.1\text{-}21)$$

Dividing Eq 9.1-19 by Eq 9.1-21 gives the transfer function $H[s]$ of the simple-hold circuit:

$$H[s] = \frac{P[s]}{S^s[s]} = \frac{1 - e^{-sT}}{s\tau} \qquad (9.1\text{-}22)$$

Setting $s = j\omega$ gives the frequency response of the simple-hold circuit:

$$H[j\omega] = \frac{1 - e^{-j\omega T}}{j\omega\tau} \qquad (9.1\text{-}23)$$

Factoring $e^{-j\omega T/2}$ from the expression gives

$$H[j\omega] = e^{-j\omega T/2}\frac{e^{j\omega T/2} - e^{-j\omega T/2}}{j\omega\tau} \qquad (9.1\text{-}24)$$

Now $\sin[\omega T/2]$ is equal to

$$\sin\left[\frac{\omega T}{2}\right] = \frac{(e^{j\omega T/2} - e^{-j\omega T/2})}{2j} \qquad (9.1\text{-}25)$$

Combining Eqs 9.1-24, -25 gives

$$H[j\omega] = \frac{T}{\tau}e^{-j\omega T/2}\frac{\sin[\omega T/2]}{\omega T/2} \qquad (9.1\text{-}26)$$

This can be expressed in the form

$$H[j\omega] = \frac{T}{\tau}M[\omega]e^{-j\omega T/2} \qquad (9.1\text{-}27)$$

where $M[\omega]$ represents the normalized frequency response magnitude of the demodulation process, which is

$$M[\omega] = \frac{\sin[\omega T/2]}{\omega T/2} = \frac{\sin[\pi f/F]}{\pi f/F} \qquad (9.1\text{-}28)$$

A plot of $M[\omega]$ is shown as curve ② in Fig 9.1-6. For comparison the dashed curve ① shows the response of an ideal sampled-data demodulator, which passes all data out to $\frac{1}{2}$ of the sampling frequency unchanged, and eliminates everything at higher frequencies.

The factor $e^{-j\omega T/2}$ in Eq 9.1-27 represents a time delay equal to $\frac{1}{2}$ of the sampling period T. This has unity magnitude response and a phase lag proportional to frequency given by

$$\text{Ang}[e^{-j\omega T/2}] = -\frac{\omega T}{2}\text{ rad} = -\frac{f}{F}180° \qquad (9.1\text{-}29)$$

This time delay of half a sampling period is equivalent to a phase lag of half a cycle (or 180°) at the sampling frequency $f = F$.

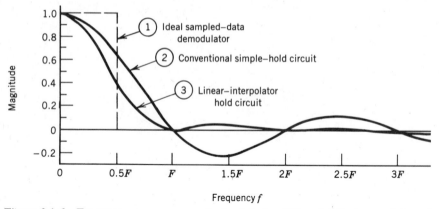

Figure 9.1-6 Frequency-response magnitude plots for different techniques of demodulating sampled data.

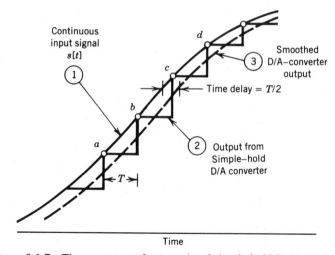

Figure 9.1-7 Time response of conventional simple-hold D/A converter.

Figure 9.1-7 illustrates this digital-to-analog conversion process in the time domain. Curve ① is the original continuous signal $s[t]$. Points a, b, c, d are the sampled-data values defining $s^s[t]$. Curve ② is the output from the simple-hold D/A converter which demodulates $s^s[t]$. When the demodulated signal is filtered to attenuate the high-frequency components, the smoothed plot shown by the dashed curve ③ remains. (The time delay of the filter is not included in the plot ③.) The smoothed signal lags the continuous signal $s^s[t]$ by $\frac{1}{2}$ of a sampling period. This illustrates the principle that the simple-hold D/A conversion causes a time delay equal to $\frac{1}{2}$ of the sampling

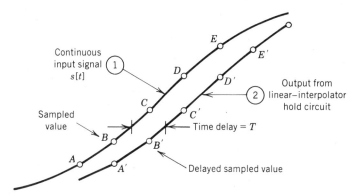

Figure 9.1-8 Time response of linear-interpolator hold circuit.

period. The filter that attenuates the harmonics of the stepped waveform ②
to form the smoothed curve ③ adds additional time delay to the signal. The
dashed curve ③ is a theoretical abstraction, which shows only the time delay
of the simple-hold demodulation process, and omits subsequent time delays
due to filtering.

9.1.4 Linear-Interpolator Hold Circuit

A much smoother output signal can be obtained by using a hold circuit that
provides straight-line interpolation between the sampled points. In Fig 9.1-8,
curve ① is the input signal $s[t]$, and points a, b, c, d are the sampled values
$s^s[t]$. These sampled values are delayed by one period T to produce the values
a', b', c', d'; and the demodulation process connects straight lines between
adjacent points. A full cycle of time delay is required in the demodulation, in
order to calculate the difference between successive values. This difference
information is used to compute the slopes of the straight-line interpolations.

Figure 9.1-9 shows a process that theoretically could act as a linear-interpolator hold circuit. The difference between successive samples ($y_n - y_{n-1}$) is
calculated and converted to analog form in the D/A converter. The output
from the D/A converter is fed to an integrator having the transfer function
$1/Ts$. The integrator output is the demodulated analog signal. This circuit can
theoretically provide linear interpolation, but is not practical, because inaccu-
racies in the integrator would produce drift in the output. This drift can be
corrected by resetting the integrator voltage to the correct output at each
sampling instant, as is done in the circuit of Fig 9.1-10.

Figure 9.1-10 shows a practical linear-interpolator hold circuit. Two D/A
converters are used: DAC1, which provides conventional simple-hold demodulation of the sampled signal, and DAC2, which converts to analog form the
differential values between successive samples of the signal. The differential

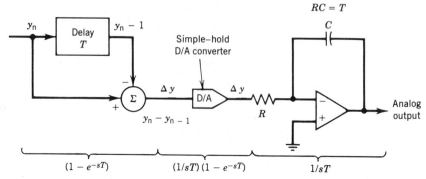

Figure 9.1-9   Theoretical linear-interpolator hold circuit.

signal from DAC2 is fed to an opamp integrator circuit, having an *RC* time constant equal to the sample period *T*. By integrating this differential information, the integrator provides linear interpolation of the sample data at its output. At each sampling instant, the integrator output is compared with the output from DAC1. The FET switch is closed at each sampling instant, and the feedback circuit adjusts the capacitor voltage to correct for drift and make the integrator output agree with the output from DAC1.

There is one sample period of time delay in the formation of the demodulated signal by the integrator. To compensate for this delay, the output from DAC1 is appropriately delayed relative to the input data. The operation of DAC1 is actually delayed by only half a sample period. This eliminates the effect of transients in DAC1, because the output from DAC1 is allowed to settle by half a sample period before it is used to reset the voltage on the integrator capacitor.

To illustrate the timing of this demodulation process, it is assumed in Fig 9.1-10 that a continuous signal $y[t]$ is sampled, and that the sampled data are converted by the linear-interpolator hold circuit back to a continuous signal equivalent to $y[t]$. As is shown, there is a time delay of one sample period between the input continuous signal $y[t]$ and the reconstruction of $y[t]$ provided at the integrator output.

The transfer function of the linear-interpolator hold circuit can be derived from the ideal process shown in Fig 9.1-9. This transfer function is

$$H[s] = (1 - e^{-sT})\left\{\frac{1}{sT}(1 - e^{-sT})\right\}\frac{1}{sT}$$

$$= \frac{T}{\tau}\left(\frac{1 - e^{-sT}}{sT}\right)^2 \tag{9.1-30}$$

The factor $(1 - e^{-sT})$ is the transfer function of the subtraction process

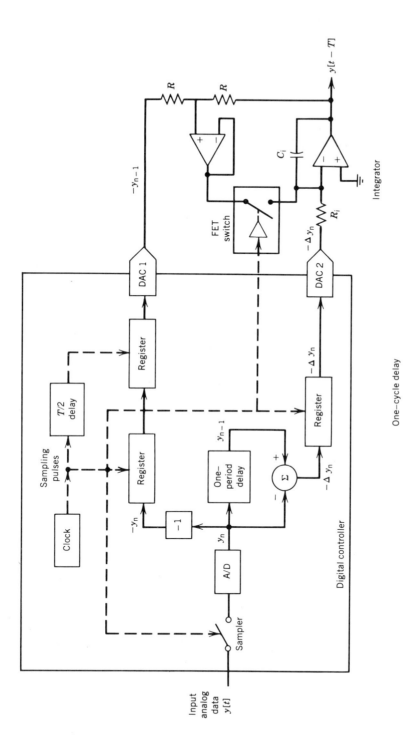

One-cycle delay

Figure 9.1-10 Practical linear-interpolator hold circuit.

$(y_n - y_{n-1})$. The factor $1/sT$ is the transfer function of the analog integrator. The factor within the braces { } is the transfer function of the D/A converter, which was given in Eq 9.1-22. Setting $s = j\omega$ in Eq 9.1-30 gives the frequency response. This can be simplified to the following form, which is similar to Eq 9.1-26:

$$H[j\omega] = \frac{T}{\tau}\left(\frac{1 - e^{-j\omega T}}{j\omega T}\right)^2$$

$$= \frac{T}{\tau}e^{-j\omega T}\left(\frac{\sin[\omega T/2]}{\omega T/2}\right)^2 \qquad (9.1\text{-}31)$$

This can be expressed as

$$H[j\omega] = \frac{T}{\tau}M[\omega]e^{-j\omega T} \qquad (9.1\text{-}32)$$

The function $M[\omega]$ is the normalized magnitude response, which is

$$M[\omega] = \left(\frac{\sin[\omega T/2]}{\omega T/2}\right)^2 = \left(\frac{\sin[\pi f/F]}{\pi f/F}\right)^2 \qquad (9.1\text{-}33)$$

The exponential $e^{-j\omega T}$ in Eq 9.1-32 represents the delay of one sample period in the process. This produces twice the phase lag of the simple-hold circuit. The magnitude response $M[\omega]$ of Eq 9.1-33 is the square of the magnitude response of the simple-hold circuit, which was given in Eq 9.1-28. A plot of this is given as curve ③ of Fig 9.1-6.

A simple-hold circuit, provided by a D/A converter, may require appreciable subsequent filtering to achieve adequate reduction of the sampling harmonics. A linear-interpolator hold circuit, with its much lower harmonic output, may not need subsequent filtering. Consequently, the overall phase lag of the simple-hold circuit and its filtering may actually be greater than the phase lag of a linear-interpolator hold circuit.

9.1.5 Oversampling

The effect of a linear-interpolator hold circuit can be approximated with a simpler circuit by using oversampling, a technique that is commonly used in compact-disk audio players. In Fig 9.1-11, diagram a shows a circuit that can provide times-4 oversampling. The corresponding waveforms for a single digital value are shown in diagram b.

The digital input is fed to a simple-hold D/A converter, which forms the analog signal A. The signal A is sampled by three sample-and-hold circuits to form signals B, C, and D. The sampling instants of the sample-and-hold

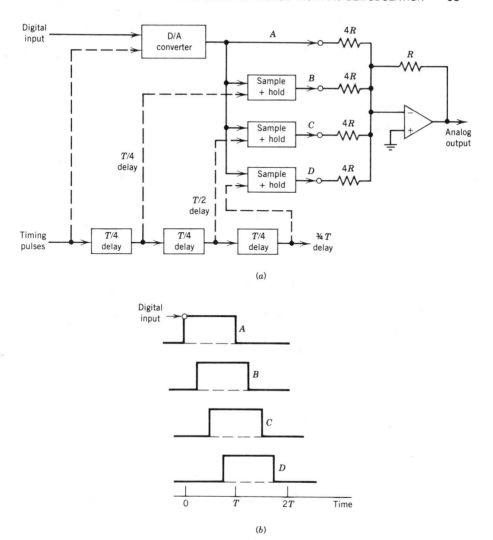

Figure 9.1-11   Times-4 oversampled hold circuit: (*a*) circuit; (*b*) waveforms.

circuits are delayed relative to those of the D/A converter by $\frac{1}{4}$, $\frac{1}{2}$, and $\frac{3}{4}$ sample period. The four analog signals *A, B, C, D* for a single digital value are shown in diagram *b*. These four signals are summed in an opamp, with a gain of $\frac{1}{4}$, to form the analog output voltage. The resultant analog output voltage (ignoring the phase reversal of the opamp) is shown in Fig 9.1-12*c*. This response (relative to the digital input value) is the sum of the four waveforms of Fig 9.1-11*b*, divided by 4.

Figure 9.1-12 shows the analog responses, for a single digital value, of the simple-hold circuit (diagram *a*), the linear-interpolator hold circuit (diagram *b*), and the times-4 oversampled hold circuit of Fig 9.1-11 (diagram *c*).

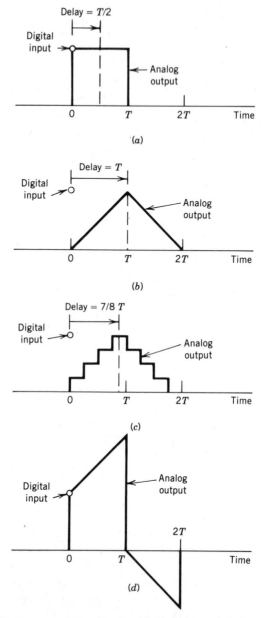

Figure 9.1-12 Analog responses of various hold circuits to a single digital value: (*a*) simple-hold circuit; (*b*) linear-interpolator hold circuit; (*c*) times-4 oversampled hold circuit; (*d*) first-order hold circuit.

(Diagram d shows the response for the first-order hold circuit, to be discussed in Section 9.1.6.) Note that the analog output with times-4 oversampling in diagram c approximates that for the linear-interpolator hold in diagram b.

Figure 9.1-12 shows that the response for the simple-hold circuit has a time delay of $T/2$; that for the linear-interpolator hold circuit has a time delay of T; and that for the times-4 oversampled hold circuit has a time delay of $\frac{7}{8}T$.

If the output from the times-4 oversampled hold circuit is fed through a lowpass filter, the resultant response can closely approximate that for the linear-interpolator hold circuit. Assume, for example, a second-order filter with the transfer function $1/(1 + \tau s)^2$, where $\tau = T/8$. This filter would provide 11 : 1 attenuation of the ripple-frequency fundamental of Fig 9.1-12c, which is the fourth harmonic of the sampling frequency. The time delay of this filter is 2τ, which is $T/4$. Hence, the total time delay of the hold circuit plus filter is $\frac{9}{8}T$. This is only 12.5% greater than that of the linear-interpolator hold circuit.

9.1.6 First-Order Hold Circuit

The sampled-data literature makes little reference to the linear-interpolator hold circuit, but frequently discusses the "first-order hold" circuit. These two circuits should not be confused. The first-order hold circuit provides linear *extrapolation* of the sampled data, not linear *interpolation*. The difference between two sampled values is measured, to set the slope of the output signal in the subsequent sample interval.

The response of the first-order hold circuit to a single digital value was shown in Fig 9.1-12d. Figure 9.1-13 shows the response of this hold circuit to

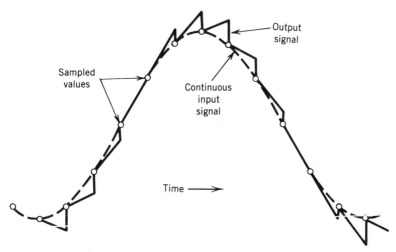

Figure 9.1-13 Response of first-order hold circuit to a sampled sinusoid at $\frac{1}{12}$ of the sampling frequency.

the sampled values of a continuous sinusoidal input, with a frequency of $\frac{1}{12}$ of the sampling frequency. The figure shows that the first-order hold circuit generates large output harmonics when the frequency content of the input signal is at all high. Consequently, this circuit has little practical value. This conclusion is consistent with the findings of Ragazzini and Franklin [9.1] (p. 39), who reported that the first-order hold circuit is "not commonly employed in feedback control systems".

In contrast, the linear-interpolator hold circuit has many potential applications. For example, Oppenheim, Willsky, and Young [9.2] discuss its use in the signal processing of digitized imagery data.

9.1.7 Polynomial-Fit Demodulation of Sampled Data

The linear-interpolator hold circuit draws a straight line between successive sampled values. As shown in Fig 9.1-14, this approach can be extended by constructing a third-order polynomial that passes through four successive sampled values a, b, c, d. The section of that polynomial shown by the solid curve is the output signal between sampled points b, c. This polynomial has the following form, where time t is zero at point b:

$$x = K_0 + K_1 t + K_2 t^2 + K_3 t^3 \tag{9.1-34}$$

To solve for the coefficients, set $x = a$ at $t = -T$, $x = b$ at $t = 0$, $x = c$ at $t = +T$, and $x = d$ at $t = +2T$. Applying these relations to Eq 9.1-34 yields

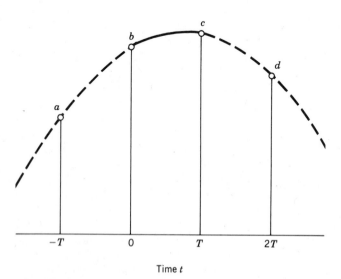

Time t

Figure 9.1-14   Four-point polynomial fit for demodulating sampled data.

the coefficients:

$$K_0 = b \tag{9.1-35}$$

$$K_1 = \frac{-2a + 5b + 6c - d}{6T} \tag{9.1-36}$$

$$K_2 = \frac{a - 2b + c}{2T^2} \tag{9.1-37}$$

$$K_3 = \frac{-a - 5b - 3c + d}{6T^3} \tag{9.1-38}$$

These coefficients are substituted into Eq 9.1-34 to provide third-order interpolation between the sample points b, c. This approach is generally not practical in the actual detection stage. However, it can be implemented within the computer to increase the data rate by generating sample values between the original points. A linear-interpolator hold circuit can be used to provide straight-line interpolation between the resultant data points. This demodulation process can achieve very accurate reconstruction of the signal, but causes a time delay of two sample periods.

9.2   DIGITAL COMPUTATION FOR SAMPLED-DATA FILTERING AND SIMULATION

This section develops algorithms for processing digital data to achieve specified frequency-response characteristics. These are used to simulate dynamic systems on a digital computer. The simulation procedure developed in this section is called *serial simulation* to distinguish it from the parallel simulation procedure using the Runge–Kutta integration routine to be described in Section 9.4.

9.2.1   Signal-Flow Diagram of Servo for Computer Simulation

As an example for computer simulation, consider the stage-positioning servo studied in Chapter 7. Figure 9.2-1 is a simplified version of the signal-flow diagram of that servo shown earlier in Fig 7.5-1. The torque disturbance and the effects of the motor back EMF have been eliminated, but these will be included later. The motor electrical break frequency ω_e has been replaced by its reciprocal, the motor electrical time constant τ_e, defined by

$$\tau_e = 1/\omega_e \tag{9.2-1}$$

To simplify the diagram, the subscript c for the controlled variable is omitted.

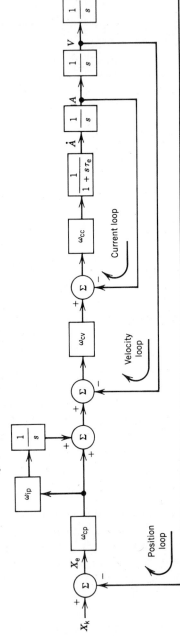

ω_{cp}, ω_{cv}, ω_{cc} = asymptotic gain crossover frequencies of position loop, velocity loop, and current loop

ω_{ip} = break frequency of position-loop integral network

τ_e = electrical time constant due to servo motor inductance

Figure 9.2-1 Simplified signal-flow diagram of stage-positioning servo to be used for simulation.

The following transform variables are defined:

$$X = X_c = \text{controlled position}$$
$$V = sX = \text{controlled velocity} \qquad (9.2\text{-}2)$$
$$A = sV = \text{controlled acceleration} \qquad (9.2\text{-}3)$$
$$\dot{A} = sA = \text{controlled rate of change of acceleration} \qquad (9.2\text{-}4)$$
$$X_k = \text{command position}$$
$$X_e = X_k - X = \text{position error} \qquad (9.2\text{-}5)$$

To simulate this signal-flow diagram on a digital computer, computer algorithms are needed to implement the transfer functions $1/s$ for an integrator and $1/(1 + s\tau)$ for a lowpass filter. A method for deriving such algorithms will now be presented.

9.2.2 Algorithm for Sampled-Data Filter

To develop a general method for designing linear sampled-data filters, let us examine a sampled-data formula for achieving integration. Assume that a variable $y[t]$ is calculated by integrating a variable $x[t]$:

$$y = \int x \, dt \qquad (9.2\text{-}6)$$

A well-known sampled-data routine for implementing this equation is trapezoidal integration, which is illustrated in Fig 9.2-2. The nth samples of x and y are defined as x_n and y_n, and the $(n-1)$th samples are defined as x_{n-1} and y_{n-1}. As shown in Fig 9.2-2, the area under the x-curve between the $(n-1)$th sample and the nth sample is approximately

$$\Delta y_n \doteq \tfrac{1}{2}(x_n + x_{n-1})T \qquad (9.2\text{-}7)$$

where T is the sample period, which is the time between the $(n-1)$th and the nth sample. In this equation, the area under the x-curve between these two samples is approximated by a trapezoid which is formed by drawing a straight line between points x_{n-1} and x_n. Since y is the integral of x, its value at the nth sample is

$$y_n = y_{n-1} + \Delta y_n \qquad (9.2\text{-}8)$$

where y_{n-1} is the integral of x up to the $(n-1)$th sample. Substituting Eq 9.2-7 into Eq 9.2-8 gives the trapezoidal integration formula:

$$y_n = y_{n-1} + \frac{T}{2}(x_n + x_{n-1}) \qquad (9.2\text{-}9)$$

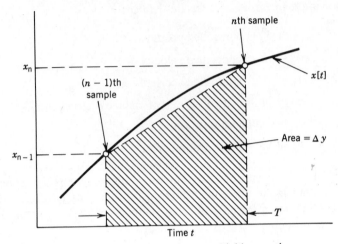

Figure 9.2-2   Illustration of trapezoidal integration.

To find the frequency response of this algorithm, the Laplace transform of the equation is taken, and s is replaced by $j\omega$. In terms of a continuous signal, x_n can be considered to represent $x[t]$, and x_{n-1} to represent $x[t - T]$. Thus

$$x_n = x[t], \qquad x_{n-1} = x[t - T] \qquad\qquad (9.2\text{-}10)$$

Hence, the Laplace transforms of x_{n-1} and x_n are related by

$$\mathscr{L}[x_{n-1}] = \mathscr{L}[x[t - T]] = e^{-sT}\mathscr{L}[x[t]] = e^{-sT}\mathscr{L}[x_n] \quad (9.2\text{-}11)$$

(Remember that e^{-sT} in the Laplace transform corresponding to a time delay T). Using this principle, the Laplace transform of Eq 9.2-11 can be taken to give

$$\mathscr{L}[y_n] = \mathscr{L}[y_{n-1}] + \frac{T}{2}\{\mathscr{L}[x_n] + \mathscr{L}[x_{n-1}]\}$$

$$= e^{-sT}\mathscr{L}[y_n] + \frac{T}{2}\{\mathscr{L}[x_n] + e^{-sT}\mathscr{L}[x_n]\} \qquad (9.2\text{-}12)$$

The Laplace transforms of the x and y sample values are designated as

$$\mathscr{L}[y_n] = Y \qquad\qquad (9.2\text{-}13)$$
$$\mathscr{L}[x_n] = X \qquad\qquad (9.2\text{-}14)$$

Hence Eq 9.2-12 becomes

$$Y = e^{-sT}Y + \frac{T}{2}(X + e^{-sT}X) \qquad\qquad (9.2\text{-}15)$$

By algebraic manipulation of Eq 9.2-15, the transfer function Y/X is calculated as follows:

$$Y(1 - e^{-sT}) = \frac{T}{2}X(1 + e^{-sT}) \qquad (9.2\text{-}16)$$

$$\frac{Y}{X} = \frac{(T/2)(1 + e^{-sT})}{1 - e^{-sT}} \qquad (9.2\text{-}17)$$

To obtain the frequency response, set $s = j\omega$. The exponentials become

$$e^{-sT} = e^{-j\omega T} = \cos[\omega T] - j\sin[\omega T] \qquad (9.2\text{-}18)$$

Substituting this into Eq 9.2-17 gives the frequency response between X and Y:

$$\frac{Y}{X} = \frac{(T/2)(1 + \cos[\omega T] - j\sin[\omega T])}{1 - \cos[\omega T] + j\sin[\omega T]} \qquad (9.2\text{-}19)$$

Multiply the numerator and denominator by $\{(1 - \cos[\omega T] - j\sin[\omega T])$ to rationalize the expression. The frequency response reduces to

$$\frac{Y}{X} = \frac{(T/2)(-2j\sin[\omega T])}{2(1 - \cos[\omega T])} = \frac{T\sin[\omega T]}{2j(1 - \cos[\omega T])} \qquad (9.2\text{-}20)$$

For small values of ωT, the sine and cosine expressions can be approximated by

$$\sin[\omega T] \doteq \omega T \qquad (9.2\text{-}21)$$

$$\cos[\omega T] \doteq 1 - \tfrac{1}{2}(\omega T)^2 \qquad (9.2\text{-}22)$$

Substituting these into Eq 9.2-20 gives the approximate frequency response

$$\frac{Y}{X} \doteq \frac{1}{j\omega} \qquad (9.2\text{-}23)$$

Thus, at low frequencies, trapezoidal integration exhibits the frequency response $1/j\omega$, which is the transfer function of an ideal integrator.

This result can be derived much more simply by employing the "W-transform" method developed by Johnson, Lindorff, and Nordling [9.3], [9.4]. However, we replace their symbol W with U, so that W can be reserved to represent a pseudo angular frequency, which approximates ω at low frequencies. The first step is to make the substitution

$$\bar{z} = e^{-sT} \qquad (9.2\text{-}24)$$

This is the transfer function for a time delay of one sample period. The sampled-data literature generally deals with the reciprocal of this (e^{sT}), which is designated z. The author uses the variable $\bar{z}$ (which is equal to $1/z$) because time shifts only occur in terms of delays. Thus, when the variable z is used, it occurs naturally in the form z^{-1}, z^{-2}, etc. This issue is discussed further in Section 9.5.2. In terms of $\bar{z}$, the transform of the basic trapezoidal integration algorithm of Eq 9.2-9 is

$$Y = \bar{z}Y + \frac{T}{2}(X + \bar{z}X) \qquad (9.2\text{-}25)$$

Note that $\bar{z}$ is substituted for each time delay. Algebraic manipulation of Eq 9.2-25 yields the following transfer function:

$$\frac{Y}{X} = \frac{(T/2)(1 + \bar{z})}{1 - \bar{z}} \qquad (9.2\text{-}26)$$

This could also be derived from Eq 9.2-17 by setting e^{-Ts} equal to $\bar{z}$.

In the Johnson–Lindorff–Nordling transform, U is defined as

$$U = \frac{1 - \bar{z}}{1 + \bar{z}} \qquad (9.2\text{-}27)$$

It can be seen by substitution that this transformation applies both ways:

$$\bar{z} = \frac{1 - U}{1 + U} \qquad (9.2\text{-}28)$$

In Eq 9.2-27, replace $\bar{z}$ by e^{-sT} and set $s = j\omega$ to obtain the frequency response. This yields

$$U = \frac{1 - e^{-sT}}{1 + e^{-sT}} = \frac{1 - e^{-j\omega T}}{1 + e^{-j\omega T}} \qquad (9.2\text{-}29)$$

Multiply numerator and denominator by $e^{j\omega T/2}$, to obtain

$$U = \frac{e^{j\omega T/2} - e^{-j\omega T/2}}{e^{j\omega T/2} + e^{-j\omega T/2}} \qquad (9.2\text{-}30)$$

Remember that $\cos[\Theta]$ and $\sin[\Theta]$ are expressed as follows in terms of complex exponentials:

$$\cos[\Theta] = \frac{e^{j\Theta} + e^{-j\Theta}}{2} \qquad (9.2\text{-}31)$$

$$\sin[\Theta] = \frac{e^{j\Theta} - e^{-j\Theta}}{2j} \qquad (9.2\text{-}32)$$

Hence, $j \tan[\Theta]$ is equal to

$$j \tan[\Theta] = j\frac{\sin[\Theta]}{\cos[\Theta]} = \frac{e^{j\Theta} - e^{-j\Theta}}{e^{j\Theta} + e^{-j\Theta}} \qquad (9.2\text{-}33)$$

Comparing Eqs 9.2-30, -33 shows that

$$U = j \tan[\omega T/2] \qquad (9.2\text{-}34)$$

For angles less than 45° (or $\pi/4$ rad) the function $\tan[\Theta]$ can be approximated by

$$\tan[\Theta] \doteq \Theta \qquad \text{for} \quad \Theta < \pi/4 \text{ rad} = 45° \qquad (9.2\text{-}35)$$

The maximum error of this approximation, which occurs at 45°, is 21%. Applying this approximation to Eq 9.2-34 gives

$$U = j \tan\left[\frac{\omega T}{2}\right] \doteq \frac{j\omega T}{2} = \frac{sT}{2} \qquad \text{for} \quad \frac{\omega T}{2} < \frac{\pi}{4} \qquad (9.2\text{-}36)$$

In terms of frequency in hertz, the limit is

$$f = \frac{\omega}{2\pi} < \frac{1}{4T} = \frac{F}{4} \qquad (9.2\text{-}37)$$

where F is the sampling frequency, which is equal to $1/T$. This shows that the approximation of Eq 9.2-36 holds with reasonable accuracy up to $\frac{1}{4}$ of the sample frequency. Since sampled data can only convey frequency information up to $\frac{1}{2}$ of the sample frequency, this approximation is good up to $\frac{1}{2}$ of the maximum frequency that can be conveyed by the sampled data.

Let us apply the U-transform to trapezoidal integration. Substituting Eq 9.2-28 into Eq 9.2-26 gives the following for the transfer function of trapezoidal integration, in terms of U:

$$\frac{Y}{X} = \frac{(T/2)(1 + \bar{z})}{1 - \bar{z}} = \frac{(T/2)[1 + (1 - U)/(1 + U)]}{1 - (1 - U)/(1 + U)}$$

$$= \frac{(T/2)[(1 + U) + (1 - U)]}{(1 + U) - (1 - U)} = \frac{(T/2)(2)}{2U} = \frac{T}{2U} \qquad (9.2\text{-}38)$$

Apply the approximation of Eq 9.2-36 by replacing U with $sT/2$. Equation 9.2-38 becomes

$$\frac{Y}{X} = \frac{T}{2U} \doteq \frac{T}{2(sT/2)} = \frac{1}{s} = \frac{1}{j\omega} \qquad (9.2\text{-}39)$$

Thus, the frequency response for trapezoidal integration approximate $1/j\omega$ quite accurately up to $\frac{1}{4}$ of the sampling frequency.

The preceding discussion can be summarized by the following equations:

$$\bar{z} = e^{-sT} = \text{transform of one-cycle time delay} \qquad (9.2\text{-}40)$$

$$U = \frac{1 - \bar{z}}{1 + \bar{z}} \qquad (9.2\text{-}41)$$

$$\bar{z} = \frac{1 - U}{1 + U} \qquad (9.2\text{-}42)$$

$$U \doteq \frac{sT}{2}, \quad s \doteq \frac{2U}{T} \qquad \text{for} \quad f < \frac{F}{4} \qquad (9.2\text{-}43)$$

To apply these equations to another example, let us calculate the sampled-data routine having the transfer function of a single-order lowpass filter of time constant τ:

$$\frac{Y}{X} = \frac{1}{1 + s\tau} \qquad (9.2\text{-}44)$$

To simplify the calculations, the time constant τ is defined to be a factor B times the sample period T:

$$\tau = BT \qquad (9.2\text{-}45)$$

Hence the lowpass-filter transfer function is

$$\frac{Y}{X} = \frac{1}{1 + sBT} \qquad (9.2\text{-}46)$$

The relation of Eq 9.2-43 is used to obtain the approximate sampled-data transfer function in terms of U:

$$\frac{Y}{X} = \frac{1}{1 + (2U/T)BT} = \frac{1}{1 + 2BU} \qquad (9.2\text{-}47)$$

Substituting Eq 9.2-41 into this gives the sampled-data transfer function in terms of $\bar{z}$:

$$\frac{Y}{X} = \frac{1}{1 + [2B(1 - \bar{z})/(1 + \bar{z})]} = \frac{1 + \bar{z}}{(1 + \bar{z}) + 2B(1 - \bar{z})}$$

$$= \frac{1 + \bar{z}}{(1 + 2B) + \bar{z}(1 - 2B)} \qquad (9.2\text{-}48)$$

Multiply both sides by the denominators to obtain

$$Y(1 + 2B) + \bar{z}Y(1 - 2B) = X + \bar{z}X \tag{9.2-49}$$

Solve this for Y:

$$Y = \frac{\bar{z}Y(2B - 1) + X + \bar{z}X}{2B + 1} \tag{9.2-50}$$

The inverse transform of this is the sampled-data computer routine:

$$y_n = \frac{y_{n-1}(2B - 1) + x_n + x_{n-1}}{2B + 1} \tag{9.2-51}$$

Assume for example that $B = 9.5$, so that $\tau = 9.5T$. Hence, $(2B + 1) = 20$, and $(2B - 1) = 18$. Equation 9.2-51 becomes

$$y_n = \frac{18y_{n-1} + x_n + x_{n-1}}{20} \tag{9.2-52}$$

When implementing these computer routines in software, it is convenient to use the subscript p to represent "the past value of." Thus, x_p is the past value of x, and y_p is the past value of y. The trapezoidal integration routine of Eq 9.2-26 is expressed as

$$y = y_p + \frac{T}{2}(x + x_p) \tag{9.2-53}$$

Before the next cycle, the following equations should be implemented, to set the past values equal to the present values:

$$x_p = x \tag{9.2-54}$$

$$y_p = y \tag{9.2-55}$$

In like fashion, the computer routine for the low-pass transfer function $1/(1 + s\tau)$ is, from Eq 9.2-51,

$$y = \frac{y_p(2(\tau/T) - 1) + x + x_p}{2(\tau/T) + 1} \tag{9.2-56}$$

9.2.3 Digital Serial Simulation of Servo

The preceding section has developed computer routines for implementing the transfer functions $1/s$ and $1/(1 + s\tau)$. These can be applied to simulate the

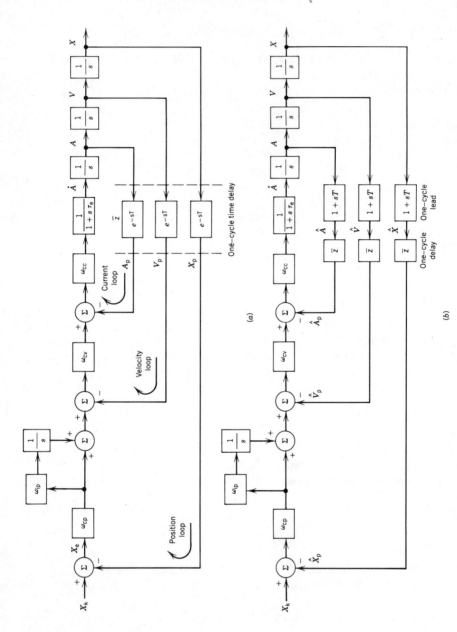

Figure 9.2-3 Method of compensating for phase lag due to one-cycle time delay around feedback loop: (*a*) computation diagram showing effect of one-cycle time delay; (*b*) system with time-delay compensation.

elements of the signal-flow diagram of Fig 9.2-1. However, the computations are performed in a feedback manner, and the feedback signals are derived from computations performed in the previous cycle. Hence, there is a one-cycle time delay around each feedback loop, which adds phase lag to the loop. To achieve accurate simulation, each loop must include lead compensation which offsets the phase lag caused by this one-cycle time delay.

Figure 9.2-3a shows the signal-flow diagram of the stage-positioning servo given in Fig 9.2-1, which is modified to include a one-cycle time delay in each of the three feedback paths. The transfer function of a one-cycle delay is e^{-sT}, which is represented as $\bar{z}$. The feedback loops provide feedback of position X, velocity V, and acceleration A. The delayed signals are the past values of X, V, and A, and so are represented as X_p, V_p, and A_p.

The frequency-response transfer function for a one-cycle time delay is $e^{-j\omega T}$. This has unity magnitude and a phase given by

$$\text{Ang}[\bar{z}] = \text{Ang}[e^{-j\omega T}] = -\omega T \qquad (9.2\text{-}57)$$

To compensate for this phase lag at low frequencies, the compensation transfer function $(1 + sT)$ is inserted in cascade with the time delay. For real frequencies $(s = j\omega)$, the magnitude and phase of this compensation transfer function is

$$|1 + sT| = |1 + j\omega T| = \sqrt{1 + (\omega T)^2} \qquad (9.2\text{-}58)$$

$$\text{Ang}[1 + sT] = \text{Ang}[1 + j\omega T] = \arctan[\omega T] \qquad (9.2\text{-}59)$$

For $\omega T \ll 1$, these can be approximated by

$$|1 + j\omega T| \doteq 1 \qquad (9.2\text{-}60)$$

$$\text{Ang}[1 + j\omega T] \doteq +\omega T \qquad (9.2\text{-}61)$$

Thus, at low frequencies, the compensation factor has approximately unity magnitude response, and has a phase lead of $+\omega T$ radian. This phase lead cancels the phase lag $-\omega T$ (given in Eq 9.2-57) produced by the one-cycle time delay.

The effect of this compensation factor at high frequencies is discussed in Section 9.3. As will be shown, the high-frequency response of the compensation factor can cause instability if the sampling rate is too low. This can be avoided by using a sampling frequency that is 16 times greater than the maximum gain crossover frequency of any feedback loop.

In Fig 9.2-3b, a compensation factor of the form $(1 + sT)$ is placed in series with each one-cycle time-delay factor in the feedback loops (represented by the transfer function $\bar{z}$). The effect of a $(1 + sT)$ compensation factor is to provide at low frequencies a phase lead equivalent to a time shift in the future of one sample period T. Passing a signal X through a $(1 + sT)$ transfer function produces a signal labeled $\hat{X}$, which (for a low-frequency signal) is equal to the future value of X, one cycle in the future. Hence, the signal $\hat{X}$ is called the "predicted value of X" or the "future value of X." The signal $\hat{X}$ is passed through the one cycle time delay $\bar{z}$ to form the signal $\hat{X}_p$, which is called the "past value of the predicted X" or the "past value of the future X." Thus, at low frequencies, $\hat{X}_p$ is equivalent to X.

To simplify the computation, it is convenient to normalize all of the $1/s$ integrations by dividing them by T, so as to provide transfer functions of the form $1/sT$. To compensate for this change, a gain constant at a prior point of each loop is multiplied by T. The resultant signal-flow diagram is shown in Fig 9.2-4a.

This normalization of the integration transfer functions changes the variables of the signal-flow diagram. The velocity V changes to TV, the acceleration A changes to T^2A, and the rate of acceleration $\dot{A}$ changes to $T^3\dot{A}$. All of these variables now have the same units as the output position X, and so are represented as follows by the symbol X with appropriate subscripts:

$$X_v = TV \qquad \text{(normalized velocity)} \qquad (9.2\text{-}62)$$

$$X_a = T^2A \qquad \text{(normalized acceleration)} \qquad (9.2\text{-}63)$$

$$X_{ra} = T^3\dot{A} \qquad \text{(normalized rate of acceleration)} \qquad (9.2\text{-}64)$$

The simulation is implemented in terms of the normalized variables X_v, X_a, X_{ra}. To obtain the actual values of velocity, acceleration, and rate of acceleration, these normalized values are divided by the sample period T raised to the appropriate power.

To simplify the calculations further, each $(1 + sT)$ compensation factor is shifted to the forward part of the loop, as indicated by the dashed lines in Fig 9.2-4a. Figure 9.2-4b shows the resultant signal-flow diagram after the factor $(1 + sT)$ of the current loop has been shifted to the forward part of that loop. The normalized integration $1/sT$ is replaced by $(1 + sT)/sT$. The factor $1/(1 + sT)$ is inserted in the velocity loop after this point to keep the signal-flow diagrams equivalent. When the $(1 + sT)$ factor in the feedback path of the velocity loop is placed in the forward path of the velocity loop, it cancels this factor $1/(1 + sT)$ and places a factor $1/(1 + sT)$ in the position loop. Moving the $(1 + sT)$ factor from the feedback path of the position loop to the forward path cancels this factor $1/(1 + sT)$ in the position loop.

The resultant signal-flow diagram used for computation is shown in Fig 9.2-5. The $(1 + sT)$ compensation factors in the three feedback paths have been replaced by a single $(1 + sT)$ factor in the forward path of the current loop. This is combined with the integration of that loop to produce the net transfer function $(1 + sT)/sT$. Also the variables X, X_v, and X_a have been replaced by their predicted values $\hat{X}$, $\hat{X}_v$, $\hat{X}_a$, which correspond to values one sample period in the future. To obtain the actual signal X, the signal $\hat{X}$ can be passed through the transfer function $1/(1 + Ts)$ as shown. However, if X is

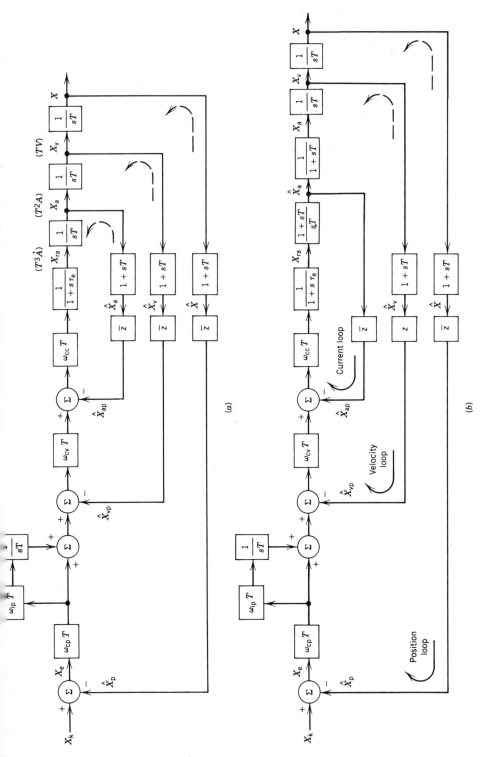

Figure 9.2-4 Simplification of simulation signal-flow diagram: (*a*) normalization of integrations; (*b*) effect of shifting time-delay compensation factor to forward path of current loop.

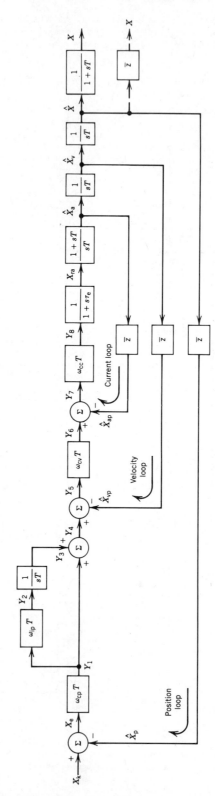

Figure 9.2-5 Final signal-flow diagram of stage-positioning servo for serial simulation.

TABLE 9.2-1   Algorithms for Simulating Common Normalized Transfer Functions

Transfer Function	Algorithm
(1) $\dfrac{Y}{X} = \dfrac{1}{sT}$	$y = y_p + \frac{1}{2}(x + x_p)$
(2) $\dfrac{Y}{X} = \dfrac{1 + sT}{sT}$	$y = y_p + \frac{3}{2}x - \frac{1}{2}x_p$
(3) $\dfrac{Y}{X} = \dfrac{1 + sT/2}{sT}$	$y = y_p + x$
(4) $\dfrac{Y}{X} = \dfrac{1}{1 + s\tau}$	$y = \dfrac{y_p(2\tau/T - 1) + x + x_p}{(2\tau/T + 1)}$
(5) $\dfrac{Y}{X} = \dfrac{1}{1 + \omega_x/s}$	$y = \dfrac{y_p(1 - \omega_x T/2) + x - x_p}{1 + \omega_x T/2}$
(6) $\dfrac{Y}{X} = \dfrac{1}{1 + sT}$	$y = \frac{1}{3}(y_p + x + x_p)$
(7) $\dfrac{Y}{X} = 1 + sT$	$y = -y_p + 3x - x_p$

not used subsequently in a feedback manner, an adequate representation of X can be derived much more simply by delaying $\hat{X}$ by one sample period. Either approach gives a good representation of X for low-frequency information.

Table 9.2-1 shows the algorithms for simulating the various transfer functions in the signal-flow diagram of Fig 9.2-5, along with others that are particularly useful. The algorithm for the transfer function $1/(1 + \omega_x/s)$ is needed to simulate the effect of the motor back EMF loop, which will be included later in Fig 9.2-6. The following simple integration formula is often used to provide sampled-data integration:

$$y = y_p + Tx \qquad \text{(Simple Integration Formula)} \qquad (9.2\text{-}65)$$

Normalizing this gives $y = y_p + x$. As shown in the table, the transfer function for this is $(1/sT)(1 + sT/2)$.

It was not absolutely necessary to convert the signal-flow diagram of Fig 9.2-4a to the form of Fig 9.2-5. This conversion was made to simplify the calculations performed in the simulation. Sometimes, it may not be desirable to shift the compensation factor $(1 + sT)$ from the feedback path to the forward path. In such a case, the transfer function $(1 + sT)$ could be simulated directly in the calculations. Therefore, Table 9.2-1 includes the algorithm for this transfer function.

One can write by inspection from the signal-flow diagram of Fig 9.2-5 the steps in a digital-computer program for simulating the system response. The variables labeled in the general form Y_1, Y_2, Y_3, etc. on the signal-flow diagram represent signals for which the physical significance is not specifically defined. All of the variables in the signal-flow diagram represent the transforms of the signals, whereas the variables used in the computer routine represent the actual time-domain signals. To be consistent with the standard symbolism of this book, the signals in the computer program should be represented by lowercase letters. However, to do this would confuse the process of translating the signal-flow diagram to a computer program. Accordingly, all of the variables in the computer program are represented by the same uppercase letters used in the signal-flow diagram, even though they represent time-domain signals in the computer program.

The following is an outline of a computer program expressed in BASIC computer-language format for simulating the response of the signal-flow diagram of Fig 9.2-5. The symbol T, which represents the sample period, is not to be confused with the variable TME, which represents the elapsed time. (The variable TIME should not be used, because it represents absolute time in BASIC.) The program steps are numbered in sequence to simplify the presentation. In the actual program, space should be left in the numbering sequence to allow for subsequent program changes.

SPECIFY PARAMETERS:

System parameters:

$$\omega_{cp} = ?, \ \omega_{ip} = ?, \ \omega_{cv} = ?, \ \omega_{cc} = ?, \ \tau_e = 1/\omega_e = ?$$

Program parameters:

$T = ?$ (sample period)
$\text{TME}_{Max} = ?$ (maximum elapsed time for computation)
$L_{Max} = ?$ (number of computation cycles per print cycle)

SPECIFY INITIAL CONDITIONS:

Set to Zero the Initial Values Required in Computation:

$$\hat{X}_p = \hat{X}_{vp} = \hat{X}_{ap} = X_{rap} = Y_{2p} = Y_{3p} = Y_{8p} = 0$$

Specify Initial Values of the Program Parameters:

$\text{TME} = 0$ (initial time)
$L = 0$ (initiate print counter)

SPECIFY INPUT SIGNAL(S):

100   TME = TME + T
101   X_k = Function[TME] (given function of time)

SYSTEM EQUATIONS (Read from Fig 9.2-5 and Table 9.2-1):

200   $X_e = X_k - \hat{X}_p$
201   $Y_1 = (\omega_{cp} * T) * X_e$
202   $Y_2 = (\omega_{ip} * T) * Y_1$
203   $Y_3 = Y_{3p} + (1/2)*(Y_2 + Y_{2p})$
204   $Y_4 = Y_1 + Y_3$
205   $Y_5 = Y_4 - \hat{X}_{vp}$
206   $Y_6 = (\omega_{cv} * T) * Y_5$
207   $Y_7 = Y_6 - \hat{X}_{ap}$
208   $Y_8 = (\omega_{cc} * T) * Y_7$
209   $X_{ra} = (X_{rap}*(2*(\tau_e/T) - 1) + Y_8 + Y_{8p})/(2*(\tau_e/T) + 1)$
210   $\hat{X}_a = \hat{X}_{ap} + (3/2)* X_{ra} - (1/2)* X_{rap}$
211   $\hat{X}_v = \hat{X}_{vp} + (1/2)*(\hat{X}_a + \hat{X}_{ap})$
212   $\hat{X} = \hat{X}_p + (1/2)*(\hat{X}_v + \hat{X}_{vp})$
213   $X = \hat{X}_p$

CHECK PRINT COUNT, PRINT DATA:

300   $L = L + 1$
301   IF $(L < L_{Max})$ GOTO 400
302   $L = 0$
303   PRINT data   (TME, X, X_e, $\hat{X}_v$, etc.)

SET PAST VALUES EQUAL TO PRESENT VALUES:

400   $\hat{X}_p = \hat{X}$
401   $Y_{2p} = Y_2$
402   $Y_{3p} = Y_3$
403   $\hat{X}_{vp} = \hat{X}_v$
404   $\hat{X}_{ap} = \hat{X}_a$
405   $Y_{8p} = Y_8$
406   $X_{rap} = X_{ra}$

CHECK TIME LIMIT, REPEAT

500   IF (TME < TME_{Max}) GOTO 100
501   END

This program prints data every L_{Max} cycles. Data for time TME = 0 are obtained by printing the initial (past) values of the variables.

To achieve accurate simulation, the sampling frequency $1/T$ must be much greater than the maximum gain crossover frequency of any feedback loop or the maximum break frequency of any transfer function. As will be shown in Section 9.3, accurate simulation is generally assured if the sampling frequency is at least a factor of 16 times greater than any gain crossover frequency or any break frequency. Thus, for any gain crossover frequency $f_c = \omega_c/2\pi$, the minimum allowable sampling frequency F is

$$\frac{1}{T} = F \geq 16f_c = 16\frac{\omega_c}{2\pi} = 2.5\omega_c \qquad (9.2\text{-}66)$$

This can be simplified to the requirement

$$\omega_c T \leq 1/2.5 = 0.4 \qquad (9.2\text{-}67)$$

For any transfer function being simulated, this becomes:

$$\omega_x T \leq 0.4 \qquad (9.2\text{-}68)$$

where ω_x is the highest break frequency of the transfer function. Setting $\omega_x = 1/\tau$ gives

$$T \leq 0.4\tau \qquad (9.2\text{-}69)$$

Thus the sampling period T should be no greater than 40% of the smallest time constant τ being simulated.

On the other hand, when a factor having low damping is simulated, an even shorter sampling period T may be needed. As will be shown in Section 9.3 (Eq 9.3-35), the simulated damping ratio ζ of an underdamped quadratic factor of natural frequency ω_n is reduced from the desired value by the following damping ratio error:

$$\Delta\zeta = 0.01(\omega_n T/0.4)^3 \qquad (9.2\text{-}70)$$

For example, if $\omega_n T = 0.4$, the simulated damping ratio is 0.01 lower than the desired value.

Two convenient command input signals for studying the system response are the unit step and the unit ramp. For a unit step input, the function of time to be specified in step 100 for the command input X_k is

100   $X_k = 1$ (for unit step input)

For a unit ramp input, the function of time is

100   $X_k = $ TME   (for unit ramp input)

In designating the variables, the following symbolism has been used to represent one-cycle time shifts in the past and future:

X_{ap} = *past value of* X_a, which is the value of X_a calculated in the previous cycle.

$\hat{X}_a$ = *future* (or *predicted*) *value of* X_a, which for a low-frequency signal is equal to the value of X_a one cycle in the future; obtained by passing X_a through the transfer function $(1 + sT)$.

$\tilde{X}_a$ = *delayed value of* X_a, which for a low-frequency signal is equal to the value of X_a one cycle in the past; obtained by passing X_a through the transfer function $1/(1 + sT)$.

The delayed value of X_a, represented by $\tilde{X}_a$, was not illustrated in Fig 9.2-5, but will be used in subsequent examples.

To simplify the computer program, the following are suggested for the names of the variables actually used in the program:

XA = X_a	(general variable X_a)
XAP = X_{ap}	(past value of X_a)
XAPP = X_{app}	(past value of X_{ap})
XAF = $\hat{X}_a$	(future, or predicted, X_a)
XAFP = $\hat{X}_{ap}$	(past value of future X_a)
XAD = $\tilde{X}_a$	(delayed X_a)
XADP = $\hat{X}_{ap}$	(past value of delayed X_a)
DXA = $\dot{X}_a$	(derivative of X_a)
DDXA = $\ddot{X}_a$	(second derivative of X_a)
TME = Time	(elapsed time)
T = T	(sample period)

The BASIC programs used in small personal computers often limit the names of variables to two alphanumeric characters. For such computers, self-explanatory variable names are generally not possible, and so a cross-reference table may be needed to relate the variable names to meaningful symbols. This problem can be minimized by using one letter to designate the present value of X, and another to designate the past value of X. For example, B2 could be the past value of A2, and BA the past value of AA.

9.2.4 Extension of Servo Simulation

The control system designed in Chapter 7 includes saturation limits which restrict the maximum absolute values of velocity and torque. These limits can readily be simulated in the following manner. Designate as follows the maximum values of the stage velocity and the force applied to the stage:

V_{Max} = maximum stage velocity,

F_{Max} = maximum force applied to stage, when friction is ignored.

The maximum stage acceleration in the absence of friction is

$$A_{\text{Max}} = F_{\text{Max}}/M_{(c)} \tag{9.2-71}$$

where $M_{(c)}$ is the total mass reflected to the stage (controlled member). Define the following normalized parameters:

$$X_{v(\text{Max})} = TV_{\text{Max}} \tag{9.2-72}$$

$$X_{a(\text{Max})} = T^2 A_{\text{Max}} \tag{9.2-73}$$

To provide the velocity limits, add the following lines to the computer program between steps 204, 205:

204a   IF $Y_4 > X_{v(\text{Max})}$ THEN $Y_4 = +X_{v(\text{Max})}$
204b   IF $Y_4 < -X_{v(\text{Max})}$ THEN $Y_4 = -X_{v(\text{Max})}$

The maximum torque can be limited by adding the following lines to the computer program between steps 206, 207:

206a   IF $Y_6 > X_{a(\text{Max})}$ THEN $Y_6 = +X_{a(\text{Max})}$
206b   IF $Y_6 < -X_{a(\text{Max})}$ THEN $Y_6 = -X_{a(\text{Max})}$

When the servo operates under saturated conditions, large overshoot and even instability can result because of excessive voltage stored on the capacitor of the position-loop integral network. This issue is discussed in Chapter 12, Section 12.2.4. As will be shown in Fig 12.2-5, it is common practice to place a diode limiter circuit across the capacitor of an integral network, in order to limit the voltage stored on the capacitor. The effect of such a limiter can be simulated in the following manner.

In the simulation signal-flow diagram of the stage-positioning servo of Fig 9.2-5, the variable Y_3 corresponds to the charge stored on the integral-network capacitor, and is equivalent to the integral-network correction signal X_{cor}, shown in Fig 2.4-7 of Chapter 2. For a low-frequency input, the integral network correction signal Y_3 is approximately equal to the velocity-loop feedback signal $\hat{X}_{vp}$. If sampled-data dynamics are ignored, this simplifies to

$$Y_3 = X_v = TV \tag{9.2-74}$$

this equation omits the caret and the subscript p on X_v.

A servo generally exhibits maximum velocity only under slew conditions. Under accurate tracking, it usually operates at much less than maximum velocity. Hence the charge stored on the integral network can correspond to a velocity command that is much less than maximum velocity. Thus, the maximum value of the variable Y_3 can be limited by

$$\text{Max}|Y_3| = \frac{\text{Max}|X_v|}{\beta} = \frac{X_{v(\text{Max})}}{\beta} = \frac{TV_{\text{Max}}}{\beta} \tag{9.2-75}$$

where β is a nondimensional factor appreciably greater than unity. For this example, a good value for β is

$$\beta = 4 \tag{9.2-76}$$

Equation 9.2-75 is equal to the expression of Eq 9.2-72, divided by β. If $|Y_3|$ is set so that it does not exceed this limit of Eq 9.2-75, the integral network does not build up excessive charge during saturated operation, provided that β is sufficiently large. This limit can be set by adding the following lines to the computer program, between steps 203, 204:

203a IF $Y_3 > X_{v(\text{Max})}/\beta$ THEN $Y_3 = X_{v(\text{Max})}/\beta$
203b IF $Y_3 < -X_{v(\text{Max})}/\beta$ THEN $Y_3 = -X_{v(\text{Max})}/\beta$

Including these steps in the computer program is equivalent to placing zener diodes across the integral-network capacitor, as shown in Fig 12.2-5.

Figure 9.2-5 was derived from Fig 9.2-1, which is a simplified version of the signal-flow diagram of the stage-positioning servo given in Fig 7.5-1. When the complete signal-flow diagram of Fig 7.5-1 is used, the resultant diagram for computer simulation is shown in Fig 9.2-6. This includes the effects of motor back EMF and a load disturbance input. It also includes two test inputs, X_{kv} and X_{ka}, which are used to test the responses of the velocity loop and the current feedback loop.

The signals $X_{a(d)}$, $X_{a(t)}$ in Fig 9.2-6 are defined as

$$X_{a(d)} = T^2 A'_d \tag{9.2-77}$$

$$X_{a(t)} = T^2 A_t \tag{9.2-78}$$

The variables A'_d and A_t were shown in Fig 7.5-1. In accordance with Eqs 7.5-3, -4, the variable A'_d is proportional to the equivalent force or torque disturbance, and is given by

$$A_d = \frac{F_{d(c)}}{M_{(c)}} \tag{9.2-79}$$

$$A'_d = A_d\left(1 - \frac{\omega_{cm}}{\omega_{cc}}\right) \tag{9.2-80}$$

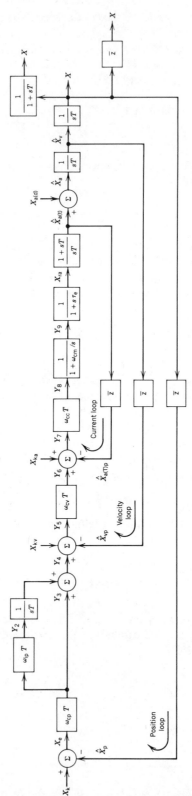

Figure 9.2-6 Serial simulation signal-flow diagram for stage-positioning servo, including effect of motor back EMF and friction disturbance signal.

TABLE 9.2-2   Parameters Used in Simulating Signal-Flow Diagram of Fig 9.2-6

Parameter	Value	Parameter	Value
(1) ω_e	1109 sec^{-1}	(6) ω_{cm}	14.0 sec^{-1}
(2) ω_{cc}	1109 sec^{-1}	(7) $F_{d(c)}$	21.25 lb
(3) ω_{cv}	457 sec^{-1}	(8) $M_{(c)}$	1.497 lb-sec^2/in.
(4) ω_{cp}	142 sec^{-1}	(9) A_d	14.2 in./sec^2
(5) ω_{ip}	32.7 sec^{-1}	(10) A'_d	14.0 in./sec^2

The variable A_d is the acceleration that the controlled member (stage) would experience if the total equivalent friction force were applied to the controlled member, with no other forces. The variable A'_d includes the effect of the back-EMF loop on the equivalent force disturbance. The variable A_t is proportional to motor torque, and is the stage acceleration that would be produced if the load force disturbance $F_{d(c)}$ were zero. The variables $\hat{X}_{a(t)}$, $\hat{X}_{a(d)}$ are the predicted values of $X_{a(t)}$, $X_{a(d)}$, which for low-frequency information are the values of these variables one cycle in the future. Note that the feedback signal for the current loop is $\hat{X}_{a(t)p}$, which is the past value of $\hat{X}_{a(t)}$, and for a low-frequency signal is equal to $X_{a(t)}$.

9.2.5   Results of Simulation

Table 9.2-2 shows the values of the parameters required to simulate the signal-flow diagrams of Figs 9.2-5, -6 for the stage-positioning servo. Items (1) to (6) were obtained from Table 7.2-1, and items (7), (8) were obtained from Table 7.1-4. The values of items (7), (8) are substituted in Eq 9.2-79 to obtain the value of A_d given in item (9). Items (2), (6), (9) are substituted in Eq 9.2-80 to obtain the value of A'_d given in item (10).

The maximum break frequency is ω_e or ω_{cc}, both of which are equal to 1109 sec^{-1}. Substituting this for ω_x in Eq 9.2-68 gives the following allowable upper value for the sampling period T:

$$T \le \frac{0.4}{\omega_x} = \frac{0.4}{\omega_{cc}} = \frac{0.4}{1109 \text{ sec}^{-1}} = 3.6 \times 10^{-4} \text{ sec} \qquad (9.2\text{-}81)$$

Accordingly, the following value is selected for the sample period:

$$T = 2.5 \times 10^{-4} \text{ sec} = 0.25 \text{ msec} \qquad (9.2\text{-}82)$$

This corresponds to a sampling frequency ($F = 1/T$) of 4000 Hz.

The signals X_{ka}, X_{kv} of Fig 9.2-6 are input command signals for measuring the responses of the current and velocity loops. The feedback response of the current loop to a unit step is measured by: (1) opening the velocity loop by setting $\omega_{cv} = 0$; (2) setting the acceleration command input X_{ka} equal to 1.0; and (3) reading the signal $\hat{X}_{a(t)p}$. The feedback response of the velocity loop to a unit step is measured by: (1) opening the position loop by setting $\omega_{cp} = 0$; (2) setting the velocity command input X_{kv} equal to 1.0; and (3) reading the signal $\hat{X}_{vp}$.

As shown in Chapter 6, the maximum error caused by static friction can be calculated by determining the peak response of the servo to a step of load torque or force equal to the friction level. The representative error response to friction is measured by: (1) setting the A'_d disturbance input to 14.0 in./sec^2 in

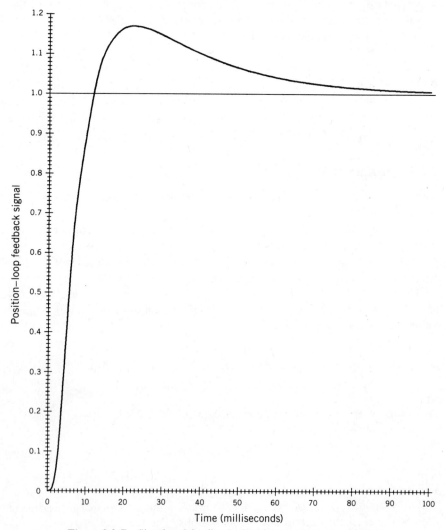

Figure 9.2-7   Simulated feedback-signal response to a unit step.

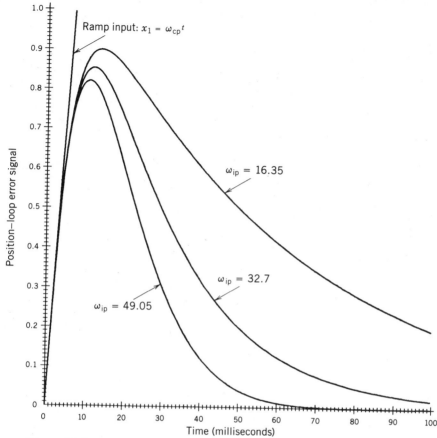

Figure 9.2-8 Simulated error response to a ramp input, for three values of ω_{ip}.

accordance with Table 9.2-2; (2) setting $X_k = 0$; and (3) reading the error signal X_e, which gives the error response in inches. The maximum value of this response is the peak error produced by friction.

Figure 9.2-7 shows the simulated feedback response to a unit step input to the servo of Fig 9.2-5, using the parameters of items (1) to (5) of Table 9.2-2. Figure 9.2-8 shows the simulated error responses to a ramp input for three specified values of ω_{ip}, including the value 32.7 sec^{-1} given in Table 9.2-2. The ramp input is normalized in terms of ω_{cp}, and is given by

$$x_i = \omega_{cp}t = \left(142 \text{ sec}^{-1}\right)t \qquad (9.2\text{-}83)$$

These responses were obtained from the serial-simulation program given in Section 9.2.3, using a sample period T of 0.25 msec. (The plots of Figs 9.2-7, -8 were prepared by Dominic Gasbarro, a student at the University of Lowell.)

9.3   PSEUDO-FREQUENCY CONCEPT FOR EVALUATING THE DYNAMIC LIMITATIONS OF SAMPLED-DATA ALGORITHMS

9.3.1   Definition of Pseudo Frequency W

In Section 9.2, sampled-data routines were developed to achieve specified frequency-response characteristics by noting that the variable U, which is defined as $(1 - \bar{z})/(1 + \bar{z})$, is approximately equal to $sT/2$ at low frequencies. In order to investigate the limitations of this approximation, it is convenient to define the complex pseudo frequency p, which is exactly related to U by

$$U = p\frac{T}{2}$$

(9.3-1)

The real and imaginary parts of p are defined as

$$p = \alpha + jW$$

(9.3-2)

Hence U is related to α and W by

$$U = \alpha\frac{T}{2} + jW\frac{T}{2}$$

(9.3-3)

As was shown in Eq 9.2-34, for real frequencies (i.e., for $s = j\omega$), U is imaginary and is equal to

$$U = j\tan\left[\frac{\omega T}{2}\right]$$

(9.3-4)

Equating the imaginary parts of U in Eqs 9.3-3, -4 gives the following expression, which holds for real frequencies ($s = j\omega$):

$$\frac{WT}{2} = \tan\left[\frac{\omega T}{2}\right]$$

(9.3-5)

For $\omega T/2 \ll 1$, $\tan[\omega T/2]$ is approximately equal to $\omega T/2$, and so W is approximately equal to ω. At higher frequencies, Eq 9.3-5 can be solved as follows to calculate one of these variables from the other:

$$WT = 2\tan[\omega T/2]$$

(9.3-6)

$$\omega T = 2\arctan[WT/2]$$

(9.3-7)

Figure 9.3-1 shows a plot of the normalized true frequency ωT versus the normalized pseudo frequency WT. The vertical axis also shows f/F, which is the ratio of real frequency f in hertz, divided by the sampling frequency F.

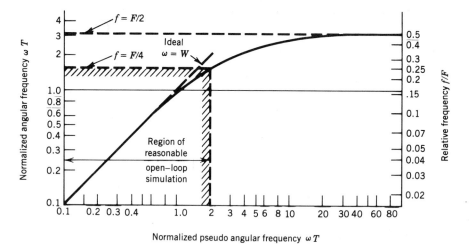

Figure 9.3-1 Plot of normalized true angular frequency (ωT) versus normalized pseudo angular frequency (WT).

Note that ωT is related as follows to f/F:

$$\omega T = 2\pi f T = 2\pi\left(\frac{f}{F}\right) \tag{9.3-8}$$

At low frequencies, ω is approximately equal to W. This approximation holds reasonably well up to the limit $WT = 2$. At this point, ωT and f/F are equal to

$$\omega T = 2\arctan\left[\frac{WT}{2}\right] = 2\arctan[1] = \frac{\pi}{2} = 1.57 \tag{9.3-9}$$

$$\frac{f}{F} = \frac{\omega T}{2\pi} = \frac{1}{4} \tag{9.3-10}$$

Thus, W can be approximated by ω with reasonable accuracy up to $\frac{1}{4}$ of the sampling frequency, which is $\frac{1}{2}$ of the maximum frequency of the sampled information. Above $\frac{1}{4}$ of the sampling frequency, the variables W and ω depart strongly. The pseudo frequency W increases to infinity as the true frequency ω increases from $\frac{1}{4}$ to $\frac{1}{2}$ of the sampling frequency. When W is infinite, ωT is equal to π, and f/F is equal to 0.5.

To apply this plot of ω versus W, consider the exact frequency response of the normalized trapezoidal integration algorithm, which is

$$y = y_{\mathrm{p}} + \tfrac{1}{2}(x + x_{\mathrm{p}}) \tag{9.3-11}$$

The Laplace transform of this is

$$Y = \bar{z}Y + \tfrac{1}{2}(X + \bar{z}X) \tag{9.3-12}$$

Solving for the ratio Y/X gives

$$\frac{Y}{X} = \frac{1 + \bar{z}}{2(1 - \bar{z})} = \frac{1}{2U} \tag{9.3-13}$$

This is expressed as follows in terms of the complex pseudo frequency p by setting $U = pT/2$:

$$\frac{Y}{X} = \frac{1}{pT} \tag{9.3-14}$$

To obtain the pseudo-frequency response, p is replaced by jW:

$$\frac{Y}{X} = \frac{1}{jWT} \tag{9.3-15}$$

In Fig 9.3-2, curve ① is a plot of the magnitude of this frequency response versus the normalized pseudo frequency WT.

The horizontal scale of Fig 9.3-2 represents both the normalized pseudo frequency WT and the normalized true frequency ωT. Curve ② shows the frequency response of the trapezoidal integration formula in terms of the true frequency ω; whereas curve ① is the corresponding response in terms of the pseudo frequency W. Curve ② is derived from curve ① by calculating for each value of W the corresponding value of ω. Consider, for example, point A on curve ①, which occurs at $WT = 2.5$. At this value of WT, the magnitude of the transfer function, which is $1/WT$, is equal to $1/2.5$ or 0.4, as shown on the vertical scale. The value of ωT at the corresponding point B on curve ② is derived from Eq 9.3-7 as follows:

$$\omega T = 2\arctan\left[\frac{WT}{2}\right] = 2\arctan\left[\frac{2.5}{2}\right] = 1.79 \tag{9.3-16}$$

The frequency response of an ideal integrator would have the same shape as curve ① of Fig 9.3-2, relative to the true frequency ω. Therefore, the departure of curve ② from the ideal curve ① represents the departure of the actual frequency response from the ideal. The figure shows that the actual response ② of the integrator follows the ideal response ① very accurately up to $\frac{1}{8}$ of the sampling frequency (up to $F/8$), and there is reasonable agreement up to $\frac{1}{4}$ of the sampling frequency (up to $F/4$). The frequency response extends only to $\frac{1}{2}$ of the sampling frequency ($F/2$), which is the limit of the information band. This limit corresponds to $\omega T = \pi$.

In a sampled-data signal, the base band extends from zero to half the sampling frequency. Signal components at frequencies above the base band

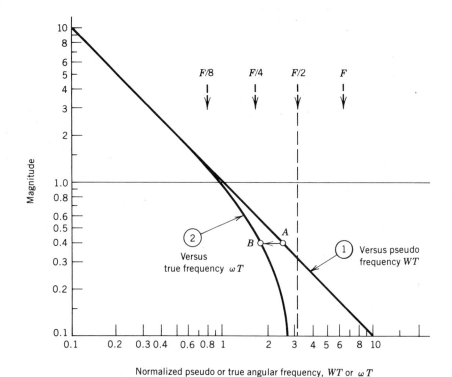

Figure 9.3-2 Magnitude of frequency response of normalized trapezoidal integration.

are merely duplicates of the base-band information, and so are usually of no concern. However, when a sampled-data signal is demodulated, any signal in a high-frequency band that passes through the filtering of the demodulation process adds distortion to the demodulated signal. To determine the effect of this distortion, it is sometimes desirable to plot the frequency response of a sampled-data signal over much more than the base frequency band. This can be done in the following manner.

Figure 9.3-3 illustrates the relationships between the magnitude and phase values of a sampled-data transfer function (or a transform) for negative frequencies and for frequencies greater than half the sampling frequency, relative to the values in the base band. The fundamental relations are

$$H[-f] = H^*[f] \qquad (9.3\text{-}17)$$

$$H[NF + f] = H[f] \qquad (9.3\text{-}18)$$

The function H is a transfer function (or a transform), the variable f is the frequency in hertz, N is an integer, and F is the sampling frequency in hertz. Equation 9.3-17 states that the value of the transfer function (or transform) H at the negative frequency $-f$ is equal to the conjugate of H at the frequency

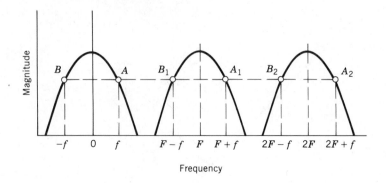

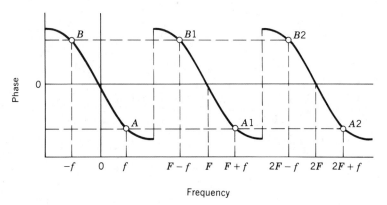

Figure 9.3-3 Magnitude and phase values of a simplified transfer function in the main band, related to the values at negative frequencies and at higher frequencies.

f. This principle is expressed as follows in terms of magnitude and phase:

$$|H[-f]| = |H[f]| \qquad (9.3\text{-}19)$$
$$\text{Ang}[H[-f]] = -\text{Ang}[H[f]] \qquad (9.3\text{-}20)$$

From Eqs 9.3-18, -19, the magnitude values in the different bands are related by

$$|H[NF + f]| = |H[f]| \qquad (9.3\text{-}21)$$
$$|H[NF - f]| = |H[-f]| = |H[f]| \qquad (9.3\text{-}22)$$

By Eqs 9.3-18, -20, the phase values in the different bands are related by

$$\text{Ang}[H[NF + f]] = \text{Ang}[H[f]] \qquad (9.3\text{-}23)$$
$$\text{Ang}[H[NF - f]] = \text{Ang}[H[-f]] = -\text{Ang}[H[f]] \qquad (9.3\text{-}24)$$

These equations are illustrated in Fig 9.3-3.

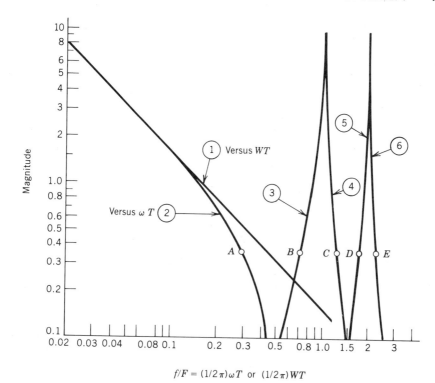

$$f/F = (1/2\pi)\omega T \text{ or } (1/2\pi)WT$$

Figure 9.3-4 Magnitude of sampled frequency response for normalized trapezoidal integration, for frequencies up to 2.5 times the sampling frequency.

Consider the magnitude plots of Fig 9.3-3. The magnitude at the frequency f is indicated by point A. This is the same value as the magnitude at the frequency $-f$, indicated by point B. The magnitude also has this same value at the frequencies $(F+f),(F-f),(2F+f),(2F-f)$, which are indicated by points A_1, B_1, A_2, B_2, respectively.

For the phase plots of Fig 9.3-3, the phase at frequency f is indicated by point A. The phase at the frequency $-f$, which is indicated by point B, is the negative of this phase at point A. The phase values at the frequencies $(F+f)$ and $(2F+f)$, indicated by points A_1 and A_2, are the same as the phase at point A at the frequency f. The phase values at the frequencies $(F-f)$ and $(2F-f)$, indicated by points B_1 and B_2, are the same as the phase at the frequency $-f$ indicated by point B, which is the negative of the phase at the frequency f, indicated by point A.

Using the principles illustrated in Fig 9.3-3, the magnitude of the frequency response of the normalized trapezoidal integration formula is plotted in Fig 9.3-4, out to 2.5 times the sampling frequency. To facilitate the construction of this plot, the frequency scale of Fig 9.3-2 is divided by 2π for both the variables ωT and WT. As shown in the following equation, this changes the

ωT scale to the frequency ratio f/F:

$$\frac{1}{2\pi}\omega T = fT = \frac{f}{F} \qquad (9.3\text{-}25)$$

Curves ① and ② of Fig 9.3-4 are derived from curves ① and ② of Fig 9.3-2 by sliding the horizontal frequency scale so that the normalized frequency $\omega T = \pi$ is shifted to $f/F = 0.5$.

Curve segments ③, ④, ⑤, and ⑥ are derived from curve ②. For example, at point A on curve ②, the magnitude is 0.36 and the frequency is $f = 0.3F$. This same magnitude (0.36) occurs at point B at the frequency $0.7F$, at point C at the frequency $1.3F$, at point D at the frequency $1.7F$, and at point E at the frequency $2.3F$.

9.3.2 Accuracy of Sampled-Data Simulation of a Feedback Loop

The preceding has shown that a sampled-data filter can provide, in an open-loop process, a very close approximation of a desired frequency response up to $\frac{1}{8}$ of the sampling frequency, and a reasonable approximation up to $\frac{1}{4}$ of the sampling frequency. On the other hand, in a feedback system, the frequency limits for good simulation are much lower, because of the one-cycle time delay around a feedback loop. This case is illustrated in Fig 9.3-5.

The transfer function for a one-cycle time delay can be expressed as follows in terms of the pseudo frequency W:

$$\bar{z} = \frac{1-U}{1+U} = \frac{1-jW(T/2)}{1+jW(T/2)} \qquad (9.3\text{-}26)$$

The magnitude of this function is unity. The phase is plotted as curve ① in Fig 9.3-5 versus the normalized pseudo frequency WT, expressed on a logarithmic scale. Scales are also shown to give the corresponding values of the normalized angular frequency ωT and the relative frequency f/F.

As was shown in Section 9.2, to compensated for the phase lag caused by the one-cycle time delay around a loop, an algorithm is used which exhibits a transfer function approximating $(1 + sT)$. The exact compensation transfer function is

$$(1 + pT) = (1 + jWT) \qquad (9.3\text{-}27)$$

Plots of the magnitude and phase of this compensation transfer function are shown as curves ② and ③ in Fig 9.3-5. Curve ④, which is the sum of the phase curves ① and ②, is the total phase. Thus, the magnitude plot ③ and phase plot ④ are the combined effect of the time delay and the time-delay compensation.

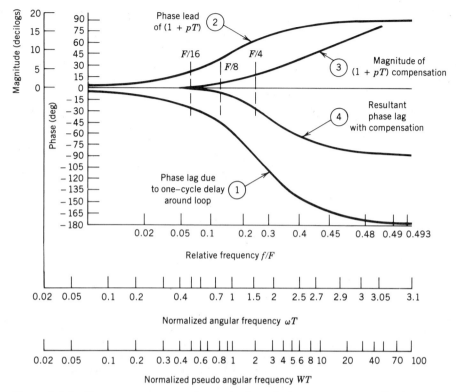

Figure 9.3-5 Frequency response of magnitude and phase versus normalized pseudo frequency WT, showing effect of $(1 + pT)$ compensation to offset phase lag of one-cycle time delay.

From Eqs 9.3-6, -7, the normalized pseudo frequency WT can be related as follows to the relative frequency f/F:

$$WT = 2 \tan[\pi f/F] \qquad (9.3\text{-}28)$$

This shows that for frequency f equal to $F/4$, $F/8$, and $F/16$, the values of WT are 1.0, 0.828, and 0.398, respectively. These frequencies are indicated by dashed vertical lines in Fig 9.3-5. At $f = F/8$ (or $WT = 0.828$), the magnitude deviation (curve ③) is $+1.1$ dg, and the phase deviation is $-5.4°$. This frequency is an upper limit to the range of acceptable simulation. At $f = F/16$ (or $WT = 0.398$), the magnitude deviation of curve ③ is $+0.3$ dg, and the phase deviation of curve ④ is $-0.8°$. These deviations are negligible for most applications.

Thus, when the $(1 + pT)$ digital compensation is added to a feedback loop to compensate for the unavoidable one-cycle time delay around the loop, the resultant loop provides very good simulation up to $\frac{1}{16}$ of the sampling frequency, and reasonable simulation up to $\frac{1}{8}$ of the sampling frequency.

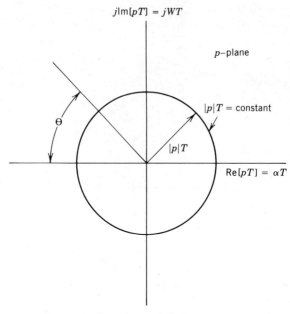

Figure 9.3-6   Complex pseudo-frequency p-plane.

9.3.3   Map of Complex Pseudo-Frequency p-Plane onto s-Plane

The relationship $U = j \tan[\omega T/2]$ given in Eq 9.3-4 holds only for real frequencies (for $s = j\omega$). An analysis of the general relationship between the variable U and the full s-plane is presented in Appendix D. The following equations are derived in that appendix, which express the real and imaginary parts of the complex frequency s (designated σ and ω) as functions of the real and imaginary parts of the complex pseudo frequency p (designated α and W):

$$\tan[\omega T] = \frac{WT}{1 - (|p|T/2)^2} \qquad (9.3\text{-}29)$$

$$\tanh[\sigma T] = \frac{\alpha T}{1 + (|p|T/2)^2} \qquad (9.3\text{-}30)$$

The variable $|p|$ is the magnitude of the complex pseudo frequency p, which is related to α and W by

$$|p|^2 = \alpha^2 + W^2 \qquad (9.3\text{-}31)$$

The complex pseudo-frequency p-plane is illustrated in Fig 9.3-6. The real part of the complex pseudo frequency is defined as α and the imaginary part is defined as W. Two sets of contours are drawn on this p-plane. One set is for constant values of $|p|$, which consists of circles about the origin. The second set is for constant values of the ratio W/α, which consists of radial lines emanating from the origin. As shown in the figure, the contours for constant values of W/α are defined in terms of the angle Θ between the radial line and the negative real p-axis.

In Fig 9.3-7, the p-plane contours of Fig 9.3-6 are mapped onto the s-plane over the following ranges: $0 < \omega T < \pi$, and $-6 < \alpha T < +6$. The frequency range in hertz corresponding to this range of ωT is $0 < f < F/2$. Thus, the plot covers the complete frequency range of the information band of the sampled data, which extends to half the sampling frequency.

The contour map of Fig 9.3-7 shows the following:

1. The p-plane maps onto the s-plane with high uniformity up to $|p|T = 1$, which corresponds to $\frac{1}{6}$ of the sampling frequency.

2. The right half of the p-plane maps onto the right half of the s-plane, and the jW axis of the p-plane maps onto the $j\omega$ axis of the s-plane.

3. As $|p|$ approaches infinity, the map of the p-plane converges to a point on the $j\omega$ axis at $\omega T = \pi$, which corresponds to $\frac{1}{2}$ of the sampling frequency ($f/F = 0.5$).

4. The contour for $|p|T = 2$ is a horizontal line of constant ωT, along which $\omega T = \pi/2$ and $f/F = 0.25$. For values of $|p|T$ greater than 2, a negative real root of p (for which $W = 0$) results in a complex root of s, for which $\omega T = \pi$, or $f = 0.5F$.

The following conclusions can be drawn from this contour map of Fig 9.3-7:

1. A stable sampled transfer function cannot have poles of p at infinity, because this results in oscillatory poles of s on the $j\omega$ axis at $\frac{1}{2}$ of the sampling frequency.

2. A pseudo Nyquist plot of the digital loop transfer function $G^s[jW]$ can be constructed relative to the pseudo frequency W, in the same manner that $G[j\omega]$ is constructed relative to the true frequency ω. If $G^s[p]$ has no poles at $p = \infty$, the loop will be stable if the $G^s[jW]$ Nyquist plot satisfies the requirements of a stable $G[j\omega]$ Nyquist plot. (The reason for this is that the right half of the p-plane maps into the right half of the s-plane.)

3. To achieve good stability, the poles of p should not exceed the limit $|p|T = 2$. If this limit is exceeded, the loop transfer function $G[j\omega]$ has underdamped high-frequency poles, even when the poles of p are real. Underdamped high-frequency poles of G generate underdamped closed-loop poles that are close to them. These closed-loop poles would tend to make the loop oscillate at half the sampling frequency.

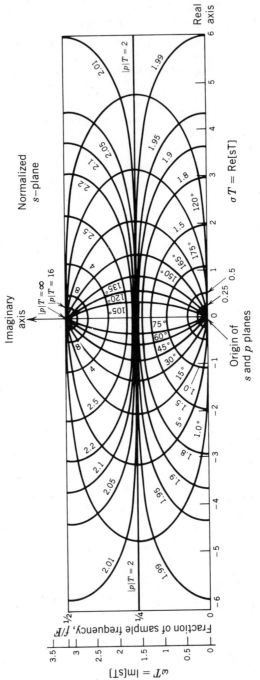

Figure 9.3-7 Map of complex pseudo-frequency p-plane onto s-plane.

132

Because of conclusion (3), the break frequency of any pole relative to the pseudo frequency W should not exceed the normalized pseudo frequency $WT = 2$, which corresponds to $\omega T = \pi/2$, or $f = 0.25F$. It is possible to reach this limit, but the limit should not be exceeded. Remember that every feedback loop has the transfer function $\bar{z}$, caused by the one-cycle time delay around the loop, which is expressed as follows in terms of p:

$$\bar{z} = \frac{1 - p(T/2)}{1 + p(T/2)} \qquad (9.3\text{-}32)$$

This has a left-half-plane pole and a right-half-plane zero with the same pseudo break frequency, $WT = 2$.

9.3.4 Application of Pseudo Frequency in Design of a Sampled-Data Feedback Loop

The map of the complex pseudo-frequency p-plane onto the s-plane shows that a stable system can have no poles at infinite p. An important implication of this requirement is that a digital feedback loop should not be closed around a differentiator; it should only be closed around an integrator. The digital transfer function of a differentiator is p, which has a pole at infinite p; while the digital transfer function of an integrator is $1/p$, which has a zero at infinite p. This principle is applied in Section 9.4, which develops general approaches to the simulation of dynamic equations.

To develop quantitative requirements for achieving good stability with digital feedback, consider the simulation of the following simple loop transfer function:

$$G = \omega_c/s \qquad (9.3\text{-}33)$$

The sampled-data transfer function for simulating this is

$$G^s = \frac{\omega_c}{p}\bar{z}(1 + pT) = \frac{\omega_c(1 - pT/2)(1 + pT)}{p(1 + pT/2)} \qquad (9.3\text{-}34)$$

The factor ω_c/p is the ideal transfer function; the factor $\bar{z}$ is the transfer function for the one-cycle time delay around the loop; and $(1 + pT)$ is the compensation for this one-cycle delay. The factor $\bar{z}$ is expressed in terms of the complex pseudo frequency p in accordance with Eq 9.3-32. Setting $p = jW$ gives the pseudo-frequency response:

$$G^s = \frac{\omega_c(1 - jWT/2)(1 + jWT)}{jWT(1 + jWT/2)} \qquad (9.3\text{-}35)$$

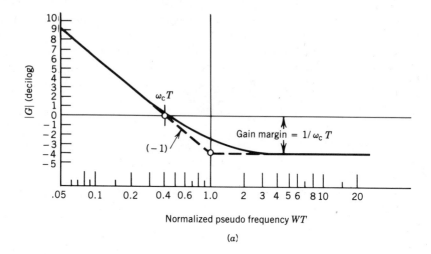

(a)

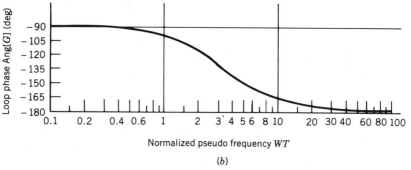

(b)

Figure 9.3-8 Pseudo-frequency response plots of sampled loop transfer function $G^s[jW]$, for simulating loop transfer function $G = \omega_c/s$: (a) magnitude plot; (b) phase plot.

The magnitude and phase of this loop transfer function are plotted in Fig 9.3-8 versus the normalized pseudo frequency WT for the particular case $\omega_c T = 0.4$. The phase plot is the same as curve ④ of Fig 9.3-5, shifted by $-90°$. The G-locus obtained from this data is plotted on a simplified Nichols chart in Fig 9.3-9. This Nichols chart shows only two contours: $|G_{ib}| = 0$ dg and $|G_{ib}| = 1.0$ dg = 2 dB.

The G-locus (curve ③) on the simplified Nichols chart of Fig 9.3-9 shows the values of the normalized pseudo frequency WT along the lower and right side of the locus. The values of the true relative frequency f/F are shown along the upper and left side of this locus. The phase of G is $-180°$ at the point $WT = \infty$, or $f/F = 0.5$. The reciprocal of $|G|$ at this point is therefore

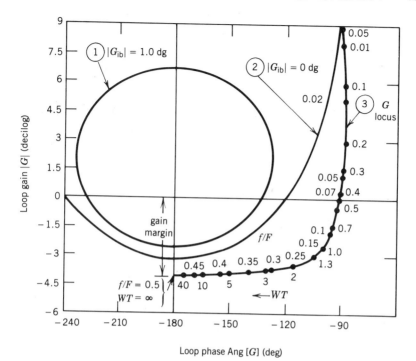

Figure 9.3-9 Locus of sampled loop transfer function of Fig 9.3-8 on abbreviated Nichols chart.

the gain margin of the loop. The plot of $|G|$ in Fig 9.3-9 shows that the magnitude of G at infinite WT is equal to the magnitude asymptote of G at $WT = 1$, which is equal to $\omega_c T$. Thus, the gain margin is

$$\text{gain margin} = 1/\omega_c T \tag{9.3-36}$$

As was stated in Chapter 3, the gain margin of a feedback loop should be at least a factor of 2.5 (or 4 dg) to assure good stability. Applying this requirement gives the following upper limit on $\omega_c T$:

$$\omega_c T \leq 1/2.5 = 0.4 \tag{9.3-37}$$

Set $\omega_c = 2\pi f_c$, and $T = 1/F$. This gives the corresponding upper limit on the gain crossover frequency f_c expressed in hertz:

$$f_c = \frac{\omega_c}{2\pi} \leq \frac{0.4}{2\pi T} = \frac{F}{15.7} \tag{9.3-38}$$

This is approximately $F/16$. Thus to assure a gain margin of at least 2.5, the gain crossover frequency should not exceed $\frac{1}{16}$ of the sample frequency F, and the value of ωT should not exceed 0.4.

Let us compare this stability requirement based on gain margin with a direct measurement of stability derived from the transient response of the digital loop. Figure 9.3-10 shows the transient error responses of this digital loop for a unit-step input for values of $\omega_c T$ from 0.3 to 0.8. These responses were obtained by simulation. Since these are sampled-data responses, the values are specified only at the sampling instants. However for clarity, straight lines are drawn between the sampled points. The response for $\omega_c T = 1.0$, which is not shown, is a constant-amplitude oscillation, with a frequency equal to half the sampling frequency, and a period equal to twice the sample period T.

As shown in Fig 9.3-10, the response is highly oscillatory for $\omega_c T = 0.8$, and is poor for $\omega_c T$ greater than 0.5. The curve for $\omega_c T = 0.4$ represents a good engineering design. This result substantiates the requirement, based on gain margin, that $\omega_c T$ should not exceed 0.4.

The preceding discussion has shown that the maximum gain crossover frequency of any feedback loop in a digital simulation program should not exceed $\frac{1}{16}$ of the sampling frequency. This is equivalent to requiring that $\omega_c T$ should not exceed 0.4. A simple way to check the stability of a digital simulation is to run the simulation a second time with the sampling period T reduced by a factor of 2. If the response does not change significantly, the sampling frequency is adequate.

9.3.5   Required Sampling Rate for Simulating Factors with Low Damping

As will be explained in Chapter 10, mechanical structural resonance in a servomechanism often exhibits poles and zeros having very low damping. When one is simulating a factor with low damping, a better criterion for the sampling rate is required. Consider the signal-flow diagram of Fig 9.3-11a. The exact loop transfer function G and feedback transfer function G_{ib} for the outer loop are

$$G = \frac{\omega_{c1}\omega_{c2}}{s(s + \omega_{c2})} \tag{9.3-39}$$

$$G_{ib} = \frac{1}{1 + G} = \frac{\omega_{c1}\omega_{c2}}{s^2 + s\omega_{c2} + \omega_{c1}\omega_{c2}} \tag{9.3-40}$$

This feedback transfer function has the form

$$G_{ib} = \frac{\omega_n^2}{s^2 + 2\zeta\omega_n s + \omega_n^2} \tag{9.3-41}$$

$$\omega_{c1} = \omega_n/2\zeta \tag{9.3-42}$$

$$\omega_{c2} = 2\zeta\omega_n \tag{9.3-43}$$

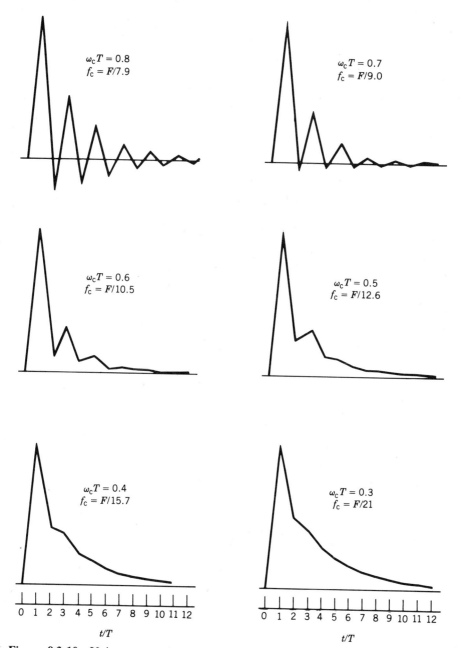

Figure 9.3-10 Unit-step transient error responses of sampled-data feedback loop having ideal transfer function ω_c/s, for different values of the parameter $\omega_c T$.

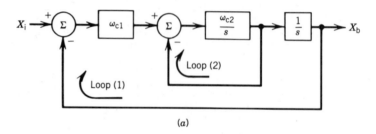

(a)

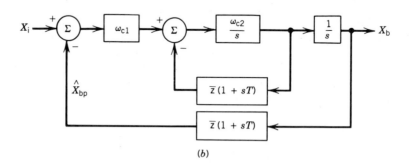

(b)

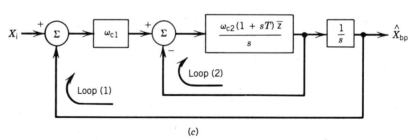

(c)

Figure 9.3-11 Serial simulation of feedback loop exhibiting underdamped response: (a) ideal signal-flow diagram; (b) serial-simulation implementation of a; (c) modification of serial-simulation diagram b to express it in the form of ideal diagram a.

Diagram b of Fig 9.3-11 is the signal-flow diagram for serial simulation of diagram a. Diagram b can be manipulated to form diagram c, which differs from the ideal diagram a in that it has the following additional transfer function in the inner loop:

$$H = (1 + sT)e^{-sT} = (1 + j\omega T)e^{-j\omega T} \qquad (9.3\text{-}44)$$

The corresponding G_{ib} transfer function can be obtained from Eq 9.3-40 by multiplying each ω_{c2} factor by H. This yields

$$G_{ib} = \frac{\omega_{c1}\omega_{c2}H}{s^2 + s\omega_{c2}H + \omega_{c1}\omega_{c2}H} = \frac{\omega_n^2 H}{s^2 + s2\zeta\omega_n H + \omega_n^2 H} \qquad (9.3\text{-}45)$$

We are concerned with the characteristics of this simulated transfer function at angular frequencies ω that are much less than $1/T$. At such low frequencies, the magnitude of H is close to unity, and so only the phase of H is of concern. At any particular low frequency, H can be approximated by

$$H \doteq 1 - j\phi \qquad (9.3\text{-}46)$$

where ϕ is the phase lag of H at that frequency. By Eq 9.3-44, ϕ is equal to

$$\phi = -\text{Ang}[H] = \omega T - \arctan[\omega T] \qquad (9.3\text{-}47)$$

Substitute the approximation for H of Eq 9.3-46 into Eq 9.3-45. This gives the following approximation of G_{ib} (for $s = j\omega$):

$$G_{ib} \doteq \frac{\omega_n^2(1 - j\phi)}{-\omega^2 + j\omega 2\zeta\omega_n(1 - j\phi) + \omega_n^2(1 - j\phi)}$$

$$\doteq \frac{\omega_n^2(1 - j\phi)}{\left(\omega_n^2 - \omega^2 + 2\zeta\omega_n\omega\phi\right) + j\left(2\zeta\omega_n\omega - \omega_n^2\phi\right)} \qquad (9.3\text{-}48)$$

The factor $(1 - j\phi)$ in the numerator merely adds a very small phase shift to the transfer function, and so can be ignored. Let us designate the resonant frequency of G_{ib} as ω_n'. This is obtained by setting the real part of the denominator of Eq 9.3-48 equal to zero, which gives

$$\omega^2 = \omega_n'^2 = \omega_n^2 + 2\zeta\omega_n\omega_n'\phi \qquad (9.3\text{-}49)$$

The last term is quite small relative to ω_n^2. Hence the quantity ω_n' in this term can be approximated by ω_n without appreciably affecting the value of the equation. This gives

$$\omega_n' = \omega_n\sqrt{1 + 2\zeta\phi} \doteq \omega_n(1 + \zeta\phi) \qquad (9.3\text{-}50)$$

Note that $\sqrt{1 + x}$ is approximately equal to $(1 + x/2)$ for $|x| \ll 1$. This simulated natural frequency ω_n' differs only slightly from the desired natural frequency ω_n. At the simulated natural frequency ω_n', G_{ib} of Eq 9.3-48 is equal to

$$G_{ib}[\omega_n'] = \frac{\omega_n^2}{j\left(2\zeta\omega_n\omega_n' - \omega_n^2\phi\right)} \doteq \frac{1}{j(2\zeta - \phi)} \qquad (9.3\text{-}51)$$

where ω_n' is approximated by ω_n. The damping ratio of this simulated resonant factor is designated ζ', and is related as follows to G_{ib} at the resonant frequency ω_n':

$$G_{ib}[\omega_n'] = 1/j2\zeta' \tag{9.3-52}$$

Setting Eqs 9.3-51, -52 equal gives for ζ'

$$\zeta' = \zeta - \frac{\phi}{2} = \zeta - \Delta\zeta \tag{9.3-53}$$

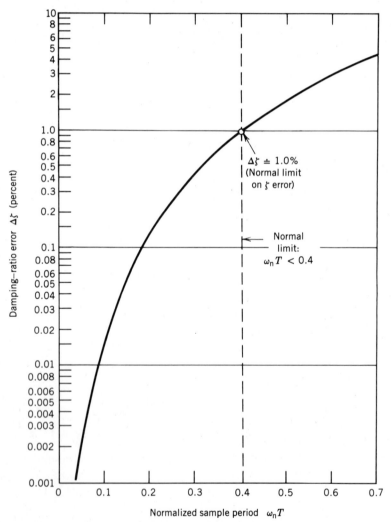

Figure 9.3-12 Plot of damping-ratio error (in percent) for serial simulation of under-damped feedback loop, versus normalized sampling period $\omega_n T$.

The parameter $\Delta\zeta$ is the shift of the simulated damping ratio ζ' from the desired damping ratio ζ. Combining Eqs 9.3-47, -53 gives for $\Delta\zeta$

$$\Delta\zeta = \frac{\phi}{2} = \tfrac{1}{2}(\omega_n'T - \arctan[\omega_n'T])$$

$$\doteq \tfrac{1}{2}(\omega_nT - \arctan[\omega_nT]) \tag{9.3-54}$$

where ω_n' is approximated by ω_n.

The damping ratio ζ' of the simulation is less than the desired damping ratio ζ, and the parameter $\Delta\zeta$ is the error in damping ratio of the simulation. The damping ratio error $\Delta\zeta$ obtained from Eq 9.3-54 is plotted in Fig 9.3-12 versus the normalized sampling period ω_nT.

According to Section 9.3.4, the parameter $\omega_{gc}T$ should not exceed 0.4, where ω_{gc} is the highest gain crossover frequency of the system. For a loop with low damping, ω_{gc} is approximately equal to ω_n. As shown in Fig 9.3-12, when $\omega_nT = 0.4$ the damping ratio of the simulated quadratic factor is in error by 1%. This damping-ratio error is reasonable if the damping ratio is high, but is quite unacceptable when the damping ratio is low. Thus, for simulating a factor with low damping, the normalized sampling period ω_nT must be appreciable less than 0.4.

For $\omega_n < 0.4$, the plot of Fig 9.3-12 can be approximated quite accurately by

$$\Delta\zeta \doteq 0.01(\omega_nT/0.4)^3 \tag{9.3-55}$$

Thus, the damping ratio error $\Delta\zeta$ is proportional to the cube of the sampling period. For example, if the normalized sampling period ω_nT is reduced by a factor of 2 from 0.4 to 0.2, the damping-ratio error $\Delta\zeta$ is reduced by a factor of 2^3, or 8: from 1% to 0.125%.

9.4 GENERALIZED APPROACHES TO SIMULATION

9.4.1 ELF Transmitter Circuit

To develop more general approaches for dynamic simulation, this section considers the simulation of an electrical circuit. Two techniques are applied: (1) the serial simulation method developed in Section 9.2, and (2) the Runge–Kutta integration method, which is formulated in terms of state variables. The discussion explains the concepts of state variables and state equations.

To give this discussion physical significance, it is applied to a circuit investigated by the author in the study of a high-power communication transmitter that would operate in the extremely-low-frequency (ELF) band, from 30 to 300 Hz. The transmitter would be driven by a device called an

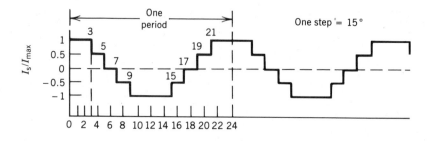

Figure 9.4-1   Waveform of source current I_s generated by current-source inverter.

inverter, which uses thyristors (silicon-controlled rectifiers), which act as switches to convert DC power to AC power with high efficiency. A "current-source inverter" would be used, powered by a constant-current source. Using appropriate electrical switching commands to the thyristors, the inverter controls the current delivered to the load, by switching the path of this constant current. The current-source inverter is used extensively in the control of large AC motors, and is described by Phillips [9.5].

To excite a sinusoidal current in the antenna, the current-source inverter would generate the current waveform shown in Fig 9.4-1. Figure 9.4-2 shows a circuit that was investigated during the transmitter study, for filtering the inverter current waveform and coupling it efficiently to the antenna. The current source I_s in Fig 9.4-2 has the inverter current waveform of Fig 9.4-1. The purpose of the circuit is to provide strong filtering of harmonics, while generating a sinusoidal antenna current I_a. The communications signal is modulated by shifting (or "keying") the frequency of the sinusoid. Communication efficiency is maximized if the frequency shift occurs, without phase change, when the antenna current passes through zero.

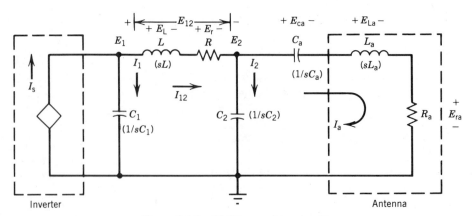

Figure 9.4-2   ELF transmitter circuit.

The antenna impedance is represented by an antenna resistance R_a in series with an antenna inductance L_a. The capacitance C_a is a tuning capacitor, which cancels the inductive antenna impedance at the mean transmitted frequency. Hence C_a is equal to

$$C_a = 1/\omega_o^2 L_a \qquad (9.4\text{-}1)$$

where ω_o is the mean transmitted frequency, expressed in rad/sec. The reactances of the inductor L and capacitors C_1 and C_2 at the frequency ω_o are set equal to the antenna resistance R_a. Hence

$$\frac{1}{\omega_o C_1} = \frac{1}{\omega_o C_2} = \omega_o L = R_a \qquad (9.4\text{-}2)$$

With these settings, the input impedance of the coupling network is resistive at the mean frequency ω_o, and is equal to the antenna resistance R_a. This input impedance is the load that is experienced by the inverter. The Q of the antenna at the mean frequency ω_o is defined as

$$Q_a = \omega_o L_a / R_a \qquad (9.4\text{-}3)$$

A typical value for Q_a is 5. The resistance R is the effective series resistance of the inductor L, which can be expressed as

$$R = \omega_o L / Q \qquad (9.4\text{-}4)$$

The parameter Q is the Q of the inductor, which is assumed to be 20.

To develop a computer program for simulating this ELF transmitter network, a signal-flow diagram of the network is constructed as shown in Fig 9.4-3. This figure presents in signal-flow form the Kirchoff circuit equations, expressed with all dynamic calculations represented as integrals. Hence, the voltage E_c across a capacitor is derived from the current I_c flowing through the capacitor, while the current I_L flowing through an inductor is derived from the voltage E_L across the inductor. Thus the calculations for a capacitor and an inductor take the following form:

Calculation for capacitor:

$$E_c = \frac{1}{sC} I_c \qquad (9.4\text{-}5)$$

Calculation for inductor:

$$I_L = \frac{1}{sL} E_L \qquad (9.4\text{-}6)$$

For a resistor, either the voltage or the current can be used as the independent

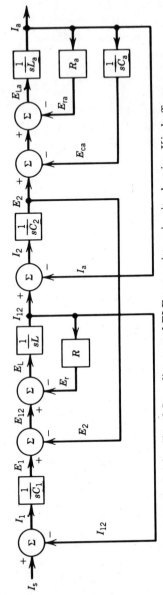

Figure 9.4-3 Signal-flow diagram of ELF transmitter circuit, showing Kirchoff equations with all dynamics expressed as integrations.

variable. With a little ingenuity, the student can readily express the Kirchoff equations for any circuit in a signal-flow diagram of this form.

9.4.2   State Variables and State Equations

In Fig 9.4-4*a*, each transfer function of Fig 9.4-3 between voltage and current variables is split, so as to separate the $1/s$ integration from the factor $1/L$ or $1/C$, with the factor $1/s$ placed last. The outputs of the integrators are called the *state variables* of the system, and the inputs to the integrators are the derivatives of the state variables. Since the signal-flow diagram has five integrators, the system has five state variables. The transfer functions of the five integrators are

$$E_1 = \frac{1}{s}\dot{E}_1 \qquad (9.4\text{-}7)$$

$$I_{12} = \frac{1}{s}\dot{I}_{12} \qquad (9.4\text{-}8)$$

$$E_2 = \frac{1}{s}\dot{E}_2 \qquad (9.4\text{-}9)$$

$$I_a = \frac{1}{s}\dot{I}_a \qquad (9.4\text{-}10)$$

$$E_{ca} = \frac{1}{s}\dot{E}_{ca} \qquad (9.4\text{-}11)$$

The variables E_1, I_{12}, E_2, I_a, and E_{ca} are the transforms of the state variables, and the variables $\dot{E}_1$, $\dot{I}_{12}$, $\dot{E}_2$, $\dot{I}_a$, and $\dot{E}_{ca}$ are the transforms of the derivatives of the state variables. Taking the inverse transforms of these equations gives

$$e_1 = \int \dot{e}_1 \, dt \qquad (9.4\text{-}12)$$

$$i_{12} = \int \dot{i}_{12} \, dt \qquad (9.4\text{-}13)$$

$$e_2 = \int \dot{e}_2 \, dt \qquad (9.4\text{-}14)$$

$$i_a = \int \dot{i}_a \, dt \qquad (9.4\text{-}15)$$

$$e_{ca} = \int \dot{e}_{ca} \, dt \qquad (9.4\text{-}16)$$

where e_1, i_{12}, e_2, i_a, and e_{ca} are the state variables, and $\dot{e}_1$, $\dot{i}_{12}$, $\dot{e}_2$, $\dot{i}_a$, and $\dot{e}_{ca}$ are the derivatives of the state variables. Hence, these equations, which

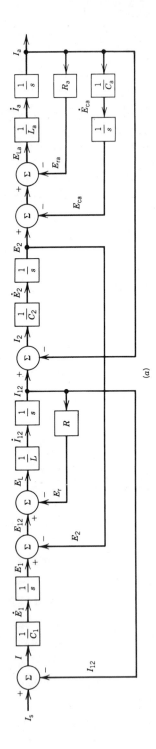

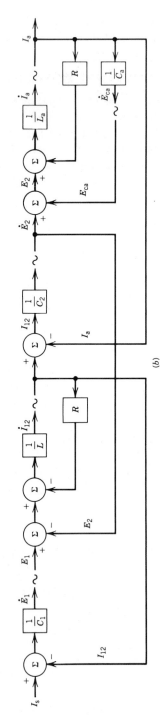

Figure 9.4.4 Modification of signal-flow diagram, with integrations expressed as separate elements, to define the state variables and state equations: (*a*) modification of Fig 9.4.3, with integrations expressed separately; (*b*) modification of (*a*), with integrations and secondary signals deleted, to simplify specification of state equations.

calculate the state variables by integrating the derivatives of the state variables, have the general form

$$y_n = \int \dot{y}_n \, dt \tag{9.4-17}$$

where y_n is the nth state variable, and $\dot{y}_n$ is the derivative of the nth state variable.

This signal-flow diagram of Fig 9.4-4a is redrawn in Fig 9.4-4b with the integrators omitted. All secondary variables are omitted, so that the diagram shows only the state variables, the derivatives of the state variables, and the input source current I_s. The following are the equations derived from that diagram, which express the transforms of the integrator inputs (the derivatives of the state variables) as functions of the transforms of the integrator outputs (the state variables) and the transform of the source current I_s (which is called a forcing function):

$$\dot{E}_i = \frac{1}{C_1}(I_s - I_{12}) \tag{9.4-18}$$

$$\dot{I}_{12} = \frac{1}{L}(-RI_{12} + E_1 - E_2) \tag{9.4-19}$$

$$\dot{E}_2 = \frac{1}{C_2}(I_{12} - I_a) \tag{9.4-20}$$

$$\dot{I}_a = \frac{1}{L_a}(E_2 - E_{ca} - I_a R_a) \tag{9.4-21}$$

$$\dot{E}_{ca} = \frac{1}{C_a}I_a \tag{9.4-22}$$

Since these equations do not contain dynamic elements, their inverse transforms can be obtained by replacing the transforms of the signals by the signals themselves. The inverse transforms of Eqs 9.4-18 to 9.4-22 are

$$\dot{e}_i = \frac{1}{C_1}(i_s - i_{12}) \tag{9.4-23}$$

$$\dot{i}_{12} = \frac{1}{L}(-Ri_{12} + e_1 - e_2) \tag{9.4-24}$$

$$\dot{e}_2 = \frac{1}{C_2}(i_{12} - i_a) \tag{9.4-25}$$

$$\dot{i}_a = \frac{1}{L_a}(e_2 - e_{ca} - i_a R_a) \tag{9.4-26}$$

$$\dot{e}_{ca} = \frac{1}{C_a}i_a \tag{9.4-27}$$

These are called the *state equations*. They express the derivatives of the state variables as functions of the state variables and the input forcing functions. (For this case, there is only one input forcing function, the source current i_s.) An important reason for expressing the system equations in terms of state variables and state equations is that this allows the system time response to be calculated by means of a Runge–Kutta computer integration routine.

9.4.3   Runge–Kutta Integration Routine for Simulating Dynamic Systems

Most computer programs for simulating dynamic systems employ the Runge–Kutta integration routine. As was shown in Section 9.2, a digital computation loop experiences a time delay of one sample cycle, which produces error in the dynamic simulation unless compensation is included to correct for it. The Runge–Kutta integration routine compensates for this time delay by extrapolating the data in such a manner as to bring the signals in time synchronism with one another. The following is a description of the Runge–Kutta routine [9.6].

To perform a Runge–Kutta integration, the system equations are expressed in the state-equation form, which has the following general format:

1. State equations:

$$\dot{y}_n[t] = \text{Function}\left[y_1[t], y_2[t], \ldots, x_1[t], x_2[t], \ldots\right] \quad (9.4\text{-}28)$$

2. Integrations:

$$y_1[t] = \int \dot{y}_1[t]\, dt, \qquad y_2[t] = \int \dot{y}_2[t]\, dt, \ldots \quad (9.4\text{-}29)$$

The functions $y_1[t]$, $y_2[t]$, etc. are the values of the state variables at a particular time t; and $x_1[t]$, $x_2[t]$, etc. are the values of the input forcing functions at that same time t. Often there is only a single input forcing function $x[t]$. The variables $\dot{y}_n[t]$ are the derivatives of the state variables $y_n[t]$. The state equations, which can be nonlinear, express the values of the derivatives $\dot{y}_n[t]$, at a particular instant of time, as functions of the values of the state variables $y_n[t]$ and the forcing functions $x_m[t]$ at that same instant of time.

The values of $y_n[t]$ at the beginning, middle, and end of a particular sample interval are designated $y_{n(1)}$, $y_{n(2)}$, $y_{n(3)}$; and the values of the input forcing functions at these same instants are designated $x_{m(1)}$, $x_{m(2)}$, $x_{m(3)}$. The Runge–Kutta algorithm operates by calculating the derivatives of the state variables at the beginning, middle, and end of each sample period. From these derivatives, designated $\dot{y}_{n(1)}$, $\dot{y}_{n(2)}$, $\dot{y}_{n(3)}$, the integral over the sample interval, designated Δy, is calculated as shown in Fig 9.4-5. For simplicity, the figure

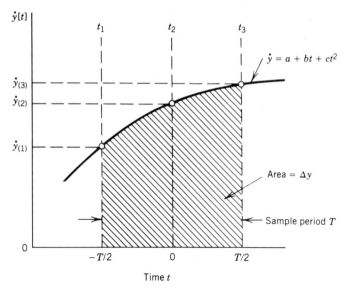

Figure 9.4-5   Calculation of integral of $\dot{y}[t]$ over sample period T for Runge–Kutta integration.

omits the n in the subscripts. The figure is a plot of the derivative $\dot{y}[t]$ over the sample interval T, which is characterized by the three values $\dot{y}_{(1)}$, $\dot{y}_{(2)}$, $\dot{y}_{(3)}$. Since only these three values are known, the best assumption one can make for the actual time function $\dot{y}[t]$ is a curve passing through the three points, described by a second-order equation of the form

$$\dot{y}[t] = a + bt + ct^2 \tag{9.4-30}$$

The values of the three coefficients a, b, c are selected so that the curve passes through the three specified points. Equation 9.4-30 is integrated over the sample interval to obtain Δy, which is indicated by the cross-hatched area in Fig 9.4-5. Appendix E shows that the resultant expression for Δy is

$$\Delta y = \frac{T}{6}\left(\dot{y}_{(1)} + 4\dot{y}_{(2)} + \dot{y}_{(3)} \right) \tag{9.4-31}$$

The calculations of the Runge–Kutta routine for one sample interval are outlined in Table 9.4-1. The following is a discussion of those calculations. The calculation for each sample interval starts with: (1) the values of the state variables $y_n[t]$ at time $t = t_i$ (at the start of that interval), and (2) the values of the input forcing functions at time t equal to t_i, $(t_i + T/2)$, and $(t_i + T)$ (at the beginning, middle, and end of that interval). Thus, at each sample interval, the calculation is started with known values of $y_{n(1)}$ for each state variable, and known values of $x_{m(1)}$, $x_{m(2)}$, $x_{m(3)}$ for each input forcing function.

TABLE 9.4-1   Outline of Calculations of Runge–Kutta Integration

---

$$y_{n(1)}, x_{n(1)} \rightarrow \boxed{\text{State equations}} \rightarrow \dot{y}_{n(1)}$$

First estimate of $y_{n(2)}$: $y_{n(2)e1} = y_{n(1)} + (T/2)\dot{y}_{n(1)}.$

$$y_{n(2)e1}, x_{n(2)} \rightarrow \boxed{\text{State equations}} \rightarrow \dot{y}_{n(2)e1}$$

Second estimate of $y_{n(2)}$: $y_{n(2)e2} = y_{n(1)} + (T/2)\dot{y}_{n(2)e1}.$

$$y_{n(2)e2}, x_{n(2)} \rightarrow \boxed{\text{State equations}} \rightarrow \dot{y}_{n(2)e2}$$

Estimate of $y_{n(3)}$: $y_{n(3)e} = y_{n(1)} + (T)\dot{y}_{n(2)e2}.$

$$y_{n(3)e}, x_{n(3)} \rightarrow \boxed{\text{State equations}} \rightarrow \dot{y}_{n(3)e}$$

Average of estimates of $\dot{y}_{n(2)}$ is designated $\dot{y}_{n(2)e}$:

$$\dot{y}_{n(2)e} = \tfrac{1}{2}(\dot{y}_{n(2)e1} + \dot{y}_{n(2)e2})$$

Integral of $\dot{y}_n$ over sampling interval is

$$\Delta y_n = \frac{T}{6}(\dot{y}_{n(1)} + 4\dot{y}_{n(2)e} + \dot{y}_{n(3)e})$$

Exact value for $y_{n(3)}$ at end of sample interval is

$$y_{n(3)} = y_{n(1)} + \Delta y_n$$

Variable $y_{n(3)}$ is the value of $y_{n(1)}$ for next sample interval.

The values of $y_{n(1)}$, $x_{m(1)}$, which are the values of the state variables and input forcing functions at time $t = t_i$ (at the start of the sample interval), are substituted into the state equations to obtain the values of the derivatives of the state variables at time $t = t_i$, which are designated $\dot{y}_{n(1)}$. From these derivatives, the following equation is used to calculate the first estimates of the state variables at time $(t_i + T/2)$, which are designated $y_{n(2)e1}$:

$$y_{n(2)e1} = y_{n(1)} + (T/2)\dot{y}_{n(1)} \qquad (9.4\text{-}32)$$

This calculation of the first estimate of $y_{n(2)}$ is illustrated in Fig 9.4-6a. These first estimates of $y_{n(2)}$, along with the values $x_{m(2)}$ of the forcing function at time $(t_i + T/2)$, are substituted into the state equations to obtain the first estimates of the derivatives at time $(t_i + T/2)$, which are designated $\dot{y}_{n(2)e1}$. From these, the following equation is used to calculate the second estimates of the state variables at time $(t_i + T/2)$, which are designated $y_{n(2)e2}$:

$$y_{n(2)e2} = y_{n(1)} + (T/2)\dot{y}_{n(2)e1} \qquad (9.4\text{-}33)$$

This calculation of the second estimate of $y_{n(2)}$ is illustrated in Fig 9.4-6b.

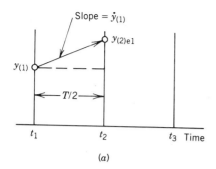

(a)

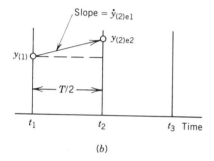

(b)

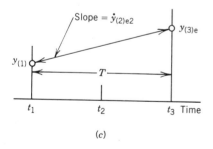

(c)

Figure 9.4-6 Runge–Kutta calculations of estimated state-variable values, at middle and end of sample interval: calculation of (a) first estimate of $y_{(2)}$; (b) second estimate of $y_{(2)}$; (c) estimate of $y_{(3)}$.

These second estimates of $y_{n(2)}$, along with the values $x_{m(2)}$ of the forcing function at time $(t_i + T/2)$, are substituted into the state equations to obtain the second estimates of the derivatives at time $(t_i + T/2)$, which are designated $\dot{y}_{n(2)e2}$. From these, the following equation is used to calculate the estimates of the state variables at time $(t_i + T)$, which are designated $y_{n(3)e}$:

$$y_{n(3)e} = y_{n(1)} + (T)\dot{y}_{n(2)e2} \qquad (9.4\text{-}34)$$

This calculation of the estimate of $y_{n(3)}$ is illustrated in Fig 9.4-6c. These estimates of $y_{n(3)}$, along with the values $x_{m(3)}$ of the forcing function at time $(t_i + T)$, are substituted into the state equations to obtain the estimates of the derivatives at time $(t_i + T)$, which are designated $\dot{y}_{n(3)e}$.

The estimates of the derivatives at point 2, at the middle of the interval, are averaged as follows to obtain the average estimated derivative, which is designated $\dot{y}_{n(2)e}$:

$$\dot{y}_{n(2)e} = \tfrac{1}{2}\left(\dot{y}_{n(2)e1} + \dot{y}_{n(2)e2}\right) \tag{9.4-35}$$

Thus, the calculations yield three derivatives for each state variable, which are the derivatives of y_n at the beginning, middle, and end of the sample interval. These three derivatives are designated $\dot{y}_{n(1)}$, $\dot{y}_{n(2)e}$, and $\dot{y}_{n(3)e}$. In accordance with Eq 9.4-31, the integral of $\dot{y}_n[t]$ over the sample interval, which is designated Δy, is calculated as follows from these three derivatives:

$$\Delta y_n = \frac{T}{6}\left(\dot{y}_{n(1)} + 4\dot{y}_{n(2)e} + \dot{y}_{n(3)e}\right) \tag{9.4-36}$$

The true value for y_n at the end of the sample interval, which is designated $y_{n(3)}$, is obtained by adding Δy_n to $y_{n(1)}$, the value at the beginning of the sample interval:

$$y_{n(3)} = y_{n(1)} + \Delta y_n \tag{9.4-37}$$

This value $y_{n(3)}$ now becomes the value of $y_{n(1)}$ for the next sample interval, and the process is repeated.

9.4.4 Matrix Representation of State Equations

To apply the Runge–Kutta integration procedure efficiently, the state equations must be expressed in a form that can be interpreted conveniently by a computer. This is achieved by writing the state equations in matrix format. The required matrix format can be simply explained by considering an example. Rules for matrix manipulation are summarized in Appendix F.

The variables of the state equations are designated by:

1. N state variables, labeled y_n.
2. N derivatives of state variables, labeled $\dot{y}_n$.
3. M forcing functions, labeled x_m.

Assume, for example, that $N = 4$ and $M = 2$, so that there are four state variables (y_1, y_2, y_3, y_4), four derivatives of state variables ($\dot{y}_1, \dot{y}_2, \dot{y}_3, \dot{y}_4$), and two forcing functions (x_1, x_2). The general form of the state equations is

$$\dot{y}_1 = a_{11}y_1 + a_{12}y_2 + a_{13}y_3 + a_{14}y_4 + b_{11}x_1 + b_{12}x_2 \qquad (9.4\text{-}38)$$

$$\dot{y}_2 = a_{21}y_1 + a_{22}y_2 + a_{23}y_3 + a_{24}y_4 + b_{21}x_1 + b_{22}x_2 \qquad (9.4\text{-}39)$$

$$\dot{y}_3 = a_{31}y_1 + a_{32}y_2 + a_{33}y_3 + a_{34}y_4 + b_{31}x_1 + b_{32}x_2 \qquad (9.4\text{-}40)$$

$$\dot{y}_4 = a_{41}y_1 + a_{42}y_2 + a_{43}y_3 + a_{44}y_4 + b_{41}x_1 + b_{42}x_2 \qquad (9.4\text{-}41)$$

These equations are expressed in matrix format as follows:

$$
\begin{bmatrix} \dot{y}_1 \\ \dot{y}_2 \\ \dot{y}_3 \\ \dot{y}_4 \end{bmatrix}
=
\begin{bmatrix}
a_{11} & a_{12} & a_{13} & a_{14} \\
a_{21} & a_{22} & a_{23} & a_{24} \\
a_{31} & a_{32} & a_{33} & a_{34} \\
a_{41} & a_{42} & a_{43} & a_{44}
\end{bmatrix}
\begin{bmatrix} y_1 \\ y_2 \\ y_3 \\ y_4 \end{bmatrix}
+
\begin{bmatrix}
b_{11} & b_{12} \\
b_{21} & b_{22} \\
b_{31} & b_{32} \\
b_{41} & b_{42}
\end{bmatrix}
\begin{bmatrix} x_1 \\ x_2 \end{bmatrix}
\qquad (9.4\text{-}42)
$$

The matrix expression of Eq 9.4-42 is equivalent to Eqs 9.4-38 to 9.4-41. The coefficients a_{pq} for the state variables are arranged in an N-by-N matrix (N rows by N columns), while the coefficients b_{pq} for the forcing functions are arranged in an N-by-M matrix (N rows by M columns). The state variables y_n, and the derivatives of the state variables $\dot{y}_n$, are arranged in N-by-1 arrays, which are called N-component (column) vectors. The forcing functions x_m are arranged in an M-by-1 array, which is an M-component vector. In order for matrix multiplication to have meaning, the number of columns of the first matrix (or vector) must be equal to the number of rows of the second matrix (or vector).

The state equations for the ELF transmitter network were given in Eqs 9.4-23 to 9.4-27. These can be expressed as

$$\dot{e}_1 = -\frac{1}{C_1}i_{12} + \frac{1}{C_1}i_s \qquad (9.4\text{-}43)$$

$$\dot{i}_{12} = \frac{1}{L}e_1 - \frac{R}{L}i_{12} - \frac{1}{L}e_2 \qquad (9.4\text{-}44)$$

$$\dot{e}_2 = \frac{1}{C_2}i_{12} - \frac{1}{C_2}i_a \qquad (9.4\text{-}45)$$

$$\dot{i}_a = \frac{1}{L_a}e_2 - \frac{R_a}{L_a}i_a - \frac{1}{L_a}e_{ca} \qquad (9.4\text{-}46)$$

$$\dot{e}_{ca} = \frac{1}{C_a}i_a \qquad (9.4\text{-}47)$$

These equations are represented in matrix format as follows:

$$
\begin{bmatrix} \dot{e}_1 \\ \dot{i}_{12} \\ \dot{e}_2 \\ \dot{i}_a \\ \dot{e}_{ca} \end{bmatrix} = \begin{bmatrix} 0 & -1/C_1 & 0 & 0 & 0 \\ 1/L & -R/L & -1/L & 0 & 0 \\ 0 & 1/C_2 & 0 & -1/C_2 & 0 \\ 0 & 0 & 1/L_a & -R_a/L_a & -1/L_a \\ 0 & 0 & 0 & 1/C_a & 0 \end{bmatrix} \begin{bmatrix} e_1 \\ i_{12} \\ e_2 \\ i_a \\ e_{ca} \end{bmatrix}
$$

$$
+ \begin{bmatrix} 1/C_1 \\ 0 \\ 0 \\ 0 \\ 0 \end{bmatrix} i_s \qquad\qquad (9.4\text{-}48)
$$

The matrix relation of Eq 9.4-42 can be condensed to the form

$$
\dot{\underline{y}} = \underline{A}\underline{y} + \underline{B}\underline{x} \qquad\qquad (9.4\text{-}49)
$$

The underlines indicate that the variables y, $\dot{y}$, and x are vectors and the transfer blocks A and B are matrices. The variable y is a four-component vector that represents the four state variables y_1, y_2, y_3, y_4. The variable $\dot{y}$ is a four-component vector that represents the four derivatives of state variables, $\dot{y}_1$, $\dot{y}_2$, $\dot{y}_3$, $\dot{y}_4$. The variable x is a two-component vector that represents the two forcing functions x_1, x_2. The transfer block A represents the matrix of the sixteen coefficients a_{pq}, and the transfer block B represents the matrix of the eight coefficients b_{pq}.

In Fig 9.4-7, the equations of the system are expressed in signal-flow diagram form, in accordance with the condensed matrix relation of Eq 9.4-49. The signals y, $\dot{y}$, and x in the signal-flow diagram represent vector signals, with four components for y and $\dot{y}$, and two components for x. The transfer blocks A and B represent matrices. The block between $\dot{y}$ and y indicates that each component of the state vector y is the integral of the corresponding component of $\dot{y}$.

Instead of using a Runge–Kutta integration to implement the integrations between $\dot{y}$ and y, an approximate digital solution can be obtained by using the simple integration formula that was stated in Eq 9.2-65. This gives for any

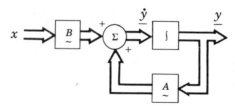

Figure 9.4-7 Signal-flow diagram representation of matrix form of state equations.

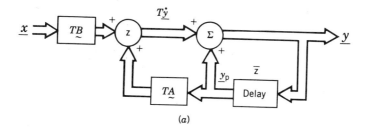

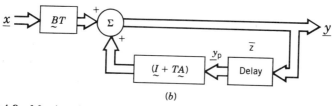

Figure 9.4-8 Matrix signal-flow diagram, which approximates Fig 9.4-7 with sampled-data computations: (a) basic diagram; (b) simplified form of (a).

one of the components y_n

$$y_n = y_{np} + T(\dot{y}_n) \qquad (9.4\text{-}50)$$

where y_{np} is the past value of the component y_n, and $\dot{y}_n$ is the derivative, which is based on the y_n-values calculated in the previous cycle. Using this relation, the system equations can be solved in a sampled-data computation according to the matrix signal-flow diagram of Fig 9.4-8a. The y_p-values represent the past values of the state variables y computed in the previous cycle. The matrices A and B are multiplied by the sample period T to form the matrices TA and TB. Since the sample period T is a scalar, all of the coefficients a_{pq}, b_{pq} of these matrices are multiplied by T.

Diagram a of Fig 9.4-8 can be simplified to form diagram b. The two feedback paths for y_p are combined by changing the matrix TA to $(I + TA)$. The matrix I is an identity matrix, and is defined as follows for this example, which has four state variables:

$$I = \begin{bmatrix} 1 & 0 & 0 & 0 \\ 0 & 1 & 0 & 0 \\ 0 & 0 & 1 & 0 \\ 0 & 0 & 0 & 1 \end{bmatrix} \qquad (9.4\text{-}51)$$

The identity matrix has unit values on the diagonal, and zero values for all other elements. The computations described by Fig 9.4-8b represent the following equations:

$$y_1 = (1 + Ta_{11})y_{1p} + Ta_{12}y_{2p} + Ta_{13}y_{3p} + Ta_{14}y_{4p} + Tb_{11}x_1 + Tb_{12}x_2$$
$$(9.4\text{-}52)$$

$$y_2 = Ta_{21}y_{1p} + (1 + Ta_{22})y_{2p} + Ta_{23}y_{3p} + Ta_{24}y_{4p} + Tb_{21}x_1 + Tb_{22}x_2$$
$$(9.4\text{-}53)$$

$$y_3 = Ta_{31}y_{1p} + Ta_{32}y_{2p} + (1 + Ta_{33})y_{3p} + Ta_{34}y_{4p} + Tb_{31}x_1 + Tb_{32}x_2$$
$$(9.4\text{-}54)$$

$$y_4 = Ta_{41}y_{1p} + Ta_{42}y_{2p} + Ta_{43}y_{3p} + (1 + Ta_{44})y_{4p} + Tb_{41}x_1 + Tb_{42}x_2$$
$$(9.4\text{-}55)$$

The sampled-data calculations indicated in Fig 9.4-8 provide only a first-order approximation of the differential equations described by Fig 9.4-7. The errors in the sampled-data calculations that are associated with the feedback loops have much more effect on the response than the errors associated with the forcing functions. Therefore, considerably better accuracy can be realized by correcting only the feedback sampled-data calculations. This can be achieved by replacing the matrix block $(I + TA)$ with a matrix designated Φ, called the *transition matrix*, which is given by

$$\Phi = I + TA + \frac{(TA)^2}{2!} + \frac{(TA)^3}{3!} + \frac{(TA)^4}{4!} + \cdots \qquad (9.4\text{-}56)$$

A matrix signal-flow diagram of the resultant sampled-data calculations is shown in Fig 9.4-9. Equation 9.4-56 has the form of the expansion of an exponential. Hence the transition matrix can be expressed as

$$\Phi = \exp[TA] \qquad (9.4\text{-}57)$$

The transition matrix can be derived as follows. The input forcing function vector x is assumed to be zero, and each state variable derivative $\dot{y}$ is

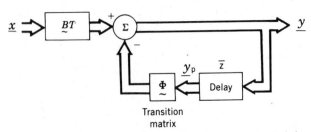

Transition
matrix

Figure 9.4-9 Matrix signal-flow diagram, showing accurate sampled-data calculations for solving matrix equations of Fig 9.4-7.

expanded in a Taylor series, over the time interval of one sample period. In order to simplify the notation, the nth derivative of $y[t]$ is designated $D^n y[t]$. The Taylor series expansion for $\dot{y}[t]$ is

$$\dot{y}[t] = Dy[t] = Dy[t_1] + D^2 y[t_1](t - t_1) + \frac{D^3 y[t_1](t - t_1)^2}{2!}$$
$$+ \frac{D^4 y[t_1](t - t_1)^3}{3!} + \cdots \qquad (9.4\text{-}58)$$

Time t_1 designates the start of the sample interval. Integrating Eq 9.4-58 gives the integral of $\dot{y}[t]$ over the sample interval:

$$\int_{t_1}^{t_1 + T} \dot{y}[t]\, dt = Dy[t_1]T + \frac{D^2 y[t_1]T}{2!} + \frac{D^3 y[t_1]T^3}{3!} + \frac{D^4 y[t_1]T^4}{4!} + \cdots$$
$$(9.4\text{-}59)$$

This is added to $y[t_1]$ to obtain the value of $y[t]$ at the end of the sample interval:

$$y[t_1 + T] = y[t_1] + \int_{t_1}^{t_1 + T} \dot{y}[t]\, dt$$

$$= y[t_1] + Dy[t_1]T + \frac{D^2 y[t_1]T}{2!} + \frac{D^3 y[t_1]T^3}{3!}$$
$$+ \frac{D^4 y[t_1]T^4}{4!} + \cdots \qquad (9.4\text{-}60)$$

The value $y[t_1 + T]$ is considered to be the present value of y. Hence $y[t_1]$ is the past value of y, designated y_p; and $D^n y[t_1]$ is the past value of the nth derivative of y, designated $D^n y_p$. Thus Eq 9.4-60 becomes

$$y = y_p + (T)Dy_p + \frac{T^2}{2!}D^2 y_p + \frac{T^3}{3!}D^3 y_p + \frac{T^4}{4!}D^4 y_p + \cdots \qquad (9.4\text{-}61)$$

Since the input forcing function vector $\underline{x}$ is set to zero, the past value of the derivative vector $\dot{\underline{y}}$ is (by Eq 9.4-49)

$$\dot{\underline{y}}_p = D\underline{y}_p = \underline{A}\underline{y}_p \qquad (9.4\text{-}62)$$

Differentiating this gives

$$D^2 \underline{y}_p = \underline{A}D\underline{y}_p = \underline{A}^2 \underline{y}_p \qquad (9.4\text{-}63)$$

$$D^n \underline{y}_p = \underline{A}D^{n-1}\underline{y}_p = \underline{A}^n \underline{y}_p \qquad (9.4\text{-}64)$$

Applying Eq 9.4-64 to Eq 9.4-61 gives the following matrix equation:

$$y = y_p + (T\underset{\sim}{A})y_p + \frac{(T\underset{\sim}{A})^2}{2!}y_p + \frac{(T\underset{\sim}{A})^3}{3!}y_p + \frac{(T\underset{\sim}{A})^4}{4!}y_p + \cdots \quad (9.4\text{-}65)$$

Factoring the vector y_p from the right-hand side gives

$$y = \left[I + T\underset{\sim}{A} + \frac{(T\underset{\sim}{A})^2}{2!} + \frac{(T\underset{\sim}{A})^3}{3!} + \frac{(T\underset{\sim}{A})^4}{4!} + \cdots \right] y_p$$

$$= \underset{\sim}{\Phi} y_p \qquad\qquad\qquad\qquad (9.4\text{-}66)$$

The expression in the brackets is the same as the equation for the transition matrix Φ given previously in Eq 9.4-56. The transition matrix is a key element in Kalman optimal-estimator theory, which is discussed in Ref [1.14] (Chapter 8).

9.4.5 Expressing Transfer Functions in Terms of Integrations

To describe the signal-flow diagram of a control system in state-variable form, all of the transfer functions must be expressed in terms of simple integrators. This can be achieved by representing the transfer functions by feedback loops, as shown in Fig 9.4-10.

Diagrams *a* to *c* of Fig 9.4-10 employ a single feedback loop with the loop transfer function $G = \omega_c/s$. The transfer functions Y/X are

Diagram *a*:

$$\frac{Y}{X} = G_{ib} = \frac{G}{1+G} = \frac{\omega_c/s}{1+\omega_c/s} = \frac{\omega_c}{s+\omega_c} \qquad (9.4\text{-}67)$$

Diagram *b*:

$$\frac{Y}{X} = G_{ie} = \frac{1}{1+G} = \frac{1}{1+\omega_c/s} = \frac{s}{s+\omega_c} \qquad (9.4\text{-}68)$$

Diagram *c*:

$$\frac{Y}{X} = K_1 G_{ie} + K_2 G_{ib} = \frac{K_1 s}{s+\omega_c} + \frac{K_2 \omega_c}{s+\omega_c} = \frac{K_1 s + K_2 \omega_c}{s+\omega_c} \qquad (9.4\text{-}69)$$

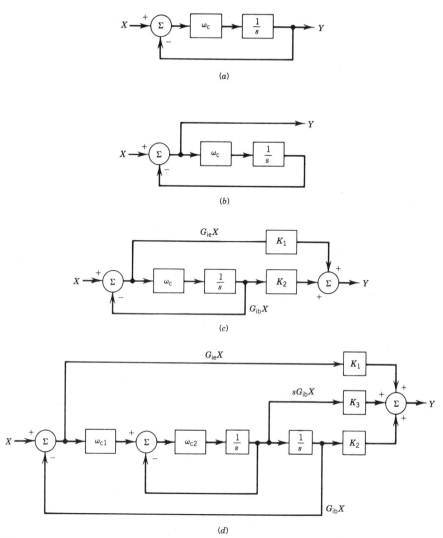

Figure 9.4-10 Signal-flow diagrams employing simple integrations to implement the following transfer functions: (a) $\omega_c/(s + \omega_c)$; (b) $s/(s + \omega_c)$; (c) $(K_1 s + K_2 \omega_c)/(s + \omega_c)$; (d) general transfer functions with underdamped poles.

To achieve an underdamped transfer function, the signal-flow diagram must have two feedback loops. A general form of such a configuration is shown in Fig 9.4-10d. The loop transfer function of the outer loop is

$$G = \frac{\omega_{c1}}{s}\frac{\omega_{c2}}{(s + \omega_{c2})} = \frac{N}{D} \qquad (9.4\text{-}70)$$

The feedback and error transfer functions of the outer loop are therefore

$$G_{ib} = \frac{G}{1 + G} = \frac{N}{N + D} = \frac{\omega_{c1}\omega_{c2}}{s^2 + \omega_{c2}s + \omega_{c1}\omega_{c2}} \tag{9.4-71}$$

$$G_{ie} = \frac{1}{1 + G} = \frac{D}{N + D} = \frac{s(s + \omega_{c2})}{s^2 + \omega_{c2}s + \omega_{c1}\omega_{c2}} \tag{9.4-72}$$

The response Y/X is

$$\begin{aligned}
Y/X &= K_1 G_{ie} + K_2 G_{ib} + K_3 s G_{ib} \\
&= \frac{K_1 s(s + \omega_{c2})}{s^2 + \omega_{c2}s + \omega_{c1}\omega_{c2}} + \frac{K_2 \omega_{c1}\omega_{c2}}{s^2 + \omega_{c2}s + \omega_{c1}\omega_{c2}} \\
&\quad + \frac{s K_3 \omega_{c1}\omega_{c2}}{s^2 + \omega_{c2}s + \omega_{c1}\omega_{c2}}
\end{aligned} \tag{9.4-73}$$

This reduces to

$$\frac{Y}{X} = \frac{K_1 s^2 + (K_1 + K_3 \omega_{c1})\omega_{c2}s + K_2 \omega_{c1}\omega_{c2}}{s^2 + \omega_{c2}s + \omega_{c1}\omega_{c2}} \tag{9.4-74}$$

The constants K_1, K_2, K_3 can have negative as well as positive values. By appropriate choice of the constants K_1, K_2, K_3, any desired transfer function can be implemented with diagrams c and d, or cascaded combinations of them.

9.4.6   Application of Serial-Simulation Procedure to ELF Transmitter Network

Now let us apply to the ELF transmitter network the simulation approach developed in Section 9.2. The Runge–Kutta integration procedure solves the system equations in a parallel manner, whereas the simulation method of Section 9.2 solves them in a serial manner. Therefore, the method of Section 9.2 is called *serial simulation*.

Figure 9.4-11a shows the computer signal-flow diagram of Fig 9.4-4a modified in accordance with the serial-simulation procedure. All of the $1/s$ integrator blocks are replaced with normalized integrations $1/sT$, and the element in the preceding block is appropriately multiplied by T. The transfer function $\bar{z}$ is placed in each feedback path to represent a one-cycle time delay, and a time-delay compensation factor $(1 + sT)$ is placed in series with each $\bar{z}$ transfer function.

To simplify Fig 9.4-11a, the compensation factors $(1 + sT)$ are shifted into the forward paths and combined with appropriate $1/sT$ integrations. Different

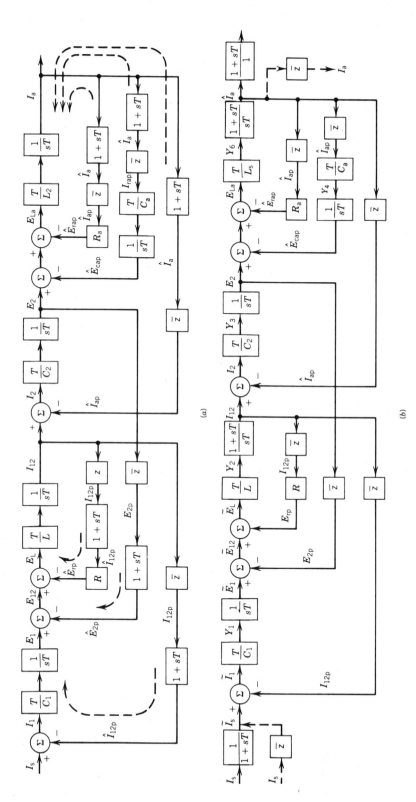

Figure 9.4-11 Signal-flow diagram of ELF transmitter circuit prepared for serial simulation: (*a*) modification of Fig 9.4-3, showing time delay, time-delay compensation, and normalization of integration transfer functions; (*b*) modification of (*a*) achieved by shifting $(1 + sT)$ transfer functions from feedback paths to forward paths.

161

signal-flow diagrams can result (each of which is valid), depending on how the factors $(1 + sT)$ are shifted. As shown by the dashed lines, the factors $(1 + sT)$ in the feedback paths of the antenna current I_a are shifted in the backward direction, and the other factors $(1 + sT)$ are shifted in the forward direction. To clarify this process, each compensation factor $(1 + sT)$ has been placed on the appropriate side of the time-delay factor $\bar{z}$.

Figure 9.4-11b shows the resultant simplified diagram. The source current I_s is fed through the lowpass transfer function $1/(1 + sT)$ to form the delayed signal $\tilde{I}_s$, which lags I_s by one sample period at low frequencies. If the signal I_s is not part of a feedback loop, the delayed signal $\tilde{I}_s$ can be approximated as the past value of I_s:

$$\tilde{I}_s \doteq I_{sp} \tag{9.4-75}$$

Similarly, the antenna current I_a can be approximated as the past value of the future antenna current $\hat{I}_a$:

$$I_a \doteq \hat{I}_{ap} \tag{9.4-76}$$

The computer program to simulate the signal-flow diagram of Fig 9.4-11b is outlined below. Remember that all of the variables in the computer equations represent the actual time-domain signals, not their transforms.

SPECIFY PARAMETERS:

> $R_a, L_a, C_a, C_1, C_2, L, R = ?$ (Circuit parameters)
> $L_{Max} = ?$ (Number of computation cycles per print cycle)
> $\text{TME}_{Max} = ?$ (Maximum time of computation)
> $T = ?$ (Sample period)

SET INITIAL VALUES TO ZERO:

> $\text{TME} = 0$ (Initial time)
> $L = 0$ (Print-cycle counter)
> $I_{sp}, I_{12p}, \tilde{E}_{1p}, Y_{1p}, E_{2p}, Y_{2p}, \hat{I}_{ap}, Y_{3p}, Y_{4p}, \hat{E}_{capp}, Y_{6p} = 0$

SPECIFY INPUT FORCING FUNCTION(S) VERSUS TIME (TME):

100    $\text{TME} = \text{TME} + T$

101    $I_s = \text{Function[TME]}$

SYSTEM EQUATIONS:

200    $\tilde{I}_s = I_{sp}$
201    $\tilde{I}_1 = \tilde{I}_s - I_{12p}$
202    $Y_1 = (T/C_1) * \tilde{I}_1$

203 $\tilde{E}_1 = \tilde{E}_{1p} + (1/2) * (Y_1 + Y_{1p})$

204 $\tilde{E}_{12} = \tilde{E}_1 - E_{2p}$

205 $E_{rp} = R * I_{12p}$

206 $\tilde{E}_L = \tilde{E}_{12} - E_{rp}$

207 $Y_2 = (T/L) * \tilde{E}_L$

208 $I_{12} = I_{12p} + (3/2) * Y_2 - (1/2) * Y_{2p}$

209 $I_2 = I_{12} - \hat{I}_{ap}$

210 $Y_3 = (T/C_2) * I_2$

211 $E_2 = E_{2p} + (1/2) * (Y_3 + Y_{3p})$

212 $Y_4 = (T/C_a) * \hat{I}_{ap}$

213 $\hat{E}_{cap} = \hat{E}_{capp} + (1/2) * (Y_4 + Y_{4p})$

214 $Y_5 = E_2 - \hat{E}_{cap}$

215 $\hat{E}_{rap} = R_a * \hat{I}_{ap}$

216 $E_{La} = Y_5 - \hat{E}_{rap}$

217 $Y_6 = (T/L_a) * \tilde{E}_{La}$

218 $\hat{I}_a = \hat{I}_{ap} + (3/2) * Y_6 - (1/2) * Y_{6p}$

219 $I_a = \hat{I}_{ap}$

CHECK PRINT COUNT, PRINT DATA:

300 $L = L + 1$

301 IF $(L < L_{Max})$ GOTO 400

302 $L = 0$

303 PRINT data (TME, X, X_e, $\hat{X}_v$, etc.)

SET PAST VALUES EQUAL TO PRESENT VALUES:

400 $I_{sp} = I_s$

401 $I_{12p} = I_{12}$

402 $\tilde{E}_{1p} = \tilde{E}_1$

403 $Y_{1p} = Y_1$

404 $E_{2p} = E_2$

405 $Y_{2p} = Y_2$

406 $\hat{I}_{ap} = \hat{I}_a$

407 $Y_{3p} = Y_3$

408 $Y_{4p} = Y_4$

409 $\hat{E}_{capp} = \hat{E}_{cap}$

410 $Y_{6p} = Y_6$

CHECK TIME LIMIT, REPEAT:

500 IF $(TME < TME_{Max})$ GOTO 100

501 END

As shown below, the variables Y_1, Y_2, Y_3, Y_4, and Y_6 are proportional to derivatives of the voltage and current variables:

$$Y_1 = T\frac{d\tilde{E}_1}{dt} \tag{9.4-77}$$

$$Y_2 = T\frac{d\tilde{I}_{12}}{dt} \tag{9.4-78}$$

$$Y_3 = T\frac{dE_2}{dt} \tag{9.4-79}$$

$$Y_4 = T\frac{d\hat{E}_{cap}}{dt} \tag{9.4-80}$$

$$Y_6 = T\frac{dI_a}{dt} \tag{9.4-81}$$

9.5 EFFECT OF SAMPLING WITHIN A CONTINUOUS-DATA FEEDBACK LOOP

9.5.1 Analysis of Effect of Sampling

9.5.1.1 *Summary of Equations.* Many feedback-control systems have feedback loops that include both digital and analog elements. The loop transfer function of the complete loop is needed to evaluate the stability and dynamic performance of the loop. This transfer function can be determined by considering input and output signals within the digital portion of the loop.

Consider the block diagram in Fig 9.5-1. The continuous signal $x[t]$ is sampled to produce the sampled signal $x^s[t]$, which is digitized and processed digitally. For simplicity it is assumed that the sampled signal is not modified in the computer. It is converted unchanged back to analog form by a digital-to-analog (D/A) converter, which acts as a simple-hold circuit. The resultant continuous analog signal is fed through a dynamic analog element, which is called the *plant*. The output from the plant, which is designated $y[t]$, is sampled to produce the sampled signal $y^s[t]$. The transfer function of the complete analog section (the D/A converter plus the plant), between the sampled signal $x^s[t]$ and the plant output $y[t]$, is designated $H[s]$. The transfer function between the sampled signal $x^s[t]$ and the sampled plant output $y^s[t]$ is designated $H^s[s]$. Thus

$$H[s] = \frac{\mathcal{L}[y[t]]}{\mathcal{L}[x^s[t]]} \tag{9.5-1}$$

$$H^s[s] = \frac{\mathcal{L}[y^s[t]]}{\mathcal{L}[x^s[t]]} \tag{9.5-2}$$

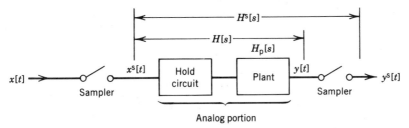

Figure 9.5-1 Diagram defining the plant transfer function $H_p[s]$, the total analog transfer function $H[s]$, and the total sampled transfer function $H^s[s]$.

In the following analyses, two different expressions are developed relating the transfer function $H^s[s]$ to $H[s]$: the first is based on the frequency response, and the second on the transient response. The frequency-response expression was derived by Linvill [9.7]. The transient-response expression, and its implementation by means of the z-transform, were derived by Ragazzini and Zadeh [9.8]. These are

$$H^s[j\omega] = \sum_{K=0}^{\infty} H[j(K\omega_s + \omega)] + \sum_{K=1}^{\infty} H^*[j(K\omega_s - \omega)] \quad (9.5\text{-}3)$$

$$H^s[s] = \sum_{K=0}^{\infty} h[KT]\bar{z}^k \quad (9.5\text{-}4)$$

where $\bar{z} = e^{-sT}$. Equation 9.5-3 shows that the sampled frequency response $H^s[j\omega]$ can be obtained by performing the following summation at specified frequencies ω:

$$H^s[j\omega] = H[j\omega] + H^*[j(\omega_s - \omega)] + H[j(\omega_s + \omega)]$$
$$+ H^*[j(2\omega_s - \omega)] + H[j(2\omega_s + \omega)] + \cdots \quad (9.5\text{-}5)$$

where $H^*[j(\omega_s - \omega)]$ is the conjugate of $H[j\omega]$ at the frequency $(\omega_s - \omega)$. Although this summation theoretically includes an infinite number of terms, a practical $H[j\omega]$ analog transfer function must have strong cutoff characteristics above half the sampling frequency, and so only a few terms are actually needed.

On the other hand, when the analog transfer function $H[s]$ is specified analytically, the time-domain expression of Eq 9.5-4 is much easier to use. As will be shown in Section 9.5.2, this infinite-series expansion can be reduced to closed form when the transfer function is expressed as a ratio of polynomials in s.

9.5.1.2 Derivation of Frequency-Response Expansion. Let us assume that the input continuous signal $x[t]$ is a sinusoid of frequency ω, given by

$$x[t] = \cos[\omega t] \quad (9.5\text{-}6)$$

Sampling this signal produces the sampled signal $x^s[t]$, which can be calculated by multiplying the continuous signal $x[t]$ by a reference signal $r[t]$ consisting of a train of impulses. This reference signal can be expressed as

$$r[t] = 1 + 2 \sum_{K=1}^{\infty} \cos[K\omega_s t] \qquad (9.5\text{-}7)$$

where ω_s is the sampling frequency in rad/sec. The sampled signal is

$$x^s[t] = r[t]x[t] = \left\{ 1 + 2 \sum_{K=1}^{\infty} \cos[K\omega_s t] \right\} x[t]$$

$$= \left\{ 1 + 2 \sum_{K=1}^{\infty} \cos[K\omega_s t] \right\} \cos[\omega t]$$

$$= \cos[\omega t] + \sum_{K=1}^{\infty} 2\cos[K\omega_s t]\cos[\omega t] \qquad (9.5\text{-}8)$$

Apply to this the following trigonometric identity:

$$2\cos[K\omega_s t]\cos[\omega t] = \cos[(K\omega_s - \omega)t] + \cos[(K\omega_s + \omega)t] \quad (9.5\text{-}9)$$

This gives the following infinite-series expansion for $x^s[t]$:

$$x^s[t] = \cos[\omega t] + \cos[(\omega_s - \omega)t] + \cos[(\omega_s + \omega)t]$$
$$+ \cos[(2\omega_s - \omega)t] + \cos[(2\omega_s + \omega)t] + \cdots \qquad (9.5\text{-}10)$$

The magnitude and phase of the transfer function $H[j\omega]$ are designated $|H[\omega]|$ and $\phi[\omega]$. Thus $H[j\omega]$ is expressed as

$$H[j\omega] = |H[\omega]|e^{j\phi[\omega]} \qquad (9.5\text{-}11)$$

Each component of $x^s[t]$ has the form $\cos[\omega_x t]$. To obtain $y[t]$, the magnitude of each of these components is multiplied by $|H[\omega_x]|$, and the cosine function is shifted in phase by $\phi[\omega]$. Hence the response $y[t]$ is

$$y[t] = |H[\omega]|\cos[\omega t + \phi[\omega]] + |H[\omega_s - \omega]|\cos[(\omega_s - \omega)t + \phi[\omega_s - \omega]]$$
$$+ |H[\omega_s + \omega]|\cos[(\omega_s + \omega)t + \phi[\omega_s + \omega]]$$
$$+ |H[2\omega_s - \omega]|\cos[(2\omega_s - \omega)t + \phi[2\omega_s - \omega]]$$
$$+ |H[2\omega_s + \omega]|\cos[(2\omega_s + \omega)t + \phi[2\omega_s + \omega]] + \cdots \qquad (9.5\text{-}12)$$

This signal $y[t]$ is sampled to form the signal $y^s[t]$, given by

$$y^s[t] = \left\{ 1 + 2 \sum_{K=1}^{\infty} \cos[K\omega_s t] \right\} y[t] \qquad (9.5\text{-}13)$$

Substitute into this the expression for $y[t]$ given in Eq 9.5-12, and expand into sum and difference frequency components, using the trigonometric identity of Eq 9.5-9. Only those frequency components in the resultant expansion that are less than half the sampling frequency need be considered, because the higher-frequency components of a sampled signal are merely repetitions of the components below half the sampling frequency. Therefore, only the difference-frequency components of the expansion of Eq 9.5-8 are included in the following expansion for $y^s[t]$:

$$y^s[t] = |H[\omega]|\cos[\omega t + \phi[\omega]]$$
$$+ |H[\omega_s - \omega]|\cos[\omega_s t - \{(\omega_s - \omega)t + \phi[\omega_s - \omega]\}]$$
$$+ |H[\omega_s + \omega]|\cos[(\omega_s + \omega)t + \phi[\omega_s + \omega] - \omega_s t]$$
$$+ |H[2\omega_s - \omega]|\cos[2\omega_s t - \{(2\omega_s - \omega)t + \phi[2\omega_s - \omega]\}]$$
$$+ |H[2\omega_s + \omega]|\cos[(2\omega_s + \omega)t + \phi[2\omega_s + \omega] - 2\omega_s t] + \cdots$$
$$(9.5\text{-}14)$$

This reduces to

$$y^s[t] = |H[\omega]|\cos[\omega t + \phi[\omega]]$$
$$+ |H[\omega_s - \omega]|\cos[\omega t - \phi[\omega_s - \omega]]$$
$$+ |H[\omega_s + \omega]|\cos[\omega t + \phi[\omega_s + \omega]]$$
$$+ |H[2\omega_s - \omega]|\cos[\omega t - \phi[2\omega_s - \omega]]$$
$$+ |H[2\omega_s + \omega]|\cos[\omega t + \phi[2\omega_s + \omega]] + \cdots \quad (9.5\text{-}15)$$

All of these components are at the frequency ω. Therefore, at this frequency ω, the frequency response between $x^s[t]$ and $y^s[t]$, which is designated $H^s[j\omega]$, is

$$H^s[j\omega] = |H[\omega]|e^{j\phi[\omega]}$$
$$+ |H[\omega_s - \omega]|e^{-j\phi[\omega_s - \omega]} + |H[\omega_s + \omega]|e^{j\phi[\omega_s + \omega]}$$
$$+ |H[2\omega_s - \omega]|e^{-j\phi[2\omega_s - \omega]} + |H[2\omega_s + \omega]|e^{j\phi[2\omega_s + \omega]} + \cdots$$
$$(9.5\text{-}16)$$

Applying the definition of Eq 9.5-11 gives

$$H[j(K\omega_s + \omega)] = |H[K\omega_s + \omega]|e^{j\phi[\omega_s + \omega]} \quad (9.5\text{-}17)$$
$$H^*[j(K\omega_s - \omega)] = |H[K\omega_s - \omega]|e^{-j\phi[\omega_s - \omega]} \quad (9.5\text{-}18)$$

where $H^*[j\omega]$ represents the conjugate of $H[j\omega]$. Applying Eqs 9.5-17 and 9.5-18 to Eq 9.5-16 gives for $H^s[j\omega]$

$$H^s[j\omega] = H[j\omega] + H^*[j(\omega_s - \omega)] + H[j(\omega_s + \omega)]$$
$$+ H^*[j(2\omega_s - \omega)] + H[j(2\omega_s + \omega)] + \cdots \quad (9.5\text{-}19)$$

This can be expressed as

$$H^s[j\omega] = \sum_{K=0}^{\infty} H[j(K\omega_s + \omega)] + \sum_{K=1}^{\infty} H^*[j(K\omega_s - \omega)] \quad (9.5\text{-}20)$$

This was given in Eq 9.5-3. A transfer function is theoretically defined for negative as well as for positive frequencies. If $j\omega$ is replaced by $-j\omega$ in a transfer function, the magnitude of the transfer function stays the same, but the sign of the phase is reversed. Thus, $H[-j\omega]$ is the conjugate of $H[j\omega]$:

$$H[-j\omega] = H^*[j\omega] \quad (9.5\text{-}21)$$

Hence, by including negative frequencies, Eq 9.5-20 can be condensed to

$$H^s[j\omega] = \sum_{K=-\infty}^{\infty} H[j(K\omega_s + \omega)] \quad (9.5\text{-}22)$$

9.5.1.3 Derivation of Time-Domain Expansion. A second expression for $H^s[j\omega]$ can be derived by examining the response in the time domain. The sampled signal $x^s[t]$ can be represented as a series of impulses, the areas of which are the sampled data. Let us consider the simple case where the sampled signal $x^s[t]$ is a single unit impulse (designated $\delta[t]$) occurring at time ($t = 0$):

$$x^s[t] = \delta[0] \quad (9.5\text{-}23)$$

The signal $y[t]$ is therefore the unit-impulse response of the transfer function $H[s]$, which is designated $h[t]$:

$$y[t] = h[t] = \mathcal{L}^{-1}[H[s]] \quad (9.5\text{-}24)$$

The signal $y[t]$ is sampled to form the signal $y^s[t]$. As shown in Fig 9.5-2, this sampled signal $y^s[t]$ consists of a series of impulses, the areas of which are equal to the values of $y[t] = h[t]$ at the sampling instants, $0, T, 2T, 3T, \ldots$. Thus

$$y^s[t] = h[0]\delta[0] + h[T]\delta[T] + h[2T]\delta[2T] + h[3T]\delta[3T] + \cdots$$
$$(9.5\text{-}25)$$

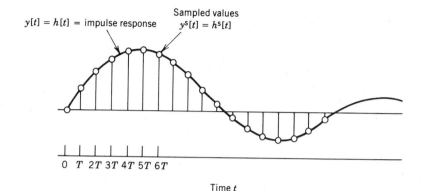

Figure 9.5-2   Sampled values of impulse response $h[t]$.

where $\delta[T]$ is a unit impulse occurring at time $t = T$, $\delta[2T]$ is a unit impulse occurring at time $t = 2T$, etc. Taking the Laplace transform of Eq 9.5-25 gives the transform of $y^s[t]$, which is designated $Y^s[s]$. This is

$$Y^s[s] = \mathcal{L}[y^s[t]]$$
$$= h[0] + h[T]e^{-sT} + h[2T]e^{-2sT} + h[3T]e^{-3sT} + \cdots \quad (9.5\text{-}26)$$

Remember that the Laplace transform of a unit impulse at time $t = 0$ is unity, and the Laplace transform of a unit impulse at time $t = t_1$ is e^{-st_1}. Equation 9.5-26 can be expressed as follows:

$$Y^s[s] = \sum_{K=0}^{\infty} h[KT]e^{-KsT} \quad (9.5\text{-}27)$$

The Laplace transform of $x^s[t]$ given in Eq 9.5-23 is

$$X^s[s] = \mathcal{L}[x^s[t]] = \mathcal{L}[\delta[0]] = 1 \quad (9.5\text{-}28)$$

Dividing Eq 9.5-27 by Eq 9.5-28 gives the transfer function $H^s[s]$ between $X^s[s]$ and $Y^s[s]$, which is

$$H^s[s] = \frac{Y^s[s]}{X^s[s]} = \sum_{K=0}^{\infty} h[KT]e^{-KsT} \quad (9.5\text{-}29)$$

It is convenient to express this in terms of the variable $\bar{z}$, which has been defined as $\bar{z} = e^{-sT}$. Thus

$$\bar{z}^K = (e^{-sT})^K = e^{-KsT} \quad (9.5\text{-}30)$$

Expressing Eq 9.5-29 in terms of $\bar{z}$ gives

$$H^s[s] = \sum_{K=0}^{\infty} h[KT]\bar{z}^K \qquad (9.5\text{-}31)$$

Note that $h[t]$ is the inverse transform of the transfer function of the analog section between the two samplers, which is designated $H[s]$. The function $h[KT]$ represents the values of $h[t]$ at the sampling instants: $t = 0, T, 2T, 3T, \ldots$. This expression was given earlier in Eq 9.5-4.

9.5.2   The z-Transform

The time-domain infinite-series expansion for $H^s[s]$ of Eq 9.3-31 can be expressed in closed form if the equation for $H[s]$ is known. For example, assume that the analog transfer function $H[s]$ is $1/s$. The impulse response $h[t]$ for this is a unit step:

$$h[t] = u[t] \qquad (9.5\text{-}32)$$

Hence Eq 9.5-31 gives for $H^s[s]$

$$H^s[s] = h[0]\bar{z}^0 + h[T]\bar{z}^1 + h[2T]\bar{z}^2 + h[3T]\bar{z}^3 + \cdots$$

$$= 1 + \bar{z} + \bar{z}^2 + \bar{z}^3 + \cdots \;\; = \;\; \frac{1}{1 - \bar{z}} \qquad (9.5\text{-}33)$$

Note that $h[0] = h[1] = h[2] = 1$. As shown, the sum of this infinite series is $1/(1 - \bar{z})$, which is the closed form of the series.

The closed-form expression for H^s is commonly called the z-*transform* of the impulse response $h[t]$. The superscript s or $*$ is often omitted, and this sampled transform is expressed as $H[z]$ to distinguish it from the analog transform, $H[s]$. Table 9.5-1 gives a general table of the z-transforms $H^s[\bar{z}]$ for basic time functions $h[t]$, along with the Laplace transforms $H[s]$.

In the sampled-data literature, the variable z is defined as e^{sT}, and this is the variable used in the z-transform tables. However, this book uses the variable $\bar{z}$, which is the reciprocal of z, because this results in more convenient transfer functions. Besides, the analyses in the preceding sections of this chapter would be much more cumbersome if the variable $z = e^{sT}$ were used instead of the variable $\bar{z} = e^{-sT}$. To convert the expressions for $H^s[\bar{z}]$ in Table 9.5-1 to the usual z-transform expressions, $\bar{z}$ should be replaced by z^{-1}.

In the Discussion associated with their basic paper, Ragazzini and Zadeh [9.8] recognized that it would be preferable from an engineering point of view to define z as e^{-sT}. However, they chose the reciprocal of this, e^{sT}, to be consistent with the earlier work by Hurewicz [9.9].

In Table 9.5-1, items (A) to (E) give general relations from which a large family of z-transforms can be developed. Item (F) is obvious, and (G) was

TABLE 9.5-1   Sampled Laplace Transforms or z-Transforms

	Impulse Response $h[t]$	Laplace Transform $H[s]$	Sampled Transform (z-Transform) $H^s[\bar{z}]$
(A)	$ah[t]$	$aH[s]$	$aH^s[\bar{z}]$
(B)	$h_1[t] + h_2[t]$	$H_1[s] + H_2[s]$	$H_1^s[\bar{z}] + H_2^s[\bar{z}]$
(C)	$h[t - NT]$	$\bar{z}^N H[s]$	$\bar{z}^N H^s[\bar{z}]$
(D)	$th[t]$	—	$T\bar{z}\dfrac{d(H^s[\bar{z}])}{dt}$
(E)	$e^{-at}h[t]$	$H[s + a]$	$H^s[\bar{z}e^{-aT}]$
(F)	$\delta[t]$	1	1
(G)	$u[t]$	$\dfrac{1}{s}$	$\dfrac{1}{1 - \bar{z}}$
(H)	t	$\dfrac{1}{s^2}$	$\dfrac{T\bar{z}}{(1 - \bar{z})^2}$
(I)	t^2	$\dfrac{2}{s^3}$	$\dfrac{T^2\bar{z}(1 + \bar{z})}{(1 - \bar{z})^3}$
(J)	t^3	$\dfrac{6}{s^4}$	$\dfrac{T^3\bar{z}(1 + 4\bar{z} + \bar{z}^2)}{(1 - \bar{z})^4}$
(K)	e^{-at}	$\dfrac{1}{s + a}$	$\dfrac{1}{1 - \bar{z}e^{-aT}}$
(L)	te^{-at}	$\dfrac{1}{(s + a)^2}$	$\dfrac{T\bar{z}e^{-aT}}{(1 - \bar{z}e^{-aT})^2}$
(M)	$\sin[\omega t]$	$\dfrac{\omega}{s^2 + \omega^2}$	$\dfrac{\bar{z}\sin[\omega T]}{1 - 2\bar{z}\cos[\omega T] + \bar{z}^2}$
(N)	$\cos[\omega t]$	$\dfrac{s}{s^2 + \omega^2}$	$\dfrac{1 - \bar{z}\cos[\omega T]}{1 - 2\bar{z}\cos[\omega T] + \bar{z}^2}$

Note: $\bar{z} = e^{-sT}$.

derived in Eq 9.5-33. Items (H), (I), and (J) are derived by applying item (D) to item (G) successively. Item (K) is derived by applying item (E) to (G). Applying item (D) to item (K) gives item (L). Items (M) and (N) are derived from item (K) by setting a equal to $j\omega$. The z-transforms for more complicated transfer functions can be derived by applying these principles along with partial-fraction expansions.

Although the Laplace transforms are uniquely paired with their time functions, the z-transforms $H^s[\bar{z}]$ are not uniquely paired with the time

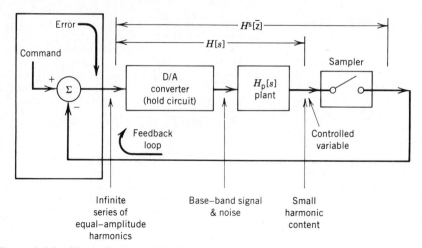

Figure 9.5-3 Block diagram of feedback-control system with sampled-data section, defining analog transfer functions $H_p[s]$ and $H[s]$, and the sampled transfer function $H^s[\bar{z}]$.

functions $h[t]$. If the values of two different time functions are the same at the sampling instants, these time functions have the same z-transform. Therefore, when the z-transform is used for transient analysis of sampled-data systems, it yields transient-response information only at the sampling instants, unless complex extensions of the approach are used.

Transient analysis using z-transforms is discussed in Section 9.5.4. The next Section, 9.5.3, shows how the z-transform can be used to perform frequency-response analysis.

9.5.3 Frequency-Response Analysis Using the z-Transform

9.5.3.1 Approximate Effects of Sampling for Different Types of Hold Circuits. Let us consider, on a physical basis, the primary effects of sampling within an otherwise analog feedback loop. Figure 9.5-3 shows the common situation where the controlled variable is measured digitally, and subtracted from the digital command signal to form the system error signal. The digital system error signal is converted to analog form in a D/A converter. To simplify the discussion, it is assumed that no dynamic processing is performed within the digital section.

As was explained in Section 9.1, the D/A converter acts as a lowpass filter, which transmits the base-band information with a small amount of harmonics. The plant also acts as a lowpass filter, to provide more attenuation of the harmonics. Ideally, there is enough lowpass filtering within the total analog section (the D/A converter plus the plant) so that the controlled variable at the input to the sampler consists only of signal components at frequencies less than half the sampling frequency. When this condition is satisfied, the sampling process itself has no dynamic effect on the loop. However, to satisfy this condition, appreciable lowpass filtering may be required in the analog section (provided by the D/A converter and the plant), and this filtering can seriously limit the dynamic performance of the feedback loop.

It often is not practical to eliminate harmonics completely at the input to the sampler. The harmonics that do pass through the sampler are converted back to the base-band frequency. The resultant distortion of the base-band signal can produce undesirable feedback around the loop that degrades its stability.

To address this issue quantitatively, let us examine the frequency-response characteristics of the sampled-data transfer functions that result from a number of plant transfer functions. Two types of D/A converters are assumed: one using a conventional simple-hold circuit, and one using a linear-interpolator hold circuit, which was described in Section 9.1. The transfer functions of these two converters are as follows:

D/A converter with a simple-hold circuit:

$$H_1 = \frac{1 - \bar{z}}{s} \tag{9.5-34}$$

D/A converter with a linear-interpolator hold circuit:

$$H_2 = \frac{(1 - \bar{z})^2}{s^2 T} \tag{9.5-35}$$

Equation 9.5-34 was derived from Eq 9.1-22, and Eq 9.5-35 was derived from Eq 9.1-30. (The transfer functions of Eqs 9.1-22, -30 were multiplied by τ to take account of the conversion of narrow pulses to impulses; and e^{-Ts} was replaced by $\bar{z}$.)

The first column of Table 9.5-2 shows a number of simple plant transfer functions labeled $H_p[s]$. The second column shows the total analog transfer function $H[s]$. This is the product of the plant transfer function $H_p[s]$ multiplied by the transfer function of the D/A converter (DAC), given in Eq 9.5-34 or -35. The first four cases (A), (B), (C), (D) assume a simple-hold circuit, and so use the D/A converter transfer function of Eq 9.5-34. The last two cases (E), (F) assume a linear-interpolator hold circuit, and so use the transfer function of Eq 9.5-35.

By applying the z-transform formulas of Table 9.5-1, along with a partial-fraction expansion for case (D), the digital transfer functions $H^s[\bar{z}]$ shown in the third column are derived from the analog transfer functions $H[s]$ given in the second column. The frequency-response characteristics of these digital

TABLE 9.5-2   Digital Transfer Functions (z-Transforms) Resulting from Various Analog Plant Transfer Functions, for D / A Converters Having Simple-Hold and Linear-Interpolator Hold Circuits

	Plant Transfer Function $H_p[s]$	Total Analog Transfer Function (Plant + DAC), $H[s]$	Digital Transfer Function in terms of $\bar{z}$, $H^s[\bar{z}]$
	Conventional DAC with Simple-Hold Circuit		
(A)	$\dfrac{1}{s}$	$\dfrac{1 - \bar{z}}{s}\dfrac{1}{s}$	$\dfrac{T\bar{z}}{1 - \bar{z}}$
(B)	$\dfrac{1}{s^2}$	$\dfrac{1 - \bar{z}}{s}\dfrac{1}{s^2}$	$\dfrac{T^2\bar{z}(1 + \bar{z})}{2(1 - \bar{z})^2}$
(C)	$\dfrac{1}{s^3}$	$\dfrac{1 - \bar{z}}{s}\dfrac{1}{s^3}$	$\dfrac{T^3\bar{z}(1 + 4\bar{z} + \bar{z}^2)}{6(1 - \bar{z})^2}$
(D)	$\dfrac{1}{s(1 + s/\omega_1)}$	$\dfrac{1 - \bar{z}}{s}\dfrac{\omega_1}{s(s + \omega_1)}$	$\dfrac{T\bar{z}}{1 - \bar{z}} - \dfrac{\bar{z}(1 - e^{-\omega_1 T})}{\omega_1(1 - \bar{z}e^{-\omega_1 T})}$
	DAC with Linear-Interpolator Hold Circuit		
(E)	$\dfrac{1}{s}$	$\dfrac{(1 - \bar{z})^2}{Ts^2}\dfrac{1}{s}$	$\dfrac{T\bar{z}(1 + \bar{z})}{2(1 - \bar{z})}$
(F)	$\dfrac{1}{s^2}$	$\dfrac{(1 - \bar{z})^2}{Ts^2}\dfrac{1}{s^2}$	$\dfrac{T^2\bar{z}(1 + 4\bar{z} + \bar{z}^2)}{6(1 - \bar{z}^2)}$

transfer functions $H^s[\bar{z}]$ are obtained by setting $\bar{z}$ equal to

$$\bar{z} = \frac{1 - U}{1 + U} = \frac{1 - p(T/2)}{1 + p(T/2)} \tag{9.5-36}$$

This is substituted for $\bar{z}$ in the relations for $H^s[\bar{z}]$, shown in the third column of Table 9.5-2, to obtain those for $H^s[p]$, listed in the second column of Table 9.5-3. These are the digital transfer functions expressed in terms of the complex pseudo frequency p. The first column of Table 9.5-3 repeats the analog transfer functions for the plant, $H_p[s]$, which were obtained from the first column of Table 9.5-2.

For frequencies below $\frac{1}{4}$ of the sampling frequency, the complex pseudo frequency p can be approximately replaced by s. The digital transfer functions

TABLE 9.5-3   Digital Transfer Functions of Table 9.5-3, Expressed in Terms of Complex Pseudo Frequency p, Compared with Analog Plant Transfer Functions

	Plant Transfer Function $H_p[s]$	Digital Transfer Function in Terms of Pseudo Frequency $H^s[p]$
	Conventional DAC with Simple-Hold Circuit	
(A)	$\dfrac{1}{s}$	$\dfrac{1 - pT/2}{p}$
(B)	$\dfrac{1}{s^2}$	$\dfrac{1 - pT/2}{p^2}$
(C)	$\dfrac{1}{s^3}$	$\dfrac{(1 - pT/2)\{1 - (pT)^2/12\}}{p^3}$
(D)	$\dfrac{1}{s(1 + s/\omega_1)}$	$\dfrac{(1 - pT/2)(1 + p/\omega_1' - p/\omega_1)}{p(1 + p/\omega_1')}$
		$\omega_1'T = 2\tanh[\omega_1 T/2]$
		$\omega_1' \doteq \omega_1 \text{ for } \omega_1 T < 1$
	DAC with Linear-Interpolator Hold Circuit	
(E)	$\dfrac{1}{s}$	$\dfrac{\bar{z}}{p}$
(F)	$\dfrac{1}{s^2}$	$\dfrac{\bar{z}\{1 - (pT)^2/12\}}{p^2}$

$H^s[p]$ of Table 9.5-3 are then approximately related as follows to the plant transfer functions $H_p[s]$:

For simple-hold D/A converter:

(A), (B):
$$H^s \doteq \left(1 - \frac{sT}{2}\right)H_p \tag{9.5-37}$$

(C):
$$H^s \doteq \left(1 - \frac{sT}{2}\right)H_p\left(1 - \frac{(sT)^2}{12}\right) \tag{9.5-38}$$

(D):
$$H^s \doteq \left(1 - \frac{sT}{2}\right)H_p\left(1 + \frac{s}{\omega_1'} - \frac{s}{\omega_1}\right) \tag{9.5-39}$$

For linear-interpolator D/A converter:

(E):
$$H^s \doteq \bar{z} H_p$$
(9.5-40)

(F):
$$H^s \doteq \bar{z} H_p \left(1 - \frac{(sT)^2}{12}\right)$$
(9.5-41)

The parameter ω_1' in Eq 9.5-39 is defined in Table 9.5-3.

For case (D), it can be assumed that the break frequency ($f_1 = \omega_1/2\pi$) is less than $\frac{1}{6}$ of the sampling frequency. For this condition, ω_1' is approximately equal to ω_1, and so the last factor of Eq 9.5-39 is approximately unity. For cases (C) and (F), the factor $(1 - (sT)^2/12)$ can also be approximated by unity. Thus, Eqs 9.5-37 to -41 simplify as follows:

For simple-hold D/A converter [(A), (B), (C), (D)]:

$$H^s \doteq \left(1 - \frac{sT}{2}\right) H_p$$
(9.5-42)

For linear-interpolator D/A converter circuit [(E), (F)]:

$$H^s \doteq \bar{z} H_p$$
(9.5-43)

As was explained in Section 9.1, a D/A converter with a linear-interpolator hold circuit produces a time delay of a full sample period (which is represented by the transfer function $\bar{z} = e^{-sT}$), and the harmonic content of the demodulated signal is very low. This explains the results of Eq 9.5-43, which shows that the sampled transfer function H^s is approximately equal to the plant transfer function H_p multiplied by the transfer function $\bar{z}$ of a one-cycle time delay.

As was shown in Section 9.5-36, a conventional simple-hold circuit produces a time delay equal to $\frac{1}{2}$ of the sample period. The low-frequency phase lag of this time delay is the same as the low-frequency phase lag of the factor $(1 - sT/2)$ of Eq 9.5-42. This $(1 - sT/2)$ factor also has a magnitude characteristic that rises with frequency, which indicates that the sampling harmonics transmitted around the loop produce high-frequency peaking in the loop transfer function. This peaking decreases the gain margin and so degrades loop stability.

9.5.3.2 Gain Crossover Frequency Achievable in Sampled-Data Feedback Loop with Simple-Hold Circuit. Let us assume that a simple-hold D/A converter is used, and that the plant has the following transfer function:

$$H_p[s] = \omega_c/s$$
(9.5-44)

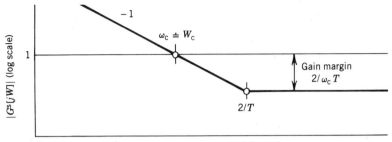

Figure 9.5-4   Magnitude asymptote plot of sampled loop transfer function for simple-hold D/A converter and plant transfer function $G_p = \omega_c / s$, with $\omega_c T = 0.8$.

As shown in Table 9.5-3, case (A), the sampled loop transfer function is

$$G^s[p] = H^s[p] = \frac{\omega_c}{p}\left(1 - \frac{pT}{2}\right) \tag{9.5-45}$$

It is assumed that there are no dynamic computations within the digital portion of the feedback loop, so that $H^s[p]$ is the total digital loop transfer function, which is designated $G^s[p]$. Setting $p = jW$ gives the pseudo-frequency response of the loop transfer function:

$$G^s[jW] = \frac{\omega_c(1 - jWT/2)}{jW} \tag{9.5-46}$$

An asymptotic magnitude plot of this pseudo-frequency response is shown in Fig 9.5-4. Just as for the transfer function of the digital loop illustrated in Fig 9.3-8, the phase reaches $-180°$ at infinite W. Hence, the gain margin is the reciprocal of the magnitude of $G^s[jW]$ at infinite W (or at $f = 0.5F$). The magnitude of G^s at infinite W is equal to the magnitude asymptote of G^s at $W = 2/T$, which is $(2/T)/\omega_c$. Hence the gain margin is

$$\text{gain margin} = \frac{(2/T)}{\omega_c} = \frac{2}{\omega_c T} \tag{9.5-47}$$

As in Section 9.3.4, a minimum gain margin of 2.5 should be assumed. The corresponding upper limit on $\omega_c T$ is

$$\omega_c T \le 2/2.5 = 0.8 \tag{9.5-48}$$

Figure 9.5-5 shows a G-locus plot of this transfer function on a simplified Nichols chart. The value for $\omega_c T$ is 0.8, and the gain margin is 4 dg, which represents a factor of 2.5. When $\omega_c T$ is equal to this limit, 0.8, the value for the

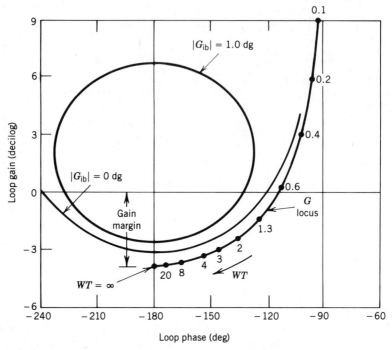

Figure 9.5-5 Abbreviated Nichols-chart plot of sampled loop transfer function for simple-hold D/A converter and plant transfer function $G_p = \omega_c/s$, with $\omega_c T = 0.8$.

asymptotic gain crossover frequency in hertz is

$$f_c = \frac{\omega_c}{2\pi} = \frac{\omega_c T}{2\pi} F = \frac{0.8}{2\pi} F = \frac{F}{7.85} \qquad (9.5\text{-}49)$$

Thus the limit on the allowable gain crossover frequency, for a loop having good stability, is approximately $\frac{1}{8}$ of the sampling frequency F. Note that ω_c is the value of the pseudo angular frequency W at the asymptotic gain crossover, not the value of the true angular frequency ω. However, this distinction is unimportant because at $\frac{1}{8}$ of the sampling frequency the values of W and ω are essentially equal.

The simple-hold D/A converter has a time delay equal to $\frac{1}{2}$ of the sample period T. This produces a phase lag that is proportional to frequency. At the sampling frequency F, this time delay represents a phase lag of $\frac{1}{2}$ cycle, which is 180°. Hence the phase shift ϕ at any frequency f is

$$\phi = -\frac{f}{F} 180° \qquad (9.5\text{-}50)$$

At $\frac{1}{8}$ of the sampling frequency, this is a phase lag of 22.5°. Thus, when the gain crossover frequency is at its maximum allowable value $(F/8)$, the simple-hold D/A converter contributes 22.5° of phase lag at gain crossover.

A lowpass filter is usually included in the plant transfer function to attenuate the sampling harmonics. The resultant loop then has much higher gain margin, but much lower phase margin, because the lowpass filter adds additional phase lag at gain crossover. To illustrate this point, assume that a single-order lowpass filter of break frequency ω_1 is added to the plant transfer function of Eq 9.5-44, to give the following plant transfer function:

$$G_p[s] = H_p[s] = \frac{\omega_c}{s(1 + s/\omega_1)} \tag{9.5-51}$$

This is the plant transfer function for case (D). Table 9.5-3 shows that the sampled loop transfer function for case (D) is

$$G^s[p] = \frac{\omega_c(1 - pT/2)(1 + p/\omega_1' - p/\omega_1)}{p(1 + p/\omega_1')} \tag{9.5-52}$$

Assume that $\omega_c T$ is set equal to 0.8, the upper limit of Eq 9.5-48. This sets f_c equal to $F/7.85$. Also assume that the filter break frequency ω_1 is set equal to 2 times the asymptotic gain crossover frequency ω_c. This gives

$$\omega_1 T = 2\omega_c T = 2(0.8) = 1.6 \tag{9.5-53}$$

As shown in Table 9.5-52, the value for $\omega_1' T$ is

$$\omega_1' T = 2 \tanh[\omega_c T] = 2 \tanh[1.6/2] = 1.33 \tag{9.5-54}$$

Substitute into Eq 9.5-52 the values of Eq 9.5-53, -54, and replace p by jW. This gives the following sampled frequency response for the loop transfer function:

$$G^s[jW] = \frac{\omega_c T(1 - jWT/2)(1 + jWT/7.8)}{jWT(1 + jWT/1.33)} \tag{9.5-55}$$

A magnitude asymptote plot of this transfer function is shown in Fig 9.5-6 versus the normalized pseudo frequency WT. As in the previous example, the phase of $G^s[jW]$ at infinite W is $-180°$, and so the gain margin is the reciprocal of $|G^s[jW]|$ at infinite W. It can be seen from Fig 9.5-6 that the gain margin is

$$\text{gain margin} = \frac{7.8}{0.8} \frac{2}{1.33} = 14.66 \tag{9.5-56}$$

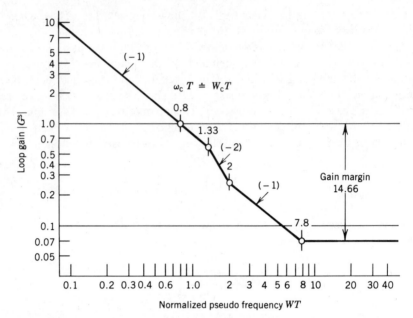

Figure 9.5-6 Magnitude asymptote plot of sampled loop transfer function for simple-hold D/A converter and plant transfer function $\omega_c/s(1 + s/\omega_1)$, with $\omega_1 = 2\omega_c$, and $\omega_c T = 0.8$.

Thus the loop has high gain margin. Its stability is limited by the phase margin. The excess phase lag of $G^s[jW]$ in Eq 9.5-55 at the asymptotic gain crossover frequency is

$$\phi_{ex}[j\omega_c] = \arctan\left[\frac{\omega_c T}{1.33}\right] + \arctan\left[\frac{\omega_c T}{2}\right] - \arctan\left[\frac{\omega_c T}{7.8}\right]$$

$$= \arctan\left[\frac{0.8}{1.33}\right] + \arctan\left[\frac{0.8}{2}\right] - \arctan\left[\frac{0.8}{7.8}\right]$$

$$= 47.0° \tag{9.5-57}$$

(Remember that the right-half-plane zero of the factor $(1 - jWT/2)$ adds phase lag at gain crossover.) This excess phase lag at gain crossover, 47.0°, is reasonably close to the ideal, 45°, usually chosen as a design criterion. Thus, the loop has reasonable stability, with a gain crossover frequency that is approximately $\frac{1}{8}$ of the sampling frequency.

The preceeding discussion has shown that it is theoretically possible to have good stability in a sampled-data feedback-control loop with a gain crossover frequency as high as $\frac{1}{8}$ of the sampling frequency. However, in most systems the gain crossover frequency is significantly lower than this, because stronger filtering is usually required to reduce sampling harmonics to acceptable levels. Under such conditions, the dynamic degradation of the sampling process is primarily the phase lag of the half-cycle delay in the D/A converter (which was given in Eq 9.5-50), plus the phase lag of the filtering required to attenuate the sampling harmonics. The total phase lag including filtering is often at least twice the phase lag of the D/A converter alone.

There are a number of reasons why strong filtering of sampling harmonics is usually required. The plant transfer functions considered in the preceding discussions are very simple. In most practical systems, the plant transfer functions are much more complicated than these in the high-frequency region. For example, as will be explained in Chapter 10, mechanical control systems usually have poorly damped high-frequency resonance peaks caused by the dynamics of the mechanical structure. If a sampling harmonic of appreciable magnitude falls at the same frequency as a structural resonance peak, instability may result. Therefore, mechanical control systems generally require strong filtering of sampling harmonics.

Besides, in high-performance control systems, sampling harmonics may saturate the power amplifier unless they are strongly filtered. For example, in a tachometer-feedback servomechanism, the tachometer feedback signal cancels most of the low-frequency portion of the amplified position error signal. Therefore, after the summation point of the tachometer signal, the control signal in a sampled-data system often consists largely of sampling harmonics. If these harmonics are not strongly filtered, they may saturate the power amplifier, and the resultant nonlinearity may severely degrade system performance.

9.5.4 Transient Analysis Using the z-Transform

This section describes the use of the z-transform to calculate the transient response of a feedback-control system that contains digital and analog data. As a first step, signal-flow diagram relations for sampled-data systems are examined.

9.5.4.1 Sampled-Data Transfer Functions for Typical Feedback Systems. Figure 9.5-7a illustrates the basic sampled-data relation, which was discussed in Section 9.5.1. The input signal X is sampled to form the sampled signal X^s. This signal is fed through the transfer function H, which has an output signal Y. Hence,

$$Y = HX^s \tag{9.5-58}$$

This is a difficult expression to apply because the responses of all of the sidebands of the sampled signal X^s must be included in the analysis. A much simpler relation occurs when Y is sampled to form the signal Y^s, which is related to X^s by

$$Y^s = H^s X^s \tag{9.5-59}$$

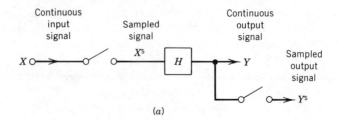

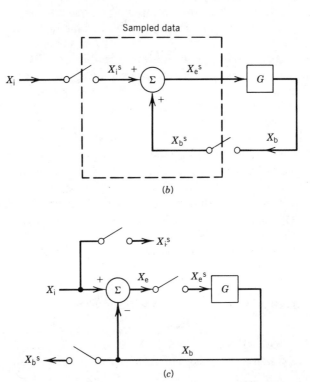

Figure 9.5-7 Sampled-data diagrams for a basic sampled system and simple feedback loops: (*a*) basic sampled-data relations; (*b*) feedback loop that samples the input and feedback signals; (*c*) feedback loop equivalent to (*b*), which samples the error signal.

The sampled transfer function H^s is the z-transform of the continuous transfer function H. The z-transform is computed by using the relations of Table 9.5-1.

Diagram *b* of Fig 9.5-7 extends the concept of diagram *a* to a simple feedback loop. The input signal X_i, feedback signal X_b, and error signal X_e are sampled. The same results occur when only the error signal is sampled, as shown in diagram *c*. In accordance with diagram *a* the sampled feedback signal X_b^s is related to the sampled error signal X_e^s by

$$X_b^s = G^s X_e^s \tag{9.5-60}$$

The transfer function G^s is the z-transform of the unsampled loop transfer function G, defined in diagram b or c by

$$X_b = G X_e^s \tag{9.5-61}$$

Since $X_e = X_i - X_b$, the corresponding sampled variables are related by

$$X_e^s = X_i^s - X_b^s \tag{9.5-62}$$

Combining Eqs 9.5-60, -62 gives

$$X_e^s = \frac{1}{1 + G^s} X_i^s \tag{9.5-63}$$

$$X_b^s = \frac{G^s}{1 + G^s} X_i^s \tag{9.5-64}$$

These have the forms

$$X_e^s = G_{ie}^s X_i^s \tag{9.5-65}$$
$$X_b^s = G_{ib}^s X_i^s \tag{9.5-66}$$

where G_{ie}^s, G_{ib}^s are the sampled error and feedback transfer functions, defined as

$$G_{ie}^s = \frac{1}{1 + G^s} \tag{9.5-67}$$

$$G_{ib}^s = \frac{G^s}{1 + G^s} \tag{9.5-68}$$

Figure 9.5-8 shows more complicated feedback configurations. These two examples are representative of a great many positional control systems that use digital positional sensors and compute the error signal digitally. In diagram a, the digital error signal is converted directly to analog form, while in diagram b dynamic digital processing is performed on the digital error signal X_e^s before it is converted to analog data.

The transfer function of the hold circuit, which converts the sampled digital data to analog form, is designated H_h; the transfer function of the plant is designated H_p; and H_f is the transfer function of the feedback circuit. In

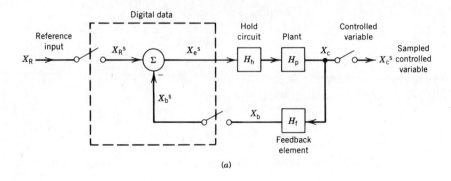

(a)

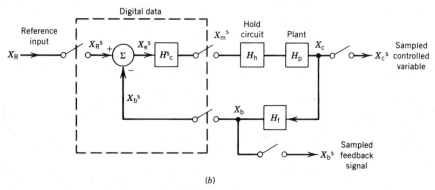

(b)

Figure 9.5-8   Typical signal-flow diagrams of control systems with digital position sensors: (a) system that converts sampled error signal directly to continuous (analog) signal; (b) system that processes sampled error signal with sampled-data compensation algorithm.

Fig 9.5-8a, the continuous loop transfer function is

$$G = H_h H_p H_f \tag{9.5-69}$$

The sampled loop transfer function G^s is

$$G^s = \left(H_h H_p H_f \right)^s \tag{9.5-70}$$

Equation 9.5-63 shows that the sampled error X_e^s is related as follows to the sampled reference input X_R^s:

$$X_e^s = G_{ie}^s X_R^s = \frac{1}{1 + G^s} X_R^s \tag{9.5-71}$$

In accordance with Eq 9.5-59, the sampled controlled variable X_c^s is related as follows to the sampled error signal:

$$X_c^s = \left(H_h H_p \right)^s X_e^s \qquad (9.5\text{-}72)$$

Combining Eqs 9.5-70 to -72 gives the transfer function relating the sampled controlled variable to the sampled reference input:

$$X_c^s = \frac{\left(H_h H_p \right)^s X_R^s}{1 + \left(H_h H_p H_f \right)^s} \qquad (9.5\text{-}73)$$

Often the feedback transfer function H_f is either a constant, or its dynamic effects are so small they can be neglected. In such a system, the transfer function H_f can be replaced by the constant K_f:

$$H_f = K_f \qquad (9.5\text{-}74)$$

For this case, the feedback signal x_b is instantaneously proportional to the controlled variable x_c, and so their sampled values are related by

$$X_b^s = K_f X_c^s \qquad (9.5\text{-}75)$$

The transfer function for X_c^s of Eq 9.5-73 now simplifies to

$$X_c^s = \frac{1}{K_f} X_b^s = \frac{1}{K_f} G_{ib}^s X_R^s$$

$$= \frac{1}{K_f} \frac{G^s}{1 + G^s} X_R^s \qquad (9.5\text{-}76)$$

The sampled loop transfer function G^s of Eq 9.5-70 simplifies to

$$G^s = \left(H_h H_p K_f \right)^s = K_f \left(H_h H_p \right)^s \qquad (9.5\text{-}77)$$

In diagram b of Fig 9.5-8, the sampled error signal X_e^s is processed digitally by a sampled-data compensation algorithm. This forms the sampled manipulated variable X_m^s, which is converted to analog data in the hold circuit. The transfer function of this sampled-data compensation is designated H_c^s, so that

$$X_m^s = H_c^s X_e^s \qquad (9.5\text{-}78)$$

The feedback signal X_b is related to X_m^s by

$$X_b = \left(H_h H_p H_f \right) X_m^s \qquad (9.5\text{-}79)$$

Hence, the sampled feedback signal is equal to

$$X_b^s = \left(H_h H_p H_f\right)^s X_m^s = \left(H_h H_p H_f\right)^s H_c^s X_e^s \qquad (9.5\text{-}80)$$

The sampled loop transfer function is

$$G^s = X_b^s / X_e^s = H_c^s \left(H_h H_p H_f\right)^s \qquad (9.5\text{-}81)$$

The sampled error is related to the reference input by

$$X_e^s = G_{ie}^s X_R^s = \frac{1}{1 + G^s} X_R^s$$

$$= \frac{1}{1 + H_c^s \left(H_h H_p H_f\right)} X_R^s \qquad (9.5\text{-}82)$$

Combining Eqs 9.5-78, -82 gives the following transfer function between the sampled reference input and the sampled manipulated variable:

$$X_m^s = H_c^s X_e^s = \frac{H_c^s X_R^s}{1 + H_c^s \left(H_h H_p H_f\right)^s} \qquad (9.5\text{-}83)$$

The controlled variable is related to the sampled manipulated variable by

$$X_c = \left(H_h H_p\right) X_m^s \qquad (9.5\text{-}84)$$

Hence, the sampled controlled variable is related to the sampled manipulated variable by

$$X_c^s = \left(H_h H_p\right)^s X_m^s \qquad (9.5\text{-}85)$$

Combining Eqs 9.5-83, -85 gives the transfer function between the sampled reference input and the sampled controlled variable:

$$X_c^s = \frac{H_c^s \left(H_h H_p\right)^s X_R^s}{1 + H_c^s \left(H_h H_p H_f\right)^s} \qquad (9.5\text{-}86)$$

When the feedback transfer function H_f can be represented by a feedback constant K_f, Eq 9.5-86 can be replaced by the simpler expression that was given in Eq 9.5-76. The sampled loop transfer function given in Eq 9.5-81 reduces to

$$G^s = H_c^s \left(H_h H_p K_f\right)^s = K_f H_c^s \left(H_h H_p\right)^s \qquad (9.5\text{-}87)$$

9.5.4.2 Examples of z-Transform Transient Analysis.

Let us consider two feedback-control systems A and B, which correspond to diagrams a, b of Fig 9.5-8. In both cases, the feedback transfer function H_f is taken to be a constant K_f. For simplicity, this feedback constant is set equal to unity:

$$H_f = K_f = 1 \tag{9.5-88}$$

A conventional simple-hold D/A converter is assumed for both cases. Hence, the hold transfer function is

$$H_h = \frac{1 - e^{-sT}}{sT} = \frac{1 - \bar{z}}{sT} \tag{9.5-89}$$

The plant is assumed to have the transfer function

$$H_p = \frac{\omega_c}{s(1 + s/\omega_f)} = \frac{\omega_c \omega_f}{s(s + \omega_f)} \tag{9.5-90}$$

For system A, which has the signal-flow diagram of Fig 9.5-8a, the unsampled loop transfer function is (from Eqs 9.5-88 to -90)

$$G_A = H_h H_p H_f = \frac{(1 - \bar{z})\omega_c \omega_f}{s^2 T(s + \omega_f)} \tag{9.5-91}$$

To simplify the calculations, define the transfer function H_A by

$$G_A = (1 - \bar{z}) H_A \tag{9.5-92}$$

Combining Eqs 9.5-91, -92 gives for H_A

$$H_A = \frac{\omega_c \omega_f}{s^2 T(s + \omega_f)} \tag{9.5-93}$$

In accordance with item (C) of the z-transform list in Table 9.5-1, the z-transform of Eq 9.5-92 is

$$G_A^s = (1 - \bar{z}) H_A^s \tag{9.5-94}$$

The expression for H_A of Eq 9.5-93 can be expanded in partial fractions as follows:

$$H_A = \frac{\omega_c}{s^2} - \frac{\omega_c/\omega_f}{s} + \frac{\omega_c/\omega_f}{s + \omega_f} \tag{9.5-95}$$

This is the standard partial-fraction method used in conventional Laplace-transform analysis. The three terms in this expansion have the same forms as items (H), (G), (K) of Table 9.5-1, in that order. Hence the z-transform of H_A is

$$H_A^s = \frac{\omega_c T \bar{z}}{(1 - \bar{z})^2} - \frac{\omega_c/\omega_f}{1 - \bar{z}} + \frac{\omega_c/\omega_f}{1 - \bar{z} \exp[-\omega_f T]} \qquad (9.5\text{-}96)$$

Combining Eqs 9.5-94, -96 gives the following expression for the sampled loop transfer function G_A^s:

$$G_A^s = \frac{\omega_c T \bar{z}}{1 - \bar{z}} - \frac{\omega_c}{\omega_f} + \frac{(\omega_c/\omega_f)(1 - \bar{z})}{1 - \bar{z} \exp[-\omega_f T]} \qquad (9.5\text{-}97)$$

Define the constant A as

$$A = \exp[-\omega_f T] \qquad (9.5\text{-}98)$$

Substitute this into Eq 9.5-97, and place G_A^s over a common denominator:

$$G_A^s = \frac{[\omega_c T - (\omega_c/\omega_f)(1 - A)]\bar{z} + [(\omega_c/\omega_f)(1 - A) - A\omega_c T]\bar{z}^2}{(1 - \bar{z})(1 - A\bar{z})} \qquad (9.5\text{-}99)$$

To proceed further with the analysis becomes very complicated unless a specific numerical example is used. Let us consider the example discussed in Section 9.5.3.2. In accordance with Eq 9.5-53, the following parameters are assumed:

$$\omega_f = 2\omega_c \qquad (9.5\text{-}100)$$

$$\omega_c T = 0.8 \qquad (9.5\text{-}101)$$

$$\omega_f T = 1.6 \qquad (9.5\text{-}102)$$

The expression for A defined in Eq 9.5-98 is equal to

$$A = \exp[-\omega_f T] = \exp[-1.6] = 0.20190 \doteq 0.2 \qquad (9.5\text{-}103)$$

As indicated, A is approximated as 0.2 in order to simplify the computations. Substituting Eqs 9.5-100 to -103 into Eq 9.5-99 gives the following the sampled loop transfer function:

$$G_A^s = \frac{0.4\bar{z} + 0.24\bar{z}^2}{(1 - \bar{z})(1 - 0.2\bar{z})} = \frac{N}{D} \qquad (9.5\text{-}104)$$

By Eq 9.5-68, the sampled feedback transfer function is

$$G_{ib(A)}^s = \frac{N}{D+N} = \frac{0.4\bar{z} + 0.24\bar{z}^2}{(1 - \bar{z})(1 - 0.2\bar{z}) + (0.4\bar{z} + 0.24\bar{z}^2)}$$

$$= \frac{0.4\bar{z} + 0.24\bar{z}^2}{1 - 0.8\bar{z} + 0.44\bar{z}^2} \qquad (9.5\text{-}105)$$

Assume that the reference input X_R is a unit step. Its transform is

$$X_R = 1/s \qquad (9.5\text{-}106)$$

In accordance with item (G) of Table 9.5-1, the z-transform of Eq 9.5-106 is

$$X_R^s = \frac{1}{1 - \bar{z}} \qquad (9.5\text{-}107)$$

Substituting Eqs 9.5-105, -107 into Eq 9.5-76 gives the following for the transform of the sampled controlled variable:

$$X_c^s = \frac{1}{K_f} G_{ib}^s X_R^s = \frac{0.4\bar{z} + 0.24\bar{z}^2}{(1 - \bar{z})(1 - 0.8\bar{z} + 0.44\bar{z}^2)}$$

$$= \frac{0.4\bar{z} + 0.24\bar{z}^2}{1 - 1.8\bar{z} + 1.24\bar{z}^2 - 0.44\bar{z}^3} \qquad (9.5\text{-}108)$$

The constant K_f was set equal to unity, in accordance with Eq 9.5-88. Using long division, divide the numerator of Eq 9.5-108 by the denominator. This yields an expansion of the form

$$X_c^s = c_0 + c_1\bar{z} + c_2\bar{z} + c_3\bar{z}^3 + c_4\bar{z}^4 + \cdots \qquad (9.5\text{-}109)$$

The coefficients derived from this long-division expansion are

$$
\begin{array}{lll}
c_0 = 0, & c_1 = 0.4, & c_2 = 0.96, \\
c_3 = 1.232, & c_4 = 1.2032 & c_5 = 1.0605, \\
c_6 = 0.9590, & c_7 = 0.9398, & c_8 = 0.9690
\end{array}
\qquad (9.5\text{-}110)
$$

In accordance with Eq 9.5-31, the expression for the transform X_c^s has the form

$$X_c^s = \sum_{k=0}^{\infty} x_c[kT] \qquad (9.5\text{-}111)$$

where $x_c[kT]$ are the values of the time function at the sampling instants:

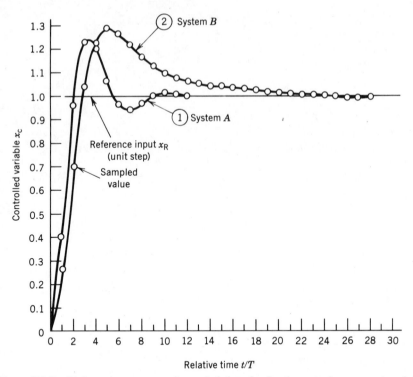

Figure 9.5-9   Unit-step responses of sampled-data feedback-control systems A and B.

$t = 0, T, 2T, 3T, \ldots$. Hence, the coefficients $c_0, c_1, \ldots, c_k$ given in Eq 9.5-110, derived from the long-division expansion of X_c^s, are the values of the controlled variable x_c at the sampling instants.

In Fig 9.5-9, curve ① is a plot of the coefficients c_k for system A obtained from Eq 9.5-110. The circles indicate the values of the controlled variable x_c at the sampling instants, in response to a unit step of the reference input x_R. This z-transform analysis does not yield any information concerning the response x_c between the sampling instants. The continuous plot is a smooth curve drawn between these points, which is a reasonable estimate of the transient response when the system response is well behaved.

System B, which has the signal-flow diagram of Fig 9.5-8b, is assumed to have the same analog transfer functions H_h, H_p, H_f as system A, except that the gain crossover frequency ω_c is different. System B also has digital compensation, represented by the transfer function H_c^s, which implements an integral-network transfer function, ideally represented by

$$H_c = 1 + \frac{\omega_i}{s} \qquad (9.5\text{-}112)$$

A magnitude asymptote plot of this integral-compensation transfer function is shown in Fig 9.5-10a. Applying the method described in Section 9.2 gives the following sampled-data algorithm to implement this transfer function:

$$y = y_p + x\left(1 + \frac{\omega_i T}{2}\right) - x_p\left(1 - \frac{\omega_i T}{2}\right) \qquad (9.5\text{-}113)$$

To obtain the corresponding sampled transfer function H_c^s, take the Laplace transform of Eq 9.5-113, which gives

$$Y = \bar{z}Y + X\left(1 + \frac{\omega_i T}{2}\right) - \bar{z}X\left(1 - \frac{\omega_i T}{2}\right) \qquad (9.5\text{-}114)$$

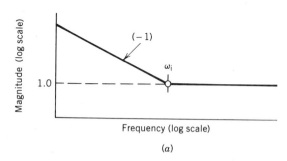

(a)

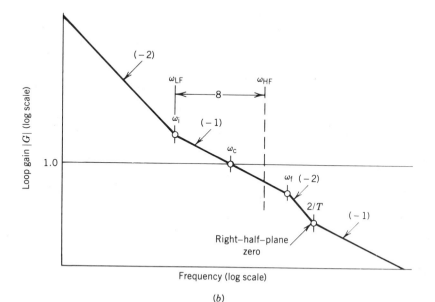

(b)

Figure 9.5-10 Approximate magnitude asymptote plots of integral-compensation transfer function and loop transfer function for sampled-data system B: (a) magnitude asymptote plot for digital integral-compensation algorithm; (b) asymptote plot of loop gain for sampled-data feedback loop with integral-compensation algorithm.

Combining X and Y terms gives

$$Y(1 - \bar{z}) = X\left\{\left(1 + \frac{\omega_i T}{2}\right) - \bar{z}\left(1 - \frac{\omega_i T}{2}\right)\right\} \qquad (9.5\text{-}115)$$

The ratio Y/X is the sampled transfer function H_c^s:

$$H_c^s = \frac{Y}{X} = \frac{(1 + \omega_i T/2) - \bar{z}(1 - \omega_i T/2)}{1 - \bar{z}} \qquad (9.5\text{-}116)$$

The sampled loop transfer-function for system B, designated G_B^s, is related as follows to the sampled loop transfer function for loop A, designated G_A^s:

$$G_B^s = H_c^s G_A^s \qquad (9.5\text{-}117)$$

where H_c^s is given by Eq 9.5-116 and G_A^s was given in general form in Eq 9.5-99.

Since system B has an integral network, which adds phase lag to the feedback loop, its gain crossover frequency ω_c must be lowered somewhat to maintain good stability. The continuous approximation of the loop transfer function for system B, designated G_B', is

$$G_B' = H_c H_p H_f\left(1 - \frac{sT}{2}\right) = \frac{\omega_c(1 + \omega_i/s)(1 - sT/2)}{s(1 + s/\omega_f)} \qquad (9.5\text{-}118)$$

In this expression, the dynamic effect of sampling (with a simple-hold D/A converter) is approximated by $(1 - sT/2)$, as explained in Section 9.5.3. An asymptotic magnitude plot of this approximate loop transfer function is shown in Fig 9.5-10b.

Let us choose optimum parameters for this transfer function, in accordance with the approximation procedure given in Chapter 3, Section 3.4, for a loop with a double integration at low frequencies. The break frequency at $\omega = 2/T$ corresponds to a right-half-plane zero, which has the same phase as a left-half-plane pole. Hence, in the optimization procedure, this break frequency is treated as if it represented a left-half-plane pole. The equation for ω_{HF} is

$$\frac{1}{\omega_{HF}} = \frac{1}{\omega_f} + \frac{T}{2} \qquad (9.5\text{-}119)$$

The value for ω_f is the same as that given for system A in Eq 9.5-102, which is

$$\omega_f = 1.6/T \qquad (9.5\text{-}120)$$

Combining Eqs 9.5-119, -120 gives

$$\omega_{HF} = 0.8889/T \qquad (9.5\text{-}121)$$

Assume that the ratio ω_{HF}/ω_{LF} is 8, which yields the following value for $\text{Max}|G_{ib}|$:

$$\text{Max}|G_{ib}| = \frac{\omega_{HF}/\omega_{LF} + 1}{\omega_{HF}/\omega_{LF} - 1}$$

$$= \frac{8+1}{8-1} = 1.286 \qquad (9.5\text{-}122)$$

Since ω_{LF} is equal to ω_i, the value for ω_i is

$$\omega_i = \omega_{LF} = \frac{\omega_{HF}}{8} = \frac{0.8889}{8T} = \frac{0.1111}{T} = \frac{1}{9T} \qquad (9.5\text{-}123)$$

The optimum value for ω_c is

$$\omega_c = \tfrac{1}{2}(\omega_{HF} + \omega_{LF}) = \tfrac{1}{2}\frac{(0.8889 + 0.1111)}{T}$$

$$= \frac{0.5000}{T} = \frac{1}{2T} \qquad (9.5\text{-}124)$$

From Eqs 9.5-120, -124, the ratio ω_c/ω_f is

$$\frac{\omega_c}{\omega_f} = \frac{0.5}{1.6} = \frac{5}{16} \qquad (9.5\text{-}125)$$

Substitute into Eq 9.5-116 the value for ω_i in Eq 9.5-123. This gives the following for the sampled integral-compensation transfer function:

$$H_c^s = \frac{\left(1 + \tfrac{1}{18}\right) - \bar{z}\left(1 - \tfrac{1}{18}\right)}{1 - \bar{z}} = \frac{19 - 17\bar{z}}{18(1 - \bar{z})} \qquad (9.5\text{-}126)$$

The general expression for the sampled loop transfer function for loop A was given in Eq 9.5-99. To obtain the corresponding transfer function for the gain crossover frequency of loop B, substitute into Eq 9.5-99 the values for $\omega_c T$ and $\omega_c \omega_f$ from Eqs 9.5-124, -125, and the values for $\omega_f T$ and $A = \exp[-\omega_f T]$ from Eqs 9.5-102, -103. (Note that ω_f is the same in systems A and B.) This gives

$$G_{A(B)}^s = \frac{\left[0.5 - \tfrac{5}{16}(1 - 0.2)\right]\bar{z} + \left[\tfrac{5}{16}(1 - 0.2) - 0.2(0.5)\right]\bar{z}^2}{(1 - \bar{z})(1 - 0.2\bar{z})}$$

$$= \frac{0.25\bar{z} + 0.15\bar{z}^2}{(1 - \bar{z})(1 - 0.2\bar{z})} \qquad (9.5\text{-}127)$$

This transfer function $G_{A(B)}^s$ is the expression for G_A^s that corresponds to the value of ω_c for system B.

Substitute into Eq 9.5-117 the expression for H_c^s of Eq 9.5-126, and substitute for G_A^s the expression for $G_{A(B)}^s$ of Eq 9.5-127. This gives the following sampled loop transfer function for loop B:

$$G_B^s = H_c^s G_{A(B)}^s = \frac{(19 - 17\bar{z})(0.25\bar{z} + 0.15\bar{z}^2)}{18(1 - \bar{z})^2(1 - 0.2\bar{z})}.$$

$$= \frac{4.75\bar{z} - 1.4\bar{z}^2 - 2.55\bar{z}^3}{18 - 39.6\bar{z} + 25.2\bar{z}^2 - 3.6\bar{z}^3} = \frac{N}{D} \qquad (9.5\text{-}128)$$

The corresponding sampled feedback transfer function is

$$G_{ib\,(B)}^s = \frac{N}{D + N}$$

$$= \frac{4.75\bar{z} - 1.4\bar{z}^2 - 2.55\bar{z}^3}{18 - 39.6\bar{z} + 25.2\bar{z}^2 - 3.6\bar{z}^3 + 4.75\bar{z} - 1.4\bar{z}^2 - 2.55\bar{z}^3}$$

$$= \frac{4.75\bar{z} - 1.4\bar{z}^2 - 2.55\bar{z}^3}{18 - 34.85\bar{z} + 23.8\bar{z}^2 - 6.15\bar{z}^3} \qquad (9.5\text{-}129)$$

As in Eq 9.5-108, the sampled controlled variable for system B, for a unit step of X_R, is

$$X_c^s = \frac{1}{K_f} G_{ib\,(B)}^s X_R^s = \frac{1}{K_f(1 - \bar{z})} G_{ib\,(B)}^s$$

$$= \frac{4.75\bar{z} - 1.4\bar{z}^2 - 2.55\bar{z}^3}{(1 - \bar{z})(18 - 34.85\bar{z} + 23.8\bar{z}^2 - 6.15\bar{z}^3)}$$

$$= \frac{4.75\bar{z} - 1.4\bar{z}^2 - 2.55\bar{z}^3}{18 - 52.85\bar{z} + 58.65\bar{z}^2 - 29.95\bar{z}^3 + 6.15\bar{z}^4} \qquad (9.5\text{-}130)$$

The expression for $G_{ib\,(B)}^s$ was obtained from Eq 9.5-129, X_R^s was obtained from Eq 9.5-107, and K_f was set equal to unity. For this case, the approximation of Eq 9.5-103, where a was set equal to 0.2, results in significant error in the tail of the transient. When a is changed to its exact value (0.2190), the expression for X_c^s in Eq 9.5-130 changes to

$$X_c^s = \frac{4.8148\bar{z} - 1.6263\bar{z}^2 - 2.3994\bar{z}^3}{18 - 52.819\bar{z} + 58.462\bar{z}^2 - 29.676\bar{z}^3 + 6.0336\bar{z}^4} \qquad (9.5\text{-}131)$$

Using long division, divide the numerator of Eq 9.5-131 by the denominator. This yields an expansion of the form of Eq 9.5-109, with the following coefficients:

$$\begin{array}{lll} c_0 = 0, & c_1 = 0.2675, & c_2 = 0.6946, \\ c_3 = 1.0361, & c_4 = 1.2253 & c_5 = 1.2860, \\ c_6 = 1.2692, & c_7 = 1.2205, & c_8 = 1.1686 \end{array} \qquad (9.5\text{-}132)$$

The resultant transient response is plotted in Fig 9.5-9 as curve ②. Note that sampled-data control systems A and B were both designed with approximate frequency-response analysis to have maximum gain crossover frequency consistent with good stability. The transient responses in Fig 9.5-9 are consistent with this criterion.

9.5.5   Serial Simulation of Sampled-Data Systems

This z-transform method for transient-response analysis of sampled-data systems yields values only at the sampling instants. When there is good high-frequency smoothing in the transfer functions H_h and H_p of the hold circuit and plant, the response between sampling instants is smooth, and no further information is needed. However, mechanical resonance and other dynamic effects in some control systems can cause high-frequency oscillations in the response. When these are present, information concerning the response between the sampling instants is needed.

The z-transform method can be extended to provide response data between the sampling instants. This approach, called the *modified z-transform*, is described by Ragazzini and Franklin [9.1] (Section 8.4, p. 212) and by Tou [9.10] (Section 6.4, p. 255).

On the other hand, z-transform transient analysis of complicated sampled-data control systems, even without the modified z-transform, is a cumbersome design method in today's world of the computer. A much more convenient approach is to apply the serial-simulation technique described in Section 9.2. Let us consider the steps for using serial simulation to compute the response of a sampled-data control system. The sampling frequency for such a simulation should be at least a factor of 8 greater than the sampling frequency of the sampled-data system.

Assume that the sampled-data feedback-control system, being modeled with serial simulation, has a sampling element followed by a simple-hold circuit, as shown in Fig 9.5-11a. Let us consider the steps in the computer program for implementing the response of this sample-and-hold process. The following sampling parameters are defined:

T_s = system sampling period

T = simulation sampling period

The simulation sampling period T is chosen such that the ratio T_s/T is an integer, which is designated J_{Max}:

$$J_{\text{Max}} = \frac{T_s}{T} = \text{integer} \qquad (9.5\text{-}133)$$

This integer J_{Max} should be no less than 8. The program initially specifies the value for J_{Max} along with the initial condition: $J = 0$. The steps in the serial simulation of a sampled-data system, for implementing the sample-and-hold

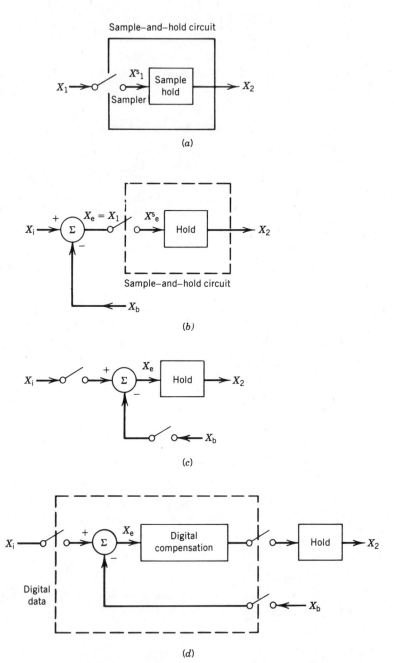

Figure 9.5-11 Signal-flow diagrams of sampled-data portion of digital systems that are simulated with serial simulation: (*a*) sample-and-hold circuit; (*b*) feedback loop that applies sample-and-hold to error signal; (*c*) feedback loop that samples input and feedback signals, and holds sampled error signal; (*d*) feedback loop that processes digital error signal with sampled-data compensation algorithm.

process of Fig 9.5-11a, are as follows:

101 IF $J \neq 0$ THEN $X_2 = X_{2p}$
102 IF $J = 0$ THEN $X_2 = X_1$
103 $X_{2p} = X_2$
104 $J = J + 1$
105 IF $J = J_{Max}$ THEN $J = 0$
106 CONTINUE AND REPEAT

Note that X_{2p} is the past value of X_2. With this routine, X_2 samples X_1 during the first cycle, and holds that value for the next $(J_{Max} - 1)$ cycles. Then X_2 samples X_1 again.

This same approach is used for the sample-and-hold configurations of diagrams b and c of Fig 9.5-11, in which the error signal of a feedback loop is sampled. The program given above is preceded with the step:

100 $X_1 = X_e = X_i - X_b$

The simulation procedure is somewhat different for the case shown in diagram d of Fig 9.5-11, which uses dynamic digital computation to process the error signal. For this case, the following initial conditions are specified:

$$J_{Max} = T_s/T = \text{integer}$$
$$J = 0, \qquad X_{ep} = 0, \qquad X_{2p} = 0$$

The program is

100 IF $J \neq 0$ THEN $X_2 = X_{2p}$ GOTO 105
101 $X_e = X_i - X_b$
102 $X_2 = c_1 X_{2p} + c_2 X_e + c_3 X_{ep}$
103 $X_{ep} = X_e$
104 $X_{2p} = X_2$
105 $J = J + 1$
106 IF $J = J_{Max}$ THEN $J = 0$
107 CONTINUE AND REPEAT

The variable X_{ep} is the past value of the error X_e, and X_{2p} is the past value of X_2. Step 102 is the sampled-data algorithm for the dynamic digital compensation. For example, suppose that integral compensation is used, having the ideal transfer function:

$$H_c = \left(1 + \frac{\omega_i}{s}\right) \qquad (9.5\text{-}134)$$

In accordance with Eq 9.5-113, given in Section 9.5.4, the corresponding sampled-data algorithm for step 102 is

$$102 \quad X_2 = X_{2p} + \left(1 + \frac{\omega_i T}{2}\right) X_e - \left(1 - \frac{\omega_i T}{2}\right) X_{ep}$$

Hence, for this integral-compensation transfer function, the coefficients c_1, c_2, c_3 of step 102 are

$$c_1 = 1, \qquad c_2 = 1 + \frac{\omega_i T}{2}, \qquad c_3 = 1 - \frac{\omega_i T}{2} \qquad (9.5\text{-}135)$$

9.5.6 Summary of Digital Control-System Design

A control system that includes digital signal processing handles signals in terms of sampled data. Exact sampled-data analysis of such a system can be very complicated if the system is at all complex. However, such analysis is generally not necessary.

Usually there is sufficient filtering in the feedback loop for the control system to be analyzed with frequency response, in which the effect of sampling is approximated by the transfer function $(1 - sT/2)$. (This assumes a conventional simple-hold digital-to-analog converter.) If one suspects that this approximation is not adequate, the control system can be accurately simulated using serial simulation, as described in Section 9.5.5.

When the digital signal processing incorporates a difference equation that provides dynamic compensation, the frequency response of the compensation can be readily determined by applying the principles of Sections 9.2 and 9.3. The Laplace transform of the difference equation is taken by representing each one-cycle time delay by $\bar{z}$, which is equal to $\exp[-sT]$. The variable $\bar{z}$ is replaced with the following to obtain the approximate frequency response of the difference equation:

$$\bar{z} = \frac{1 - pT/2}{1 + pT/2} \qquad (9.5\text{-}136)$$

This yields a transfer function expressed in terms of the pseudo complex frequency p, which approximates the true complex frequency s for values of $|s|$ up to $1/4$ of the sampling frequency.

Chapter 10

Structural Dynamics

This chapter introduces the issue of structural dynamics. Generally, mechanical structural resonance is the primary limitation on the dynamic performance of systems that control mechanical elements. The dynamic effect of compliance in the gear train between a servo motor and an inertial load is analyzed, and reduced to a computer model. This computer model is applied to the servo studied in Chapter 7, and simulated in Chapter 9, to provide a very realistic serial simulation of a practical servomechanism.

10.1 EFFECT OF STRUCTURAL RESONANCE ON FEEDBACK-CONTROL LOOPS

For feedback-control systems that drive mechanical devices, the dynamics of the mechanical structure is usually the most important constraint on system performance. When tests are made on such a control system, the dynamic limitations caused by structural resonance are usually very clear. However, at that time it is often too late to change the structure.

Although analysis of structural dynamics in a feedback-control system is highly desirable in the design phase, this can be very difficult. A realistic structure can have many independent nodes, each of which can resonate independently in six degrees of freedom (three in translation and three in rotation). Another complication is that the dynamic characteristics of the motor (particularly its output impedance) are integrally tied with those of the structure. For these reasons, dynamic structural models are often gross simplifications of reality.

In recent years mechanical engineers have developed sophisticated techniques for modeling structural dynamics. However, these techniques are so complicated it is often prohibitively expensive to apply them thoroughly. The problem is further confused by the fact that the mechanical engineers who are skilled in the use of these mechanical models often do not know how to include the effects of the drive motor and other control elements, or to analyze

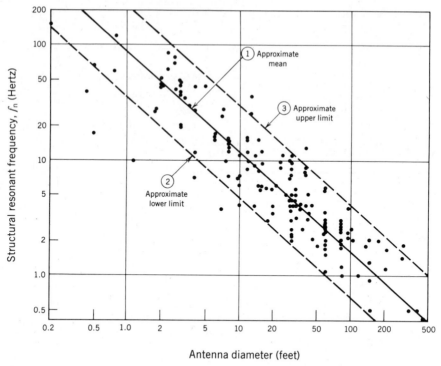

Figure 10.1-1   Servo structural resonant frequencies of servo-driven parabolic antennas, plotted as a function of antenna diameter (compiled by Pidhayny).

the problem so as to define properly the resonant transfer functions experienced by the control loops.

For a servo-driven parabolic antenna, the servo structural resonant frequency is a parameter than can strongly affect both dynamic performance and cost. This is defined as the lowest structural resonant frequency experienced in the antenna servo loops. It is often called the locked-rotor resonant frequency, because it is the frequency at which the structure would resonate if the motor rotor were locked. A good method for measuring this resonant frequency is to oscillate the antenna with a fixed torque amplitude, and find the frequency at which the motor velocity amplitude is a minimum.

Figure 10.1-1 shows values of the servo structural resonant frequency of 160 servo-controlled parabolic antennas, expressed as a function of antenna aperture diameter. Each point on the chart represents a different antenna. These data have been compiled over the past 25 years by Denny D. Pidhayny of Aerospace Corp. Appendix H gives a more detailed figure, which shows the names of the antennas that correspond to the points on Fig 10.1-1, and the companies that built them.

Line ① in Fig 10.1-1 is a plot of the approximate mean structural resonant frequency for the antennas. This mean is characterized by

$$\text{Mean: } f_n = 90/(D/\text{ft})^{7/8} \text{ Hz} \tag{10.1-1}$$

The variable D is the antenna aperture diameter expressed in feet (ft), and f_n is the mean structural resonant frequency in hertz. Dashed lines ② and ③ are drawn a factor of 2.5 below the mean plot ①, and a factor of 2.5 above the mean plot. As the figure shows, almost all of the antennas fall between lines ② and ③.

Thus, for estimating, one can use Eq 10.1-1 to obtain a mean practical value of the structural resonant frequency for any specified antenna diameter. The structural resonant frequency for a practical antenna can generally be assumed to be greater than 0.4 times this mean value, and less than 2.5 times this mean value.

To increase the structural resonant frequency of a large antenna appreciably can be very expensive, because structural stiffness is proportional to the square of the resonant frequency. The antennas in Fig 10.1-1 have many different purposes. Hence for a given diameter, some antennas are designed to have higher resonant frequencies than others, and so are more expensive.

This chapter presents an introduction to the very complex problem of structural dynamics. One should always treat structural dynamics with a great deal of respect, and be very conservative in predicting performance. Above all, dynamic measurements of the actual responses of the control system with its structural load are essential.

10.2   ELECTRICAL ANALOGS OF MECHANICAL MODELS

A convenient method for developing the transfer function of a mechanical model is to represent the model by an equivalent electrical circuit. Figure 10.2-1a shows for illustration a mechanical model that moves in a single dimension x. There are two masses, M_1 and M_2. These are coupled together by a spring of stiffness k_2 in parallel with a damper (otherwise known as a dashpot or shock absorber) of damping constant b_2. Mass M_1 is coupled to a fixed member (ground) by a spring of stiffness k_1 and a damper of damping constant b_1 while mass M_2 is coupled to ground by a spring of stiffness k_3 and damping constant b_3. The displacement of mass M_1 is designated x_1, and the displacement of mass M_2 is designated x_2. The velocity of mass M_1 is V_1, and the velocity of mass M_2 is V_2.

A forcing device attached to ground exerts a force F_1 on mass M_1. A second forcing device attached to mass M_1 exerts a force F_2 on mass M_2, and on mass M_1 exerts the reaction to this force, which is $-F_2$.

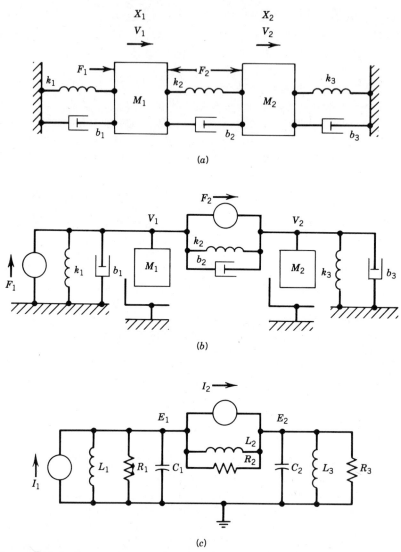

Figure 10.2-1 Derivation of analogous electric circuit from dynamic mechanical model: (*a*) linear-motion mechanical model; (*b*) mechanical circuit; (*c*) analogous electric circuit.

Diagram *b* of Fig 10.2-1 shows a mechanical circuit that is equivalent to the mechanical model of diagram *a*. Note that each mass is represented as a two-terminal circuit element, where one of these terminals is ground. The ground node indicates that the acceleration of the mass is measured relative to inertial space, or "ground."

The equations for the force across a mass, spring, and damper are expressed as follows in terms of the relative velocity V between the two terminals of the mechanical circuit element:

Mass:
$$F = M\frac{d^2x}{dt^2} = M\frac{dV}{dt} \qquad (10.2\text{-}1)$$

Spring:
$$F = kx = k\int V\,dt \qquad (10.2\text{-}2)$$

Damper:
$$F = bV \qquad (10.2\text{-}3)$$

To construct the electrical analog of the mechanical circuit, the velocity V is represented by a voltage E, and the force F is represented by a current I. The current I through a capacitor, inductor, or resistor is related as follows to the voltage E across the element:

Capacitor:
$$I = C\frac{dE}{dt} \qquad (10.2\text{-}4)$$

Inductor:
$$I = \frac{1}{L}\int E\,dt \qquad (10.2\text{-}5)$$

Resistor:
$$I = \frac{1}{R}E \qquad (10.2\text{-}6)$$

Equations 10.2-4 to 10.2-6 for the electrical circuit elements are equivalent to Eqs 10.2-1 to 10.2-3, respectively, for the mechanical circuit elements, if the following substitutions are made:

$$\text{(force)} \quad F \rightarrow I \quad \text{(current)} \qquad (10.2\text{-}7)$$
$$\text{(velocity)} \quad V \rightarrow E \quad \text{(voltage)} \qquad (10.2\text{-}8)$$
$$\text{(mass)} \quad M \rightarrow C \quad \text{(capacitance)} \qquad (10.2\text{-}9)$$
$$\text{(compliance)} \quad 1/k \rightarrow L \quad \text{(inductance)} \qquad (10.2\text{-}10)$$
$$\text{(damping constant)} \quad b \rightarrow 1/R \quad \text{(conductance)} \qquad (10.2\text{-}11)$$

Note that the compliance of a spring is the reciprocal of the spring stiffness k.
By making the substitutions of Eqs 10.2-7 to 10.2-11 in the mechanical circuit of diagram b of Fig 10.2-1, the mechanical circuit is converted to the equivalent electrical circuit of diagram c. The values of the signals and parameters of the electrical circuit are set equal to the corresponding mechani-

cal quantities as follows:

$$I_1 = F_1 \tag{10.2-12}$$

$$I_2 = F_2 \tag{10.2-13}$$

$$E_1 = V_1 = \frac{dx_1}{dt} \tag{10.2-14}$$

$$E_2 = V_2 = \frac{dx_2}{dt} \tag{10.2-15}$$

$$C_1 = M_1 \tag{10.2-16}$$

$$C_2 = M_2 \tag{10.2-17}$$

$$L_1 = 1/k_1 \tag{10.2-18}$$

$$L_2 = 1/k_2 \tag{10.2-19}$$

$$L_3 = 1/k_3 \tag{10.2-20}$$

$$R_1 = 1/b_1 \tag{10.2-21}$$

$$R_2 = 1/b_2 \tag{10.2-22}$$

$$R_3 = 1/b_3 \tag{10.2-23}$$

This approach can be extended to apply to rotary motion, by replacing force F with torque T, and by replacing linear velocity V with angular velocity Ω. The mechanical rotation equations corresponding to Eqs 10.2-1 to 10.2-3 are

Inertia:
$$T = J\frac{d^2\Theta}{dt^2} = J\frac{d\Omega}{dt} \tag{10.2-24}$$

Torsional spring:
$$T = K\Theta = K\int \Omega \, dt \tag{10.2-25}$$

Torsional damper:
$$T = B\Omega \tag{10.2-26}$$

The uppercase symbols K and B are used to represent torsional stiffness and torsional damping constant, while lowercase k and b represent linear stiffness and linear damping constant. Thus, the following substitutions are made to create an electrical analog of a rotary mechanical circuit:

(torque)	$T \rightarrow I$	(current)	(10.2-27)
(angular velocity)	$\Omega \rightarrow E$	(voltage)	(10.2-28)
(inertia)	$J \rightarrow C$	(capacitance)	(10.2-29)
(torsional compliance)	$1/K \rightarrow L$	(inductance)	(10.2-30)
(torsional damping constant)	$B \rightarrow 1/R$	(conductance)	(10.2-31)

It is also possible to develop an electrical analog of a mechanical circuit that is the dual of the above equations, by replacing force with voltage, and velocity by current. For the basic circuit elements, the following are the differential equations expressing voltage as a function of current:

Capacitor: $$E = \frac{1}{C} \int I \, dt \qquad \qquad (10.2\text{-}32)$$

Inductor: $$E = L \frac{dI}{dt} \qquad \qquad (10.2\text{-}33)$$

Resistor: $$E = RI \qquad \qquad (10.2\text{-}34)$$

Comparing these with the expressions in Eqs 10.2-1 to 10.2-3, for the mechanical circuit elements, gives the following substitutions to create the dual electrical analog:

$$\text{(force)} \quad F \rightarrow E \quad \text{(voltage)} \qquad (10.2\text{-}35)$$
$$\text{(velocity)} \quad V \rightarrow I \quad \text{(current)} \qquad (10.2\text{-}36)$$
$$\text{(mass)} \quad M \rightarrow L \quad \text{(inductance)} \qquad (10.2\text{-}37)$$
$$\text{(compliance)} \quad 1/k \rightarrow C \quad \text{(capacitance)} \qquad (20.2\text{-}38)$$
$$\text{(damping constant)} \quad b \rightarrow R \quad \text{(resistance)} \qquad (10.2\text{-}39)$$

This approach can be extended to apply to rotary motion.

This dual electrical analog may seem physically more reasonable, because voltage seems physically more akin to force than to velocity, and current seems physically more akin to velocity than to force. However, the problem with this dual analog is that mechanical elements that are in series result in electrical elements that are in parallel, and vice versa. Consequently, it is much more difficult to derive the electrical circuit from the mechanical circuit when this dual electrical analog is used.

For example, if two springs are in series, they both carry the same force. For the recommended analog, these series springs are replaced with inductors in series, which carry the same current. If two springs are in parallel, the relative velocities across the terminals are the same. For the recommended analog, these parallel springs are replaced with two inductors in parallel, which are subjected to the same voltage. Thus, to maintain the same parallel and series connections of elements in the mechanical and electrical circuits, force must be replaced with current, and velocity must be replaced with voltage.

10.3 DYNAMIC STRUCTURAL MODEL OF A MOTOR DRIVING AN INERTIAL LOAD THROUGH A COMPLIANT GEAR TRAIN

This section derives the structural frequency-response equations experienced in the position and velocity loops closed around a servo motor driving an inertial load through a compliant gear train. General frequency-response plots of these responses are derived.

10.3.1. Analysis

Diagram *a* of Fig 10.3-1 shows a simplified mechanical model of a motor driving an inertial load through a compliant gear train. To simplify the discussion, the motor is reflected through the gear ratio, to provide the equivalent motor torque, motor velocity, and motor inertia reflected to the output load. The motor inertia, load inertia, and interconnecting gearing can be represented by two inertias coupled by a compliant shaft. The signals and parameters of the mechanical model are

$T_m = N \times$ (motor torque) = motor torque reflected to output

$\Omega_m = (1/N) \times$ (motor angular velocity)
    = motor angular velocity reflected to output

Ω_o = output angular velocity

$J_m = N^2 \times$ (motor inertia)
    = motor inertia reflected to output

J_o = output load inertia

K = stiffness of gear train relative to output

B_p = damping constant of gear train relative to output (expressed as an equivalent parallel damper across the gear train)

The sum of the reflected motor inertia J_m and the output inertia J_o is designated J:

$$J = J_m + J_o$$
$$= \text{total inertia reflected to output} \qquad (10.3\text{-}1)$$

Diagram *b* of Fig 10.3-1 shows the mechanical circuit for this mechanical model. Diagram *c* shows the corresponding electric analog, where the following substitutions have been made:

$$I_m = T_m \qquad (10.3\text{-}2)$$
$$E_m = \Omega_m \qquad (10.3\text{-}3)$$
$$E_o = \Omega_o \qquad (10.3\text{-}4)$$
$$C_m = J_m \qquad (10.3\text{-}5)$$
$$C_o = J_o \qquad (10.3\text{-}6)$$
$$L = 1/K \qquad (10.3\text{-}7)$$
$$R_p = 1/B_p \qquad (10.3\text{-}8)$$

As will be shown in Section 10.4, a high-Q circuit, consisting of a resistor R_p in parallel with an inductor L, can be replaced approximately with a resistor R_s in series with the inductor, where R_s is

$$R_s = \frac{(\omega L)^2}{R_p} \qquad (10.3\text{-}9)$$

Total inertia $J = J_m + J_o$

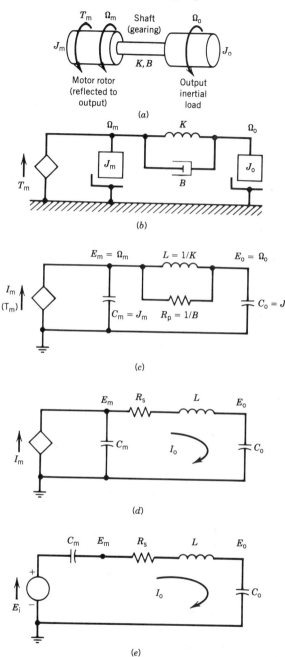

(a)

(b)

(c)

(d)

(e)

Figure 10.3-1 Electric circuit analog for motor driving inertial load through compliant gear train: (a) simplified mechanical model of motor, gearing, and inertial load; (b) mechanical circuit; (c) analogous electric circuit; (d) circuit with parallel resistor R_p replaced with equivalent series resistor R_s; (e) circuit with I_m, C_m replaced with Thevenin equivalent.

By applying this principle, diagram c of Fig 10.3-1 simplifies to diagram d. The Q of the circuit is related as follows to the series and parallel resistances:

$$Q = \frac{\omega L}{R_s} = \frac{R_p}{\omega L} \qquad (10.3\text{-}10)$$

Structural damping is nonlinear, and consequently the damping factor B_p varies with frequency. Generally, the resultant Q of a mechanical circuit is approximately a constant, independent of frequency. Therefore, it is convenient to characterize structural damping in terms of Q. For a gear-driven structure, the primary source of structural damping is friction in the gear train. A reasonable range of Q for a gear-driven structure appears to be 10–20.

Diagram d is simplified to diagram e by replacing the current source I_m and parallel capacitor C_m with the Thevenin equivalent, which consists of a voltage source E_i in series with the same capacitor C_m. This Thevenin equivalent voltage source E_i is equal to the source current I_m multiplied by the impedance $1/sC_m$ of the capacitor:

$$E_i = I_m \frac{1}{sC_m} \qquad (10.3\text{-}11)$$

This results in a simple series circuit. The output current I_o flowing in this circuit is analogous to the output torque delivered to the output inertia J_o. Diagram e shows that the equation for the current I_o is

$$I_o = \frac{E_i}{R_s + sL + (1/sC_m) + (1/sC_o)} \qquad (10.3\text{-}12)$$

The capacitance of the series combination of capacitors C_o and C_m is defined as C_{mo}, given by

$$\frac{1}{C_{mo}} = \frac{1}{C_m} + \frac{1}{C_o} \qquad (10.3\text{-}13)$$

$$C_{mo} = \frac{C_m C_o}{C_m + C_o} \qquad (10.3\text{-}14)$$

Equation 10.3-13 is applied to Eq 10.3-12, and E_i is replaced by the expression of Eq 10.3-11. This gives

$$I_o = \frac{I_m(1/sC_m)}{R_s + sL + (1/sC_{mo})} = \frac{(C_{mo}/C_m)I_m}{1 + sC_{mo}R_s + s^2 LC_{mo}} \qquad (10.3\text{-}15)$$

(The numerator and denominator were multiplied by sC_{mo} to obtain the second expression.) Equation 10.3-14 is substituted for C_{mo} in the numerator to give

$$I_o = \frac{[C_o/(C_m + C_o)]I_m}{1 + sC_{mo}R_s + s^2LC_{mo}} \qquad (10.3\text{-}16)$$

Figure 10.3-1d shows that the output voltage E_o and the motor voltage E_m are related as follows to the output current I_o:

$$E_o = I_o\frac{1}{sC_o} \qquad (10.3\text{-}17)$$

$$E_m = I_o\left(\frac{1}{sC_o} + sL + R\right) \qquad (10.3\text{-}18)$$

Substitute the expression for I_o of Eq 10.3-16 into Eqs 10.3-17, -18, and solve for the transfer functions E_o/I_m and E_m/I_m:

$$\frac{E_o}{I_m} = \frac{1}{s(C_m + C_o)(1 + sC_{mo}R_s + s^2LC_{mo})} \qquad (10.3\text{-}19)$$

$$\frac{E_m}{I_m} = \frac{1 + sR_sC_o + s^2LC_o}{s(C_m + C_o)(1 + sC_{mo}R_s + s^2LC_{mo})} \qquad (10.3\text{-}20)$$

Equations 10.3-19, -20 are converted to the following analogous mechanical equations:

$$\frac{\Omega_o}{T_m} = \frac{1}{s(J_m + J_o)[1 + s(J_{mo}/B_s) + s^2(J_{mo}/K)]} \qquad (10.3\text{-}21)$$

$$\frac{\Omega_m}{T_m} = \frac{1 + s(J_o/B_s) + s^2(J_o/K)}{s(J_m + J_o)[1 + s(J_{mo}/B_s) + s^2(J_{mo}/K)]} \qquad (10.3\text{-}22)$$

In accordance with Eq 10.3-14, the inertia parameter J_{mo} is defined as

$$J_{mo} = \frac{J_mJ_o}{J_m + J_o} \qquad (10.3\text{-}23)$$

The parameter B_s is the damping factor of the equivalent damper in series with the gearing compliance K, and is equal to the reciprocal of the analogous series resistance R_s. By Eq 10.3-10, R_s is equal to $\omega L/Q$. Hence B_s is equal to

$$B_s = \frac{1}{R_s} = \frac{Q}{\omega L} = \frac{QK}{\omega} \qquad (10.3\text{-}24)$$

where L is replaced by $1/K$ in accordance with Eq 10.3-7. The damping factor B_p for an equivalent damper in parallel with the spring is equal to the reciprocal of the analogous parallel resistance R_p. By Eq 10.3-10, R_p is equal to ωLQ. Hence the parallel damping factor B_p is equal to

$$B_p = \frac{1}{R_p} = \frac{1}{\omega LQ} = \frac{K}{\omega Q} \qquad (10.3\text{-}25)$$

The inertia sum $(J_m + J_o)$ is replaced with the total inertia J. Thus

$$J = J_m + J_o \qquad (10.3\text{-}26)$$

Both sides of Eqs 10.3-21, -22 are multiplied by sJ. This gives the following normalized responses:

$$\frac{J\alpha_o}{T_m} = \frac{Js\Omega_o}{T_m} = \frac{1}{1 + s(J_{mo}/B_s) + s^2(J_{mo}/K)} \qquad (10.3\text{-}27)$$

$$\frac{J\alpha_m}{T_m} = \frac{Js\Omega_m}{T_m} = \frac{1 + s(J_o/B_s) + s^2(J_o/K)}{1 + s(J_{mo}/B_s) + s^2(J_{mo}/K)} \qquad (10.3\text{-}28)$$

where α_o and α_m are the transforms of the output acceleration and the motor acceleration, which are equal to $s\Omega_o$ and $s\Omega_m$, respectively. The quadratic factors are expressed as follows in terms of the general quadratic parameters:

$$\frac{J\alpha_o}{T_m} = \frac{Js\Omega_o}{T_m} = \frac{1}{1 + 2\zeta_2(s/\omega_{n2}) + (s/\omega_{n2})^2} \qquad (10.3\text{-}29)$$

$$\frac{J\alpha_m}{T_m} = \frac{Js\Omega_m}{T_m} = \frac{1 + 2\zeta_1(s/\omega_{n1}) + (s/\omega_{n1})^2}{1 + 2\zeta_2(s/\omega_{n2}) + (s/\omega_{n2})^2} \qquad (10.3\text{-}30)$$

where

$$\omega_{n1}^2 = \frac{K}{J_o} \qquad (10.3\text{-}31)$$

$$\omega_{n2}^2 = \frac{K}{J_{mo}} \qquad (10.3\text{-}32)$$

$$\frac{2\zeta_1}{\omega_{n1}} = \frac{J_o}{B_s} \qquad (10.3\text{-}33)$$

$$\frac{2\zeta_2}{\omega_{n2}} = \frac{J_{mo}}{B_s} \qquad (10.3\text{-}34)$$

The expression for ω_{n2}^2 of Eq 10.3-32 can be simplified as follows:

$$\omega_{n2}^2 = \frac{K}{J_{mo}} = K\frac{J_m + J_o}{J_m J_o} = \frac{J}{J_m}\frac{K}{J_o}$$

$$= \frac{J}{J_m}\omega_{n1}^2 \qquad (10.3\text{-}35)$$

Taking the square roots of Eqs 10.3-31, -32, and -35 gives the two natural frequencies:

$$\omega_{n1} = \sqrt{K/J_o} \qquad (10.3\text{-}36)$$

$$\omega_{n2} = \sqrt{K/J_{mo}} = \omega_{n1}\sqrt{J/J_M} \qquad (10.3\text{-}37)$$

As will be shown in Section 10.4, the damping ratio ζ is equal to $1/2Q$. With mechanical damping, the Q can be assumed to be constant, and so the damping ratio ζ is constant, independent of frequency. Hence ζ_1 and ζ_2 are equal, and so can be replaced by

$$\zeta_1 = \zeta_2 = 1/2Q \qquad (10.3\text{-}38)$$

With this substitution, Eqs 10.3-29, -30 become

$$\frac{J\alpha_o}{T_m} = \frac{Js\Omega_o}{T_m} = \frac{1}{1 + (1/Q)(s/\omega_{n2}) + (s/\omega_{n2})^2} \qquad (10.3\text{-}39)$$

$$\frac{J\alpha_m}{T_m} = \frac{Js\Omega_m}{T_m} = \frac{1 + (1/Q)(s/\omega_{n1}) + (s/\omega_{n1})^2}{1 + (1/Q)(s/\omega_{n2}) + (s/\omega_{n2})^2} \qquad (10.3\text{-}40)$$

10.3.2 Plots of Structural Frequency Response

Setting $s = j\omega$ in Eqs 10.3-39, -40 gives the following normalized frequency responses, which represent the responses produced by structural dynamics in the position and velocity loops:

Position loop:

$$\frac{J\alpha_o}{T_m} = \frac{Js\Omega_o}{T_m} = \frac{1}{\left[1 - (\omega/\omega_{n2})^2\right] + j(\omega/\omega_{n2})(1/Q)} \qquad (10.3\text{-}41)$$

Tachometer velocity loop:

$$\frac{J\alpha_m}{T_m} = \frac{Js\Omega_m}{T_m} = \frac{\left[1 - (\omega/\omega_{n1})^2\right] + j(\omega/\omega_{n1})(1/Q)}{\left[1 - (\omega/\omega_{n2})^2\right] + j(\omega/\omega_{n2})(1/Q)} \qquad (10.3\text{-}42)$$

The output position signal θ_o used for position feedback is derived from the output acceleration signal α_o, while the motor velocity signal Ω_m is derived from the motor acceleration α_m. Hence, the transfer function $J\alpha_o/T_m$ is an element within the position loop, and the transfer function $J\alpha_m/T_m$ is an element within the tachometer velocity loop.

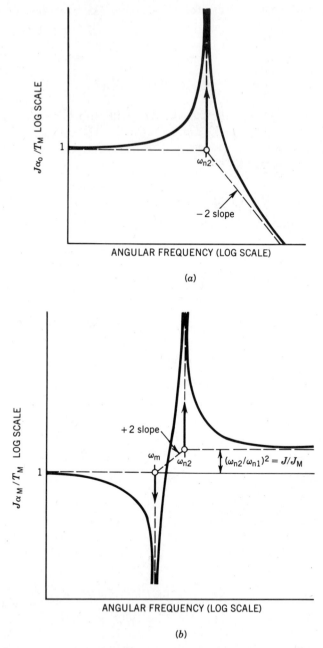

Figure 10.3-2 Magnitude of structural resonant frequency responses for infinite Q: (*a*) position loop; (*b*) velocity loop.

Let us consider the case of zero damping. Setting $1/Q$ equal to zero in Eqs 10.3-41, -42 gives

Position loop (zero damping):

$$\frac{J\alpha_o}{T_m} = \frac{1}{1 - (\omega/\omega_{n2})^2} \qquad (10.3\text{-}43)$$

Velocity loop (zero damping):

$$\frac{J\alpha_m}{T_m} = \frac{1 - (\omega/\omega_{n1})^2}{1 - (\omega/\omega_{n2})^2} \qquad (10.3\text{-}44)$$

Magnitude plots of these transfer functions are shown in Fig 10.3-2. Both responses have unity magnitude at low frequency and a peak at the frequency

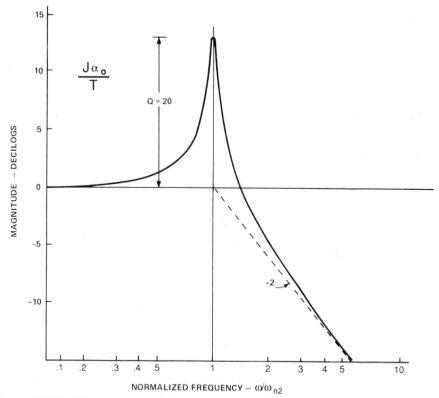

Figure 10.3-3 Effect of mechanical resonance in position loop: normalized magnitude of frequency response between torque and output acceleration.

ω_{n2}. The motor acceleration response has a zero at the frequency ω_{n1}. At high frequencies, the motor acceleration response given by Eq 10.3-44 has the following value:

$$\frac{J\alpha_m}{T_m} = \frac{(\omega/\omega_{n1})^2}{(\omega/\omega_{n2})^2} = \frac{(\omega_{n1})^2}{(\omega_{n2})^2} = \frac{J}{J_m} \qquad (10.3\text{-}45)$$

This has used the relation of Eq 10.3-37. Thus, the high-frequency magnitude of the normalized velocity loop response $J(\alpha_m/T_m)$ is equal to the inertia ratio J/J_m.

Good approximations of the magnitude response for reasonably high values of Q can be obtained by determining the magnitude values at the peak resonant frequency ω_{n2} and at the dip resonant frequency ω_{n1}. Setting $\omega = \omega_{n2}$ in the position loop response of Eq 10.3-41 gives

Position loop at peak ($\omega = \omega_{n2}$):

$$\frac{J\alpha_o}{T_m} = \frac{1}{j(1/Q)} = -jQ \qquad (10.3\text{-}46)$$

Thus, the peak magnitude of the position loop response is equal to Q. This is illustrated in Fig 10.3-3, which gives a plot of $|J\alpha_o/T_m|$. The Q is assumed to be 20, which is a conservative value for a gear-train-driven servo.

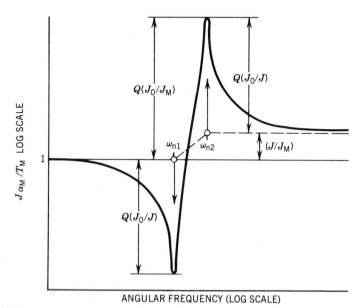

Figure 10.3-4 Approximate magnitude of structural resonant frequency response in velocity loop.

The value of the velocity-loop response at the peak frequency ω_{n2} is

Velocity loop at peak ($\omega = \omega_{n2}$):

$$\frac{J\alpha_m}{T_m} = \frac{\left[1 - (\omega_{n2}/\omega_{n1})^2\right] + j(\omega_{n2}/\omega_{n1})(1/Q)}{j(1/Q)}$$

$$= \frac{\omega_{n2}}{\omega_{n1}} - j\left[1 - \left(\frac{\omega_{n2}}{\omega_{n1}}\right)^2\right]Q \qquad (10.3\text{-}47)$$

Applying Eq 10.3-35 simplifies this to

$$\frac{J\alpha_m}{T_m} = \sqrt{J/J_m} - j\left(1 - \frac{J}{J_m}\right)Q$$

$$= \sqrt{J/J_m} + j\frac{J_o}{J_m}Q \qquad (10.3\text{-}48)$$

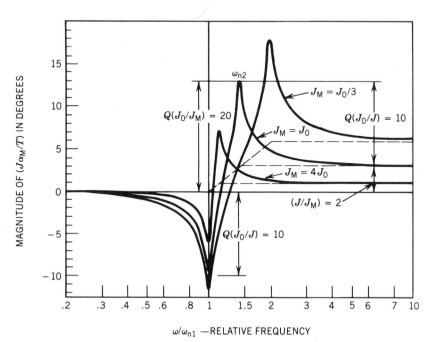

Figure 10.3-5 Magnitude of structural resonant frequency response in velocity loop for small J_m/J_o ratios ($Q = 20$).

The value of the velocity-loop frequency response at the dip frequency ω_{n1} is

Velocity loop at dip (ω_{n1}):

$$
\frac{J\alpha_m}{T_m} = \frac{j(1/Q)}{\left[1 - (\omega_{n1}/\omega_{n2})^2\right] + j(\omega_{n1}/\omega_{n2})(1/Q)}
$$

$$
= \frac{1}{(\omega_{n1}/\omega_{n2}) - jQ\left\{1 - (\omega_{n1}/\omega_{n2})^2\right\}} \tag{10.3-49}
$$

Applying Eq 10.3-35 simplifies this to

$$
\frac{J\alpha_m}{T_m} = \frac{1}{\sqrt{J_m/J} - jQ\{1 - (J_m/J)\}}
$$

$$
= \frac{1}{\sqrt{J_m/J} - jQ(J_o/J)} \tag{10.3-50}
$$

If Q is reasonably large, and if the motor inertia J_m is not very much larger than the output inertia J_o, the responses at the peak and dip frequencies given

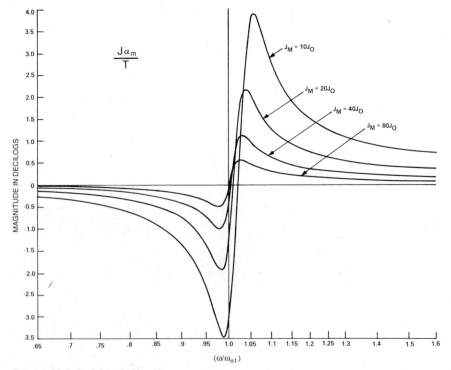

Figure 10.3-6 Magnitude of structural resonant frequency response in velocity loop for large J_m/J_o ratios $(Q = 20)$.

in Eqs 10.3-48, -50 can be approximated by

Velocity loop at resonant peak (ω_{n2}):

$$J\frac{\alpha_m}{T_m} \doteq j\frac{QJ_o}{J_m} \tag{10.3-51}$$

Velocity loop at resonant dip (ω_{n1}):

$$J\frac{\alpha_m}{T_m} \doteq j\frac{1}{Q(J_o/J)} \tag{10.3-52}$$

The frequency response corresponding to these approximations is shown in Fig 10.3-4. Note that the low-frequency dip departs from the low-frequency asymptote of the response by the factor $Q(J_o/J)$, which is the same as the departure of the high-frequency peak from the high-frequency asymptote.

Figures 10.3-5 and 10.3-6 are exact magnitude plots of the structural frequency responses of the velocity loop, for different values of the inertia

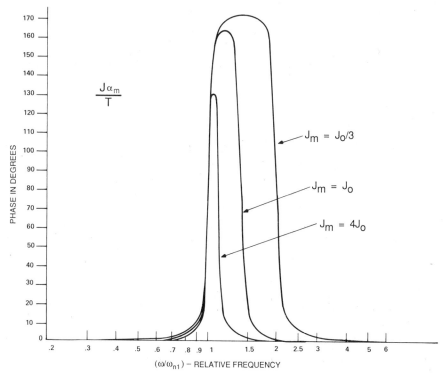

Figure 10.3-7 Phase of structural resonant frequency response in velocity loop for small J_m/J_o ratios ($Q = 20$).

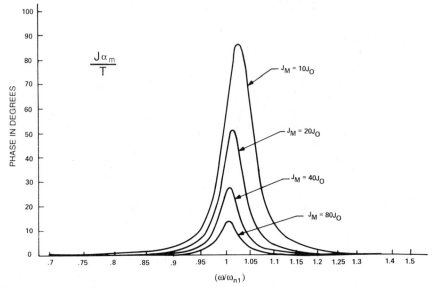

Figure 10.3-8 Phase of structural resonant frequency response in velocity loop for large J_m/J_o ratios ($Q = 20$).

ratio J_m/J_o, with Q equal to 20. Figures 10.3-7 and 10.3-8 are the corresponding phase plots. These show that when the reflected motor inertia J_m is much larger than the output inertia J_o, the effect of structural resonance in the velocity loop is small. The reason for this is that the inertia of the motor isolates the tachometer signal from the resonance effects of the load inertia and the gearing compliance.

10.4 STRUCTURAL DAMPING

This section develops some general relations pertaining to structural damping. It also investigates the question of what damping ratio (or Q) should be assumed when analyzing the structural transfer functions of a control system.

10.4.1 Equivalent Series and Parallel Representations of Damping

Let us analyze the damping of electric circuits in terms of Q. For a series RL circuit, consisting of an inductance L in series with a resistance R_s, the Q at any frequency ω is the ratio of the reactance ωL of the inductance at that frequency to the series resistance R_s:

$$Q = \omega L / R_s \qquad (10.4\text{-}1)$$

For a parallel RL circuit, consisting of an inductance L in parallel with a resistance R_p, the Q is the ratio of the resistance R_p to the inductive reactance ωL:

$$Q = R_p/\omega L \tag{10.4-2}$$

These two definitions of Q are consistent, because high-Q parallel and series RL circuits having the same inductance and same Q at a given frequency also have approximately the same impedance. To prove this, note that the impedance of an inductor L in parallel with a resistor R_p is

$$Z = \frac{R_p(j\omega L)}{R_p + j\omega L} = \frac{j\omega L}{1 + j(\omega L/R_p)} \tag{10.4-3}$$

Now $1/(1 + a)$ can be approximated as follows when $|a|$ is much less than unity:

$$\frac{1}{1 + a} \doteq 1 - a \qquad \text{for} \quad |a| \ll 1 \tag{10.4-4}$$

When the Q is much greater than unity, Eq 10.4-2 shows that $\omega L/R_p$ is much less than unity:

$$\omega L/R_p \ll 1 \qquad \text{for} \quad Q \gg 1 \tag{10.4-5}$$

Hence Eq 10.4-3 can be approximated by

$$Z \doteq j\omega L \left\{ 1 - j\frac{\omega L}{R_p} \right\} = j\omega L + \frac{(\omega L)^2}{R_p} \tag{10.4-6}$$

This impedance is equivalent to the same inductance L in series with the resistance R_s having the value

$$R_s = \frac{(\omega L)^2}{R_p} \tag{10.4-7}$$

Combining Eqs 10.4-1, -7 shows that the two expressions for Q in Eqs 10.4-1, -2 are equivalent.

Thus, over a frequency band where the Q of the circuit is high, a resistance R_s in series with an inductance L is approximately equivalent to a resistance R_p in parallel with the inductance L, where R_p and R_s are related by Eq 10.4-7. If the parallel resistance R_p is a constant, independent of frequency, the equivalent series resistance R_s varies as the square of the frequency; and if the series resistance R_s is a constant, the equivalent parallel resistance R_p varies as the square of the frequency.

The winding resistance of an inductor acts like a constant series resistance, independent of frequency. However, because of core losses, the total equivalent series resistance of an inductor is greater than the winding resistance, and this total equivalent series resistance varies somewhat with frequency.

The resistors R_s and R_p in diagrams c and d of Fig 10.3-1 are used to model the effect of damping in the gear train. Except when a shock absorber (or damper) is added to a mechanical system, the structural damping is generally nonlinear hysteresis damping. With hysteresis damping, the energy dissipated per cycle is independent of frequency, and depends only on the amplitude of the motion. For this to occur, the Q of the circuit must be a constant, independent of frequency. It can be seen from Eqs 10.4-1, -2 that when Q is constant, the values of the equivalent series resistance R_s and parallel resistance R_p must both vary in proportion to the frequency ω.

The Q can be expressed in terms of the ratio of energy dissipated per cycle divided by the stored energy. If the current through an inductor L is sinusoidal, the peak energy stored in an inductor is

$$U_{\text{store}} = \tfrac{1}{2}L(I_{\text{Max}})^2 \qquad (10.4\text{-}8)$$

where I_{Max} is the maximum value of the sinusoidal current through the inductor, which is equal to the RMS value of the current multiplied by $\sqrt{2}$. If this inductor is in a resonant circuit, oscillating with a capacitor, the stored energy is transferred back and forth between the inductor and the capacitor. Therefore, the peak energy stored in the inductor is equal to the total energy stored in the inductor–capacitor circuit, which stays constant. The average power dissipated in a resistor R_s in series with the inductor L is

$$P_{\text{dis}} = \tfrac{1}{2}R_s(I_{\text{Max}})^2 \qquad (10.4\text{-}9)$$

Multiplying this by the period T of the sinusoid gives the energy dissipated per cycle, which is

$$U_{\text{dis}} = P_{\text{dis}}T = \frac{P_{\text{dis}}}{f} = 2\pi\frac{P_{\text{dis}}}{\omega} \qquad (10.4\text{-}10)$$

where $f = 1/T$ is the frequency of the sinusoid in hertz, and ω is the frequency in rad/sec. Combining Eqs 10.4-8 to -10 gives the following for the ratio of stored energy to energy dissipated per cycle:

$$\frac{U_{\text{store}}}{U_{\text{dis}}} = \frac{L}{(2\pi/\omega)R_s} = \frac{\omega L}{(2\pi)R_s} = \frac{Q}{2\pi} \qquad (10.4\text{-}11)$$

Therefore the Q can be expressed as

$$Q = 2\pi\frac{U_{\text{store}}}{U_{\text{dis}}} \qquad (10.4\text{-}12)$$

Thus the Q of an inductor, or the Q of a spring, is equal to 2π times the ratio: the stored energy divided by the energy dissipated per cycle.

The higher the Q of a resonant circuit, the lower is the damping ratio ζ of the transient response of that circuit. To relate Q to damping ratio ζ, consider a series RLC circuit. Its impedance is

$$Z = sL + R_s + \frac{1}{sC} = \frac{L}{s}\left(s^2 + s\frac{R_s}{L} + \frac{1}{LC}\right) \qquad (10.4\text{-}13)$$

This impedance can be expressed as follows in terms of the general quadratic parameters (the natural frequency ω_n and the damping ratio ζ):

$$Z = \frac{L}{s}\left[s^2 + 2\zeta\omega_n s + \left(\frac{s}{\omega_n}\right)^2\right] \qquad (10.4\text{-}14)$$

Comparing Eqs 10.4-13, -14 shows that

$$\omega_n^2 = 1/LC \qquad (10.4\text{-}15)$$

$$2\zeta\omega_n = R_s/L \qquad (10.4\text{-}16)$$

By Eq 10.4-1, the Q of the series circuit at the natural frequency ω_n is $\omega_n L/R_s$. Therefore, by Eq 10.4-16, Q is related as follows to the damping ratio ζ:

$$Q = \frac{\omega_n L}{R_s} = \frac{1}{2\zeta} \qquad (10.4\text{-}17)$$

This relation $Q = 1/2\zeta$ applies to any resonant circuit of high Q.

10.4.2 Structural-Damping Values for Practical Structures

A number of different parameters are commonly used to characterize structural damping. As shown by Lazan [10.1], these are defined as follows, along with the symbols generally applied to them:

ζ = *damping ratio* (or percentage of critical damping)
η = *loss factor* (or loss tangent)

$$\eta = 2\zeta \qquad (10.4\text{-}18)$$

Q = *amplification factor*

$$Q = \frac{1}{\eta} = \frac{1}{2\zeta} \qquad (10.4\text{-}19)$$

Ψ = *specific damping capacity*, the ratio of vibrational strain energy dissipated during one cycle of vibration, divided by the vibration strain energy at the beginning of the cycle (normally measured on solid cylinders stressed in torsion),

$$\Psi = 2\pi\eta = \pi\zeta = 2\pi/Q \qquad (10.4\text{-}20)$$

δ = *logarithmic decrement*, defined as follows in terms of the peak values of two successive cycles of the damped oscillation, labeled x_n and x_{n+1}:

$$\delta = \ln[x_n/x_{n+1}] \qquad (10.4\text{-}21)$$

For $Q > 5$, δ is approximated within 3% by

$$\delta \doteq \Delta x/\bar{x} \qquad (10.4\text{-}22)$$

where Δx is the difference in amplitudes of two successive peaks and $\bar{x}$ is the average of the two peaks. The parameter δ is related as follows to the other damping parameters:

$$\delta = 2\pi\zeta = \frac{\Psi}{2} = \pi\eta = \frac{\pi}{Q} \qquad (10.4\text{-}23)$$

Structural damping is generally measured by determining the logarithmic decrement δ of a damped oscillation. Damping can be included in calculations of the frequency response of a complicated structure by replacing Young's modulus E in the structural equations by the following complex quantity, which is a function of the loss factor η:

$$E \rightarrow E(1 - j\eta) \qquad (10.4\text{-}24)$$

Smithells [10.2] gives a list of the specific damping capacities Ψ measured on a number of commercial alloys at room temperature. This information is summarized in the first two columns of Table 10.4-1, which shows the minimum and maximum values of specific damping capacity Ψ for each class of alloy. The subsequent columns show the corresponding minimum and maximum values of the loss factor $\eta = \Psi/2\pi$ and amplification factor $Q = 1/\eta$.

Table 10.4-1 shows that the values of the amplification factor Q for the usual structural materials, steel and aluminum, are extremely high (125 to 6300). However, the effective values of Q actually measured on structures are generally much lower than these values. One reason for this is that appreciable structural damping can result from bending at the joints of riveted or bolted structures, where metal-to-metal rubbing occurs. Another reason is that com-

TABLE 10.4-1   Materials Damping Values for Commercial Alloys at Room Temperature*

Material	Specific Damping Capacity Ψ (Percent)		Loss Factor $\eta = \Psi/2\pi$ (Percent)		Amplification $Q = 1/\eta$	
	Min.	Max.	Min.	Max.	Min.	Max.
Cast irons	1.4	19.3	0.2	3.1	33	450
Steels	0.15	3.8	0.02	0.6	170	4200
Copper alloys	0.25	1.35	0.04	0.2	470	2500
Aluminum alloys	0.1	5.0	0.016	0.8	125	6300
Magnesium alloys	0.4	7.4	0.06	1.2	85	1600
Maganese alloys	21	42	3.3	6.7	15	30
Nickel alloys	9.4	26	1.5	4.1	24	67

*From Smithells [10.2].

plicated structures vibrate in many different modes, and nonlinear coupling between the modes transfers energy out of a vibrating mode.

In 1965, Robert E. Pike and the author [10.3] conducted a program for measuring the frequency responses of a hydraulic servo driving a simulated radar-antenna load. The simulated load was an I-beam (100 in. long, weighing 234 lb) pivoted at its center, and loaded with 450 lb of steel plate at each end. The beam was rotated by a pair of hydraulic linear actuators, driven by a hydraulic control valve. The angular rate was measured by a tachometer and a gyro.

Compliance in the hydraulic fluid produced a primary resonance at 6 Hz. This had an amplification (Q) of 20, the damping of which was established primarily by throttling in the control valve. By sensing hydraulic pressure, a pressure feedback loop was closed around the control valve, which eliminated this resonance. Other resonance dips and peaks occurred at 17 Hz and higher, and were due to resonance in the structure itself. The rate-of-change of phase in these resonance regions was consistent with a Q of 100. However, the maximum peaking of the magnitude response was about 20 : 1. Close to each peak (pole) there was a dip (zero) which partially canceled the peaking due to the pole.

One can generally assume that the actual frequency-response peaking measured on a structure will not exceed 25 : 1. However, this does not necessarily mean that the maximum value of the true Q is 25. Usually the structure is subject to multiple modes which interact to keep the amplification for a single mode from exceeding 25 : 1, even though the actual Q may be higher.

On the other hand, higher resonance peaks are sometimes encountered. The author experienced a structural mode with a measured Q of 170 in the design of the flexure-supported stage of a step-and-repeat camera for making integrated circuits. This is discussed in Chapter 15.

10.4.3 Damping in a Structure Driven by a Gear Train

10.4.3.1 Analytical Model. For a structure driven through a gear train, the friction in the gear train is a strong source of structural damping. Appendix I presents an analysis of the energy dissipated in a gear train, which assumes the mechanical gear-train model shown in Fig 10.4-1. This model normalizes the gear ratio to unity, just as was done in Section 10.3. The compliance and friction are assumed to be distributed evenly along the gear train. The gear-train compliance is separated into four elements of equal compliance, and the friction torque is separated into three equal friction torques, designated ΔT, which are represented in the diagram by dampers.

At resonance, the structural compliance is placed under stress during one quarter cycle of the oscillation, as energy is transformed from kinetic energy stored in the inertia to stress energy stored in the compliance. During the next quarter cycle, energy is released from the compliance and transferred back into momentum. Because of friction, energy is dissipated when the compliance is placed under stress, and an equal amount of energy is dissipated when the energy is released from the compliance. (This assumes that the oscillation is forced, so that the oscillation amplitude remains constant.) Thus, the energy that is dissipated, when the compliance is placed under stress, is $\frac{1}{4}$ of the total energy dissipated during an oscillation cycle.

The analysis in Appendix I assumes that the structure is resonating at the frequency of null of the tachometer-feedback loop (ω_{n1} in Fig 10.3-2). Motion of the motor is very small at this frequency, and so the analysis assumes that the motor angular displacement Θ_m during the oscillation is zero. In the model, energy is transferred back and forth between the load inertia and the gear-train compliance.

The efficiency of the gear train is designated η. The torque T_m, applied by the gear train to the motor shaft, is related as follows to the load torque T_L, applied by the load inertia to the other end of the gear train:

$$T_m = \eta T_L \qquad (10.4\text{-}25)$$

Hence, the friction torque dissipated in the gear train is

$$T_f = T_L - T_m = (1 - \eta)T_L \qquad (10.4\text{-}26)$$

The model assumes that the friction is separated into three equal torques, designated ΔT, equal to

$$\Delta T = \frac{T_f}{3} = (1 - \eta)\frac{T_L}{3} \qquad (10.4\text{-}27)$$

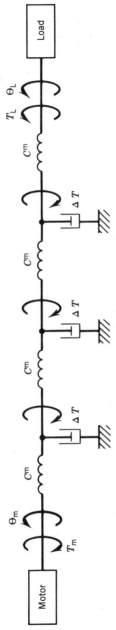

Figure 10.4-1 Assumed mechanical model of gear train, including effects of gear-train compliances and friction torques.

225

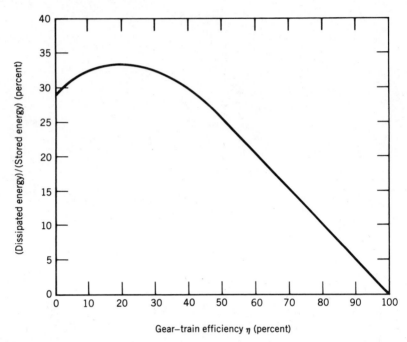

Figure 10.4-2   Ratio of dissipated energy to stored energy for compliant gear train.

Appendix I derives the following formula for the energy dissipated by friction (U_d), divided by the energy stored in the gear-train compliance (U_{st}), during the quarter cycle in which the gear train compliance is placed under stress:

$$\frac{U_d}{U_{st}} = \frac{(1 - \eta)(4 + 14\eta)}{14 + 8\eta + 14\eta^2} \tag{10.4-28}$$

where η is the gear-train efficiency, defined by Eq 10.4-25. A plot of Eq 10.4-28 is shown in Fig 10.4-2.

As shown in Section 10.4.2, the specific damping capacity Ψ is the energy dissipated per cycle of an oscillation divided by the stored energy. This is 4 times the ratio of Eq 10.4-28:

$$\Psi = 4\frac{U_d}{U_{st}} \tag{10.4-29}$$

The amplification factor Q is related to specific damping capacity Ψ in Eq 10.4-20, which gives

$$Q = 2\pi/\Psi \tag{10.4-30}$$

Combining Eqs 10.4-29, -30 gives the amplification factor Q:

$$Q = \frac{\pi}{2} \frac{U_{st}}{U_d}$$

(10.4-31)

The data for U_d / U_{st} given in Fig 10.4-2 are substituted into Eq 10.4-31 to obtain the plot of Fig 10.4-3, which shows the amplification factor Q as a function of the efficiency η of the gear train.

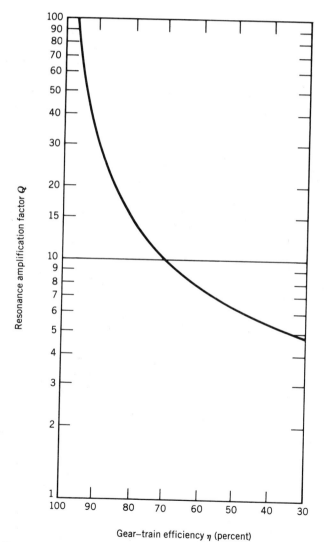

Figure 10.4-3 Plot of structural-resonance amplification factor Q versus the efficiency η of a gear train.

A typical gear-train efficiency for a well-designed antenna drive is 80%. According to Fig 10.4-3, the corresponding amplification factor Q of the oscillation is 15. Thus, it would appear that a Q of 15 should be a reasonable estimate for a gear-train driven antenna.

The bearings of the antenna add additional damping to the structural resonance, which should reduce the amplification factor Q below 15. On the other hand, the gear-train model analyzed in Appendix I, on which Fig 10.4-3 is based, is merely a first-cut estimate for gear-train friction, and so can be appreciably in error.

10.4.3.2 Experimental Data on Structural Damping.
Let us compare this analytical result with some experimental data. In Fig 10.4-4, curve ① shows the structural frequency response experienced by the tachometer velocity loop of an antenna servo driving a 46-ft tracking, telemetry, and control (TT and C) antenna. This is a plot of the magnitude of the frequency response from the motor current (or motor torque) to the motor angular velocity. The response was supplied by Denny D. Pidhayny, and was measured primarily to determine the lowest structural resonance dip, which occurs at 4.5 Hz. (This dip is followed by a resonant peak at 7.5 Hz.) This is a typical test for measurement of structural resonant frequency.

Curve ① in Fig 10.4-4 was not measured to determine damping accurately, and so gives poor amplitude resolution in the vicinity of the dip at 4.5 Hz. Besides, harmonics of the frequency response tend to fill in the null. Consequently, the true null at 4.5 Hz may be considerably deeper than indicated by the data.

Curves ②, ③ of Fig 10.4-4 are theoretical plots of the transfer function between motor torque and motor velocity, which were obtained from Eq 10.3-42, for $Q = 5$ and 10. The values for $\omega_{n1}/2\pi$ and $\omega_{n2}/2\pi$ in Eq 10.3-42 were set equal to 4.5 and 7.5 Hz, the corresponding values for the measured data. Equation 10.3-42 gives the transfer function from the motor torque T_m to the motor acceleration α_m. This transfer function is divided by $j\omega$ to obtain the transfer function from motor torque to motor velocity, which is plotted. The theoretical curves ②, ③ have been arbitrarily displaced vertically so that they agree with the measured data at 3 Hz. The dashed line ④ is the corresponding frequency response from motor torque to motor angular velocity for a pure inertial load. The measured data did not have an absolute scale, and so absolute signal-level comparisons cannot be made.

Figure 10.4-4 shows that a Q of at least 10 is required to characterize the steep magnitude slope near the dip at 4.5 Hz. The ratio from the peak at 7.5 Hz to the null at 4.5 Hz is only 13.7 dg for the measured data, whereas it is 20.4 dg for the $Q = 10$ curve and 14.5 dg for the $Q = 5$ curve. On the other hand, a more accurate measurement of the null at 4.5 Hz would probably have increased this ratio by at least 3 dg. Thus, the ratio from peak to null should correspond to an amplification factor Q of at least 7.

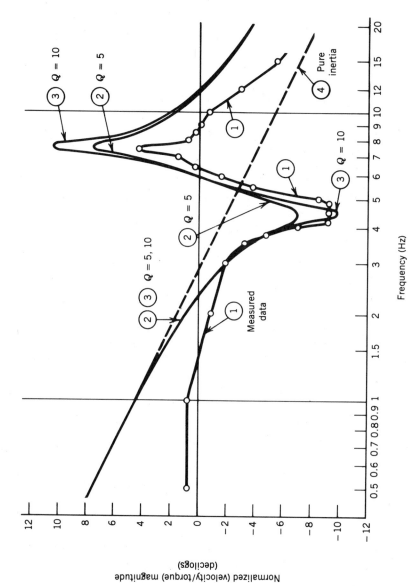

Figure 10.4-4 Frequency response from motor current (or torque) to motor velocity for azimuth loop of 46-ft antenna.

The measured frequency response of curve ① in Fig 10.4-4 does not follow the theoretical curve, but instead approaches a constant velocity amplitude at low frequencies. This indicated strong damping at high velocity amplitudes. This might be caused by saturation in the motor, or by strong windage torques in the motor and antenna. At a lower amplitude, the damping may be considerably lower. Hence, at a lower amplitude, the low-frequency response might follow the theoretical curve, and the high-frequency peak at 7.5 Hz might be appreciably higher, indicating a higher Q.

The data in Fig 10.4-4 give the structural frequency response measured on the azimuth axis of the 46-ft antenna, at zero elevation angle. Figure 10.4-5 shows the corresponding frequency response measured on the elevation axis. The dashed line shows for reference the corresponding frequency response for a pure inertial load. The vertical location of the dashed line is arbitrary. This elevation-axis frequency response is much more complicated than that for the azimuth axis in Fig 10.4-4. The elevation axis has resonant nulls at 4.4 and 8 Hz, and resonant peaks at 5 and 10 Hz. Clearly a more complicated structural model is required to characterize this frequency response than the one developed in this chapter. Nevertheless, the model of this chapter can still provide a first-order analysis of the elevation loop.

The preceeding illustrates the fact that structural resonance is a very complicated phenomenon. There can be many modes of structural resonance, and the damping effects associated with these nodes can be highly nonlinear. Besides, it is difficult to make good structural-resonance measurements on large structures, because the frequency-response tests can exert strong stresses in the structure.

Undoubtedly, various control-system engineers have performed measurements of structural dynamics that provide more definitive data than have been presented here. However, this technical information is not available in the open literature, partly for proprietary reasons, but primarily because the control field has shown little interest in this very critical issue.

A few comments from specialists in the control field may be helpful. Several years ago, the author learned from an engineer involved in antenna-servo simulation that he generally assumed a Q of 20 in his structural simulations, and found that this was consistent with experimental results. Hobart Cress, a mechanical engineer at Aerospace Corp, who has had extensive mechanical-engineering experience in antenna design, generally has assumed a damping ratio ζ of 0.02 in his analyses, which corresponds to a Q of 25. On the other hand, Denny D. Pidhayny, an electrical control-systems engineer, has generally assumed that a damping ratio of 0.08 was reasonable, which corresponds to a Q of 6.25.

Hobart Cress also provided the following comments concerning structural friction. Breakaway friction is greater than running Coulomb friction. There is also a significant component of viscous friction in bearings, and at high speeds the viscous bearing friction typically is approximately equal to the Coulomb, or static, friction. Some bearings have much more static friction than others.

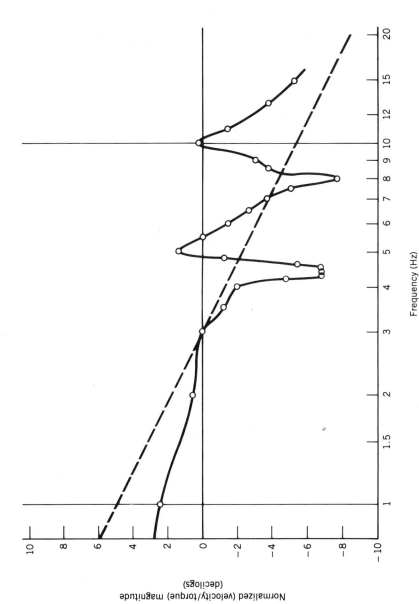

Figure 10.4-5 Frequency response from motor current (or torque) to motor velocity for elevation loop of 46-ft antenna.

231

Tapered roller bearings are generally undesirable for accurate control applications, because the shoulders on the bearings result in sliding contact, rather than rolling contact, and hence produce high Coulomb friction. Since wind resistance is proportional to the square of the wind speed, it is usually negligible at low antenna angular velocity, but can be appreciable at high velocity.

Charles Fassnacht is a mechanical engineer at Datron, who has designed many servo-controlled antennas. He has found a Q of 20 to be a good general conservative value for the design of antenna structures. Somewhat higher values of Q can sometimes be experienced for modes associated with isolated structural members. For modes that are damped by a gear train, the Q will be less than 20. Nevertheless he feels that it is still desirable to assume $Q = 20$ in the design analysis.

This information concerning damping pertains to antenna control systems. However, the same principles should apply at least approximately to any mechanical control system (or servo) that uses a gear train in the drive. (An example of particular interest today is the servo-controlled arm of a mechanical robot.) Unless one has information to the contrary, it seems prudent to assume a Q of at least 10 in the analysis of any gear-driven structure and a Q of 20 should generally give a conservative estimate.

10.5 LIMITATIONS OF GAIN CROSSOVER FREQUENCY CAUSED BY STRUCTURAL RESONANCE

10.5.1   The Effect of Structural Resonance on the Position-Loop Frequency Response

To illustrate the effect of the structural frequency response on a control system, let us design the loop transfer function of a position loop that controls a structural resonant load. To simplify the discussion, the reflected motor inertia is assumed to be much larger than the output inertia, so that the velocity loop is able to achieve close to ideal performance.

With this assumption, to first approximation the effect of the velocity loop can be ignored in designing the position loop. The loop transfer function of the position loop, in its simplest form, is

$$G_{(p)} = \frac{\omega_{cp}}{s\left[1 + (1/Q)(s/\omega_{n2}) + (s/\omega_{n2})^2\right]} \qquad (10.5\text{-}1)$$

A sketch of the magnitude plot of $G_{(p)}$ is shown in Fig 10.5-1. This has 180° of phase lag at the structural resonant frequency ω_{n2}, and so ω_{n2} is the phase crossover frequency. The loop gain at the phase crossover frequency is then

$$\left|G_{(p)}[\omega_{n2}]\right| = Q\frac{\omega_{cp}}{\omega_{n2}} \qquad (10.5\text{-}2)$$

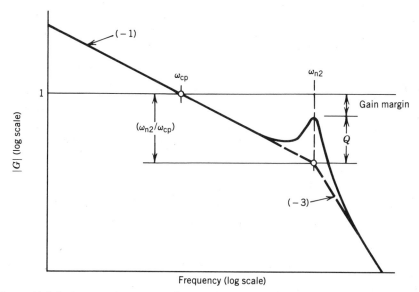

Figure 10.5-1 Loop-gain response for position loop with structural resonance and no resonance compensation.

The reciprocal of this is the gain margin, which is

$$\text{gain margin} = \frac{1}{\left|G_{(p)}[\omega_{n2}]\right|} = \frac{\omega_{n2}}{Q\omega_{cp}} \qquad (10.5\text{-}3)$$

Assume that $Q = 20$, and that the minimum allowable gain margin is 2.5. The following is the maximum allowable value for the gain crossover frequency:

$$\omega_{cp} \leq \tfrac{1}{50}\omega_{n2} \qquad (10.5\text{-}4)$$

Thus, with this approach, the gain crossover frequency must be a factor of 50 below the structural resonant frequency. A much higher gain crossover frequency can be achieved by adding a lowpass filter to the feedback loop. The lowpass filter improves the response by (1) attenuating the resonant peak and (2) by adding phase lag at the frequency of the resonant peak, so that the peak occurs well beyond the phase crossover frequency.

Let us consider for example the loop transfer function shown by the magnitude plot of Fig 10.5-2 and the phase plot of Fig 10.5-3. The loop transfer function is

$$G_{(p)} = \frac{\omega_{cp}\left(1 + \dfrac{\omega_{ip}}{s}\right)}{s\left(1 + \dfrac{s}{\omega_f}\right)^2 \left[1 + \dfrac{1}{Q}\dfrac{s}{\omega_{n2}} + \left(\dfrac{s}{\omega_{n2}}\right)^2\right]} \qquad (10.5\text{-}5)$$

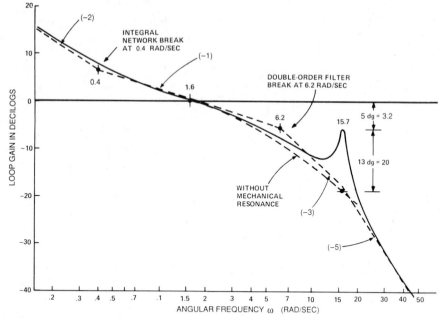

Figure 10.5-2 Loop-gain response for position loop with structural resonance and double-order lowpass filter ($Q = 20$).

The structural resonant frequency is 2.5 Hz, which corresponds to $\omega_{n2} = 15.7$ rad/sec. The gain crossover frequency ω_{cp} is set approximately a decade below this resonant frequency (at 1.6 rad/sec), and the integral break frequency ω_{ip} is set a factor of 4 below the gain crossover frequency ω_{cp} (at 0.4 rad/sec). The parameters of this frequency response are

$$\omega_{cp} = 1.6 \text{ rad/sec} \tag{10.5-6}$$

$$\omega_{ip} = 0.4 \text{ rad/sec} \tag{10.5-7}$$

$$\omega_f = 6.2 \text{ rad/sec} \tag{10.5-8}$$

$$\omega_{n2} = 15.7 \text{ rad/sec} \tag{10.5-9}$$

$$Q = 1/2\zeta = 20 \tag{10.5-10}$$

The Nichols-chart plot of this transfer function is shown in Fig 10.5-4. This plot shows that the resonant peak at 15.7 rad/sec occurs at a point well beyond 180° of phase lag, and the loop has excellent gain margin. By readjusting the parameters, a higher gain crossover frequency can be achieved with good stability.

This example is presented to give insight into the design of feedback loops having structural resonant peaks. Frequency-response analysis helps to pro-

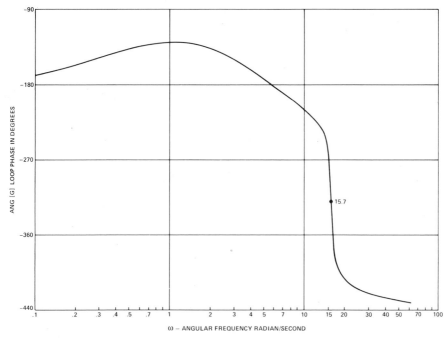

Figure 10.5-3   Loop-phase response for position loop with structural resonance and double-order lowpass filter ($Q = 20$).

vide the understanding needed to configure the loop design. However, to optimize the design parameters, it is often simpler to use simulation, as described in Section 10.6.

10.5.2   Limitations on Gain Crossover Frequency due to Structural Resonance in Practical Servo-Driven Antennas

To obtain estimates of the limitation that structural resonance places on the dynamic performance of a control system, let us consider two practical examples: the 46-ft antenna discussed in Section 10.4.3, and the 60-ft Advent antenna, the control system for which was designed by the author.

10.5.2.1   46-ft Antenna. Figure 10.5-5 shows the measured step responses, for successive positive and negative steps, made on the azimuth velocity loop of the 46-ft antenna. (These should be compared with the simulated response presented later in Section 10.6, Fig 10.6-6.) Unfortunately, the time scale is very compressed, and so the values of delay time T_d for these step responses cannot be measured very accurately. This problem is complicated further by the secondary oscillation in the response. The secondary oscillation should be averaged to obtain the delay time corresponding to the main component of the

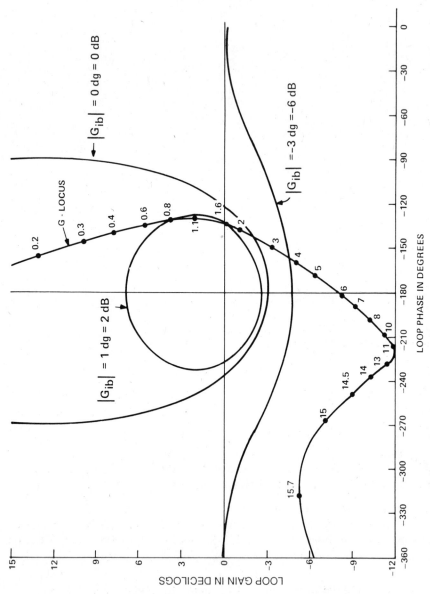

Figure 10.5-4 Nichols-chart plot for loop transfer function of position loop corresponding to plots of Figs 10.5-2, -3.

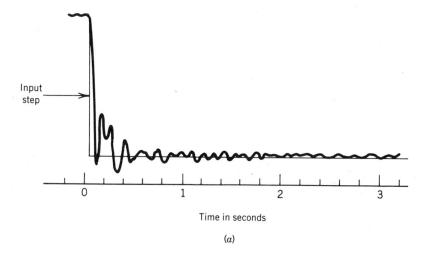

(a)

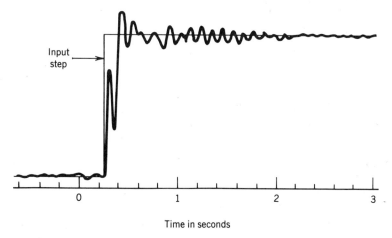

Time in seconds

(b)

Figure 10.5-5   Step responses of velocity loop for 46-ft antenna.

response. The author estimates from these responses a delay time T_d of 0.06 sec. The corresponding estimate for the gain crossover frequency of the velocity loop is

$$\omega_{cv} \doteq \frac{1}{T_d} = \frac{1}{0.06 \text{ sec}}$$

$$= 16.7 \text{ sec}^{-1} = 2\pi(2.65 \text{ Hz}) \qquad (10.5\text{-}11)$$

As was shown in Fig 10.4-4, for this antenna the lowest structural resonant

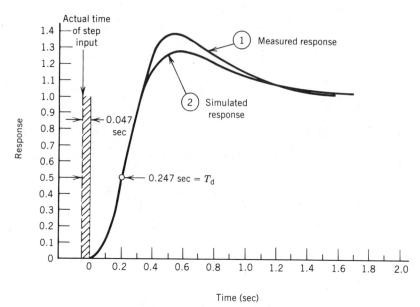

Figure 10.5-6 Measured step response of position loop for 46-ft antenna, compared with simulated response.

frequency f_n in azimuth is 4.5 Hz. The ratio of the resonant frequency to the gain crossover frequency is

$$\frac{\omega_n}{\omega_{cv}} = \frac{f_n}{f_{cv}} = \frac{4.5 \text{ Hz}}{2.65 \text{ Hz}} = 1.8 \qquad (10.5\text{-}12)$$

In Fig 10.5-6, curve ① shows the measured step response of the azimuth position loop of the 46-ft antenna, and curve ② shows the corresponding response derived from simulation. This measured response was not obtained from a direct recording, and so its zero time reference may well be in error. (Unfortunately, it is common practice in the control field to ignore the time of the input step when recording step-response data.)

The position loop has a double integration at low frequency, and so has a velocity error coefficient equal to zero (an infinite velocity constant). Figure 10.5-7 shows the unit step response of such a loop, which indicates that the area of the overshoot must be equal to the area of $(1 - x_b)$ up to the first point where x_b exceeds unity. (This issue was explained in Chapter 5, Fig 5.2-8.) When the principle of Fig 10.5-7 is applied to Fig 10.5-6, it can be shown that the input step for the measured response must preceed the $t = 0$ axis by 0.047 sec.

With the corrected time scale, the delay time T_d of the measured step response of Fig 10.5-6 is 0.247 sec. Hence, the estimated value for the

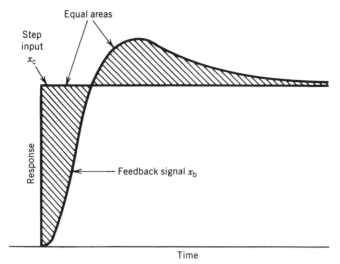

Figure 10.5-7 General relation for step response of loop with low-frequency double integration.

position-loop gain crossover frequency is

$$\omega_{cp} \doteq \frac{1}{T_d} = \frac{1}{0.247 \text{ sec}} = 4.0 \text{ sec}^{-1} \qquad (10.5\text{-}13)$$

The information supplied with this step response indicates that the acceleration constant K_2 of the position loop was

$$K_2 = 7.0 \text{ sec}^{-2} \qquad (10.5\text{-}14)$$

As was explained in Chapter 3, Section 3.2, a loop with a low-frequency double integration has a transfer function of the following form at frequencies below ω_c:

$$G_p = \frac{\omega_{cp}\left(1 + \omega_{ip}/s\right)}{s} = \frac{\omega_{cp}\omega_{ip}\left(1 + s/\omega_{ip}\right)}{s^2}$$

$$= \frac{K_2\left(1 + s/\omega_{ip}\right)}{s^2} \qquad (10.5\text{-}15)$$

Hence, the acceleration constant K_2 is equal to

$$K_2 = \omega_{ip}\omega_{cp} \qquad (10.5\text{-}16)$$

Combining Eqs 10.5-13, -14, -16 gives for ω_{ip}

$$\omega_{ip} = \frac{K_2}{\omega_{cp}} = \frac{7.0 \text{ sec}^{-2}}{4.0 \text{ sec}^{-1}}$$

$$= 1.75 \text{ sec}^{-1} = \omega_{cp}/2.3 \qquad (10.5\text{-}17)$$

The measured response in Fig 10.5-6 has excessive overshoot. However, if ω_{ip} is lowered by a moderate amount, the overshoot can probably be reduced below 30% without reducing the gain crossover frequency ω_{cp} appreciably.

By Eq 10.5-13, the ratio of the structural resonant frequency (4.5 Hz) to the gain crossover frequency of the position loop is

$$\frac{f_n}{f_{cp}} = \frac{2\pi f_n}{\omega_{cp}} = \frac{2\pi(4.5 \text{ Hz})}{4.0 \text{ sec}^{-1}} = 7.1 \qquad (10.5\text{-}18)$$

To measure the delay time T_d, the transient response must be recorded when the feedback loop is operating in its linear range; otherwise saturation can appreciably increase the delay time of the response. A convenient way to determine the saturation limits of a servo is to command a large step of velocity. The azimuth velocity loop of this 46-ft antenna was given successive step commands of velocity of ± 15 deg/sec. During the resultant transient, the antenna decelerated from -9 deg/sec to zero velocity in 1.20 sec, and then accelerated from zero velocity to $+9$ deg/sec in 1.90 sec. The motor current was saturated at its maximum value during this period. Hence the maximum values of deceleration (dec) and acceleration (acc) of the azimuth servo are

$$\text{Max[dec]} = \frac{9 \text{ deg/sec}}{1.20 \text{ sec}} = 12.5 \text{ deg/sec}^2 \qquad (10.5\text{-}19)$$

$$\text{Max[acc]} = \frac{9 \text{ deg/sec}}{1.90 \text{ sec}} = 7.9 \text{ deg/sec}^2 \qquad (10.5\text{-}20)$$

The mean of these two values is

$$\text{Mean}[\alpha_{\text{Max}}] = \tfrac{1}{2}(12.5 + 7.9) \text{ deg/sec}^2$$

$$= 10.2 \text{ deg/sec}^2 \qquad (10.5\text{-}21)$$

During deceleration, the friction torque increases the acceleration, whereas during acceleration it decreases the acceleration. Hence, the acceleration component due to friction, which is the ratio of friction torque to inertia, is equal to

$$\frac{T_{f(c)}}{J_{(c)}} = \tfrac{1}{2}(12.5 - 7.9) \text{ deg/sec}^2$$

$$= 2.3 \text{ deg/sec}^2 \qquad (10.5\text{-}22)$$

The amplitude of the velocity step that just begins to saturate the servo is approximately

$$\Delta\Omega_{sat} = \frac{\text{mean}[\alpha_{Max}]}{\omega_{cv}} = \frac{10.2 \text{ deg/sec}^2}{16.7 \text{ sec}^{-1}}$$

$$= 0.61 \text{ deg/sec} \qquad (10.5\text{-}23)$$

The amplitude of the step of position that just begins to saturate the servo is

$$\Delta\Theta_{sat} = \frac{\Delta\Omega_{sat}}{\omega_{cp}} = \frac{0.61 \text{ deg/sec}}{4.0 \text{ sec}^{-1}} = 0.15° \qquad (10.5\text{-}24)$$

Hence, to obtain linear transient responses, the velocity loop should be tested with a step of velocity not exceeding 0.61 deg/sec, and the position loop should be tested with a step of position not exceeding 0.15°.

The accuracy of the azimuth servo in response to friction can be estimated from the data that have been given. As was shown in Chapter 6, Eq 6.3-12, the peak error caused by friction is approximately

$$\text{Max}|\Theta_e| = \frac{T_{f(c)}}{J_{(c)}\omega_{cv}\omega_{cp}} \qquad (10.5\text{-}25)$$

The ratio $T_{f(c)}/J_{(c)}$ is given in Eq 10.5-22, and the values for ω_{cv}, ω_{cp} are given in Eqs 10.5-11, -13. Substituting these into Eq 10.5-25 gives

$$\text{Max}|\Theta_e| = \frac{2.3 \text{ deg/sec}^2}{(16.7 \text{ sec}^{-1})(4.0 \text{ sec}^{-1})} = 0.034° \qquad (10.5\text{-}26)$$

This indicates that the antenna azimuth servo should have a peak error due to friction of approximately ±0.034°.

10.5.2.2 60-Foot Advent Antenna.

Now let us consider the 60-ft Advent antenna, which operated at 2 GHz and 9 GHz. Two models of this antenna were installed by GTE Sylvania Electric Products, Inc., in 1962, at Fort Dix, New Jersey and Camp Roberts, California, for the U.S. Army Signal Research and Development Laboratory. These antennas were to be used for communication and control of low-flying communication satellites, and were designed to have very high accuracy, and high acceleration and velocity, to expedite station-keeping. However, after the antennas were constructed, satellites were placed in high-altitude synchronous orbits, and so the high-performance capabilities of the antenna servo were not needed.

During slew, the antenna could accelerate at 8 deg/sec² to a maximum velocity of 8 deg/sec. To achieve very high positional accuracy, a flywheel was coupled through a clutch to each motor shaft, and was declutched during slew. The flywheel had 3 times the motor inertia, and so increased the effective motor inertia by a factor of 4. With the flywheel coupled to the motor, the gain of the rate loop could be increased, to allow greater stiffness, and hence reduced error from friction and wind gusts.

Two motors were used in each axis (azimuth and elevation), which drove the antenna through separate gear trains. Gear backlash was eliminated by applying a preload torque between the motors.

With such high motor inertia, a common-mode oscillation could occur between the two motor inertias of a given axis, through the compliance of the interconnecting gear trains. To avoid such an oscillation, the two motors were mounted back-to-back, and their rear shafts were coupled together through a hydraulic damper.

The motors were 40-horsepower (hp) DC motors, which were overdriving to 80 hp during transient conditions. Each motor was controlled by a metadyne, which is a two-stage rotary amplifier with current feedback. To achieve high dynamic performance, the metadyne was designed to have comparatively low power gain. The control field of the metadyne was driven by a high-performance 1-kilowatt magnetic amplifier. The metadyne could deliver twice rated motor current, with 60° of phase lag in the current feedback loop at 2.5 Hz, and no peaking in the frequency response.

The following performance data were obtained from the test report on the Altair antenna [10.4]. The antenna had a position resolution (peak friction error) of $\pm 0.002°$. (The specification was $\pm 0.0075°$.) Breakaway friction was

TABLE 10.5-1   Dynamic Parameters Measured in August 1982 on Advent Antenna, Installed at Fort Dix, New Jersey

Parameter	Azimuth		Elevation	
Flywheels:	Off	On	Off	On
Structural resonant frequency f_n (Hz)	3.5	3.2	2.8	2.8
Velocity Loop:				
f_{cv} (Hz)	1.95	1.90	1.85	2.2
Phase margin (deg)	63	63	63	64
Gain margin	5.0	6.3	?	?
Manual position loop:				
f_{cp} (Hz)	0.8	0.8	0.55	0.55
Phase margin (deg)	65	65	70	75
Gain margin	2.8	2.8	2.8	3.2
Track position loop:				
f_{cp} (Hz)		0.83		
Phase margin (deg)		58		
Gain margin		2.5		
Frequency ratios:				
f_n / f_{cv}	1.8	1.7	1.5	1.3
f_n / f_{cp}	4.4	4	5.1	5.1

21,000 ft-lb in azimuth, and 36,000 ft-lb in elevation. High winds were not encountered during the tests. To simulate the effect of a 45-mph wind load in azimuth, the preload on the gear train was increased to the equivalent torque (180,000 ft-lb). At this high preload, the peak error increased to $\pm 0.006°$.

Table 10.5-1 shows dynamic parameters measured on the Advent azimuth and elevation servos. The structural natural frequency was found by measuring the sharp dip in the velocity-loop frequency response. As indicated by the phase-margin and gain-margin values, the loops had good stability. Peak overshoot of the step responses of the position loops (manual and tracking) did not exceed 30%.

Thus, the Advent antenna achieved an f_n/f_{cp} ratio of 4 to 5.1. As shown in Eq 10.5-18, the 46-ft antenna achieved an f_n/f_{cp} ratio of 7.1. Since the Advent antenna had a sophisticated drive, it should not be used as a typical example. The ratio 7.1 for the 46-ft antenna is based on an estimate, and so may be somewhat in error. A reasonable general estimate for the f_n/f_{cp} ratio for a well-designed antenna servo appears to be

$$f_n/f_{cp} = 8 \quad \text{(general estimate)} \tag{10.5-27}$$

10.6   COMPUTER SIMULATION OF STRUCTURAL RESONANCE

10.6.1   Signal-Flow Diagram of Structural Resonance for Computer Simulation

Diagram a of Fig 10.6-1 shows the equivalent electric circuit for the structural resonance model shown earlier in Fig 10.3-1d. The basic signal-flow diagram of this circuit for computer simulation is shown in diagram b. All of the dynamic elements are expressed in the form of integrators. Thus, the voltage across a capacitor is calculated from the current through the capaitor, and the current in an inductor is calculated from the voltage across the inductor. Diagram c shows the equivalent signal-flow diagram for the mechanical circuit. This is obtained by making the following substitutions in the electric signal-flow diagram b:

$$E_m = \Omega_m \qquad \text{(motor angular velocity)} \tag{10.6-1}$$

$$E_{gt} = \Omega_g \qquad \text{(differential angular}$$
$$\text{velocity across gear train)} \tag{10.6-2}$$

$$1/L = K_g \qquad \text{(stiffness of gear train)} \tag{10.6-3}$$

$$C_o = J_o \qquad \text{(output inertia)} \tag{10.6-4}$$

$$C_m = J_m \qquad \text{(motor inertia)} \tag{10.6-5}$$

$$I_m = T_m \qquad \text{(motor torque)} \tag{10.6-6}$$

$$I_o = T_o \qquad \text{(output torque)} \tag{10.6-7}$$

$$R_s = \frac{1}{B_s} = \frac{\omega}{QK_g} \qquad \text{(gear-train damping)} \tag{10.6-8}$$

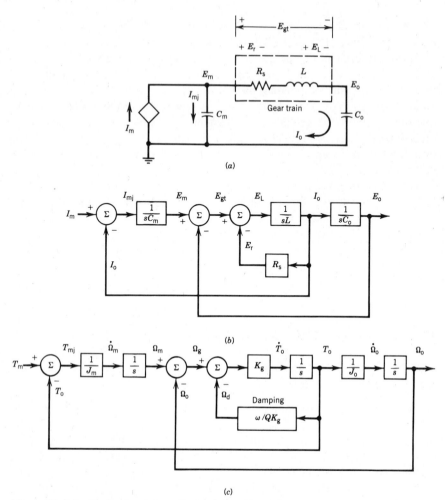

Figure 10.6-1 Basic signal-flow diagram for structural resonance model of Fig 10.3-1: (*a*) analogous electric circuit; (*b*) signal-flow diagram of circuit; (*c*) signal-flow diagram of structural model.

The expression for B_s given in Eq 10.3-24 of Section 10.3 is used to relate R_s to the spring constant K_g of the gear train and the Q of the damping in the gear train.

Figure 10.6-1*c* could be used directly for simulation. However, a much more convenient simulation model can be obtained by simplifying the diagram in the following manner. The signal-flow diagram of Fig 10.6-1*c* is modified to form Fig 10.6-2*a*. The feedback path that characterizes gear-train damping is moved after the K_g block, by multiplying that path by K_g, thereby changing the feedback element from $\omega/K_g Q$ to ω/Q. Since Ω_o is equal to $(1/sJ_o)T_o$, the output torque T_o can be derived from Ω_o in accordance with

$$T_o = (sJ_o)\Omega_o \tag{10.6-9}$$

Thus, in Fig 10.6-2a T_o is derived by feeding the signal Ω_o through the block sJ_o. The velocity Ω_g across the gearing is equal to the motor velocity minus the output velocity:

$$\Omega_g = \Omega_m - \Omega_o \tag{10.6-10}$$

Solving for Ω_m gives

$$\Omega_m = \Omega_g + \Omega_o \tag{10.6-11}$$

Hence Ω_m can be derived, as shown, by adding Ω_g and Ω_o.

The next simplification is to feed back the T_o signal to a point after the $1/sJ_m$ block. As shown by the dashed line, this is achieved by multiplying the T_o signal by $1/sJ_m$ and adding the result to the Ω_o signal. This eliminates the Ω_m signal. However, as shown in Eq 10.6-11, an alternative computation of the Ω_m signal is obtained by adding Ω_g and Ω_o.

Diagram a of Fig 10.6-2 is simplified to form diagram b. The block $1/J_o$ is moved across the gear-damping feedback loop, and combined with the K_g block to produce the block K_g/J_o. By Eq 10.3-31 of Section 10.3, this is equal to ω_{n1}^2:

$$\omega_{n1}^2 = K_g/J_o \tag{10.6-12}$$

The transfer function of the T_o feedback path of diagram a, from Ω_o to the point where it is added to Ω_o (the dashed path), is equal to

$$sJ_o \frac{1}{sJ_m} = \frac{J_o}{J_m} \tag{10.6-13}$$

Hence, as shown in diagram b, the net transfer function in the feedback path, from the Ω_o signal to the summation point, is $(1 + J_o/J_m)$, which is equal to

$$1 + \frac{J_o}{J_m} = \frac{J_m + J_o}{J_m} = \frac{J}{J_m} \tag{10.6-14}$$

To generate diagram c, this block J/J_m is moved from the feedback path to the forward path, and the reciprocal of this, which is J_m/J, is inserted in front of the summation point. In the forward path, the blocks J/J_m and ω_{n1}^2 are combined to form the following transfer function:

$$\frac{J}{J_m} \omega_{n1}^2 = \omega_{n2}^2 \tag{10.6-15}$$

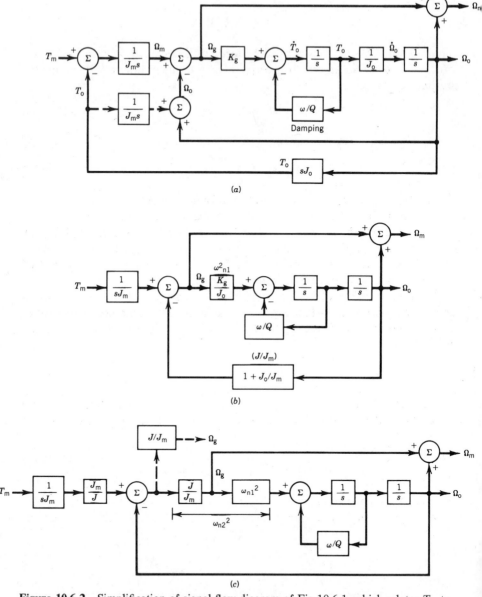

Figure 10.6-2 Simplification of signal-flow diagram of Fig 10.6-1, which relates T_m to Ω_o and Ω_m: (*a*) modification of Fig 10.6-1, which derives T_o from Ω_o signal; (*b*) modification of (*a*), which combines T_o and Ω_o signals to form single feedback path; (*c*) simplification of (*b*), which shifts J/J_m transfer function to forward path.

This relation was given in Section 10.3, Eq 10.3-35. Combining these two blocks eliminates the Ω_g signal. Hence, Ω_g is reconstructed by feeding the output from the summation point through the block J/J_m, as shown by the dashed line.

Thus, diagram c simplifies to Fig 10.6-3a. The input elements to the feedback loop are reduced to $1/sJ$. Two new variables are defined, which are called the stiff-body (or rigid-body) angular acceleration α_s and the stiff-body angular velocity Ω_s, which are defined as

$$\alpha_s = T_m/J = \text{stiff-body angular acceleration} \qquad (10.6\text{-}16)$$

$$\Omega_s = \alpha_s/s = \text{stiff-body angular velocity} \qquad (10.6\text{-}17)$$

The stiff-body angular acceleration α_s and the stiff-body angular velocity Ω_s are the acceleration and velocity that would be produced by the given motor torque if the structure were infinitely stiff.

Fig 10.6-3a has an outer feedback loop, which contains an inner feedback loop that characterizes the structural damping. The figure shows the loop transfer function of the outer loop, which is designated G, and the corresponding G_{ib} and G_{ie} transfer functions. It can be seen that the motor velocity Ω_m is expressed as follows in terms of G_{ie} and G_{ib}:

$$\Omega_m = \Omega_o + \frac{J}{J_m} G_{ie}\Omega_s = \left(G_{ib} + \frac{J}{J_m} G_{ie}\right)\Omega_s \qquad (10.6\text{-}18)$$

Now G_{ib} is related to G_{ie} by

$$G_{ib} = 1 - G_{ie} \qquad (10.6\text{-}19)$$

Substituting this into Eq 10.6-18 gives

$$\Omega_m = \left((1 - G_{ie}) + \frac{J}{J_m} G_{ie}\right)\Omega_s = \left[1 + \left(\frac{J}{J_m} - 1\right)G_{ie}\right]\Omega_s$$

$$= \left(1 + \frac{J_o}{J_m} G_{ie}\right)\Omega_s \qquad (10.6\text{-}20)$$

By means of Eq 10.6-20, the signal-flow diagram of Fig 10.6-3a is reduced to that shown in Fig 10.6-3b. The variable ω in the block ω/Q is approximated by the constant ω_{n2}. This expresses the dynamic relations in a convenient normalized form for simulation. In the servo signal-flow diagram, the motor torque signal T_m may not be expressed directly. To apply this dynamic model, one should use, as the input, the velocity that would exist without structural dynamics, and feed this input into the model as the stiff-body velocity Ω_s.

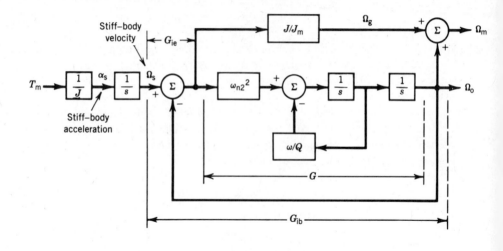

(a)

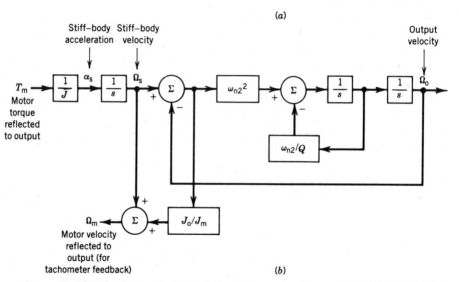

(b)

Figure 10.6-3 Final simplified signal-flow diagram for structural model: (*a*) simplification of Fig 10.6-2*c*; (*b*) final form.

10.6.2. Serial Computer Simulation of a Stage-Positioning Servo Including Structural Resonance

To illustrate the use of the structural resonance signal-flow diagram of Fig 10.6-3*b*, let us apply it to the signal-flow diagram of the stage-positioning servo that was given in Fig 9.2-3 of Chapter 9. Figure 10.6-4*a* shows the resultant signal-flow diagram of the servo, which includes the structural signal-flow diagram of Fig 10.6-3*b*, along with a lowpass filter in the position loop to attenuate the structural-resonance peak. To simplify the diagram, the effect of the motor back-EMF loop is ignored, and the subscript c is omitted from the X_c, V_c, and A_c symbols. The controlled velocity signal $V = V_c$ of Fig

9.2-3 is taken to be the stiff-body velocity V_s, and the signal-flow diagram of Fig 10.6-3b is placed between V_s and the integration $1/s$, which forms the output controlled position signal $X = X_o = X_c$.

Diagram a of Fig 10.6-4 is modified in diagram b as follows to express it in the form for serial simulation: (1) a time-delay transfer function $\bar{z}$ is placed in each feedback path along with a compensation transfer function $(1 + sT)$, and (2) the gain parameters are normalized so that all integrations are of the form $1/sT$. Each $(1 + sT)$ compensation factor is moved to the forward part of the loop in the direction shown by the arrow, and combined with the appropriate integration, to produce diagram c. The result of this is to change the integration factors $1/sT$ in the current loop and gear-damping loop to $(1 + sT)/sT$, and to add the delay factor $1/(1 + sT)$ prior to the velocity loop. Two test inputs X_{kv}, X_{ka} are added to diagram c to test the responses of the velocity and current loops.

The serial simulation program is expressed in terms of the inverse transforms of the variables in the diagram, which are defined as follows:

X_k = input command position

X_e = position error signal

$X = X_o = X_c$ = output (controlled) position

$V = V_o = V_c$ = output (controlled) velocity

$X_v = TV$

$A = A_o = A_c$ = output (controlled) acceleration

$X_a = T^2A$

$\dot{A}$ = Rate-of-Change of output acceleration

$X_{ra} = T^2\dot{A}$

V_m = motor velocity reflected to output

$X_{v(m)} = TV_m$

V_s = stiff-body velocity

$X_{v(s)} = TV_s$

A_s = stiff-body acceleration

$X_{a(s)} = T^2A_s$

A_t = Motor torque normalized in terms of output acceleration

$X_{a(t)} = T^2A_t$

$\dot{A}_t$ = Rate-of-change of motor torque, normalized in terms of output acceleration

$X_{ra(t)} = T^2\dot{A}_t$

A_d = motor disturbance (load) torque normalized in terms of output acceleration

$X_d = T^2A_d$

X_{kv}, X_{ka} = Command inputs for testing the responses of the velocity and current loops

$Y_1, Y_2, \ldots$ = unspecified signals

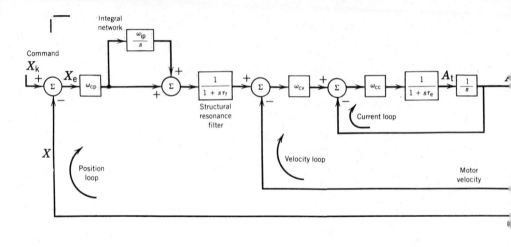

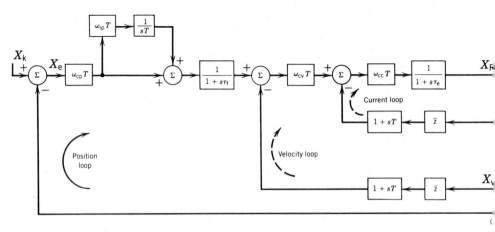

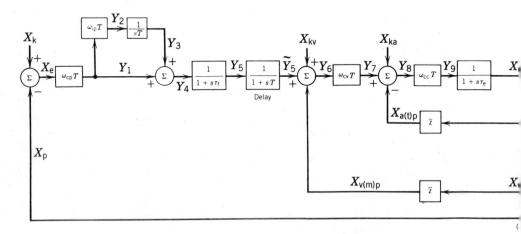

Figure 10.6-4 Signal-flow diagram for simulating stage-positioning servo with structural resonance: (*a*) signal-flow diagram of servo, including structural resonance; (*b*) serial-simulation form of (*a*), including loop delay, delay compensation, and normalization; (*c*) final diagram for simulation.

250

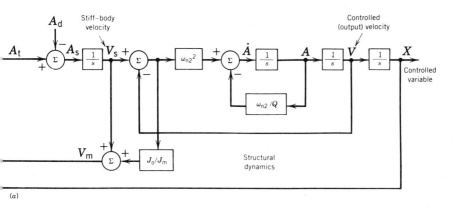

(a)

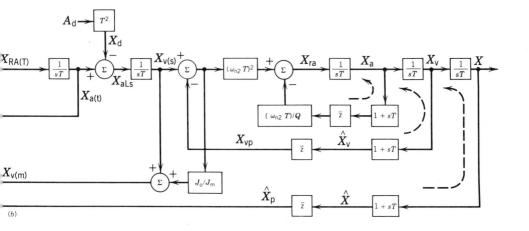

(b)

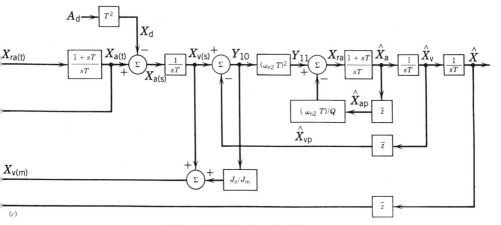

(c)

Figure 10.6-4 Continued.

These signals are modified by the following:

A subscript ending in p denotes the past value, which is the value calculated in the preceding cycle.

The caret ($\hat{}$) over a signal denotes the predicted value, which leads the variable by one cycle period T.

The tilde ($\tilde{}$) over a signal denotes the delayed value, which lags the variable by one cycle period T.

The signal Y_5 is fed into the transfer function $1/(1 + sT)$, which provides a one-cycle delay at low frequencies. Hence the output from this transfer function is designated $\tilde{Y}_5$.

From Fig 10.6-4c one can write the following computer simulation program for the control system. Note that $X**2$ means X^2.

SPECIFY PARAMETERS

System Parameters

$\omega_{cp} = ?$ (position-loop gain crossover frequency)
$\omega_{ip} = ?$ (position-loop integral break frequency)
$\omega_{cv} = ?$ (velocity-loop gain crossover frequency)
$\omega_{cc} = ?$ (current-loop gain crossover frequency)
$\tau_f = 1/\omega_f = ?$ (time constant of position-loop resonance filter)
$\tau_e = 1/\omega_e = ?$ (electrical time constant of motor circuit)
$\omega_{n2} = ?$ (frequency of structural-resonance peak)
$J_o/J_m = ?$ (load/motor inertia ratio)
$Q = 20$ (damping Q of gear train)

Program Parameters

$T = ?$ (sample period)
$\text{TME}_{Max} = ?$ (maximum time)
$L_{Max} = ?$ (computation cycles per print cycle)

SPECIFY INITIAL CONDITIONS:

Set to Zero The Initial Values of Variables Required in Computation

$$\hat{X}_p = Y_{2p} = Y_{3p} = Y_{4p} = Y_{5p} = \tilde{Y}_{5p} = X_{v(m)p} = X_{a(t)p} = 0$$
$$Y_{9p} = \tilde{X}_{ra(t)p} = X_{v(s)p} = X_{a(s)p} = \hat{X}_{ap} = \hat{X}_{vp} = X_{rap} = 0$$

Specify Initial Values of Program Parameters

$\text{TME} = 0$ (initial time)
$L = 0$ (computation cycle index)

SPECIFY INPUTS

100    $\text{TME} = \text{TME} + T$

101   X_k = function (1) of time (TME)

102   A_d = function (2) of time (TME)

103   X_{kv} = function (3) of time (TME)

104   X_{ka} = function (4) of time (TME)

SYSTEM EQUATIONS

200    $X_e = X_k - \hat{X}_p$

201    $Y_1 = (\omega_{cp} * T) * X_e$

202    $Y_2 = (\omega_{ip} * T) * Y_1$

203    $Y_3 = Y_{3p} + (1/2)(Y_2 + Y_{2p})$

204    $Y_4 = Y_3 + Y_1$

205    $Y_5 = (Y_{5p} * (2 * \tau_f / T - 1) + Y_4 + Y_{4p}) / (2 * \tau_f / T + 1)$

206    $\tilde{Y}_5 = (1/3)(\tilde{Y}_{5p} + Y_5 + Y_{5p})$

207    $Y_6 = \tilde{Y}_5 + X_{kv} - X_{v(m)p}$

208    $Y_7 = (\omega_{cv} * T) * Y_6$

209    $Y_8 = Y_7 + X_{ka} - X_{a(t)p}$

210    $Y_9 = (\omega_{cc} * T) * Y_8$

211    $\tilde{X}_{ra(t)} = (\tilde{X}_{ra(t)p} * (2 * \tau_e / T - 1) + Y_9 + Y_{9p}) / (2 * \tau_e / T + 1)$

212    $X_{a(t)} = X_{a(t)p} + (3/2) * \tilde{X}_{ra(t)} - (1/2) * \tilde{X}_{ra(t)p}$

213    $X_{a(s)} = X_{a(t)} - (T * * 2) * A_d$

214    $X_{v(s)} = X_{v(s)p} + (1/2) * (X_{a(s)} + X_{a(s)p})$

215    $Y_{10} = X_{v(s)} - \hat{X}_{vp}$

216    $X_{v(m)} = X_{v(s)} + (J_o / J_m) * Y_{10}$

217    $Y_{11} = ((\omega_{n2} * T) * * 2) * Y_{10}$

218    $X_{ra} = Y_{11} - ((\omega_{n2} * T) / Q) * \hat{X}_{ap}$

219    $\hat{X}_a = \hat{X}_{ap} + (3/2) * X_{ra} - (1/2) * X_{rap}$

220    $\hat{X}_v = \hat{X}_{vp} + (1/2) * (\hat{X}_a + \hat{X}_{ap})$

221    $\hat{X} = \hat{X}_p + (1/2) * (\hat{X}_v + \hat{X}_{vp})$

222    $X = \hat{X}_p$

CHECK PRINT COUNT, PRINT DATA:

300    $L = L + 1$

301   IF $(L < L_{\text{Max}})$ GOTO 400

302    $L = 0$

303   PRINT data (TME, X, X_e, $\hat{X}_v$, etc.)

SET PAST VALUES EQUAL TO PRESENT VALUES:

400    $\hat{X}_p = \hat{X}$

401    $Y_{2p} = Y_2$

402    $Y_{3p} = Y_3$

403 $Y_{4p} = Y_4$
404 $Y_{5p} = Y_5$
405 $\tilde{Y}_{5p} = \tilde{Y}_5$
406 $X_{v(m)p} = X_{v(m)}$
407 $X_{a(t)p} = X_{a(t)}$
408 $Y_{9p} = Y_9$
409 $\tilde{X}_{ra(t)p} = \tilde{X}_{ra(t)}$
410 $X_{v(s)p} = X_{v(s)}$
411 $X_{a(s)p} = X_{a(s)}$
412 $\hat{X}_{ap} = \hat{X}_a$
413 $\hat{X}_{vp} = \hat{X}_v$
414 $X_{rap} = X_{ra}$

CHECK TIME LIMIT, REPEAT:

500 IF $(\text{TME} < \text{TME}_{\text{Max}})$ GOTO 100
501 END

10.6.3 Application of Simulation Program

10.6.3.1 System Parameters. Table 10.6-1 shows the parameters for simulating the stage-positioning servo, including the effects of structural dynamics. Items (1), (2), (3), and (6) (τ_e, ω_{cc}, ω_{cv}, and A_d) are assumed to be the same as in the system without structural dynamics, and so are obtained from Table 9.2-2 of Chapter 9. The gear train is assumed to have a structural natural frequency ω_{n2} of 1700 sec^{-1} (270 Hz) with a Q of 20, as shown in items (9) and (7). To attenuate the resonant peak in the position loop, a filter is added of break frequency ω_f assumed to be 400 sec^{-1}, as shown in item (10). Since this filter adds phase lag to the position loop, the gain crossover frequency ω_{cp} of the position loop and the position-loop integral break frequency ω_{ip}, should be reduced below the values (142 sec^{-1}, 32.7 sec^{-1})

TABLE 10.6-1 Parameters for Simulating the Stage-Positioning Servo, Including the Effects of Structural Dynamics

(1) $\tau_e = 1/\omega_e = \frac{1}{1109}$ sec	(7) $Q = 20$
(2) $\omega_{cc} = 1109$ sec^{-1}	(8) $J_o/J_m = 0.209$
(3) $\omega_{cv} = 457$ sec^{-1}	(9) $\omega_{n2} = 1700$ sec^{-1}
(4) $\omega_{cp} = 87$ sec^{-1}	(10) $\tau_f = 1/\omega_f = \frac{1}{400}$ sec
(5) $\omega_{ip} = 20.0$ sec^{-1}	(11) $T = 0.125 \times 10^{-3}$ sec
(6) $A_d = 14.2$ in./sec^2	

given in Table 9.2-2. The new value for ω_{HF} for the position loop is given by

$$\frac{1}{\omega_{HF}} = \frac{1}{\omega_f} + \frac{1}{\omega_{cv}} + \frac{1}{\omega_{cc}} + \frac{1}{\omega_e}$$

$$= \frac{1}{400} + \frac{1}{457} + \frac{1}{1109} + \frac{1}{1109} \text{ sec} \qquad (10.6\text{-}21)$$

which yields $\omega_{HF} = 154 \text{ sec}^{-1}$. For Max $|G_{ib}| = 1.30$, the ratio ω_{HF}/ω_{LF} is 7.67 as before. Hence the new values for ω_{ip}, ω_{cp} are

$$\omega_{ip} = \omega_{LF} = \omega_{HF}/7.67$$

$$= (154/7.67) \text{ sec}^{-1} = 20.0 \text{ sec}^{-1} \qquad (10.6\text{-}22)$$

$$\omega_{cp} = \tfrac{1}{2}(\omega_{HF} + \omega_{LF})$$

$$= \tfrac{1}{2}(154 + 20) \text{ sec}^{-1} = 87.0 \text{ sec}^{-1} \qquad (10.6\text{-}23)$$

These are shown in items (4), (5) of Table 10.6-1.

The inertia ratio J_o/J_m is obtained in the following manner. As shown in items (8), (9) of Table 7.1-4 in Section 7.1, the inertia parameters of the stage-positioning servo, relative to the motor shaft, are

$$J_m = 0.0055 \text{ in.-oz-sec}^2 \qquad (10.6\text{-}24)$$

$$J_{(m)} = 0.00665 \text{ in.-oz-sec}^2 \qquad (10.6\text{-}25)$$

For the parameters defined for the model, J is the reflected value of $J_{(m)}$ at the controlled member, and J_m is the reflected value of J_m at the controlled member. Hence the ratio J/J_m is

$$\frac{J}{J_m} = \frac{J_{(m)}}{J_m} = \frac{0.00665}{0.0055} = 1.209 \qquad (10.6\text{-}26)$$

Therefore, the inertia parameters J, J_m, and J_o for the model are related by

$$J = 1.209 J_m = J_o + J_m \qquad (10.6\text{-}27)$$

$$J_o = 0.209 J_m \qquad (10.6\text{-}28)$$

$$\frac{J_o}{J} = \frac{0.209}{1.209} = 0.173 \qquad (10.6\text{-}29)$$

Item (8) in Table 10.6-1 gives the ratio J_o/J_m, which is obtained from Eq 10.6-28. By Eq 10.3-60, the frequency ω_{n1} of the zero is

$$\omega_{n1} = \omega_{n2}\sqrt{J_m/J} = \omega_{n2}\sqrt{1/1.209} = 1546 \text{ sec}^{-1} \qquad (10.6\text{-}30)$$

The highest break frequency is $\omega_n = 1700 \sec^{-1}$. A sampling period T of 0.125 msec is assumed, as shown in item (11) of Table 10.6-1. The quantity $\omega_n T$ is equal to

$$\omega_n T = (1700 \sec^{-1})(0.125 \times 10^{-3} \sec) = 0.213 \qquad (10.6\text{-}31)$$

From Section 9.3.4, Eq 9.3-55, the structural-resonance damping-ratio error caused by sampling is

$$\Delta\zeta = 0.01(0.213/0.4)^3 = 0.00151 \qquad (10.6\text{-}32)$$

Since $Q = 20$, the desired damping ratio for the structural resonance is

$$\zeta = \frac{1}{2Q} = \frac{1}{2(20)} = 0.025 \qquad (10.6\text{-}33)$$

The simulated damping ratio is

$$\zeta' = \zeta - \Delta\zeta = 0.025 - 0.0015 = 0.0235 \qquad (10.6\text{-}34)$$

The effective Q of the simulation is $1/\zeta'$, which is 21. This effective Q is reasonably close to the desired Q, which is 20.

10.6.3.2 Measurement of Mechanical Compliance.
A gear-train natural frequency ω_{n2} of 1700 $\sec^{-1}$ is assumed. In a real system the actual value of this natural frequency could be determined from compliance measurements on the gear train. Let us calculate the gear-train deflection that corresponds to this natural frequency. By Eq 10.3-35, the torsional stiffness of the gear train, relative to the motor shaft, is

$$K = \omega_{n2}^2 J_m \frac{J_o}{J} = (1700 \sec^{-1})^2(0.0055 \text{ in.-oz-sec}^2)(0.173)$$

$$= 15{,}900 \text{ in.-oz/rad} \qquad (10.6\text{-}35)$$

The values for J_m and J_o/J are obtained from Eqs 10.6-24, -29; and the value for ω_{n2} is obtained from item (9) of Table 10.6-1. The mechanical gearing compliance, reflected to the motor shaft, is the reciprocal of K in Eq 10.6-35, which is

$$C^m = \frac{1}{K} = \frac{1}{15{,}900} \text{ rad/in.-oz} = 6.29 \times 10^{-5} \text{ rad/in.-oz}$$

$$= 0.00360 \text{ deg/in.-oz} \qquad (10.6\text{-}36)$$

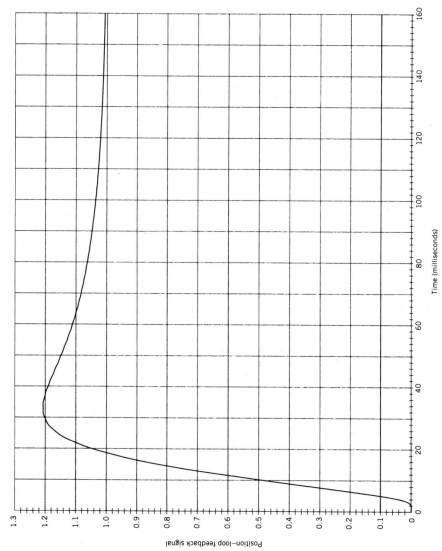

Figure 10.6-5 Feedback step response of position loop for $\omega_n = 1700$ rad/sec.

257

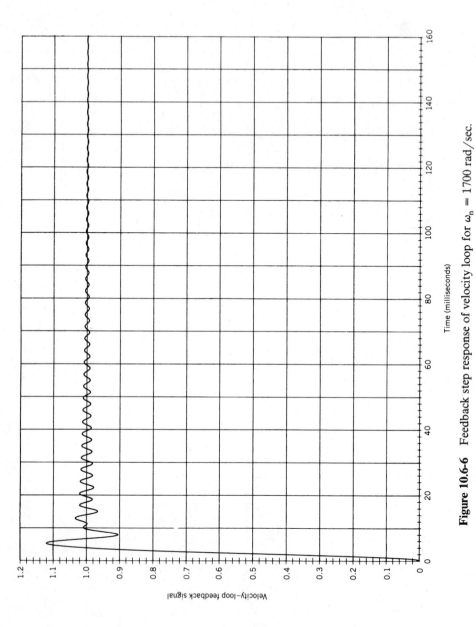

Figure 10.6-6 Feedback step response of velocity loop for $\omega_n = 1700$ rad/sec.

258

This mechanical compliance can be measured by locking the stage and observing the deflection of the motor shaft at different commanded motor torques. Adjust the current for a motor torque of 5 in.-oz, which is enough to overcome friction, and measure the angle of the motor shaft. Then increase the torque to maximum (30 in.-oz) and measure the change in the motor-shaft angle. This change $\Delta\Theta_m$ of motor angle is equal to

$$\Delta\Theta_m = C^m \Delta T_m = C^m(30 \text{ in.-oz} - 5 \text{ in.-oz}) = C^m(25 \text{ in.-oz}) \quad (10.6\text{-}37)$$

where ΔT_m is the change in motor torque. For the assumed mechanical compliance of Eq 10.6-36, the motor-shaft deflection would be

$$\Delta\Theta_m = C^m \Delta T_m = (0.00360 \text{ deg/in.-oz})(25 \text{ in.-oz}) = 0.09° \quad (10.6\text{-}38)$$

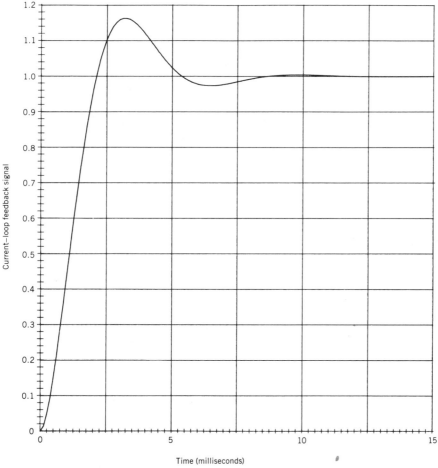

Figure 10.6-7 Feedback step response of current loop for $\omega_n = 1700$ rad/sec.

TABLE 10.6-2   Estimated Values of Delay Time T_d (Equal to $1/\omega_c$) for Current, Velocity, and Position Loops, Compared with Actual Values Measured from Step Responses

Loop	ω_c (sec^{-1})	Delay Time T_d (msec)	
		Estimated	Actual
Current	1109	0.90	1.08
Velocity	1109	2.19	2.10
Position	87	11.5	10.2

Since the torque sensitivity of the motor is 5.8 in.-oz/A [Table 7.1-4, item (3)], the required motor currents for the two torques are 5.17 A at 30 in.-oz and 0.86 A at 5 in.-oz.

10.6.3.3   Simulation Results. Figures 10.6-5 to 10.6-8 show the transient responses computed by simulation from the system defined by Table 10.6-1. These are

a. *Feedback response of position loop to unit step* (Fig 10.6-5): $X_k =$ unit step; read $X = X_p$.

b. *Feedback response of velocity loop to unit step* (Fig 10.6-6): Open the position loop by setting $\omega_{cp} = 0$. $X_{kv} =$ unit step; read $X_{v(m)p}$.

c. *Feedback response of current loop to unit step* (Fig 10.6-7): Open the velocity loop by setting $\omega_{cv} = 0$. $X_{ka} =$ unit step; read $X_{a(t)p}$.

d. *Error response to friction step disturbance* (Fig 10.6-8): $A_d = 14.2$ in./sec; read X_e (inches). (Note $X_k = 0$.)

(Figures 10.6-5 to -10 were prepared by Richard P. Rousseau, a student at University of Lowell, Massachusetts.) The feedback responses should record the proper variables. For example, the velocity feedback response should record the velocity $X_{v(m)}$ (or $X_{v(m)p}$) actually sensed by the tachometer, not $\hat{X}_v$ or $X_{v(s)}$.

A very important parameter of the step response is its delay time T_d, which is the time after the input step for the response to reach 50% of the final value. As was shown in Chapters 2 and 5, T_d is approximately equal to $1/\omega_c$. Table 10.6-2 gives the values of this estimated delay time $1/\omega_c$, and compares them with the actual values of T_d measured from the transient responses. The agreement is very good.

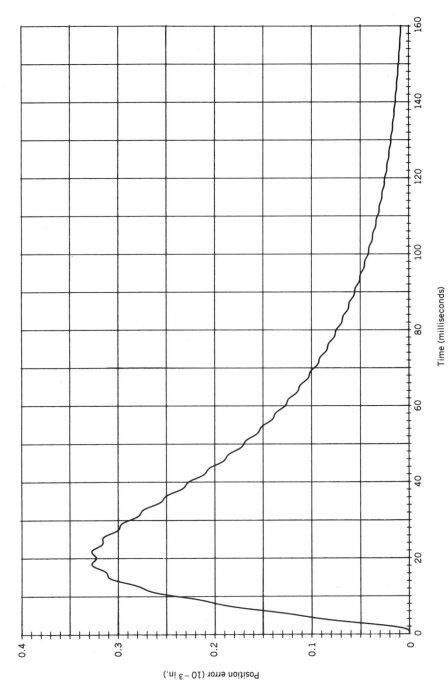

Figure 10.6-8 Error response to equivalent step of friction torque for $\omega_n = 1700$ rad/sec.

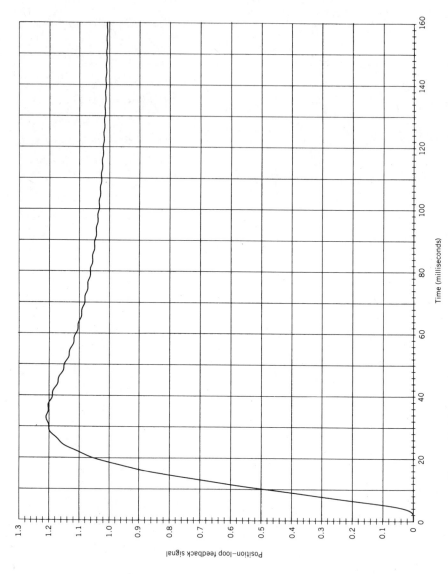

Figure 10.6-9 Feedback step response of position loop for $\omega_n = 1400$ rad/sec.

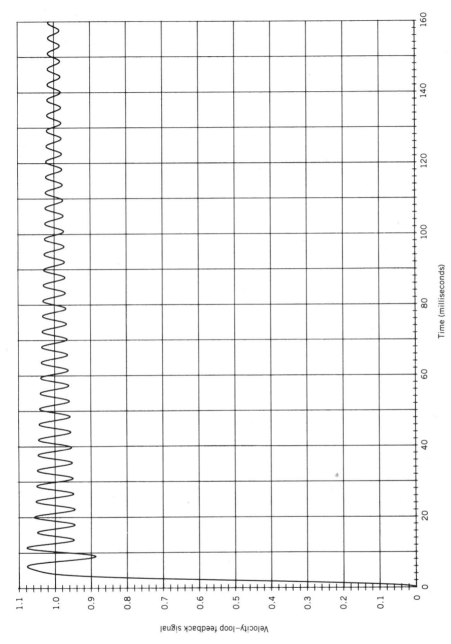

Figure 10.6-10 Feedback step response of velocity loop for $\omega_n = 1400$ rad/sec.

In accordance with Eq 7.2-62 of Chapter 7, the approximate maximum value of the error response for the friction torque disturbance is

$$\text{Max}[x_e] = \frac{F_{d(c)}}{M_{(c)}\omega_{cp}\omega_{cv}} = \frac{A_d}{\omega_{cp}\omega_{cv}}$$

$$= \frac{14.2 \text{ in./sec}^2}{(87 \text{ sec}^{-1})(457 \text{ sec}^{-1})} = 0.357 \times 10^{-3} \text{ in.} \quad (10.6\text{-}39)$$

As shown in Fig 10.6-8, the actual maximum error is 0.325×10^{-3} in., which is in good agreement with this approximation.

The step responses of the position loop and current loop in Figs 10.6-5 and -7 are quite stable. However, the step response of the velocity loop in Fig 10.6-6 exhibits a small but poorly damped oscillatory transient. Thus, as this example illustrates, it is essential that the responses of the individual feedback loops be examined independently when the dynamic characteristics of a control system are being evaluated.

Figures 10.6-9 and -10 show the effect of reducing the structural resonant frequency from 1700 sec^{-1} (270 Hz) to 1400 sec^{-1} (223 Hz), without changing the gains in the feedback loops. (The rotor shaft deflection due to compliance shown in Eq 10.6-38 would increase from 0.09° to 0.13°.) The step response of the position loop in Fig 10.6-9 superficially appears to be reasonable, but the step response of the velocity loop in Fig 10.6-10 shows that the system is on the verge of oscillation. Therefore, the amplifier gains would have to be reduced to achieve acceptable performance.

Chapter 11

Response to a Random Signal

Certain inputs to control systems are random, and can only be characterized quantitatively by statistical means. These inputs include in particular the disturbances caused by receiver noise and by wind gusts. This chapter summarizes the application of statistical techniques for handling such random input signals.

The chapter describes thermal noise and shot noise, the concept of noise bandwidth, and the Gaussian distribution. Autocorrelation and cross-correlation functions are presented, along with a general method of computing noise bandwidth.

Section 11.3 examines the processes of range and angle tracking in a radar system, showing how statistical signal concepts are used to determine the tracking errors due to receiver noise. Section 11.4 shows how wind forces are characterized statistically to calculate the tracking error caused by wind on a large satellite communication antenna.

11.1 CHARACTERISTICS OF RECEIVER NOISE SIGNALS

This section derives the basic equations for characterizing the major types of receiver noise: thermal (or Johnson) noise, which is due to thermal agitation in resistive elements, and shot (or Schottky) noise, which is due to the discrete nature of electron flow.

11.1.1 Thermal (or Johnson) Noise

Thermal agitation in a resistor generates electrical noise power that is proportional to the absolute temperature. If, as shown in Fig 11.1-1a, a resistor is connected to the input terminals of an amplifier having an input resistance matched to that of the resistor, the resistor delivers to the amplifier a noise power, per hertz of frequency, that is given by

$$\frac{dP}{df} = KT \tag{11.1-1}$$

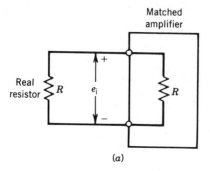

(a)

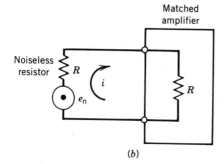

(b)

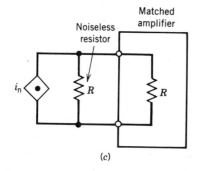

(c)

Figure 11.1-1   Definition of thermal noise: (a) basis for thermal noise definition; (b) representation of thermal noise by equivalent voltage noise source e_n; (c) representation of thermal noise by equivalent current noise source i_n.

where, T is the absolute temperature in °K (Kelvin) and K is Boltzmann's constant given by

$$K = 1.374 \times 10^{-23} \text{ joule/°K} \qquad (11.1\text{-}2)$$

At room temperature (25°C), $T = 298$ °K, so that

$$\frac{dP}{df} = 4.10 \times 10^{-21} \text{ Watt/Hz} \qquad (11.1\text{-}3)$$

At the amplifier input, there is a signal to be amplified. What is important is the ratio of signal to noise in the amplified output. To evaluate this ratio, it is convenient to express the signal and noise outputs from the amplifier relative to the input, by normalizing the voltage gain and power gain of the amplifier

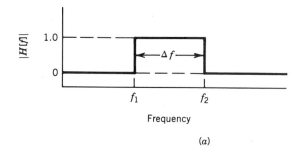

(a)

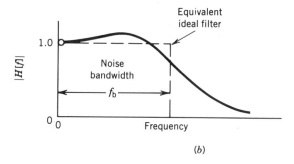

(b)

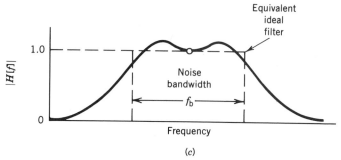

(c)

Figure 11.1-2 Definition of noise bandwidth: (a) magnitude of frequency response for ideal filter; (b) normalized magnitude of frequency response for practical lowpass filter; (c) normalized magnitude of frequency response for practical bandpass filter.

to unity at the signal frequency. Consider an ideal amplifier with a frequency response that is flat between frequencies f_1 and f_2 and is zero outside that region, as shown in diagram a of Fig 11.1-2. The output noise power, normalized relative to the input, is

$$P_o = KT\Delta f \qquad (11.1\text{-}4)$$

$$\Delta f = f_2 - f_1 \qquad (11.1\text{-}5)$$

Frequency responses of practical amplifiers are illustrated in diagrams b and c. As shown, the amplifier gain is normalized to unity at zero frequency for a lowpass response, and it is normalized to unity at the carrier frequency for a bandpass response. The following are defined:

E_i = input voltage to amplifier

E_o = normalized output voltage from amplifier

$|H| = |E_o/E_i|$ = normalized magnitude of frequency response

The normalized amplifier output power ΔP_o within a small frequency band Δf is related as follows to the input power ΔP_i in that band:

$$\frac{\Delta P_o}{\Delta P_i} = \left| \frac{E_o}{E_i} \right|^2 = |H|^2 \tag{11.1-6}$$

Since $\Delta P_i = KT\Delta f$, the output power in the band Δf is

$$\Delta P_o = KT|H|^2 \Delta f \tag{11.1-7}$$

Hence the total output power is

$$P_o = KT \int_0^\infty |H[f]|^2 \, df \tag{11.1-8}$$

It is convenient to relate this power to the power from an ideal amplifier with a flat response. Setting Eq 11.1-8 equal to Eq 11.1-4 gives

$$\Delta f = \int_0^\infty |H[f]|^2 \, df = f_b = \text{noise bandwidth} \tag{11.1-9}$$

This represents the bandwidth of an ideal flat-response amplifier that has the same normalized output noise power as the practical amplifier. It is called the *noise bandwidth* f_b of the practical amplifier.

As shown in Fig 11.1-1b, the noise power generated in a resistor can be expressed in terms of a noise voltage of RMS value e_n in series with an ideal noiseless resistor. The RMS current flowing into the matched amplifier is

$$i = e_n/2R \tag{11.1-10}$$

Hence, the power delivered to the amplifier is

$$P = i^2R = e_n^2/4R \tag{11.1-11}$$

By Eq 11.1-4, this power P is set equal to KTf_b. Hence the noise voltage is

$$e_n = \sqrt{4KTRf_b} \tag{11.1-12}$$

TABLE 11.1-1   Noise Bandwidths of Simple Filters

Filter	Transfer Function $H[\omega]$	Noise Bandwidth ω_b (rad/sec)
A	$\dfrac{\omega_1}{s + \omega_1}$	$\dfrac{\pi}{2}\,\omega_1$
B	$\dfrac{\omega_1\omega_2}{(s + \omega_1)(s + \omega_2)}$	$\dfrac{(\pi/2)\,\omega_1\omega_2}{\omega_1 + \omega_2}$
C	$\dfrac{\omega_n^2}{s^2 + 2\zeta\omega_n s + \omega_n}$	$\dfrac{\pi}{2}\dfrac{\omega_n}{2\zeta}$

At room temperature (25°C) this is equal to

$$e_n = 0.128 \times 10^{-9}\sqrt{Rf_b}\ \text{V}/\sqrt{\Omega\text{-Hz}} \tag{11.1-13}$$

This can be approximated within an error of 2% by

$$e_n \doteq \tfrac{1}{8}\sqrt{Rf_b}\ \text{nV}/\sqrt{\Omega\text{-Hz}} \tag{11.1-14}$$

As shown in Fig 11.1-1b, the noise can also be expressed in terms of a noise current of RMS value i_n in parallel with the resistor, where

$$i_n R = e_n \tag{11.1-15}$$

The RMS noise current at room temperature is

$$i_n = \sqrt{(4KT/R)f_b} \doteq \tfrac{1}{8}\sqrt{f_b/R}\ \text{nA-}\sqrt{\Omega/\text{Hz}} \tag{11.1-16}$$

11.1.2 Noise-Bandwidth Examples

11.1.2.1 Noise Bandwidths of Simple Lowpass Filters. Table 11.1-1 shows the values of noise bandwidth in rad/sec, designated ω_b, for some simple lowpass filters. Frequency-response plots for these filters are shown in Fig 11.1-3. A general method for calculating the noise bandwidth is presented in Section 11.2.4. The following is a derivation of the noise bandwidth of filter A. This is a single-order low-pass filter of break frequency $\omega_f = 2\pi f_f$, or time constant $\tau_f = 1/\omega_f$. Its transfer function in terms of angular frequency ω is

$$H[\omega] = \frac{1}{1 + j\omega/\omega_f} = \frac{\omega_f}{j\omega + \omega_f} \tag{11.1-17}$$

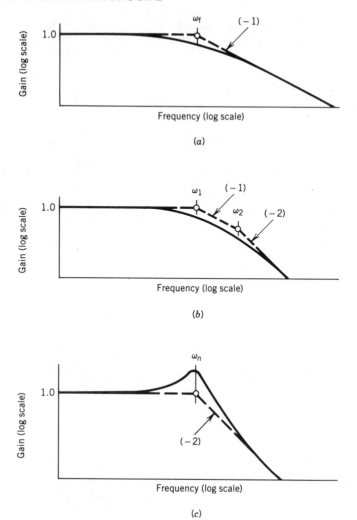

Figure 11.1-3 Frequency-response magnitude plots for filters considered in Table 11.1-1: (a) filter A; (b) filter B; (c) filter C.

The square of the magnitude of any transfer function $H[\omega]$ can be obtained from

$$|H[\omega]|^2 = H[\omega]H[\omega]^* = H[\omega]H[-\omega] \qquad (11.1\text{-}18)$$

where $H[\omega]*$ is the complex conjugate of $H[\omega]$. Applying Eq 11.1-18 to Eq 11.1-17 gives for filter A

$$|H[\omega]|^2 = \frac{\omega_f}{(\omega_f + j\omega)}\frac{\omega_f}{(\omega_f - j\omega)} = \frac{\omega_f^2}{\omega_f^2 + \omega^2} \qquad (11.1\text{-}19)$$

From Eq 11.1-9, the noise bandwidth ω_b expressed in rad/sec is

$$\omega_b = \int_0^\infty |H[\omega]|^2 \, d\omega = \int_0^\infty \frac{\omega_f^2}{\omega_f^2 + \omega^2} \, d\omega \qquad (11.1\text{-}20)$$

Now,

$$\int_0^\infty \frac{d\omega}{\omega^2 + \omega_f^2} = \frac{1}{\omega_f} \arctan \frac{\omega}{\omega_f} \bigg|_{\omega=0}^\infty = \frac{1}{\omega_f} \frac{\pi}{2} \qquad (11.1\text{-}21)$$

Hence the noise bandwidth in rad/sec is

$$\omega_b = \omega_f^2 \frac{1}{\omega_f} \frac{\pi}{2} = \frac{\pi}{2} \omega_f \qquad (11.1\text{-}22)$$

The noise bandwidth in hertz is

$$f_b = \frac{\pi}{2} f_f = \frac{\pi}{2} \frac{\omega_f}{2\pi} = \frac{\omega_f}{4} = \frac{1}{4\tau_f} \qquad (11.1\text{-}23)$$

11.1.2.2 Noise Bandwidth of Ideal Averaging Integrator. The next example is an ideal averaging integrator, which integrates the input signal over a fixed time interval T_i, to obtain the average of the input over that interval. Figure 11.1-4 shows its unit-impulse response, which has a constant value over the time interval T_i, and is zero outside that interval. Since the output is equal to the average of the input, the zero-frequency gain of the process must be unity. Hence the area under the impulse response is unity, and so the amplitude of the unit-impulse response is $1/T_i$. This impulse response was considered in Chapter 2, Fig 2.2-4a, along with its Fourier transform, which is the frequency response of the filter.

To calculate the noise bandwidth of this filter, it is convenient to use the following general relation of Fourier-transform theory (from Chapter 2, Eq

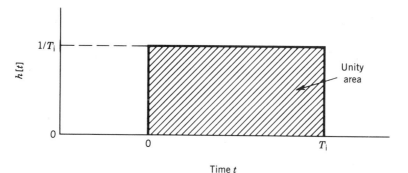

Time t

Figure 11.1-4 Impluse response of ideal integrator.

2.2-38):

$$\int_{-\infty}^{\infty} h[t]^2 \, dt = \int_{-\infty}^{\infty} |H[f]|^2 \, df \qquad (11.1\text{-}24)$$

where $H[f]$ is the frequency response of the filter and $h[t]$ is the unit impulse response. Since $|H[f]| = |H[-f]|$, the frequency domain integral from $-\infty$ to $+\infty$ is twice the integral from 0 to $+\infty$:

$$\int_{-\infty}^{\infty} |H[f]|^2 \, df = 2\int_{0}^{\infty} |H[f]|^2 \, df \qquad (11.1\text{-}25)$$

Applying Eqs 11.1-24, -25 to the definition of noise bandwidth in Eq 11.1-9 gives

$$f_b = \int_{0}^{\infty} |H[f]|^2 \, df = \frac{1}{2}\int_{-\infty}^{\infty} h[t]^2 \, dt \qquad (11.1\text{-}26)$$

Equation 11.1-26 allows the noise bandwidth to be calculated either in the frequency domain or in the time domain. For the ideal integrator, the time-domain calculation is simpler. Since the impulse response $h[t]$ is equal to $1/T_i$ over the integration interval T_i, and is zero elsewhere, the noise bandwidth is

$$f_b = \frac{1}{2}\int_{-\infty}^{\infty} h[t]^2 \, dt = \frac{1}{2}\int_{0}^{T_i} \left(\frac{1}{T_i}\right)^2 dt = \frac{1}{2T_i} \qquad (11.1\text{-}27)$$

11.1.2.3 Noise Bandwidths of Simple Feedback Loops. The noise bandwidth of a feedback loop is defined in terms of its transfer function G_{ib}

$$f_b = \int_{0}^{\infty} |G_{ib}|^2 \, df \qquad (11.1\text{-}28)$$

Table 11.1-2 shows the noise-bandwidth values for three useful loop transfer functions. Asymptotic frequency-response plots of loop gain are shown in Fig 11.1-5.

TABLE 11.1-2 Noise Bandwidths of Feedback Loops in Fig 11.1-5

	Transfer Function		Noise Bandwidth
Loop	G	G_{ib}	ω_b (rad/sec)
D	$\dfrac{\omega_c}{s(1 + s/\omega_f)}$	$\dfrac{\omega_c \omega_f}{s^2 + \omega_f s + \omega_c \omega_f}$	$\dfrac{\pi}{2}\omega_c$
E	$\dfrac{\omega_c(1 + \omega_i/s)}{s}$	$\dfrac{\omega_c(s + \omega_i)}{s^2 + \omega_c s + \omega_c \omega_i}$	$\dfrac{\pi}{2}(\omega_c + \omega_i)$
F	$\dfrac{\omega_c(1 + \omega_i/s)}{s(1 + s/\omega_f)}$	$\dfrac{\omega_c \omega_f(s + \omega_i)}{s^3 + \omega_f s^2 + \omega_c \omega_f s + \omega_c \omega_f \omega_i}$	$\dfrac{\pi \omega_f(\omega_c + \omega_i)}{2(\omega_f - \omega_i)}$

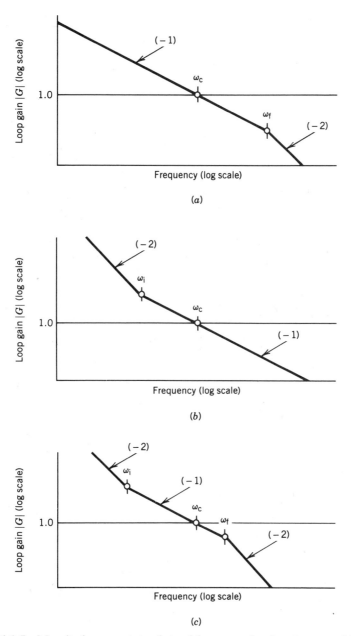

Figure 11.1-5 Magnitude asymptote plots of loop transfer functions considered in Table 11.1-2: (*a*) loop *D*; (*b*) loop *E*; (*c*) loop *F*.

For loop D, the noise bandwidth, expressed in terms of the parameters of the loop transfer function, is independent of the value of the break frequency ω_f. The corresponding G_{ib} transfer function of this loop is the same as the transfer function for filters A, B, and C shown in Table 11.1-1. If ω_f is infinite, G_{ib} is the same as $H[\omega]$ for filter A; if $4\omega_c < \omega_f < \infty$, G_{ib} has two real poles and is the same as H for filter B; and if $\omega_f < 4\omega_c$, G_{ib} is underdamped and the same as H for filter C.

11.1.3 Shot (or Schottky) Noise

11.1.3.1 Coin-Flipping Probability Experiment. Another major type of noise is shot (or Schottky) noise, which is caused by the discrete nature of electron flow in an electric current. The statistical characteristics of this noise can be derived from simple probability considerations. Let us consider a coin-flipping experiment.

If one flips a coin $2N$ times, on the average there are N heads, but the actual number of heads deviates from N in a random manner. To a good approximation, the RMS value of this deviation, σN, is as follows for reasonably large values of N:

$$\sigma N = \sqrt{N} \qquad (11.1\text{-}29)$$

Assume that the coin is flipped 200 times; the average number of heads is $N = 100$. The RMS deviation (or standard deviation) from this average is $\sqrt{N} = \sqrt{100} = 10$.

The probability of achieving a specific number of heads is described approximately by the Gaussian, or normal, distribution. The derivative of that distribution is called the Gaussian (or normal) probability density function, which is defined as

$$\frac{dP_g[x]}{dx} = \frac{1}{\sqrt{2\pi}} e^{-x^2/2\sigma^2} \qquad (11.1\text{-}30)$$

where σ is the RMS value (or standard deviation), and x is the deviation from the mean value. The normalized deviation x/σ is designated z. A plot of the Gaussian probability density function $dP_g[z]/dz$ is shown in Fig 11.1-6, versus the normalized deviation

$$z = x/\sigma. \qquad (11.1\text{-}31)$$

The total area under the curve is unity. The probability that the number of heads falls within a given range of x is the area under the curve over the corresponding normalized range of x/σ. When normalized in terms of the

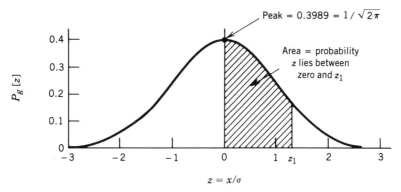

Figure 11.1-6   Plot of normal (or Gaussian) probability density function.

variable z, the Gaussian probability density is

$$\frac{dP_{\mathrm{g}}[z]}{dz} = \frac{1}{\sqrt{2\pi}} e^{-z^2/2} \qquad (11.1\text{-}32)$$

This Gaussian probability density $dP_{\mathrm{g}}[z]/dz$ cannot be integrated analytically, and so its integral, the probability distribution, is commonly expressed in tabular form. Values of this integral, obtained from Feller [11.1], are shown in Table 11.1-3, where the limits of the integral are from 0 to z. We designate this integral from zero to z as $P_{\mathrm{g}}[z]$. The integral of the Guassian density function from $-\infty$ to z is commonly designated $\Phi[z]$. This is related as follows to the function $P_{\mathrm{g}}[z]$ given in Table 11.1-3:

$$\Phi[z] = 0.5 + \mathrm{sgn}[z]\, P_{\mathrm{g}}[z] \qquad (11.1\text{-}33)$$

where $\mathrm{sgn}[z]$ is the sign of z, which is $+1$ for $z > 0$, and -1 for $z < 0$.

For large values of z, it is more convenient to consider the integral from z to infinity, which is equal to $(0.5 - P_{\mathrm{g}}[z])$. This integral can be approximated by

$$\left(0.5 - P_{\mathrm{g}}[z]\right) \doteq \frac{1}{\sqrt{2\pi}\, z} e^{-z^2/2} \qquad \text{(for large } z\text{)} \qquad (11.1\text{-}34)$$

As shown by Feller [11.1], the exact value of $(0.5 - P_{\mathrm{g}}[z])$ is less than this approximation, but is greater than the approximation multiplied by $(1 - 1/z^2)$. To solve for z as a function of $P_{\mathrm{g}}[z]$, Eq 11.1-34 can be expressed as

$$z \doteq \sqrt{-2\ln\left[\sqrt{2\pi}\, z \left(0.5 - P_{\mathrm{g}}[z]\right)\right]} \qquad \text{(for large } z\text{)} \qquad (11.1\text{-}35)$$

This expression is a weak function of the value of z on the right-hand side,

TABLE 11.1-3 Values of Gaussian Probability Integral from 0 to z, Designated $P_g[z]$, where $z = x/\sigma$

$z =$ x/σ	$P_g[z]$	$z =$ x/σ	$P_g[z]$	$z =$ x/σ	$P_g[z]$
0.0	0.000 000	1.5	0.433 193	3.0	0.498 650
0.1	0.039 828	1.6	0.445 201	3.1	0.499 032
0.2	0.079 260	1.7	0.455 439	3.2	0.499 313
0.3	0.117 911	1.8	0.464 070	3.3	0.499 517
0.4	0.155 422	1.9	0.471 283	3.4	0.499 663
0.5	0.191 462	2.0	0.477 250	3.5	0.499 767
0.6	0.225 747	2.1	0.482 136	3.6	0.499 841
0.7	0.258 036	2.2	0.486 097	3.7	0.499 892
0.8	0.288 145	2.3	0.489 276	3.8	0.499 928
0.9	0.315 940	2.4	0.491 802	3.9	0.499 952
1.0	0.341 345	2.5	0.493 790	4.0	0.499 968
1.1	0.364 334	2.6	0.495 339	4.1	0.499 979
1.2	0.384 930	2.7	0.496 533	4.2	0.499 987
1.3	0.403 200	2.8	0.497 445	4.3	0.499 991
1.4	0.419 243	2.9	0.498 134	4.4	0.499 995

and so can be solved iteratively by guessing the value of z in the right-hand side of the equation, and using the calculated value of z to update the guess. For example, assume that $(0.5 - P_g[z])$ is 10^{-6}, and guess that $z = 3$ (a poor guess). The first calculation is

$$z = \sqrt{-2\ln\left[\sqrt{2\pi}\,(3)(10^{-6})\right]} = 4.86 \qquad (11.1\text{-}36)$$

The value $z = 4.86$ is used as the next guess. The iteration proceeds as follows:

guess $z = 4.86$;   calculate $z = 4.757$

guess $z = 4.757$;   calculate $z = 4.7617$

guess $z = 4.7617$;   calculate $z = 4.7615$ $\qquad (11.1\text{-}37)$

Let us apply Table 11.1-3 to the coin-flipping example. What is the probability that the number of heads lies between 80 and 115? The mean value is 100, and so the deviation from the mean is

$$x = -20 \text{ to } +15 \qquad (11.1\text{-}38)$$

Since the RMS deviation from the mean (the standard deviation) σ is 10, the

TABLE 11.1-4 Approximate Probability that the Number of Heads Lies within a Particular Range

Range	Probability
100 to 115	43.32%
80 to 100	47.73%
80 to 115	91.05%
Less than 80	50% − 47.73% = 2.27%
Greater than 115	50% − 43.32% = 6.68%

normalized deviation x/σ from the mean is

$$z = \frac{x}{\sigma} = -2.0 \text{ to } +1.5 \tag{11.1-39}$$

The integral from $z = 0$ to $z = 1.5$ is obtained by reading from Table 11.1-3 the value for $z = 1.5$, which is

$$P_g[z] = P_g[1.5] = 0.4332 \tag{11.1-40}$$

Since the normal probability curve is symmetric about $z = 0$, the integral from $z = -2.0$ to 0 is the same as the integral from $z = 0$ to $z = +2.0$. This probability is read from Table 11.1-3 at $z = 2.0$, which is

$$P_g[-2.0] = P_g[2.0] = 0.4773 \tag{11.1-41}$$

The probability that the number of heads lies between 80 and 115 is the sum of Eqs 11.1-40, -41, which is 0.9105, or 91.05%. From this data, the probabilities shown in Table 11.1-4 can be obtained. These results are only approximate, because the distribution of number of heads is quantized, and so is only approximately Gaussian. On the other hand, the greater the average number of heads N, the more closely the distribution approximates a Gaussian distribution.

11.1.3.2 Derivation of Shot-Noise Formula.

Let us relate this to the noise associated with the flow of electrons, which is called shot noise. A constant current I has the following average rate of electron flow:

$$\langle \dot{N} \rangle = I/q \tag{11.1-42}$$

The dot (˙) above a variable represents differentiation, and the angular brackets $\langle \ \rangle$ around a variable represent average value. The parameter q is the charge on an electron. If this current is fed into an ideal averaging integrator of integration time T_i, the average number of electrons detected by the integrator is

$$\langle N \rangle = \langle \dot{N} \rangle T_i \qquad (11.1\text{-}43)$$

The actual number that is detected deviates in a random manner from this average value. The RMS value of this deviation is

$$\sigma N = \sqrt{\langle N \rangle} = \sqrt{\langle \dot{N} \rangle T_i} = \sqrt{\left(\frac{I}{q}\right)T_i} \qquad (11.1\text{-}44)$$

This deviation of electron count can be interpreted as a variation of the rate of flow of electrons. The equivalent RMS variation of the electron flow rate is

$$\sigma \dot{N} = \frac{\sigma N}{T_i} = \sqrt{\frac{I}{qT_i}} \qquad (11.1\text{-}45)$$

The apparent RMS variation of the electric current is then

$$\sigma i = q \, \sigma \dot{N} = \sqrt{\frac{qI}{T_i}} \qquad (11.1\text{-}46)$$

From Eq 11.1-27, the noise bandwidth f_b of the averaging integrator is $1/2T_i$. Replacing $1/T_i$ by $2f_b$ in Eq 11.1-46 gives the basic equation for shot noise, which expresses the RMS noise current in terms of the noise bandwidth of the integrator:

shot noise:
$$\sigma i = \sqrt{2qIf_b} \qquad (11.1\text{-}47)$$

Although this shot-noise formula was derived for an ideal integrator, it applies to any kind of filter. The value of the charge on an electron is

$$q = 1.59 \times 10^{-19} \text{coulomb} \qquad (11.1\text{-}48)$$

Substituting this into Eq 11.1-47 gives the following quantitative formula for shot noise:

$$\sigma i = 0.566\sqrt{I_{(\mu A)}f_b} \text{ pA}/\sqrt{\mu\text{A-Hz}} \qquad (11.1\text{-}49)$$

where $I_{(\mu A)}$ represents the average current in microamperes (μA).

A noisy signal displayed on an oscilloscope exhibits fluctuations about an average value. It is convenient to characterize the magnitude of the fluctuations in terms of a "peak-to-peak" value for the deviation of the noise.

However, what one calls "peak-to-peak" for a noise signal depends on the criterion that one sets to define peak-to-peak limits.

Assume that one chooses limits which are exceeded 1% of the time (the signal is within these limits 99% of the time). Table 11.1-3 shows that this corresponds to the range $\pm 2.6\sigma$. Hence for this 99% criterion, the peak-to-peak deviation is 5.2 times the RMS value σ. Alternatively, one might choose limits that include 95% of the signal. Since this corresponds to $\pm 2.0\sigma$, the peak-to-peak value for this criterion is 4 times the RMS value σ. The peak-to-peak value that one measures from observing a noisy signal is usually between 4 and 6 times the RMS value, depending on what criterion the observer selects in making his judgement.

This point is illustrated in Fig 11.1-7, which shows a photograph of a noisy waveform, compared with the corresponding plot of the Gaussian amplitude distribution. (This is obtained from an excellent practical discussion of noise in opamp circuits presented by Ryan and Scranton [11.2].)

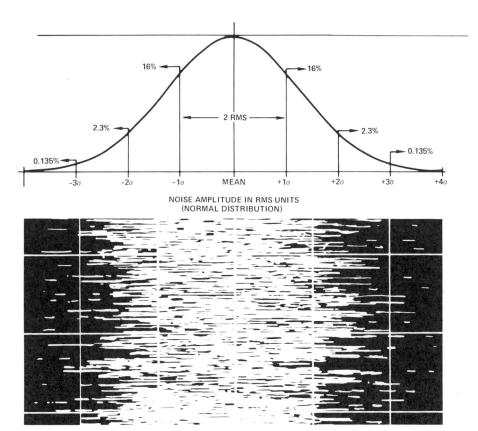

Figure 11.1-7 Gaussian amplitude distribution, compared with random noise waveform. Reprinted from Ref [11.2] with permission from Analog Devices.

11.1.4 Other Types of Noise

Thermal noise and shot noise have a flat noise spectrum at the point where they are generated, and so are called "white" noise. Besides thermal noise and shot noise, amplifiers also exhibit "flicker" noise and sometimes "popcorn" noise, which are particularly important at low frequencies.

Flicker (or $1/f$) noise has a spectrum that decreases with increasing frequency. The power density (watt/Hz) is inversely proportional to frequency, and so the noise voltage density (volt/$\sqrt{\text{Hz}}$) is inversely proportional to the square root of frequency. Flicker noise in semiconductors is due to random fluctuations in the number of surface recombinations. The contact between granules in carbon resistors also generates flicker noise as well as shot noise. For this reason, low-noise wirewound or metal-film resistors are used instead of carbon resistors in low-noise circuits. Note that a "low-noise" resistor generates just as much thermal noise as a carbon resistor.

Popcorn noise is characterized by random jumps between two or more levels. In semiconductors, it is due to random on–off recombination action in the semiconductor material, leading to erratic switching of the gain of the device. This noise has the sound of popcorn when displayed audibly. Figure 11.1-8 shows typical waveforms produced by white noise, flicker noise, and popcorn noise, which was presented by Ryan and Scranton [11.2].

11.1.5 Application to Error Components of a Control System

Suppose there are a number of independent noise sources contributing to the noise in a particular voltage E. Since the noise sources are independent, the noise powers add. Noise power is proportional to the square of the voltage, and so the total RMS voltage e is related as follows to the separate noise-voltage components e_1, e_2:

$$e^2 = e_1^2 + e_2^2 \tag{11.1-50}$$

$$e = \sqrt{e_1^2 + e_2^2} \tag{11.1-51}$$

This principle can be applied to any number of noise components. The combined RMS noise voltage is the RSS (root-sum-squared) combination of the individual noise-voltage components.

The noise contribution in a controlled variable is generally only one of many error components, produced by effects that are independent of one another. These can include errors due to static friction, tachometer imperfections, encoder quantization and inaccuracy, wind torque, etc. It is convenient to treat all of these error components as if they were noise components. The times at which they occur are independent of one another, and so they add in a random fashion. Hence, the combined error is the square root of the sum of the squares of the individual error components.

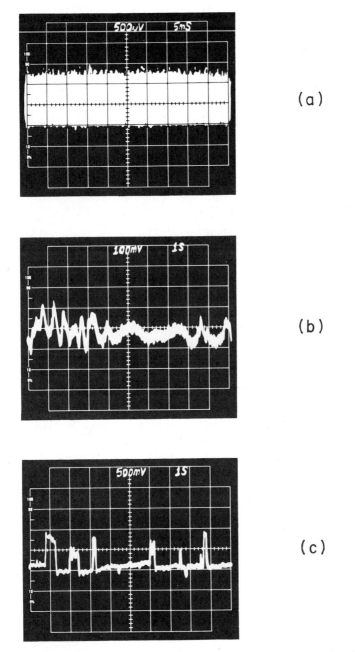

Figure 11.1-8 Noise signatures: (*a*) white noise; (*b*) flicker (or $1/f$) noise; (*c*) popcorn noise. Reprinted from Ref [11.2] with permission from Analog Devices.

Many error components are characterized by peak values rather than RMS values. How does one combine the error of a position sensor, which is usually characterized by its peak value, with error components due to thermal and shot noise, which have Gaussian amplitude distributions? By convention, peak value is generally considered to be equivalent to an error of $\pm 3\sigma$. Hence $\frac{1}{3}$ of the peak error of the position sensor is considered to be its effective RMS value σ (or *sigma*).

As was shown in Fig 11.1-7, the value that one regards as the peak (or peak-to-peak) noise is somewhat arbitrary. Nevertheless, limits of ± 3 sigma (σ) are commonly used to define the peak value of noise, for the following reasons:

1. The probability of exceeding ± 3 sigma is only 0.37%.
2. The probability of exceeding ± 3 sigma by 10% (i.e., the probability of exceeding ± 3.3 sigma) is only 0.1%.
3. The probability of exceeding ± 3 sigma by 30% (i.e., the probability of exceeding ± 3.9 sigma) is only 0.01%.

Thus the limits of ± 3 sigma are not exceeded very often, and it is rare that these limits are exceeded by an appreciable factor. As a result, ± 3 sigma has been found to be a good practical definition of peak value in many engineering applications.

11.1.6   Relation Between Rise Time and Noise Bandwidth

Chapter 5 showed that the rise time T_r of a lowpass filter is approximately related as follows to the half-power frequency ω_{hp}:

$$T_R \doteq \frac{2}{\omega_{hp}} = \frac{2}{2\pi f_{hp}} = \frac{1}{\pi f_{hp}} \tag{11.1-52}$$

where f_{hp} is the half-power frequency in hertz. Table 11.1-1 shows that for a single-order lowpass filter (A) the noise bandwidth f_b is related as follows to the half-power frequency f_{hp}:

$$f_b = \frac{\pi}{2} f_{hp} \tag{11.1-53}$$

This same relation holds approximately for all lowpass filters that are reasonably well damped. Hence Eqs 11.1-52, -53 can be combined to give the following general approximation relating the rise time of a lowpass filter to its noise bandwidth:

$$T_R \doteq \frac{1}{2f_b} \tag{11.1-54}$$

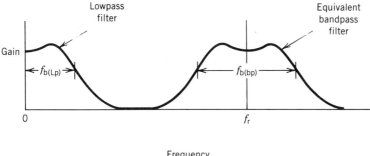

Figure 11.1-9 Relation between frequency-response magnitude plots of equivalent lowpass and bandpass filters.

As was shown in Chapter 8, the bandpass frequency response that is equivalent to a given lowpass frequency response is obtained by shifting the frequency response of the lowpass filter (including its response at negative frequencies) upward in frequency by the carrier (or reference) frequency f_r. This is illustrated in Fig 11.1-9. The figure shows that the noise bandwidth of the bandpass filter $f_{b(bp)}$ is twice the noise bandwidth of the equivalent lowpass filter $f_{b(Lp)}$:

$$f_{b(bp)} = 2f_{b(Lp)} \qquad (11.1\text{-}55)$$

Figure 11.1-10 compares the transient responses of equivalent lowpass and bandpass filters. Diagram a shows the step response of the lowpass filter. Diagram b shows a step-modulated carrier signal of frequency f_r that is fed into the equivalent bandpass filter, and diagram c shows the resultant response. The envelope of the response of the bandpass filter in diagram c is the same as the step response in diagram a of the equivalent lowpass filter. Hence, the rise time T_R of the step response of the lowpass filter is the same as that of the envelope of the response of the equivalent bandpass filter to a step-modulated carrier signal. This rise time is related approximately as follows to the noise bandwidth of the bandpass filter:

$$T_R \doteq \frac{1}{f_{b(bp)}} \qquad (11.1\text{-}56)$$

This is obtained by combining Eqs 11.1-54, -55.

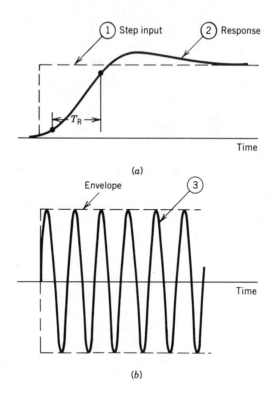

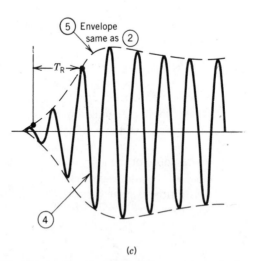

Figure 11.1-10 Waveforms showing effect of the response of a bandpass filters on the envelope of a pulse-modulated carrier signal: (*a*) step response of equivalent lowpass filter; (*b*) step-modulated carrier input to bandpass filter; (*c*) response of bandpass filter to step-modulated carrier.

11.2 BASIC MATHEMATICAL TOOLS FOR STATISTICAL ANALYSIS OF SIGNALS

This section presents a summary of basic mathematical tools used in the statistical analysis of signals. Detailed derivations of these concepts are given by Truxal [11.3] (Chapters 7, 8), by Phillips [11.4], and by Newton, Gould, and Kaiser [11.5].

11.2.1 The Convolution Integral

The convolution integral is an equation for calculating the response of a system in the time domain. It is not a statistical analysis tool per se, but is very important in statistical analysis. It can be derived by physical reasoning in the following manner.

As shown in Fig 11.2-1, assume that an input signal $x_i[t]$ is composed of a large number of narrow pulses of width Δt and magnitude $x_i[t]$. The time interval Δt is made sufficiently small, relative to the response time of the

$$h(\tau)$$

(a) UNIT - IMPULSE RESPONSE

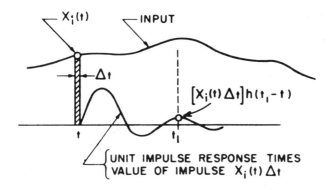

$$X_i(t) \qquad INPUT$$

$$\Delta t$$

$$[X_i(t)\Delta t]h(t_1-t)$$

$$\begin{cases} \text{UNIT IMPULSE RESPONSE TIMES} \\ \text{VALUE OF IMPULSE } X_i(t)\Delta t \end{cases}$$

(b) CALCULATION OF RESPONSE TO INPUT

Figure 11.2-1 Waveforms for deriving the convolution integral. Reprinted with permission from Ref [11.6] © 1957 IRE (now IEEE).

system, for each pulse to have the effect of an impulse of the same area $(x_i[t] \Delta t)$. Diagram a shows the unit-impulse response of the system, which is designated $h[t]$. Diagram b shows the input $x_i[t]$ fed into the system, and one of the pulses into which $x_i[t]$ is divided. The response of the pulse is obtained by multiplying the unit impulse response $h[t]$ by $x_i[t] \Delta t$ to obtain the response curve shown in diagram b. At time t_1 the response to this pulse has the value

$$(x_i[t] \Delta t) h[t_1 - t] \tag{11.2-1}$$

The total response $x_o[t_1]$ to the input $x_i[t]$ at time t_1 is obtained by adding the responses for all of the pulses into which $x_i[t]$ is divided:

$$x_o[t_1] = \sum (x_i[t] \Delta t) h[t_1 - t] \tag{11.2-2}$$

This summation includes all of the pulses between $t = -\infty$ and $t = t_1$. Passing to the limit converts the summation to an integration:

$$x_o[t_1] = \int_{-\infty}^{t_1} dt \, x_i[h] h[t_1 - t] \tag{11.2-3}$$

The convolution process is illustrated in Fig 11.2-2 in a different manner. The time t_1 at which the output is being calculated can be considered to represent the time of observation. The unit-impulse response $h[\tau]$ is shown projected from the time of observation t_1 in the direction of past (or negative) time, and hence represents $h[t_1 - t]$. Multiplying the input curve $x_i[t]$, point by point, by the impulse response $h[t_1 - t]$ gives the product curve $x_i[t] h[t_1 - t]$. By Eq 11.2-3, the response x_o at present time t_1 is the integral of this product curve from $-\infty$ up to the present time t_1, and hence represents the net area under the product curve.

As shown in Fig 11.2-2, the time interval $(t_1 - t)$ is designated τ, which represents elapsed time: a time variable projected from present time t_1 into the past. To express the convolution integral in terms of τ, note that $d\tau$ is equal to $-dt$. Changing the variable from t to τ in Eq 11.2-3 gives

$$x_o[t_1] = \int_{t_1 - \tau = -\infty}^{t_1 - \tau = t_1} (-d\tau) x_i[t_1 - \tau] h[\tau] \tag{11.2-4}$$

Expressing the limits in terms of τ gives

$$x_o[t_1] = \int_{\tau=0}^{\infty} d\tau \, h[\tau] x_i[t_1 - \tau] \tag{11.2-5}$$

This is a more convenient form of the convolution integral. The construction of Fig 11.2-2 shows why the unit impulse response $h[\tau]$ is termed a *weighting function*. The impulse response $h[\tau]$ multiplies, or "weights," the past values

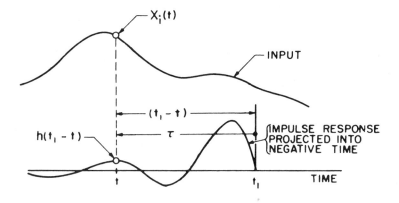

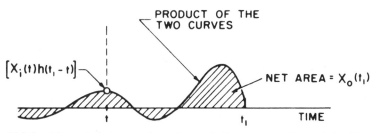

Figure 11.2-2   Alternate interpretation of convolution process. Reprinted with permission from Ref [11.6] © 1957 IRE (now IEEE).

of the input. The impulse response $h[\tau]$ decays to zero at infinite τ, and hence the values of the input $x_i[t]$ far in the past have little effect on the present output $x_o[t_1]$. The more recent values of the input are weighted much more than long-past values. Since the impulse response is zero for negative values of time, the part of the input that has not yet occurred cannot affect the present value of the output.

11.2.2   Correlation Functions

11.2.2.1   *Definition.* The correlation functions are fundamental tools for measuring the statistical properties of signals. Consider a system with an input $x_i[t]$ and an output $x_o[t]$. There are four correlation functions pertaining to these signals, which are

$$\phi_{ii}[\tau] = \text{autocorrelation function of input } x_i[t]$$

$$\phi_{oo}[\tau] = \text{autocorrelation function of output } x_o[t]$$

$$\phi_{io}[\tau], \phi_{oi}[\tau] = \text{cross-correlation functions between input } x_i[t] \text{ and output}$$
$$x_o[t]$$

These correlation functions are defined by

$$\phi_{ii}[\tau] = \lim_{T \to \infty} \frac{1}{2T} \int_{-T}^{+T} dt \, x_i[t] x_i[t + \tau] \qquad (11.2\text{-}6)$$

$$\phi_{oo}[\tau] = \lim_{T \to \infty} \frac{1}{2T} \int_{-T}^{+T} dt \, x_o[t] x_o[t + \tau] \qquad (11.2\text{-}7)$$

$$\phi_{io}[\tau] = \lim_{T \to \infty} \frac{1}{2T} \int_{-T}^{+T} dt \, x_i[t] x_o[t + \tau] \qquad (11.2\text{-}8)$$

$$\phi_{oi}[\tau] = \lim_{T \to \infty} \frac{1}{2T} \int_{-T}^{+T} dt \, x_o[t] x_i[t + \tau] \qquad (11.2\text{-}9)$$

Some important characteristics of the correlation functions can be obtained by inspection. The autocorrelation functions have the same value if τ is replaced by $-\tau$. Hence

$$\phi_{ii}[\tau] = \phi_{ii}[-\tau] \qquad (11.2\text{-}10)$$

and similarly for $\phi_{oo}[\tau]$. Thus, an autocorrelation function is an even function of τ, but the cross-correlation functions are generally not. In Eq 11.2-8, if the variable τ in $\phi_{io}[\tau]$ is replaced by $-\tau$, the relation is the same as that for $\phi_{oi}[\tau]$ of Eq 11.2-9. Thus,

$$\phi_{io}[\tau] = \phi_{oi}[-\tau] \qquad (11.2\text{-}11)$$

The value of the autocorrelation function $\phi_{ii}[\tau]$ of Eq 11.2-6 at $\tau = 0$ is

$$\phi_{ii}[0] = \lim_{T \to \infty} \frac{1}{2T} \int_{-T}^{+T} dt \, x_i[t]^2 \qquad (11.2\text{-}12)$$

By definition, this is the average of x_i^2, which is the mean square value of the input, designated $\langle x_i^2 \rangle$. Thus

$$\phi_{ii}[0] = \langle x_i^2 \rangle \qquad (11.2\text{-}13)$$

11.2.2.2   *Calculation of Correlation Functions.* The process of computing an autocorrelation function is illustrated in Fig 11.2-3. Diagram *a* shows a plot of an input function $x_i[t]$ and the function $x_i[t + \tau_1]$, which is the same plot shifted to the right by the time shift τ_1. These two curves, $x_i[t]$ and $x_i[t + \tau_1]$, are multiplied together, point by point, to obtain the product curve $x_i[t] x_i[t + \tau_1]$ shown in diagram *b*. The average value of this product curve is

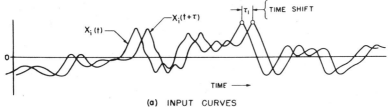

(a) INPUT CURVES

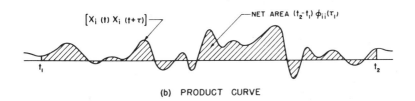

(b) PRODUCT CURVE

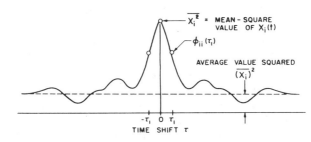

(c) AUTOCORRELATION FUNCTION $\phi_{ii}(\tau)$

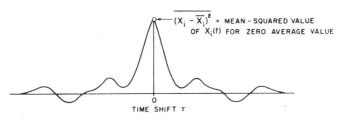

(d) AUTOCORRELATION FUNCTION FOR SIGNAL WITH
ZERO AVERAGE VALUE

Figure 11.2-3 Description of autocorrelation process. Reprinted with permission from Ref [11.6] © 1957 IRE (now IEEE).

the autocorrelation function value $\phi_{ii}[\tau_1]$. Thus, $\phi_{ii}[\tau_1]$ is

$$\phi_{ii}[\tau_1] = \big\langle x_i[t] x_i[t + \tau_1] \big\rangle$$

$$= \lim_{t_2 - t_1 \to \infty} \frac{1}{t_2 - t_1} \int_{t_1}^{t_2} dt\, x_i[t] x_i[t + \tau_1] \qquad (11.2\text{-}14)$$

In practice the averaging is performed over a finite interval of time $(t_2 - t_1)$.

However, if this time interval is much greater than the maximum time shift τ_1, the value $\phi_{ii}[\tau_1]$ is approximated with high accuracy. The expression for $\phi_{ii}[\tau_1]$ given in Eq 11.2-14 is equivalent to the more common form given in Eq 11.2-6.

The averaging process of the product curve described in diagram b gives one point $\phi_{ii}[\tau_1]$ on the convolution curve, as shown in diagram c. If the interval $(t_2 - t_1)$ can be considered infinite, it makes no difference in diagram a if the second curve is shifted in the positive or negative direction by the amount τ_1. In either case there are two identical curves shifted from one another by the amount τ_1. Thus, the autocorrelation function $\phi_{ii}[\tau]$ must have the same value at $-\tau$ that it has at $+\tau$, and so $\phi_{ii}[\tau]$ is symmetric about the zero-τ axis.

If the input $x_i[t]$ does not have any periodic components, the variations in $x_i[t]$ are uncorrelated for large time shifts. Consequently, as τ approaches infinity the autocorrelation function $\phi_{ii}[\tau]$ approaches a constant equal to the square of the average value of the input x_i, which is designated $\langle x_i \rangle^2$.

The final value of $\phi_{ii}[\tau]$ is almost always of no interest. Therefore it is generally removed, either by subtracting from the original input curve $x_i[t]$ its average value $\langle x_i \rangle$, or by subtracting from the autocorrelation function the square of the average value of the input, $\langle x_i \rangle^2$. Consequently, the autocorrelation function that is actually used has the form of Fig 11.2-3d, which has a final value of zero.

The cross-correlation function $\sigma_{io}[\tau]$ given in Eq 11.2-8 is calculated in the same manner as the autocorrelation function, except that the input $x_i[t]$ is multipled by the shifted output $x_o[t + \tau]$, and the average of the resultant product is calculated. To calculate the other cross-correlation function, $\phi_{oi}[\tau]$, the output curve $x_o[t]$ is held fixed, and the input $x_i[t]$ is shifted forward to obtain $x_i[t + \tau]$. The average of the resultant product curve $x_o[t]x_i[t + \tau]$ is taken to obtain the value of the cross-correlation function.

The same value results by shifting the output curve forward in time by τ_1 or by shifting the input curve backward in time by τ_1. Hence $\phi_{io}[\tau]$ is equal to $\phi_{oi}[-\tau]$, as was shown in Eq 11.2-11. On the other hand, the sence of the relative shift between the two curves usually makes a difference. Therefore, $\phi_{io}[\tau]$ is generally not equal to $\phi_{io}[-\tau]$.

In calculating cross-correlation functions, average values of the input $x_i[t]$ and output $x_o[t]$ are generally removed by subtraction. Hence, the final values of the cross-correlation functions are generally zero.

11.2.2.3 Transient Approach for Relating Correlation Functions.

Figure 11.2-4 shows that the correlation functions can be related to one another in a simple fashion by regarding them as transient input and output signals of the system. (See Ref [11.6].) The input autocorrelation function $\phi_{ii}[\tau]$ is assumed to be a transient function of time, which is fed into the system $H[\omega]$. As shown, the resultant transient output from the system is the cross-correlation function $\phi_{io}[\tau]$. From this, the other cross-correlation function $\phi_{oi}[\tau]$ is formed by rotating the $\phi_{io}[\tau]$ transient about the zero-time axis. This $\phi_{oi}[\tau]$

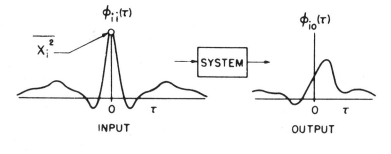

(a) RESPONSE TO $\phi_{ii}(\tau)$

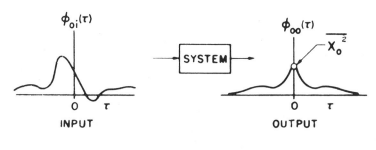

(b) RESPONSE TO $\phi_{oi}(\tau)$

Figure 11.2-4   Transient response of system to correlation-function input. Reprinted with permission from Ref [11.6] © 1957 IRE (now IEEE).

transient is fed into the system, and the resultant system output is the output autocorrelation function $\phi_{oo}[\tau]$.

The relations illustrated in Fig 11.2-4 can be readily derived by using the convolution integral along with the definitions of the correlation functions. The following is a derivation of the transient relationship shown in diagram *a*. The convolution integral of Eq 11.2-5 gives the following for the output response $y[t]$ of the system to the transient input $\phi_{ii}[t]$:

$$y[t] = \int_0^\infty d\tau\, h[\tau]\phi_{ii}[t - \tau]$$

$$= \int_0^\infty d\tau\, h[\tau]\left\{ \lim_{T \to \infty} \frac{1}{2T} \int_{-T}^{T} dt_1\, x_i[t_1]x_i[t_1 + (t - \tau)]\right\} \quad (11.2\text{-}15)$$

where $h[\tau]$ is the impulse response of the system, and the function within the braces { } is the expression for $\phi_{ii}[t - \tau]$ obtained from Eq 11.2-6. Note that t_1 is the time variable of the autocorrelation integral. The order of integration of Eq 11.2-15 can be reversed, which gives

$$y[t] = \lim_{T \to \infty} \frac{1}{2T} \int_{-T}^{T} dt_1 \, x_i[t_1] \left\{ \int_0^\infty d\tau \, h[\tau] x_i[(t_1 + t) - \tau] \right\} \quad (11.2\text{-}16)$$

The expression within the braces is a convolution of the input $x_i[t_1 + t]$ with the system impulse response $h[\tau]$, and so represents the output $x_o[t_1 + t]$. Thus, the convolution integral of Eq 11.2-5 gives

$$x_o[t_1 + t] = \int_0^\infty d\tau \, h[\tau] x_i[(t_1 + t) - \tau] \quad (11.2\text{-}17)$$

Substituting Eq 11.2-17 into Eq 11.2-16 gives

$$y[t] = \lim_{T \to \infty} \frac{1}{2T} \int_{-T}^{T} dt_1 \, x_i[t_1] x_o[t_1 + t] = \phi_{io}[t] \quad (11.2\text{-}18)$$

As shown, this represents $\phi_{io}[t]$, in accordance with the definition of the cross-correlation function in Eq 11.2-8.

11.2.3 Spectral Densities

Since the correlation functions are not zero at negative values of time, they do not have a Laplace transform. However, if the average values of the time functions are subtracted, so that the final values of the correlation functions are zero, the correlation functions do have Fourier transforms. The Fourier transform of the autocorrelation function is called the *spectral density*, or sometimes the *power spectrum*. It is designated $\Phi_{ii}[\omega]$ and $\Phi_{oo}[\omega]$ for the input and output signals. Applying the Fourier-transform relations given in Chapter 2 (Eqs 2.2-20, -21) yields

$$\Phi_{ii}[\omega] = \int_{-\infty}^{\infty} \phi_{ii}[\tau] e^{-j\omega\tau} \, d\tau \quad (11.2\text{-}19)$$

$$\phi_{ii}[\tau] = \frac{1}{2\pi} \int_{-\infty}^{\infty} \Phi_{ii}[\omega] e^{j\omega\tau} \, d\omega \quad (11.2\text{-}20)$$

Since an autocorrelation function is an even function of time (e.g., $\phi_{ii}[\tau]$ is equal to $\phi_{ii}[-\tau]$, the spectral densities $\Phi_{ii}[\omega]$ and $\Phi_{oo}[\omega]$ have zero phase. The Fourier transforms of the cross-correlation functions $\phi_{io}[\tau]$ and $\phi_{oi}[\tau]$ are designated $\Phi_{io}[\omega]$ and $\Phi_{oi}[\omega]$. Unlike the spectral densities, these transforms generally have a phase response.

The spectral density can also be calculated by transforming the time function directly, rather than by transforming its autocorrelation function.

Designate the Fourier transform of the time function $x_i[t]$ measured over the time interval $-T < t < +T$ as $X_{i(T)}[\omega]$. The spectral density of $x_i[t]$ can be shown to be

$$\Phi_{ii}[\omega] = \lim_{T \to \infty} \frac{1}{2T} X_{i(T)}[\omega] X_{i(T)}[\omega]^*$$

$$= \lim_{T \to \infty} \frac{1}{2T} |X_{i(T)}[\omega]|^2 \qquad (11.2\text{-}21)$$

The spectral density can be measured directly from the signal by means of a narrow-bandwidth variable-frequency filter. If the input signal $x_i[t]$ is passed through a narrow-bandwidth filter centered at the frequency ω_1, the average value of the output from the filter, divided by the bandwidth of the filter, approaches $\Phi_{ii}[\omega_1]$ as the bandwidth approaches zero.

As was shown in Fig 11.2-4, the time response of the system to the transient input $\phi_{ii}[t]$ is the function $\phi_{io}[t]$. Hence, the Fourier transforms of these correlation functions are related by

$$\Phi_{io}[\omega] = H[\omega]\Phi_{ii}[\omega] \qquad (11.2\text{-}22)$$

where $H[\omega]$ is the frequency response of the system. Similarly, the time response of the system to the transient input $\phi_{oi}[t]$ is $\phi_{oo}[t]$. Hence, the transforms of these correlation functions are related by

$$\Phi_{oo}[\omega] = H[\omega]\Phi_{oi}[\omega] \qquad (11.2\text{-}23)$$

If a time function $f_2[t]$ is equal to $f_1[-t]$, their Fourier transforms are related by

$$F_2[\omega] = F_1[-\omega] = F_1[\omega]^* \qquad (11.2\text{-}24)$$

Remember that $F[-\omega]$ is equal to $F[\omega]^*$, which is the complex conjugate of $F[\omega]$. Thus, the cross-correlation transforms are related to each other by

$$\Phi_{oi}[\omega] = \Phi_{io}[-\omega] = \Phi_{io}[\omega]^* \qquad (11.2\text{-}25)$$

Substitute Eq 11.2-25 into Eq 11.2-23:

$$\Phi_{oo}[\omega] = H[\omega]\Phi_{io}[\omega]^* \qquad (11.2\text{-}26)$$

Substitute for $\Phi_{io}[\omega]^*$ the conjugate of Eq 11.2-22:

$$\Phi_{oo}[\omega] = H[\omega]H[\omega]^*\Phi_{ii}[\omega]^*$$

$$= H[\omega]H[\omega]^*\Phi_{ii}[\omega] = |H[\omega]|^2\Phi_{ii}[\omega] \qquad (11.2\text{-}27)$$

Since $\Phi_{ii}[\omega]$ has no phase, it is equal to $\Phi_{ii}[\omega]^*$. This equation shows that the output spectral density is equal to the input spectral density multiplied by the square of the magnitude of the system frequency response. This principle was derived earlier in Section 11.1 from physical reasoning, to develop the definition of noise bandwidth.

By Eq 11.2-20, the value of $\phi_{oo}[\tau]$ for $\tau = 0$ can be calculated as follows from the spectral density $\Phi_{oo}[\omega]$:

$$\phi_{oo}[0] = \frac{1}{2\pi} \int_{-\infty}^{\infty} \Phi_{oo}[\omega] \, d\omega = \int_{-\infty}^{\infty} \Phi_{oo}[f] \, df \qquad (11.2\text{-}28)$$

where $d\omega$ is replaced by $2\pi \, df$. Since $\Phi_{oo}[-f]$ is equal to $\Phi_{oo}[f]$, the integral of $\Phi_{oo}[f]$ from $-\infty$ to 0 is equal to the integral from 0 to $+\infty$. Also, by Eq 11.2-13, $\phi_{oo}[0]$ is equal to $\langle x_o^2 \rangle$, the mean square value of the output x_o. Hence

$$\langle x_o^2 \rangle = \phi_{oo}[0] = 2 \int_0^{\infty} \Phi_{oo}[f] \, df \qquad (11.2\text{-}29)$$

Substituting Eq 11.2-29 into Eq 11.2-27 gives the following expression for the mean square value of the output in terms of the spectral density of the input:

$$\langle x_o^2 \rangle = 2 \int_0^{\infty} \Phi_{ii}[f] |H[f]|^2 \, df$$

$$= \frac{1}{\pi} \int_0^{\infty} \Phi_{ii}[\omega] |H[\omega]|^2 \, d\omega \qquad (11.2\text{-}30)$$

11.2.4 Calculation of Noise Bandwidth

Frequency-response calculations of noise bandwidth and mean square error can be greatly simplified by using tables of integrals of the general form

$$I_n = \frac{1}{2\pi j} \int_{-j\infty}^{+j\infty} H[s] H[-s] \, ds \qquad (11.2\text{-}31)$$

The transfer function $H[s]$ is defined as follows in terms of a ratio of two polynomials in s, where the order of the numerator is less than that of the denominator:

$$H[s] = \frac{c_{n-1} s^{n-1} + \cdots + c_1 s + c_0}{d_n s^n + d_{n-1} s^{n-1} + \cdots + d_1 s + d_0} \qquad (11.2\text{-}32)$$

The parameter n is the order of s in the denominator of $H[s]$. Appendix G gives equations for the integral I_n for values of n from 1 to 6, as functions of the coefficients c_m, d_m of the numerator and denominator of $H[s]$. These were obtained from Appendix E of Newton, Gould, and Kaiser [11.5]. That reference shows how these integrals are derived, and gives expressions for the integrals for values of n from 1 to 10. Somewhat different and more cumbersome forms of these integrals were given earlier in the Appendix of James, Nichols, and Phillips [1.2], which listed the integrals for $n = 1$ to $n = 7$.

The noise-bandwidth formula of Eq 11.1-9 can be expressed as follows in rad/sec:

$$\omega_b = \int_0^\infty |H[\omega]|^2 \, d\omega = \int_0^\infty H[\omega]H[-\omega] \, d\omega$$

$$= \frac{1}{2}\int_{-\infty}^\infty H[\omega]H[-\omega] \, d\omega \qquad (11.2\text{-}33)$$

The following shows that ω_b/π is equal to the integral expression I_n defined by Eq 11.2-31:

$$\frac{\omega_b}{\pi} = \frac{1}{2\pi}\int_{-\infty}^\infty H[\omega]H[-\omega] \, d\omega = \frac{1}{2\pi j}\int_{-\infty}^\infty H[j\omega]H[-j\omega] \, d(j\omega)$$

$$= \frac{1}{2\pi j}\int_{-j\infty}^{+j\infty} H[s]H[-s] \, ds = I_n \qquad (11.2\text{-}34)$$

Thus, the noise bandwidth of any transfer function $H[s]$ which does not exceed sixth order in s in the denominator, can be obtained from the integral expressions of Appendix G. The noise bandwidth for a higher order of s in the denominator (up to 10th order) can be obtained from Ref [11.5]. The noise bandwidth ω_b in rad/sec is equal to the tabulated integral expression I_n multiplied by π. Hence, the noise bandwidth f_b in hertz is equal to

$$f_b = I_n/2 \qquad (11.2\text{-}35)$$

For example, the G_{ib} transfer function for loop F shown in Table 11.1-2 is

$$G_{ib} = \frac{s\omega_c\omega_f + \omega_c\omega_f\omega_i}{s^3 + \omega_f s^2 + \omega_c\omega_f s + \omega_c\omega_f\omega_i} \qquad (11.2\text{-}36)$$

In accordance with Eq 11.2-32, the polynomial coefficients are

$$d_3 = 1, \qquad d_2 = \omega_f, \qquad d_1 = \omega_c\omega_f, \qquad d_0 = \omega_c\omega_f\omega_i \quad (11.2\text{-}37)$$

$$c_2 = 0, \qquad c_1 = \omega_c\omega_f, \qquad c_0 = \omega_c\omega_f\omega_i \qquad (11.2\text{-}38)$$

Substitute $c_2 = 0$, $d_3 = 1$ in the expression for I_3 shown in Eqs G-5, G-6 of

Appendix G. This gives

$$I_3 = \frac{c_2^2 d_0 d_1 + \left(c_1^2 - 2c_0 c_2\right) d_0 d_3 + c_0^2 d_2 d_3}{2 d_0 d_3 \left(d_1 d_2 - d_0 d_3\right)}$$

$$= \frac{c_1^2 d_0 + c_0^2 d_2}{2 d_0 \left(d_1 d_2 - d_0\right)} \qquad (11.2\text{-}39)$$

Substitute into this the remaining coefficients of Eqs 11.2-37, -38. This gives

$$I_3 = \frac{\left(\omega_c \omega_f\right)^2 \left(\omega_c \omega_f \omega_i\right) + \left(\omega_c \omega_f \omega_i\right)^2 \left(\omega_f\right)}{2 \left(\omega_c \omega_f \omega_i\right) \left[\left(\omega_c \omega_f\right)\left(\omega_f\right) - \left(\omega_c \omega_f \omega_i\right)\right]}$$

$$= \frac{\omega_c \omega_f + \omega_i \omega_f}{2 \left(\omega_f - \omega_i\right)} = \frac{\omega_f \left(\omega_c + \omega_i\right)}{2 \left(\omega_f - \omega_i\right)} \qquad (11.2\text{-}40)$$

Hence, for G_{ib} of Eq 11.2-36, the noise bandwidth in hertz is

$$f_b = \frac{I_3}{2} = \frac{1}{4} \frac{\omega_f \left(\omega_c + \omega_i\right)}{\omega_f - \omega_i} \qquad (11.2\text{-}41)$$

The noise bandwidth in rad/sec is

$$\omega_b = 2 \pi f_b = \frac{\pi}{2} \frac{\omega_f \left(\omega_c + \omega_i\right)}{\omega_f - \omega_i} \qquad (11.2\text{-}42)$$

11.3 ERROR IN A RADAR TRACKING SYSTEM DUE TO RECEIVER NOISE

A good example of a feedback-control application where statistical analysis of noise is very important is a tracking radar system. The detection and tracking processes of a tracking radar system are analyzed in a simple manner by the author in Ref [1.14] (Chapter 7). The following is a summary of the equations derived in that reference for calculating the tracking errors in range and angle.

11.3.1 Transmission-Path Loss of Radar Example

Figure 11.3-1 is a simplified drawing of a tracking radar antenna, which has a parabolic reflector, in front of which is a feed supported on spars. Energy radiating from the feed is reflected by the paraboloid and projected in the direction of the antenna axis (the boresight), but the beam diverges somewhat because of diffraction. If there is a target in the transmitted beam, it reflects a

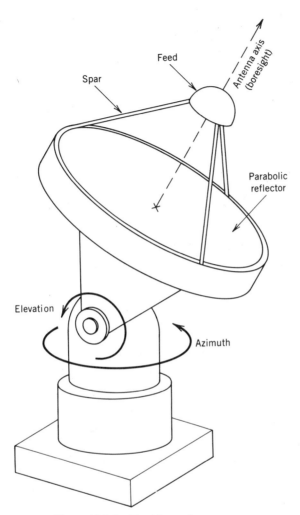

Figure 11.3-1 Tracking radar antenna.

small amount of energy back to the antenna, which is intercepted by the reflector and projected back to the feed. The direction of the antenna beam is controlled by servos, which drive the antenna gimbal structure in two axes, called azimuth and elevation. The azimuth servo rotates the whole structure about a vertical axis, thereby changing its heading relative to north. The elevation gimbal rides on the rotating azimuth carriage, and tilts the antenna boresight upward and downward relative to the horizontal plane.

The analysis in Ref [1.24] (Chapter 7) is based on a specific radar example, the basic parameters of which are listed in Table 11.3-1. The radar operates at a frequency f of 10 GHz [item (1)]. The wavelength λ is related to the

TABLE 11.3-1   Calculation of Signal Power into Receiver

Symbol	Parameter	Value
(1) f	Frequency	10 GHz
(2) λ	Wavelength	3.0 cm
(3) D	Reflector diameter	4.0 ft = 122 cm
(4) η	Antenna efficiency	0.60
(5) G_m	Peak antenna gain	9390
(6) Θ_b	Half-power bandwidth	1.60°
(7) R	Range to target	50 km
(8) A_t	Radar cross section	1.0 m²
(9) P_r/P_t	Transmission loss^{-1}	6.398×10^{-18}
(10) $1/L_w$	Waveguide loss^{-1}	-2 dB = 0.631
(11) P_{gen}	Transmitter power	100 kW
(12) P_s	Receiver signal power	4.03×10^{-13} W
(13) τ_p	Radar pulse width	0.1 μsec
(14) E_s	Receiver signal energy	4.03×10^{-20} J
(15) ΔR	Range resolution	15 m

frequency f by

$$\lambda = c/f \qquad (11.3\text{-}1)$$

where c is the speed of light (3×10^8 m/sec). As shown in item (2), the corresponding wavelength λ is 3 cm. The antenna diameter D is assumed to be 4 ft, or 122 cm [item (3)].

The peak (or maximum) gain G_m of the antenna gain pattern is equal to

$$G_m = \eta \left(\frac{\pi D}{\lambda} \right)^2 \qquad (11.3\text{-}2)$$

where η is the antenna efficiency. A reasonable value for the antenna efficiency for a radar tracking antenna is 0.6, as shown in item (4). Applying the values of items (2), (3), (4) to Eq 11.3-2 gives the maximum antenna gain G_m shown in item (5), which is 9390.

The half-power beamwidth of an antenna gain pattern is designated Θ_b. This is the angle between points 3 dB below peak gain. The approximate value for Θ_b is

$$\Theta_b \doteq 65(\lambda/D) \text{ deg} \qquad (11.3\text{-}3)$$

Applying the values of items (2), (3) to Eq 11.3-3 gives an approximate 3-dB beamwidth Θ_b of 1.60°, as shown in item (6).

Near the peak of the antenna beam, any practical tracking antenna has a gain pattern that is approximately Gaussian, and so can be expressed as

$$\frac{G}{G_m} = \exp\left[-\left(\frac{\phi}{\Psi}\right)^2\right]$$

(11.3-4)

where ϕ is the angle of the target signal from the peak of the antenna beam (the boresight), and Ψ is a constant that is proportional to the beamwidth. The ratio G/G_m should be $\frac{1}{2}$ when $\phi = \pm\Theta_b/2$. Setting this condition in Eq 11.3-4 gives

$$\Psi = 0.601\Theta_b$$

(11.3-5)

Take 10 times the logarithm of the reciprocal of Eq 11.3-4. This gives the following expression for the loss in dB, relative to peak antenna gain, due to beam pointing error ϕ:

$$\text{dB loss} = 12(\phi/\Theta_b)^2 \text{ dB}$$

(11.3-6)

Note that this loss is 3 dB when $\phi = \pm\Theta_b/2$.

It is assumed that the radar is tracking a target at a range R of 50 kilometers [item (7)]. The assumed radar cross section A_t of the target is 1.0 square meter [item (8)], which is a typical value for a small aircraft. The ratio of receive power P_r to transmit power P_t is given by the basic "radar-range" equation:

$$\frac{P_r}{P_t} = \frac{G^2\lambda^2 A_t}{(4\pi)^3 R^4}$$

(11.3-7)

where G is the antenna gain in the direction of the target. (This equation neglects atmospheric attenuation, which will be ignored.)

It is assumed that the angle tracking error is relatively small, so that the target is close to the peak of the radar beam. Hence antenna gain G in Eq 11.3-7 can be set equal to G_m given in item (5). Using the values of λ, R, and A_t of items (2), (7), (8) gives the transmission ratio P_r/P_t in item (9). The transmission loss is the reciprocal of this value, and so item (9) is called the reciprocal loss, which is designated loss^{-1}. Waveguide elements between the transmitter tube and the antenna, and between the antenna and the receiver, produce a signal loss, which is called the waveguide loss L_w. The reciprocal waveguide loss L_w^{-1} is assumed to be -2 dB, which represents a ratio of 0.631 [item (10)]. The peak power generated in the transmitter is designated P_{gen}, and is assumed to be 100 kilowatt [item (11)]. Multiplying items (9), (10), (11) gives the received signal power P_s at the input to the receiver. As shown in item (12), the signal power P_s is 4.03×10^{-13} watt.

The radar generates a pulsed waveform. The pulse width is designated τ_p, and is assumed to be 0.1 microsecond, as shown in item (13). The received signal energy is designated E_s, and is equal to the received signal power P_s multiplied by the pulse width τ_p:

$$E_s = P_s\tau_p \tag{11.3-8}$$

The received signal power P_s of item (12) is multiplied by the pulse width τ_p of item (13) to obtain the received signal energy E_s shown in item (14), which is 4.03×10^{-20} joule.

The time t for the radar pulse to travel from the radar to the target at range R, and return, is equal to

$$t = 2R/c \tag{11.3-9}$$

where c is the speed of light (3×10^8 m/sec). Hence, the range increment ΔR corresponding to a time increment Δt is given by

$$\frac{\Delta R}{\Delta t} = \frac{c}{2} = 1.5 \times 10^8 \text{ m/sec} = 150 \text{ m/}\mu\text{sec} \tag{11.3-10}$$

The 0.1-μsec pulse width τ_p of item (13) is multiplied by this factor 150 m/μsec to obtain the corresponding range increment ΔR shown in item (15), which is 15 m. This is the range resolution of the radar.

11.3.2   Calculating the RMS Range-Tracking Noise Error for a Single Radar Pulse

Figure 11.3-2 shows a block diagram of the radar receiver, including the range-tracking loop and the automatic-gain-control (AGC) loop. The receiver signal consists of a 10-GHz carrier, modulated with rectangular pulses of

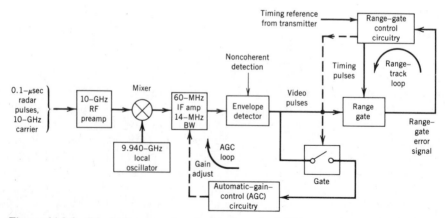

Figure 11.3-2   Block diagram of radar receiver, including range-gate and AGC circuitry.

0.1-μsec duration. This signal is fed through a 10-GHz RF preamplifier, and then to a mixer. (In early vacuum-tube radar receivers, RF amplification was not practical, and so the received radar signal was fed directly to the mixer.) The mixer multiples (or heterodynes) the received radar signal with a local-oscillator signal at 9.940 GHz, to produce sum and difference frequency components at 19.940 GHz and at 0.060 GHz. (This is mathematically the same as the modulation process discussed in Chapter 8, Section 8.1.) The sum signal at 19.940 GHz is rejected by filtering, and the difference signal at 0.060 GHz (or 60 MHz) is fed into the IF amplifier.

The 60-MHz IF amplifier has a 14-MHz bandwidth. As will be explained, this bandwidth is chosen to optimize the signal/noise ratio of the detected radar pulses. In accordance with Eq 11.1-56 in Section 11.1.5, the resultant rise time of the pulse envelope at the output of the IF amplifier is approximately

$$T_R \doteq \frac{1}{f_{b(bp)}} = \frac{1}{14\text{ MHz}} = 0.070\ \mu\text{sec} \qquad (11.3\text{-}11)$$

Since this rise time is significant relative to the 0.1-μsec pulse width of the radar signal, the pulse modulation at the output of the IF amplifier is smoothed appreciably.

The output signal from the IF amplifier is fed through an envelope detector, which detects the peak value of each cycle of the waveform. This generates a video pulse, which (except for noise) has the shape of the envelope of the IF signal. This video pulse is fed into a range-gate circuit, which generates a range gate that tracks the video pulse, and provides a signal proportional to the range-gate tracking error.

An early–late range-gate circuit is assumed, with the block diagram shown in Fig 11.3-3. The gating function provided by the circuit is shown in Fig 11.3-4. The parameter τ_g is defined as

τ_g = range-gate width = total duration of early gate plus late gate.

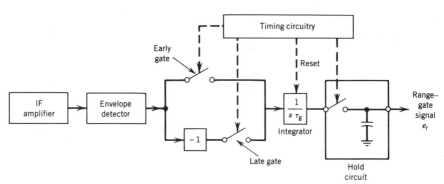

Figure 11.3-3   Basic design of early–late range-gate detector.

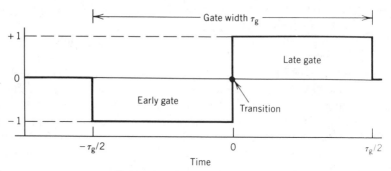

Figure 11.3-4   Range-gate function.

The early gate is closed for the time interval $\tau_g/2$ preceding the estimated time of the center of the received pulse, and the late gate is closed for an equal time after that estimated time. The range-gate circuit generates a range-error signal equal to the time difference between the actual center of the pulse and the estimated center of the pulse. This error signal is used in a feedback manner to vary the timing of the range gate, to keep the received pulse centered in the range gate.

Figure 11.3-5 shows the error response characteristic provided by the early–late gate circuit. This is a plot of the range-gate error signal, versus the timing error between the center of the radar pulse and the center of the range gate. The solid curve shows the error characteristic when the gate width τ_g is greater than $2\tau_p$. The dashed curves show the conditions for $\tau_g = 2\tau_p$ and for $\tau_g < 2\tau_p$.

The range tracker normally operates in the central (linear) region of the characteristic of Fig 11.3-5, between points 3 and 5. It is generally desirable that this central linear region be as wide as possible. Hence, τ_g should not be

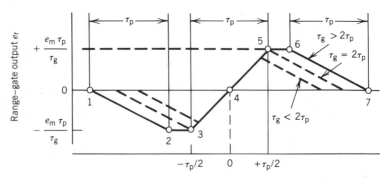

Figure 11.3-5   Plot of output voltage from range gate versus range-gate timing error.

less than $2\tau_{\mathrm{p}}$. When τ_{g} is greater than $2\tau_{\mathrm{p}}$ the range-tracking noise error is increased. Therefore, a reasonable operating condition is a range-gate width τ_{g} that is twice the pulse width:

$$\tau_{\mathrm{g}} = 2\tau_{\mathrm{p}} \tag{11.3-12}$$

This condition is assumed in the example.

The IF amplifier generally consists of several stagger-tuned stages. The resultant bandpass response can be approximated quite well by an ideal rectangular bandpass filter. As shown in Ref [1.14] (Chapter 7), the optimum noise bandwidth of a rectangular bandpass filter for the assumed range gate is

$$f_{\mathrm{b(bp)}} = 1.4/\tau_{\mathrm{p}} \tag{11.3-13}$$

For the assumed pulse width ($\tau_{\mathrm{g}} = 0.1$ μsec), the optimum noise bandwidth of the bandpass IF amplifier is 14 MHz.

The RMS range-tracking error per pulse for the example is calculated in Table 11.3-2. The receiver noise power density at the input to the IF amplifier is designated p_{n}, and is equal to

$$p_{\mathrm{n}} = KT \tag{11.3-14}$$

where K is Boltzmann's constant, and T is the noise temperature of the receiver. As shown in item (1), the receiver noise temperature is assumed to be 700 °K. Item (2) gives the Boltzmann constant K, which was obtained from Section 11.1, Eq 11.1-2. By Eq 11.3-14, the temperature T in item (1) is multiplied by K in item (2) to obtain the noise power density p_{n} of item (3).

TABLE 11.3-2   Calculation of RMS Range Tracking Error per Pulse, σR_1

Symbol	Parameter	Value
(1) T	Receiver noise temperature	700 K
(2) K	Boltzmann's constant	1.374×10^{-23} J/°K
(3) $p_{\mathrm{n}} = KT$	Noise power density	9.62×10^{-21} W/Hz
(4) E_{s}	Receiver signal energy	4.03×10^{-20} J
(5) $E_{\mathrm{s}}/p_{\mathrm{n}}$	Matched-filter signal/noise	4.189
(6) L_{m}	Receiver matching loss	1.216 = 0.85 dB
(7) $P_{\mathrm{s}}/P_{\mathrm{n}}$	Signal/noise into detector	3.44
(8) $P_{\mathrm{n}}/P_{\mathrm{s}}$	Noise/signal into detector	0.291
(9) L_{d}	Noncoherent detection loss	1.291
(10) p_{nd}	Detected noise power density	1.24×10^{-20} W/Hz
(11) $E_{\mathrm{s}}/p_{\mathrm{nd}}$		3.25
(12) $\sigma t_1/\tau_{\mathrm{p}}$	Rel. RMS time error per pulse	0.261
(13) τ_{p}	Pulse width	0.1 μsec
(14) σt_1	RMS time error per pulse	0.0261 μsec
(15) σR_1	RMS range error per pulse	3.92 m

In calculating signal/noise ratios, the signal gain of the receiver is ignored, and all signal and noise levels are referenced to the input to the receiver. Hence the noise power density p_n of item (3) holds at the output of the IF amplifier as well as at the input.

The ideal receiver for any signal is called the matched filter for that signal. For practical reasons, a matched filter is usually not implemented exactly. However, it provides an ideal reference against which actual receivers are compared. The signal/noise power ratio at the output of a matched-filter receiver is

Signal/noise at output of matched filter:

$$\text{Matched}\left[\frac{P_s}{P_n}\right] = \frac{E_s}{p_n} \qquad (11.3\text{-}15)$$

The signal/noise ratio P_s/P_n is the ratio of signal power to noise power at the time of the peak signal. In Table 11.3-2, item (4) is the received signal energy E_s, which was obtained from Table 11.3-1, item (14). This value for E_s in item (4) is divided by p_n in item (3), to obtain the ratio E_s/p_n of item (5). By Eq 11.3-12, this is the signal/noise power ratio at the output of an ideal matched-filter receiver. Note that this signal/noise power ratio, which is 4.189, is nondimensional.

As shown in Ref [1.14] (Chapter 7), the IF amplifier for the assumed amplifier and range gate has a matching loss L_m of 0.85 dB, which is a factor of 1.216. This matching loss, shown in item (6), describes the reduction of actual signal/noise ratio relative to that of an ideal matched-filter receiver. Thus, the actual signal/noise power ratio at the output of the IF amplifier is

Signal/noise at IF amplifier output:

$$\frac{P_s}{P_n} = \frac{1}{L_m}\text{ matched}\left[\frac{P_s}{P_n}\right] = \frac{E_s/p_n}{L_m} \qquad (11.3\text{-}16)$$

Therefore, the ratio E_s/p_n of item (5) is divided by the matching loss L_m of item (6) to obtain the signal/noise power ratio at the output of the actual IF amplifier, shown in item (7). This is the signal/noise power ratio at the input to the envelope detector.

There are two signal-detection processes in Fig 11.3-2: the mixing process, which converts the 10-GHz carrier to 60 MHz; and the envelope detector, which converts the pulse-modulated 60-MHz carrier to video pulses. The mixer cross-correlates the received signal with the local oscillator, and so provides coherent detection. This coherent detection process maintains phase information as well as amplitude information. In contract, the envelope detector destroys phase information in the detection process, and so is a noncoherent detector.

With coherent detection, the signal/noise power ratio is unaffected by the detection process. However, with noncoherent detection, there is a loss of signal/noise ratio when the signal/noise ratio at the input to the noncoherent detector is not much greater than unity. The effect of noncoherent detection can be approximated by including in the analysis the following noncoherent detection loss L_d:

Noncoherent detection loss:

$$L_d = 1 + P_n/P_s \qquad (11.3\text{-}17)$$

The term P_n/P_s is the noise/signal power ratio at the input to the noncoherent detector, which is the reciprocal of the signal/noise power ratio.

For example, Eq 11.3-17 shows that, if the signal/noise power ratio P_s/P_n at the input to the envelope detector (or any noncoherent detector) is unity, the noncoherent detection loss L_d is 2, and so there is a 3-dB loss in the detector. If this signal/noise ratio is 10, the noncoherent detection loss L_d is 1.1 (or 0.4 dB). On the other hand, if the signal/noise ratio is 0.1, the noncoherent detection loss L_d is 11 (or 10.4 dB).

This illustrates the principle that the signal/noise ratio at the input to a noncoherent detection process should, if possible, be appreciably greater than unity. If this is achieved, the noncoherent detection loss is small, and the noncoherent detection process is essentially as good as coherent detection. When the signal/noise ratio is less than unity, there is considerable loss in noncoherent detection.

In Table 11.3-2, item (7) gives the signal/noise power ratio P_s/P_n at the input to the envelope detector. The reciprocal of this is the noise/signal ratio P_n/P_s shown in item (8), which is 0.291. Adding unity to this ratio gives the noncoherent detection loss L_d given in item (9). This loss L_d is 1.291, or 1.1 dB.

The noncoherent detection loss increases the effective noise density at the output of the envelope detector, which is designated p_{nd}. This detected noise power density p_{nd} is related as follows to the noise power density p_n at the input to the envelope detector:

Effective noise power density at detector output:

$$p_{nd} = p_n L_d = p_n\left(1 + \frac{P_n}{P_s}\right) \qquad (11.3\text{-}18)$$

The noncoherent detection loss L_d is multipled by the IF-amplifier noise power density p_n of item (3) to obtain the detected noise power density p_{nd}, given in item (10). The signal energy per pulse, E_s of item (4), is divided by p_{nd} of item (10) to obtain the ratio E_s/P_{nd}, shown in item (11), which is 3.25.

The RMS range-gate noise timing error per pulse is designated σt_1, where the subscript 1 indicates that this is the error for a single pulse measurement. Reference [1.14] (Chapter 7) shows that the relative RMS timing error is equal to

Relative RMS range-gate timing error per pulse:

$$\frac{\sigma t_1}{\tau_p} = \frac{1}{1.5\sqrt{2\,E_s/p_{nd}}} \tag{11.3-19}$$

This relation applies to the early–late range-gate circuit, for $\tau_g = 2\tau_p$ (as given by Eq 11.3-12), when the IF amplifier has an optimized rectangular bandpass of noise bandwidth given by Eq 11.3-13. Substitute into Eq 11.3-19 the value $E_s/p_{nd} = 3.25$ given in item (11). This gives $\sigma t_1/\tau_p = 0.261$, as shown in item (12).

The pulse width τ_p is 0.1 μsec, as was shown in Table 11.3-1, item (13). This is repeated in Table 11.3-2, item (13). Multiplying the ratio $\sigma t_1/\tau_p = 0.261$ in item (12) by this pulse width. This gives the RMS range-track timing error per pulse σt_1, shown in item (14). This is multiplied by 150 m/μsec (Eq 11.3-10) to obtain the RMS range error per pulse, σR_1, which is 3.92 m, as shown in item (15).

Thus, the range gate derives from each pulse a measure of the range to the target. Because of radar receiver noise, there is a random error in each pulse measurement, which has a Gaussian distribution with an RMS value of 3.92 m.

The timing of the range gate is controlled by digital range-gate control circuitry, which uses the range-gate error signal to keep the range gate centered over the received target pulse. The range-gate control loop is illustrated in the receiver block diagram that was shown in Fig 11.3-2. A timing reference pulse is received from the transmitter at the instant that the radar pulse is transmitted. The range-gate control circuitry counts the time following each transmitted pulse, for an interval equal to the estimated elapsed time between the transmitted pulse and the received target pulse, which is proportional to the estimated target range. By means of a digital feedback loop, this estimated elapsed time is controlled so that the range-gate error signal is minimized. The circuit provides a measure of the estimated target range, which is proportional to the estimated elapsed time.

In order for the range-gate control to work properly, the gain of the 60-MHz IF amplifier must be controlled to keep the video pulse level at the output of the envelope detector approximately constant. This is achieved by means of the automatic-gain-control (AGC) feedback loop illustrated in Fig 11.3-2.

To achieve automatic gain control, the video pulses from the envelope detector are fed through a gate, which is timed by the range-gate timing circuitry to be open for the full gate width of the range gate. The video pulses passing through this gate are integrated, and fed into the automatic gain control (AGC) circuitry. This AGC circuitry controls the IF amplifier gain to keep the pulse amplitudes approximately constant.

When the signal/noise power ratio at the output of the envelope detector is close to unity, or less than unity, this AGC loop is strongly affected by the noise level. Consequently, at low signal levels, gain control is degraded, which in turn degrades the dynamic performance of the range-gate tracking loop.

11.3.3   Response of Range-Tracking Feedback Loop

The digital range-gate tracking loop has a loop transfer function of the following form:

$$G_{(r)} = \frac{\omega_{c(r)}\left(1 + \dfrac{\omega_{i(r)}}{s}\right)}{s} \tag{11.3-20}$$

This transfer function ignores sampled-data effects. As was shown in Section 11.1, Table 11.1-2 (loop E), the noise bandwidth in rad/sec for this loop transfer function is

$$\omega_{b(r)} = \frac{\pi}{2}\left(\omega_{c(r)} + \omega_{i(r)}\right) \tag{11.3-21}$$

Hence, the noise bandwidth in hertz is

$$f_{b(r)} = \frac{\omega_{b(r)}}{2\pi} = \tfrac{1}{4}\left(\omega_{c(r)} + \omega_{i(r)}\right) \tag{11.3-22}$$

Reference [1.14] (Chapter 8) shows that the optimum value of the ratio ω_c/ω_i for a feedback loop of this type is

$$\text{Optimum}[\omega_c/\omega_i] = 1.84 \tag{11.3-23}$$

This condition yields the minimum peak error due to target acceleration, for a given noise bandwidth. It is convenient to round off this 1.84 ratio to 2.0, which gives the following for the range track loop:

$$\omega_{i(r)} = \tfrac{1}{2}\omega_{c(r)} \tag{11.3-24}$$

As shown in item (1) of Table 11.3-3, the range-tracking gain crossover frequency $\omega_{c(r)}$ is assumed to be 100 sec^{-1}. By Eq 11.3-24, $\omega_{i(r)}$ is 50 sec^{-1} [item (2)]. Substituting these values for $\omega_{c(r)}$, $\omega_{i(r)}$ of items (1), (2) into Eq 11.3-22 gives a noise bandwidth $f_{b(r)}$ of 37.5 Hz, as shown in item (3). By Eq 11.1-27 of Section 11.1.2.2, the equivalent integration time corresponding to

TABLE 11.3-3   Calculation of Errors of Range-Tracking Feedback Loop

Symbol	Parameter	Value
(1) $\omega_{c(r)}$	Gain crossover frequency	100 $\sec^{-1}$
(2) $\omega_{i(r)}$	Integral break frequency	50 $\sec^{-1}$
(3) $f_{b(r)}$	Noise bandwidth	37.5 Hz
(4) $T_{i(r)}$	Effective integration time	13.33 msec
(5) F_p	Pulse repetition frequency	1.50 kHz
(6) T_p	Pulse repetition period	0.667 msec
(7) N	Number of pulses integrated	20
(8) σR_1	RMS range noise error per pulse	3.92 m
(9) σR	RMS tracking-range noise error	0.877 m
(10)	Peak (3-sigma) range noise error	2.63 m
(11) V_t	Maximum target velocity	600 m/sec
(12)	Maximum error at lock-on	± 6.0 m
(13)	Linear region of range gate	± 7.5 m

this 37.5-Hz noise bandwidth is

$$T_{i(r)} = \frac{1}{2f_{b(r)}} = \frac{1}{2(37.5 \text{ Hz})} = 13.33 \text{ msec} \qquad (11.3\text{-}25)$$

This is shown in item (4).

The pulse repetition frequency F_p of the radar pulses is assumed to be 1.5 kHz, as shown in item (5). The reciprocal of this is the pulse repetition period T_p, which is 0.667 msec, as shown in item (6). The feedback action of the range-tracking loop effectively integrates the number of pulses that occur within the integration time $T_{i(r)}$ of the range-gate feedback loop. Hence, the effective number of pulses integrated is

$$N_{i(r)} = \frac{T_{i(r)}}{T_p} = \frac{13.33 \text{ msec}}{0.667 \text{ msec}} = 20.0 \qquad (11.3\text{-}26)$$

Item (8) shows the RMS range error per pulse, 3.92 m, obtained from Table 11.3-2, item (15). The averaging action of the range-tracking loop reduces the range error by the square root of the number of pulses integrated. Thus, the resultant RMS range noise error is

$$\sigma R = \frac{\sigma R_1}{\sqrt{N_{i(r)}}} = \frac{3.92 \text{ m}}{\sqrt{20}} = 0.877 \text{ m} \qquad (11.3\text{-}27)$$

This is shown in item (9). The peak limits of error are generally considered to be ± 3 sigma. Hence, the peak 3-sigma error is 3 times the 0.877-m RMS error, which is 2.63 m, as shown in item (10).

The range-tracking loop operates on sampled data, with a sampling frequency equal to the pulse repetition frequency, which is 1.5 kHz. This sampling frequency is nearly 100 times as large as the gain crossover frequency, 100 rad/sec or 16 Hz. Hence, the dynamic effects of sampling are small.

The range-tracking noise error can be reduced by decreasing the gain crossover frequency of the range-tracking loop, but this results in increased error due to target motion. Generally, the most important range-tracking error is the peak error occurring during lock-on. If this is too large, the signal may move outside the limits of the range gate and be lost. At lock-on, the range-tracking error experiences a ramp input equal to the radial velocity of the target. As shown in item (11) of Table 11.3-3, the maximum target velocity is assumed to be 600 m/sec, which is nearly Mach 2 at high altitudes. Assuming that this velocity is in the radial direction, the peak range-tracking error due to this ramp input is approximately

$$\text{Max}|R_e| = \frac{V_t}{\omega_{c(r)}} = \frac{600 \text{ m/sec}}{100 \text{ sec}^{-1}} = 6.0 \text{ m} \qquad (11.3\text{-}28)$$

As was shown in Fig 11.3-4, the error response characteristic of the range-tracking loop has a linear range equal to $\pm \tau_p/2$ in time, which is ± 0.05 μsec. Multiplying this by 150 m/μsec gives an equivalent variation of range of ± 7.5 m, which is shown in item (13). This is not much greater than the peak error ± 6.0 m, shown in item (12). Hence, the radar tracker may lose the target during lock-on if the initial range-gate error is excessive. To avoid this problem, the range gate may be widened during lock-on, and the gain crossover frequency of the range-gate tracking loop may be increased.

After the range gate has locked onto the target, the range-tracking error due to target motion is much smaller. The range-tracking error R_e after the lock-on transient has settled is approximately equal to

$$R_e \doteq c_2 \frac{d^2 R}{dt^2} = \frac{1}{\omega_{c(r)}\omega_{i(r)}} \frac{d^2 R}{dt^2} \qquad (11.3\text{-}29)$$

where $d^2 R/dt^2$ is the range acceleration of the target, and c_2 is the acceleration error coefficient. For the assumed parameters, this error reduces to

$$|R_e| \doteq 0.0020 g_t \text{ meter} \qquad (11.3\text{-}30)$$

where g_t is the acceleration of the target expressed in g's, in the radial direction. Even for a $10g$ acceleration, this error is only 0.02 m.

11.3.4 Measurement of Angle-Tracking Error

11.3.4.1 *Conical Scan.* There are two primary methods for measuring angle-tracking error in a radar tracking system: conical scan and monopulse. Most early tracking radars used conical scan, in which the antenna beam is

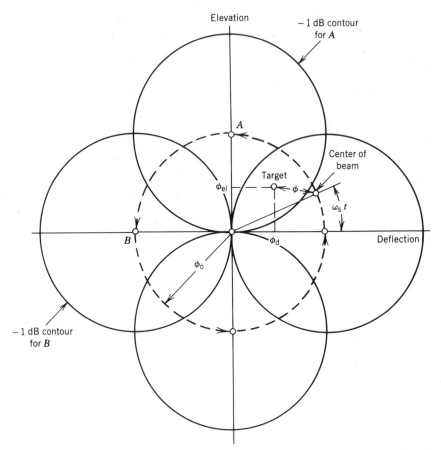

Figure 11.3-6 Typical conical scan pattern, showing gain contours at four points in scan.

scanned in a conical pattern. If the target is at the center of the scan pattern, the amplitudes of the received radar pulses are constant during the scan. When the target is displaced from the scan center, the pulse amplitudes are modulated sinusoidally at the scan frequency. The sinusoid amplitude variation is proportional to the displacement of the target from the scan center, and the phase of the sinusoid characterizes the direction from the center.

Figure 11.3-6 shows a typical conical scan pattern. The dashed circle is the path of the peak of the beam during the scan. The solid circles are the contours of 1 dB gain below peak gain, corresponding to four points on the scan separated by 90°. The angle ϕ_o is the offset angle of the scan, which is the amount that the center of the beam is offset from the center of the scan. For this case, the beam offset angle is $0.289\Theta_b$, where Θ_b is the 3-dB bandwidth of the antenna. With this offset, the gain at the center of the scan pattern is 1 dB below peak antenna gain.

As was shown in Fig 11.3-1, antenna position is controlled in terms of the azimuth and elevation angles of the antenna pedestal. However, target error is measured in terms of axes that rotate with the antenna dish. The coordinates of these tracking-error axes are designated *elevation* and *deflection*, but sometimes the term "traverse" is used in place of "deflection." The conical angle scan provides signals proportional to the elevation and deflection components of target error, measured relative to the center of the scan pattern.

If the target lies at the center of the scan pattern (zero tracking error), the amplitudes of the radar pulses are constant during the scan. When the target is displaced from the center of the scan (as indicated in Fig 11.3-6), the amplitudes of the video radar pulses at the output of the envelope detector are modulated sinusoidally as follows:

$$E = E_0 \left(1 + k_s \frac{\phi_d}{\Theta_b} \cos[\omega_s t] + k_s \frac{\phi_{eL}}{\Theta_b} \sin[\omega_s t] \right) \qquad (11.3\text{-}31)$$

where

E_0 = pulse voltage amplitude when the target is at the center of the scan
k_s = conical-scan slope parameter
ϕ_d = deflection tracking error
ϕ_{eL} = elevation tracking error
Θ_b = 3-dB beamwidth of antenna gain pattern
ω_s = angular frequency of conical scan

As shown in Ref [1.14] (Chapter 7), the conical-scan slope is equal to

$$k_s = 5.54 \frac{\phi_o}{\Theta_b} \qquad (11.3\text{-}32)$$

ϕ_o = offset angle of conical scan

In accordance with Eq 11.3-6, the conical-scan offset angle ϕ_o reduces the effective antenna gain, relative to peak gain. The gain reduction in decibels due to angle scan is

$$10 \log[L_{as}] = 12(\phi_o/\Theta_b)^2 \text{ dB} \qquad (11.3\text{-}33)$$

A typical conical-scan loss L_{as} = 1 dB. For this value, the corresponding offset angle ϕ_o and conical-scan slope k_s, obtained from Eqs 11.3-32, -33, are

$$\phi_o = 0.289 \Theta_b \qquad (11.3\text{-}34)$$

$$k_s = 1.60 \qquad (11.3\text{-}35)$$

The radar system experiences angle-scan loss in both transmission and reception. Hence, there is a total angle-scan "crossover loss," designated L_k, which is equal to

$$L_k = L_{as}^2 \qquad (11.3\text{-}36)$$

For this typical scan, which has a 1-dB angle-scan loss, the total angle-scan crossover loss L_k is 2 dB.

For small antennas, the conical angle scan may be achieved by rapidly oscillating the antenna dish, but for larger antennas the feed is oscillated. Figure 11.3-7 shows the block diagram of a conical-scan radar receiver showing the circuitry for demodulating the elevation and deflection error signals. This diagram also indicated the functions for range-gate control and automatic gain control (AGC) of the IF amplifier gain. Automatic gain control is needed to derive good angle error signals, just as it is needed to derive a good range-tracking error signal. (To simplify the diagram, the 10-GHz RF preamplifier stage is omitted.)

As shown in Fig 11.3-7, the signal at the output of the envelope detector is fed through the angle-tracking gate, of pulse width τ_{ga}. The angle-tracking gate is synchronized with the range-tracking gate. However, the angle-tracking gate width τ_{ga} may be smaller than the range-tracking gate width τ_g, in order to reduce angle-tracking noise error. The range gate can be narrower for angle-tracking, because the range-tracking loop is much faster than the angle-tracking loop, and generally keeps the target pulse close to the center of the range gate.

The signal that passes through the angle-tracking gate is fed to a pulse integrator, which integrates the energy occurring within the angle-tracking gate. The resultant integrated value is sampled by a sample-and-hold circuit, which holds the value fixed between radar pulses.

The conical-scan mechanism oscillates the feed of the antenna to produce the required conical scan pattern. Usually, the scan rate is no greater than $\frac{1}{10}$ of the pulse repetition frequency (PRF), although higher rates up to $\frac{1}{4}$ of the PRF can be used if the conical scan is synchronized with the pulse rate. For the example, the PRF is 1.5 kHz. Let us assume that the conical scan frequency is 125 Hz, which is $\frac{1}{12}$ of the PRF. For this scan rate, one radar pulse occurs every 30° of the conical-scan cycle.

Figure 11.3-8 shows the resultant video waveform. The circles (a) show the amplitudes of the radar pulses, which are obtained from the pulse integrator. The sample-and-hold circuit holds the signal constant between pulses, thereby producing the waveform (b). This waveform (b) is fed through the highpass filter shown in Fig 11.3-7, which discards the average value E_0. The resultant AC scan modulation is fed to phase-sensitive detectors, which are excited by sine and cosine reference signals obtained from the conical-scan mechanism. (The operation of phase-sensitive detectors was discussed in Chapter 8, Section 8.1.3.) The outputs from these phase-sensitive detectors are DC voltages proportional to the deflection and elevation errors ϕ_d and ϕ_{eL}.

The radar signal consists of sampled data at the pulse repetition frequency. The circuit that holds the error signal fixed between radar pulses produces a

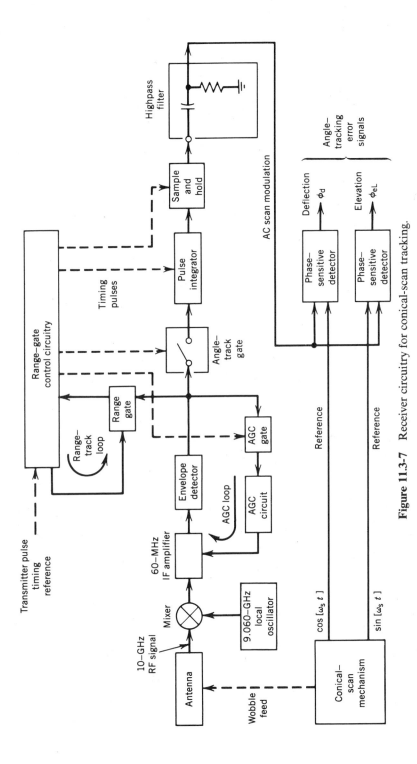

Figure 11.3-7 Receiver circuitry for conical-scan tracking.

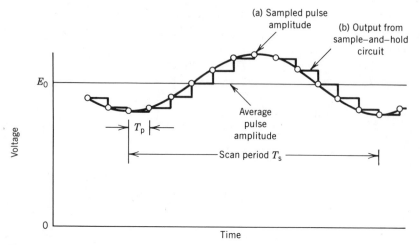

Figure 11.3-8   Derivation of conical-scan modulation from radar pulse amplitudes.

time delay of the AC scan modulation equal to $\frac{1}{2}$ of a pulse repetition period T_p. This time delay is compensated for by applying a corresponding time shift of $T_p/2$ to the sine and cosine reference signals from the conical-scan mechanism.

The elevation error signal ϕ_{eL} is used as an error signal for the servo that drives the elevation axis of the antenna pedestal; and the deflection error signal ϕ_d is used as the error signal for the servo that drives the azimuth axis. To compensate for the fact that the deflection tracking error is not measured in the same coordinates as the azimuth pedestal error, the deflection error signal is fed through a gain compensation that is proportional to $1/\cos[\Theta_{eL}]$, where Θ_{eL} is the elevation angle of the pedestal. Since $1/\cos[\Theta_{eL}]$ is equal to $\sec[\Theta_{eL}]$, this gain variation is called *secant compensation*.

11.3.4.2   Monopulse.   A serious deficiency with conical scan is that fluctuation of the radar reflectivity of the target (which is called "scintillation") modulates the amplitudes of the target pulses, and so adds error to the tracking signal. This problem is eliminated in a monopulse tracking system, which derives tracking information from a single pulse. Early tracking radars developed during World War II used conical scan. Modern high-performance radars generally use monopulse, but conical scan is sometimes used today because the system is much simpler.

Early monopulse tracking antennas used a four-horn feed to illuminate the antenna, but modern monopulse feeds often use more complicated structures to achieve better performance. With a four-horn feed, the signals from the horns illuminate four beams, which are squinted relative to one another in a manner similar to that indicated by the four solid beam contours in Fig 11.3-6. The signals received by the four horns are fed through microwave circuitry,

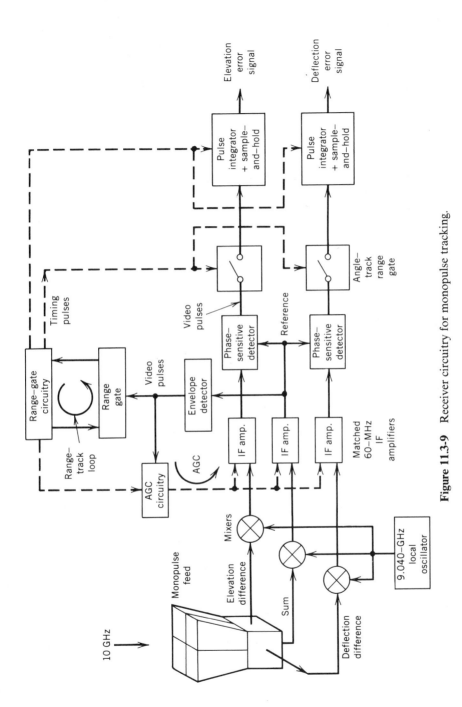

Figure 11.3-9 Receiver circuitry for monopulse tracking.

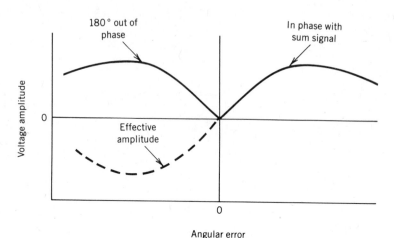

Figure 11.3-10 Amplitude of difference signal in monopulse receiver versus angle error.

which combines them to derive the following three signals: (1) the sum signal, (2) the elevation difference signal, and (3) the deflection difference signal. The transmitter power is fed into the channel for the sum signal, and excites the four horns equally.

A block diagram of the monopulse receiver circuitry is shown in Fig 11.3-9. The three signals from the feed (sum, elevation difference, and deflection difference) are fed into three matched receiver amplifiers. These consist of identical RF preamplifiers (not shown), identical mixers, and identical 60-MHz IF amplifiers. An important problem in the design of a monopulse receiver is the task of keeping the gains and phase shifts of the three amplifiers matched to one another.

The output from the sum-channel IF amplifier is fed to an envelope detector to derive video pulses. These video pulses are fed to a range gate to provide range tracking, just as for the conical-scan radar. The video pulses are also used for automatic gain control (AGC). In this case, the AGC circuitry must keep the gains of the three amplifiers matched as the radar signal level changes.

In Fig 11.3-10, the solid curve shows the amplitude of the IF signal in the elevation or deflection IF amplifier, plotted as a function of the angular tracking error in that axis. For positive error, the IF difference signal is in phase with the IF signal of the sum channel; while for negative error it is 180° out of phase.

The IF difference signal is equivalent to the position signal derived from an AC sensor, which was described in Chapter 8, Section 8.1. Hence, the corresponding DC error signal can be derived in the same manner, by feeding the AC signal from the IF amplifier into a phase-sensitive detector, which uses the sum-channel IF signal as a reference. By this means, the IF difference signal produces a positive DC error signal when it is in phase with the sum signal, and a negative DC error signal when it is out of phase with the sum signal. The dashed curve in Fig 11.3-10 shows the effective negative error signal for the latter case.

As shown in Fig 11.3-9, the signals from the elevation and deflection difference channels are fed to phase-sensitive detectors, which use the sum-channel signal as a reference. The output from each phase-sensitive detector is a train of bipolar (positive and negative) video pulses. These video pulses are passed through an angle-scan range gate, a pulse integrator, and a sample-and-hold circuit, which are the same as those used for conical scan. The final outputs are bipolar DC voltages proportional to the elevation and deflection angular errors ϕ_{eL} and ϕ_d. These error signals are used to drive the elevation and azimuth antenna pedestal servos, in the same manner as the error signals derived from conical scan.

In the region near zero angle error, the effective voltage-amplitude plot of Fig 11.3-10 is nearly linear. Hence, the elevation and deflection IF amplifier voltages (E_{eL} or E_d) can be expressed as

$$E_{eL} = k_m \frac{\phi_{eL}}{\Theta_b} E_{sum} \qquad (11.3\text{-}37)$$

$$E_d = k_m \frac{\phi_d}{\Theta_b} E_{sum} \qquad (11.3\text{-}38)$$

E_{sum} = voltage amplitude in sum channel
k_m = monopulse slope

A monopulse feed causes a reduction of antenna gain in the sum channel relative to that for a simple feed. As explained in Ref [1.14] (Chapter 7), the following are typical values of the monopulse slope k_s and the resultant sum-channel gain degradation of the antenna:

$$k_m = 1.6 \qquad (11.3\text{-}39)$$

$$\text{gain degradation} = 1.0 \text{ dB} \qquad (11.3\text{-}40)$$

This 1-dB gain degradation for a monopulse feed results in a two-way signal loss of 2 dB, relative to that of a comparable antenna with a simple feed. Note that these values of tracking slope and gain degradation are the same as the corresponding typical parameters for conical scan given in Eq 11.3-35.

11.3.5 Equations for RMS Angle-Tracking Noise Error

The angle-tracking errors for conical-scan and monopulse tracking are analyzed in Ref [1.14] (Chapter 7). The width of the range gate used to process the angle-tracking data is designated τ_{ga}:

τ_{ga} = angle-track range-gate width

This can be made appreciably narrower than τ_g, the range-gate width used for range tracking. It is assumed that the angle-tracking range-gate width τ_{ga} is greater than the radar pulse width τ_p. For this condition, the following are the effective RMS angle-tracking errors per pulse, $\sigma\Theta_1$, achieved by conical scan and monopulse:

Conical scan (for $\tau_{ga} > \tau_p$):

$$\frac{\sigma\Theta_1}{\Theta_b} = \frac{\sqrt{(1 + P_n/2P_s)(\tau_{ga}/\tau_p)}}{k_s\sqrt{E_s/P_n}} \tag{11.3-41}$$

Monopulse (for $\tau_{ga} > \tau_p$):

$$\frac{\sigma\Theta_1}{\Theta_b} = \frac{\sqrt{(1 + P_n/P_s)(\tau_{ga}/\tau_p)}}{k_s\sqrt{2E_s/P_n}} \tag{11.3-42}$$

Equation 11.3-42 gives the actual RMS tracking error of the monopulse system for a single pulse. However, the conical-scan radar cannot derive a tracking-error signal from a single pulse, and so Eq 11.3-41 merely gives an effective RMS tracking error per pulse for conical scan. For both cases, the actual angular-tracking error $\sigma\Theta$ is equal to

$$\sigma\Theta = \sigma\Theta_1/\sqrt{N_{i(a)}} \tag{11.3-43}$$

where $N_{i(a)}$ is the effective number of radar pulses integrated by the angle-tracking servo. The effective integration time of an angle-tracking servo is

$$T_{i(a)} = \frac{1}{2f_{b(a)}} \tag{11.3-44}$$

where $f_{b(a)}$ is the noise bandwidth of the angle-tracking servo. Hence, the number of radar pulses integrated by the angle-tracking servo is

$$N_{i(a)} = \frac{T_{i(a)}}{T_p} = \frac{1}{2f_{b(a)}T_p} \tag{11.3-45}$$

where T_p is the pulse repetition period of the radar.

In Eqs 11.3-41, -42 the ratio P_n/P_s is the noise/signal power ratio at the output of the IF amplifier for the conical-scan system, or at the output of the

IF sum-channel amplifier for the monopulse system. This ratio is the reciprocal of the signal/noise power ratio, which is equal to

$$\frac{P_s}{P_n} = \frac{E_s/p_n}{L_m} \qquad (11.3\text{-}46)$$

where L_m is the matching loss of the IF amplifier relative to that of an ideal
matched-filter receiver. It is assumed that L_m is the same as that given for the
range tracker, which is 0.85 dB [Table 11.3-2, item (6)].

Equations 11.3-41, -42 show that the monopulse and conical scan radars
have different noncoherent detection losses:

Conical-scan noncoherent detection loss:

$$L_{d(s)} = 1 + \frac{P_n}{2P_s} \qquad (11.3\text{-}47)$$

Monopulse noncoherent detection loss:

$$L_{d(m)} = 1 + \frac{P_n}{P_s} \qquad (11.3\text{-}48)$$

The monopulse noncoherent detection loss $(1 + P_n/P_s)$ is the same as that
shown previously in Eq 11.3-17 for range-gate tracking. However, for conical
scan, the noncoherent detection loss $(1 + P_n/2P_s)$ is lower. The reason for the
reduced loss is that the conical-scan error signal is derived from the AC
modulation of the video pulses, rather than from the pulse amplitude.

Note that there is an extra factor of $\sqrt{2}$ in the denominator of the
monopulse tracking equation. Hence, at high values of the signal/noise ratio
P_s/P_n, the monopulse error is lower than that for conical scan (assuming that
$k_m = k_s$, and the tracking gain degradations of the systems are the same).
However, when P_s/P_n drops below unity, the decreased noncoherent detection
loss for conical scan compensates for this factor of $\sqrt{2}$, and so the RMS
angular errors are nearly equal.

Thus, at high signal/noise ratios, the RMS tracking error for a monopulse
system is typically $1/\sqrt{2}$, or 70%, of the error for a conical-scan system.
However, this is generally not an important consideration in the choice of
monopulse over the much less complicated conical scan. The primary reason
for using monopulse is that target scintillation can cause large tracking errors
with conical scan, which do not occur with monopulse.

In the monopulse tracking radar, the phase-sensitive detectors that detect
the IF signals in the difference channels are coherent detection processes.
Hence, the signal/noise ratios at the outputs of the difference-signal IF
amplifiers can be much less than unity, without degrading the signal/noise
ratio in the phase-sensitive detectors.

On the other hand, these phase-sensitive detectors are coherent detection processes only when the single/noise ratio of the reference signal derived from the IF sum channel is much greater than unity. When the sum-channel signal/noise ratio is low, the difference-channel signal/noise ratio is degraded in the phase-sensitive detectors. Hence, there is a noncoherent detection loss in the phase-sensitive detectors equal to $(1 + P_n/P_s)$, where P_n/P_s is the noise/signal ratio of the reference signal from the IF sum channel.

The effective single-pulse angle-tracking errors in Eqs 11.3-41, -42 can be expressed as follows in terms of the noncoherent detection loss values $L_{d(s)}$, $L_{d(m)}$ given in Eqs 11.3-47, -48:

Conical scan (for $\tau_{ga} > \tau_p$):

$$\frac{\sigma\Theta_1}{\Theta_b} = \frac{\sqrt{L_{d(s)}(\tau_{ga}/\tau_p)}}{k_s\sqrt{E_s/P_n}} \qquad (11.3\text{-}49)$$

Monopulse (for $\tau_{ga} > \tau_p$):

$$\frac{\sigma\Theta_1}{\Theta_b} = \frac{\sqrt{L_{d(m)}(\tau_{ga}/\tau_p)}}{k_m\sqrt{2E_s/P_n}} \qquad (11.3\text{-}50)$$

11.3.6   Computation of Angle-Tracking Errors for a Monopulse Example

The example assumes a monopulse tracking antenna. The 60% antenna efficiency given in Table 11.3-1, item (4), is consistent with a gain degradation of 1.0 dB in the monopulse tracking feed, as was assumed in Eq 11.3-40.

Table 11.3-4 shows the steps for calculating the relative RMS angle-tracking noise error per pulse. Item (1) is the ratio E_s/P_n for the monopulse sum channel, and item (2) is the sum-channel IF filter matching loss L_m. These are the same as the values given in items (5), (6) of Table 11.3-2. Dividing item (1) by item (2) gives the signal/noise ratio P_s/P_n shown in item (3). This signal/noise ratio occurs at the output of the sum-channel IF amplifier, which is the reference signal fed to the monopulse phase-sensitive detectors. This ratio $P_s/P_n = 3.44$ of item (3) is the same as the P_s/P_n ratio of item (7) in Table 11.3-2.

In accordance with Eq 11.3-48, the noncoherent detection loss produced in the monopulse phase-sensitive detectors is

$$L_{d(m)} = 1 + \frac{P_n}{P_s} = 1 + \frac{1}{3.44} = 1.291 \qquad (11.3\text{-}51)$$

This is the same as the range-gate noncoherent-detection loss given in item (9) of Table 11.3-2 for range tracking.

TABLE 11.3-4   Calculation of RMS Angular-Tracking Noise Error per Pulse

Symbol	Parameter	Value
(1) E_s/p_n	Sum-channel matched-filter signal/noise ratio	4.189
(2) L_m	IF-amplifier matching loss	1.216 (0.85 dB)
(3) P_s/P_n	Signal/noise ratio of sum channel	3.44
(4) $L_{d(m)}$	Noncoherent detection loss in phase-sensitive detectors	1.291
(5) k_m	Monopulse slope	1.6
(6) τ_{ga}/τ_p	Relative angle-track gate width	1.5
(7) $\sigma\Theta_1/\Theta_b$	Relative RMS angle noise error per pulse	0.300

In accordance with Eq 11.3-39, the monopulse slope k_m is assumed to be 1.6, as shown in item (5) of Table 11.3-4. Item (6) shows that the angle-track range-gate pulse width τ_{ga} is 1.5 times greater than the RF radar pulse width τ_p, which is 0.1 μsec. This width is sufficient to accommodate range-gate tracking error and smearing of the pulse by the IF amplifier, with only a small amount of energy falling outside the gate.

The values of items (1), (4), (5), (6) of Table 11.3-4 are substituted into Eq 11.3-50. This gives the value of $\sigma\Theta_1/\Theta_b$ shown in item (7). Thus, the RMS angle error per pulse, $\sigma\Theta_1$, is 0.300 times the antenna beamwidth Θ_b.

The actual RMS angle-tracking error depends on the noise bandwidths of the angle-tracking servos. This error is calculated in Table 11.3-5. As shown in item (1), the gain crossover frequency $\omega_{c(a)}$ of the angle-tracking loop is assumed to be equal to 15 rad/sec. This loop has an integral network with a break frequency $\omega_{i(a)}$ that is set equal to $\frac{1}{3}$ of the gain crossover frequency, or 5 rad/sec, as shown in item (2). For the range-tracking loop, the integral-network break frequency ω_i was set equal to $\frac{1}{2}$ of the gain crossover frequency, which is optimum for such a loop that has essentially ideal performance. However, a greater ratio ω_c/ω_i is needed in an angle-track loop, because of the dynamic limitations of the antenna drive. Although a ratio ω_c/ω_i of 3 is assumed for the angle-tracking loop, a somewhat larger ratio may sometimes be desirable.

It is assumed that the loop transfer function of the angle-tracking loop can be approximated by

$$G_{(a)} = \frac{\omega_c(1 + \omega_i/s)}{s(1 + s/\omega_f)} \qquad (11.3-52)$$

The magnitude asymptote plot of this transfer function was shown in Section 11.1, Fig 11.1-5c (loop F). The noise bandwidth ω_b of loop F was given in

TABLE 11.3-5   Calculation of Peak Two-Axis Angle-Tracking Error Due to Receiver Noise

Symbol	Parameter	Value
(1) $\omega_{c(a)}$	Angle-tracking gain crossover frequency	$15\ \text{sec}^{-1}$
(2) $\omega_{i(a)}$	Angle-tracking integral break frequency	$5\ \text{sec}^{-1}$
(3) ω_b/ω_c	Noise bandwidth ratio	2.36
(4) $f_{b(a)}$	Angle-tracking noise bandwidth	5.63 Hz
(5) $T_{i(a)}$	Angle-tracking integration time	88.8 msec
(6) T_p	Pulse repetition period	0.667 msec
(7) $N_{i(a)}$	Number of pulses integrated by angle-tracking servos	133
(8) $\sigma\Theta_1/\Theta_b$	Relative RMS angle noise error per pulse	0.300
(9) $\sigma\Theta/\Theta_b$	Relative RMS angle noise error	0.0260
(10) Θ_b	Half-power beamwidth	1.60°
(11) $\sigma\Theta$	RMS one-axis noise error	0.0416°
(12)	Peak two-axis noise error	$\pm 0.176°$ ± 3.08 mrad

Table 11.1-2 as

$$\frac{\omega_b}{\omega_c} = \frac{\pi\omega_f(\omega_c + \omega_i)}{2\omega_c(\omega_f - \omega_i)} = \frac{\pi(1 + \omega_i/\omega_c)}{2(1 - \omega_i/\omega_f)} \tag{11.3-53}$$

It is assumed that $\omega_f/\omega_c = \omega_c/\omega_i = 3$. For this setting, the loop is the same as loop D studied in Chapter 2, Section 2.4.2. Its transient responses were plotted in Fig 2.4-11, which show that the step response has 25% peak overshoot. Figure 2.4-16 shows that the maximum value of $|G_{ib}|$ is 1.30. Hence the assumed loop parameters result in a good practical design. With this setting, the ratio ω_b/ω_c of Eq 11.3-53 is 2.36, as shown in item (3).

Thus the noise bandwidth ω_b for this case is 2.36 times greater than the gain crossover frequency ω_c. (This is actually an optimistic ratio for a practical antenna servo. When an antenna servo has strong structural resonance, the ratio ω_b/ω_c can be appreciably greater than 2.36.) The gain crossover frequency 15 rad/sec, shown in item (1), is multiplied by 2.36 to obtain the noise bandwidth ω_b of the angle tracking loop, which is 35.4 rad/sec. Dividing this by 2π gives the noise bandwidth f_b in hertz, which is 5.63 Hz, as shown in item (4).

Substitute the noise bandwidth $f_{b(a)} = 5.63$ Hz of item (4) into Eq 11.3-44. This gives the effective integration time $T_{i(a)}$ of the angle-tracking servos, shown in item (5) as 88.8 msec. The pulse repetition period T_p (0.667 msec) is shown in item (6), and was obtained from item (6) of Table 11.3-3. Dividing the integration time $T_{i(a)}$ by the pulse repetition period T_p gives the number of pulses integrated by the antenna servos, which is

$$N_{i(a)} = \frac{T_{i(a)}}{T_p} = \frac{88.8 \text{ msec}}{0.667 \text{ msec}} = 133 \qquad (11.3\text{-}54)$$

This is shown in item (7) of Table 11.3-5. Item (8) gives the relative RMS angle noise per pulse (0.300) obtained from item (7) of Table 11.3-4. This is divided by the square root of the number of pulses integrated (133) to obtain the relative RMS angle noise error:

$$\sigma\Theta/\Theta_b = \frac{\sigma\Theta_1/\Theta_b}{\sqrt{N_{i(a)}}} = \frac{0.300}{\sqrt{133}} = 0.0260 \qquad (11.3\text{-}55)$$

This is shown in item (9). The half-power beamwidth Θ_b of $1.60°$, shown in item (10), was obtained from item (6) of Table 11.3-1. The values of items (9), (10) are multiplied together to obtain the actual RMS noise error $\sigma\Theta$, shown in item (11) to be $0.0416°$.

There are two components of angle-tracking error: elevation and deflection. The total beam pointing error ϕ_e is related as follows to the elevation error ϕ_{eL} and deflection error ϕ_d:

$$\phi_e^2 = \phi_{eL}^2 + \phi_d^2 \qquad (11.3\text{-}56)$$

Since the RMS error components for the two axes are equal, the total RMS beam tracking error is $\sqrt{2}$ times the single-axis RMS error. The peak error is considered to be 3 times the RMS error. Hence, the peak 2-axis beam error Max$[\phi_e]$ is related as follows to the RMS single-axis error $\sigma\Theta$:

$$\text{Max}[\phi_e] = 3\sqrt{2}\,\sigma\Theta = 4.24\sigma\Theta \qquad (11.3\text{-}57)$$

Thus, the RMS single-axis noise error of item (11) in Table 11.3-5 is multiplied by 4.24 to obtain the peak two-axis error shown in item (12). The error in degrees is multiplied by 17.45 mrad/deg to obtain the error in milliradians, which is ± 3.08 mrad.

The angle error due to target motion is calculated in Table 11.3-6. It is assumed that the target is following a straight-line constant-velocity course. The tracking errors for this course were calculated in Chapter 5, Section 5.4, using error coefficients. The analysis showed that the maximum angular velocity Ω_{Max} and the maximum angular acceleration α_{Max} are

$$\Omega_{\text{Max}} = V_t/R_{\text{min}} \qquad (11.3\text{-}58)$$

$$\alpha_{\text{Max}} = 0.65\Omega_{\text{Max}}^2 \qquad (11.3\text{-}59)$$

where V_t is the target velocity, and R_{min} is the minimum target range.

TABLE 11.3-6   Calculation of Peak Angular Error Due to Target Motion

Symbol	Parameter	Value
(1) V_t	Maximum target velocity	600 m/sec
(2) R_{min}	Minimum range	800 m
(3) Ω_{Max}	Maximum angular velocity	0.75 rad/sec
(4) α_{Max}	Maximum angular acceleration	0.366 rad/sec^2
(5) c_2	Acceleration error coefficient	$\frac{1}{75}$ sec^2
(6)	Peak target-motion error	± 4.88 mrad

As was shown in item (11) of Table 11.3-3, the maximum target velocity V_t is assumed to be 600 m/sec, which is nearly Mach 2 at high altitudes. This is shown in item (1) of Table 11.3-6. The minimum range R_{min} is assumed to be 800 m, as shown in item (2). By Eq 11.3-58, the maximum angular velocity, shown in item (3), is

$$\Omega_{Max} = \frac{V}{R_{min}} = \frac{600 \text{ m/sec}}{800 \text{ m}} = 0.75 \text{ rad/sec} \qquad (11.3\text{-}60)$$

Substituting this into Eq 11.3-59 gives the maximum angular acceleration, shown in item (4):

$$\alpha_{Max} = 0.65\Omega_{Max}^2 = 0.65(0.75 \text{ rad/sec})^2$$
$$= 0.366 \text{ rad/sec}^2 = 366 \text{ mrad/sec}^2 \qquad (11.3\text{-}61)$$

It is assumed that the integral network has infinite gain at zero frequency, and so the velocity error coefficient is zero. Hence the maximum angle-tracking error due to target motion, after the initial lock-on transient has decayed, is approximately equal to the maximum angular acceleration multiplied by the acceleration error coefficient c_2. Item (5) gives the approximate acceleration error coefficient c_2, which is equal to $1/\omega_c\omega_i$:

$$c_2 = \frac{1}{\omega_{c(a)}\omega_{i(a)}} = \frac{1}{(15 \text{ sec}^{-1})(5 \text{ sec}^{-1})} = \frac{1}{75} \text{ sec}^2 \qquad (11.3\text{-}62)$$

Hence the maximum error due to target motion is approximately

$$\text{Max}[\Theta_e] = c_2\alpha_{Max} = \left(\frac{1}{75} \text{ sec}^2\right)(366 \text{ mrad/sec}^2)$$
$$= 4.88 \text{ mrad} \qquad (11.3\text{-}63)$$

This is shown in item (6).

Thus, the angle-tracking radar experiences a peak angle error due to target motion of ± 4.88 mrad, as shown in Table 11.3-6, and a peak two-axis error due to radar noise of ± 3.08 mrad, as shown in Table 11.3-5.

This example illustrates the statistical principles used in the calculation of radar tracking errors due to receiver noise. It shows that optimum setting of the dynamic parameters of the feedback loops require a compromise between radar noise error and error due to target motion.

11.4 STATISTICAL ANALYSIS OF TRACKING ERROR DUE TO WIND TORQUE

11.4.1 General Analysis

Besides receiver noise, another variable commonly described statistically is wind. This section shows how the statistical properties of wind can be used to calculate the tracking accuracy of a large parabolic reflector antenna. It is assumed that antenna angular velocities are sufficiently low that wind pressure due to antenna rotation is negligible.

The general expression for the pressure p exerted by wind is

$$p = \tfrac{1}{2} C_d \rho_a V^2 \qquad (11.4\text{-}1)$$

where V is the wind velocity, ρ_a is the mass density of air, and C_d is the drag coefficient of the structure, which varies with shape. For a solid structure, it is of the order of unity. The mass density of air is

$$\rho_a = 0.0024 \text{ lb-sec}^2/\text{ft}^4 \qquad (11.4\text{-}2)$$

For $C_d = 1.0$, the pressure at a wind velocity V of 60 mph (88 ft/sec) is

$$p = 9.29 \text{ lb/ft}^2 \quad (\text{at } 60 \text{ mph}) \qquad (11.4\text{-}3)$$

The torque produced by wind on an antenna is proportional to the wind pressure, and so is proportional to the square of the wind velocity. Since the tracking error due to wind is proportional to wind torque, the statistical properties of the wind torque, or wind pressure, are needed. These are not the same as the statistical properties of the wind velocity. If the wind pressure has a Gaussian distribution, the wind velocity cannot have a Gaussian distribution. The spectrum of the wind velocity is different from the spectrum of the wind torque.

On the other hand, if the relative deviation of wind velocity from mean wind velocity is small, the deviation of wind torque (or wind pressure) from mean wind torque (or mean wind pressure) is approximately proportional to the deviation of wind velocity from mean wind velocity. For this condition, the

wind torque (or wind pressure) has approximately the same statistical properties as the wind velocity. Let us represent the wind velocity V and wind torque
T as the sum of an average value plus a deviation:

$$V = V_0 + \Delta V \tag{11.4-4}$$

$$T = T_0 + \Delta T \tag{11.4-5}$$

where V_0 is the mean wind velocity, T_0 is the wind torque at mean wind
velocity, and $\Delta V, \Delta T$ are the variations from these values. Since the wind
torque T is proportional to V^2, then

$$\frac{T}{T_0} = \left(\frac{V}{V_0}\right)^2 = \left(\frac{V + \Delta V}{V_0}\right)^2$$

$$= 1 + 2\frac{\Delta V}{V_0} + \left(\frac{\Delta V}{V_0}\right)^2 \tag{11.4-6}$$

By Eq 11.4-5, T/T_0 is equal to $(1 + \Delta T/T_0)$. Hence

$$\frac{\Delta T}{T_0} = 2\frac{\Delta V}{V_0} + \left(\frac{\Delta V}{V_0}\right)^2 \doteq 2\frac{\Delta V}{V_0} \tag{11.4-7}$$

This approximation holds for $\Delta V/V_0 \ll 1$. Thus, for a small relative variation
of wind velocity from mean wind velocity V_0, the relative variation of wind
torque, $\Delta T/T_0$, is twice the relative variation of wind velocity, $\Delta V/V_0$. For
example, a 10% variation of wind velocity from the mean velocity results in a
20% variation of wind torque from the mean wind torque.

The wind spectrum commonly used to specify wind torque on antennas is
based on statistical measurements performed by Titus [11.7] of the wind
torque exerted on a 60-ft antenna. He found that the following equation fitted
the measured spectral data to reasonable accuracy:

$$\Phi_{wt}[\omega] = \frac{K\omega_1^2\omega_2^2}{(\omega^2 + \omega_1^2)(\omega^2 + \omega_2^2)} \tag{11.4-8}$$

where

$$\omega_1 = 0.12 \text{ rad/sec}, \qquad \omega_2 = 2.0 \text{ rad/sec} \tag{11.4-9}$$

and K is a constant. It is convenient to choose the value of the constant K to
normalize the spectrum so that its integral over all angular frequencies ω is
unity. If the constant K were unity, this spectrum would be the same as
$|H[\omega]|^2$ for the lowpass filter B of Table 11.1-1, and so the integral of Eq

11.4-8 would be equal to the noise bandwidth of filter B. This is

$$\int_0^\infty \frac{\omega_1^2 \omega_2^2}{\left(\omega^2 + \omega_1^2\right)\left(\omega^2 + \omega_2^2\right)} \, d\omega = \frac{\pi}{2} \frac{\omega_1 \omega_2}{\omega_1 + \omega_2} \qquad (11.4\text{-}10)$$

Hence, to make the integral of the spectrum unity, the constant K should be set equal to the reciprocal of the right-hand expression, which is

$$K = \frac{\omega_1 + \omega_2}{(\pi/2)\,\omega_1 \omega_2} \qquad (11.4\text{-}11)$$

Substituting Eq 11.4-11 into -8 gives the normalized wind-torque spectrum, the integral of which is unity:

$$\Phi_{wt(n)}[\omega] = \frac{(2/\pi)\,\omega_1 \omega_2 \left(\omega_1 + \omega_2\right)}{\left(\omega^2 + \omega_1^2\right)\left(\omega^2 + \omega_2^2\right)} \qquad (11.4\text{-}12)$$

A statistical wind specification frequency commonly used by U.S. government agencies gives the wind-torque spectrum of Eq 11.4-8 (with $\omega_1 = 0.10$ $\sec^{-1}$ rather than 0.12 $\sec^{-1}$) along with the statement that the standard deviation (sigma) of wind gust velocity is equal to 25% of the mean wind velocity. For such a large standard deviation, the torque deviation is not approximately proportional to the velocity deviation, particularly for a 3-sigma variation. To have a practical analysis tool, the wind torque must be assumed to have a Gaussian distribution, and consequently the wind velocity does not. Nevertheless, the approximation of Eq 11.4-7 can still be used to calculate the standard deviation of the torque. If the standard deviation of the wind velocity, σV, is 25% of the mean wind velocity V_0, then the standard deviation of the torque, σT, is 50% of the torque T_0 at mean wind velocity:

$$\frac{\sigma T}{T_0} = 2\frac{\sigma V}{V_0} = 2(0.25) = 0.50 \qquad (11.4\text{-}13)$$

As was shown in Chapter 6, the angle tracking error produced by a load torque T_d is given by

$$\Theta_e = T_d C^m G_{ie(p)} + \frac{T_d G_{ie(r,\,p)}}{s^2 J_{(m)}\left(1 + \omega_{cm}/s\right)} \qquad (11.4\text{-}14)$$

The first term is due to compression of the mechanical compliance C^m of the antenna structure and the gear train, while the second term is due to coercion of the servo motor by the load torque. For wind torques, which are relatively low in frequency, the second term can usually be made appreciably smaller than the first, and so is neglected in this analysis.

Equation 11.2-30 gave a general expression for the mean square output from a system, expressed in terms of the spectral density of the input signal and the system transfer function. In accordance with that equation, the RMS tracking error $\sigma\Theta$ due to wind torque is given by

$$(\sigma\Theta)^2 = \frac{1}{\pi}(C^m)^2 \int_0^\infty |G_{ie(p)}[\omega]|^2 \Phi_{wt}[\omega]\, d\omega \qquad (11.4\text{-}15)$$

where $\Phi_{wt}[\omega]$ is the spectral density of the wind torque. It is convenient to normalize the calculation by considering the RMS tracking error that would be produced under "locked-rotor" conditions, when the servo motor rotor is locked and the position loop is not operating. This locked-rotor error, which is designated $\sigma\Theta_{LR}$, can be obtained by setting $|G_{ie(p)}|$ equal to unity in Eq 11.4-15, which gives

$$(\sigma\Theta_{LR})^2 = \frac{1}{\pi}(C^m)^2 \int_0^\infty \Phi_{wt}[\omega]\, d\omega \qquad (11.4\text{-}16)$$

Dividing Eq 11.4-16 by Eq 11.4-17 gives

$$\left(\frac{\sigma\Theta}{\sigma\Theta_{LR}}\right)^2 = \int_0^\infty |G_{ie(p)}[\omega]|^2 \Phi_{wt(n)}[\omega]\, d\omega \qquad (11.4\text{-}17)$$

The function $\Phi_{wt(n)}[\omega]$ is the normalized wind-torque spectrum given in Eq 11.4-12, and defined by

$$\Phi_{wt(n)}[\omega] = \frac{\Phi_{wt}[\omega]}{\int_0^\infty \Phi_{wt}[\omega]\, d\omega} \qquad (11.4\text{-}18)$$

A general solution of Eq 11.4-17 can be obtained by assuming a tracking (position) loop having the loop transfer function of loop F, shown in Table 11.1-2 and Fig 11.1-15. The loop and error transfer functions are

$$G_{(p)} = \frac{\omega_c(1 + \omega_i/s)}{s(1 + s/\omega_f)} \qquad (11.4\text{-}19)$$

$$G_{ie(p)} = \frac{s^2(s + \omega_f)}{s^3 + \omega_f s^2 + \omega_c \omega_f s + \omega_c \omega_f \omega_i} \qquad (11.4\text{-}20)$$

The wind spectrum of Eq 11.4-12 can be related to the following lowpass transfer function:

$$H_{wt}[s] = \frac{\omega_1 \omega_2}{(s + \omega_1)(s + \omega_2)} \qquad (11.4\text{-}21)$$

For $s = j\omega$, the square of the magnitude of this is

$$|H_{wt}[\omega]|^2 = H_{wt}[\omega]H_{wt}[-\omega] = \frac{\omega_1^2\omega_2^2}{(\omega^2 + \omega_1^2)(\omega^2 + \omega_2^2)} \quad (11.4\text{-}22)$$

Comparing this with Eq 11.4-12 shows that

$$\Phi_{wt(n)}[\omega] = \frac{2}{\pi}\frac{\omega_1 + \omega_2}{\omega_1\omega_2}|H_{wt}[\omega]|^2 \quad (11.4\text{-}23)$$

Hence, Eq 11.4-17 can be expressed as

$$\left(\frac{\sigma\Theta}{\sigma\Theta_{LR}}\right)^2 = \frac{2}{\pi}\frac{\omega_1 + \omega_2}{\omega_1\omega_2}\int_0^\infty |G_{ie(p)}[\omega]H_{wt}[\omega]|^2 \, d\omega \quad (11.4\text{-}24)$$

Define the transfer function $H[s]$ as

$$H[s] = \frac{1}{\omega_1\omega_2}G_{ie(p)}[s]H_{wt}[s] \quad (11.4\text{-}25)$$

This is equal to

$$\begin{aligned} H[s] &= \frac{s^2(s + \omega_f)}{(s^3 + \omega_f s^2 + \omega_c\omega_f s + \omega_c\omega_f\omega_i)(s + \omega_1)(s + \omega_2)} \\ &= \frac{s^3 + \omega_f s^2}{D} \end{aligned} \quad (11.4\text{-}26)$$

where the denominator D is

$$\begin{aligned} D = {}&s^5 + s^4(\omega_f + \omega_1 + \omega_2) + s^3[\omega_c\omega_f + \omega_1\omega_2 + \omega_f(\omega_1 + \omega_2)] \\ &+ s^2\omega_f[\omega_c\omega_i + \omega_1\omega_2 + \omega_c(\omega_1 + \omega_2)] \\ &+ s\omega_c\omega_f[\omega_i(\omega_1 + \omega_2) + \omega_1\omega_2] + \omega_c\omega_f\omega_i\omega_1\omega_2 \end{aligned} \quad (11.4\text{-}27)$$

Equation 11.4-24 can be expressed as follows in terms of the square of the magnitude of $H[s]$ for $s = j\omega$:

$$\begin{aligned} \left(\frac{\sigma\Theta}{\sigma\Theta_{LR}}\right)^2 &= \frac{(2/\pi)(\omega_1 + \omega_2)}{\omega_1\omega_2}(\omega_1\omega_2)^2\int_0^\infty |H[\omega]|^2 \, d\omega \\ &= \omega_1\omega_2(\omega_1 + \omega_2)\frac{2}{\pi}\int_0^\infty |H[\omega]|^2 \, d\omega \end{aligned} \quad (11.4\text{-}28)$$

The integral in Eq 11.4-28 is equal to πI_5, where I_5 is the integral expression given in Appendix G for $n = 5$. From Eqs 11.4-26, -27, the coefficients of the numerator and denominator polynomials of $H[s]$ are

$$c_4 = c_1 = c_0 = 0 \qquad (11.4\text{-}29)$$

$$c_3 = 1 \qquad (11.4\text{-}30)$$

$$c_2 = \omega_f \qquad (11.4\text{-}31)$$

$$d_5 = 1 \qquad (11.4\text{-}32)$$

$$d_4 = \omega_f + \omega_1 + \omega_2 \qquad (11.4\text{-}33)$$

$$d_3 = \omega_c\omega_f + \omega_1\omega_2 + \omega_f(\omega_1 + \omega_2) \qquad (11.4\text{-}34)$$

$$d_2 = \omega_f\left[\omega_c\omega_i + \omega_1\omega_2 + \omega_c(\omega_1 + \omega_2)\right] \qquad (11.4\text{-}35)$$

$$d_1 = \omega_c\omega_f\left[\omega_i(\omega_1 + \omega_2) + \omega_1\omega_2\right] \qquad (11.4\text{-}36)$$

$$d_0 = \omega_c\omega_f\omega_i\omega_1\omega_2 \qquad (11.4\text{-}37)$$

In the expression for I_5 of Appendix G, given in Eqs G-13 to -19, set $d_5 = 1$, and substitute the values of the numerator coefficients (c_0 to c_4) of Eqs 11.4.29 to -31. This gives

$$2I_5 = \frac{m_{51} + \omega_f^2 m_{52}}{D_5} \qquad (11.4\text{-}38)$$

$$m_{51} = d_1d_2 - d_0d_3 \qquad (11.4\text{-}39)$$

$$m_{52} = d_1d_4 - d_0 \qquad (11.4\text{-}40)$$

$$m_{53} = \frac{d_2m_{52} - d_4m_{51}}{d_0} \qquad (11.4\text{-}41)$$

$$m_{54} = \frac{d_3m_{53} - d_4m_{52}}{d_0} \qquad (11.4\text{-}42)$$

$$D_5 = d_0(d_1m_{54} - d_3m_{53} + m_{52}) \qquad (11.4\text{-}43)$$

Equations 11.4-41, -42, -43 can be combined as the following single equation:

$$D_5 = \frac{d_2}{d_0}m_{51}m_{52} - \frac{d_4}{d_0}m_{51}^2 - m_{52}^2 \qquad (11.4\text{-}44)$$

Setting the integral in Eq 11.4-24 equal to πI_5 gives

$$\left(\frac{\sigma\Theta}{\sigma\Theta_{LR}}\right)^2 = \omega_1\omega_2(\omega_1 + \omega_2)\frac{2}{\pi}(\pi I_5) = \omega_1\omega_2(\omega_1 + \omega_2)(2I_5) \qquad (11.4\text{-}45)$$

The expression for $2I_5$ can be obtained from Eq 11.4-38 using the definitions for m_{51}, m_{52}, D_5 of Eqs 11.4-39, -40, -44, along with the values for d_4 to d_0 in Eqs 11.4-33 to -37.

Let us apply this expression to three examples. The first is the "optimum" loop transfer function discussed in Section 3.4, in which the gain is set so that $\text{Max}|G_{ib}|$ occurs at the frequency of minimum phase lag of G. The values of the general break frequencies are $\omega_{HF} = \omega_f$, $\omega_{LF} = \omega_i$. Hence the optimum settings are

$$\omega_c = \tfrac{1}{2}(\omega_i + \omega_f) \tag{11.4-46}$$

$$\frac{\omega_f}{\omega_i} = \frac{\text{Max}|G_{ib}| + 1}{\text{Max}|G_{ib}| - 1} \tag{11.4-47}$$

Solving for ω_i, ω_f gives

$$\omega_i = \left(1 - \frac{1}{\text{Max}|G_{ib}|}\right)\omega_c \tag{11.4-48}$$

$$\omega_f = \left(1 + \frac{1}{\text{Max}|G_{ib}|}\right)\omega_c \tag{11.4-49}$$

Curve ① in Fig 11.4-1 is the plot of the square root of Eq 11.4-45 versus the position-loop asymptotic gain crossover frequency ω_c, using Eqs 11.4-48, -49 to relate ω_i, ω_f to ω_c. The value for $\text{Max}|G_{ib}|$ is 1.30. The parameters ω_1, ω_2 are set equal to 0.12 sec^{-1}, 2 sec^{-1} in accordance with Eq 11.4-9.

For curves ②, ③ in Fig 11.4-1, the ratio ω_f/ω_c, which is designated β, is set equal to ω_c/ω_i:

$$\beta = \omega_f/\omega_c = \omega_c/\omega_i \tag{11.4-50}$$

This gives the loop transfer function for loop D that was discussed in Section 2.4.2. As was shown, the transfer function G_{ie} is

$$G_{ie} = \frac{s^2(s + \beta\omega_c)}{(s + \omega_c)(s^2 + 2\zeta\omega_c s + \omega_c^2)} \tag{11.4-51}$$

$$\zeta = \frac{\beta - 1}{2} \tag{11.4-52}$$

For curve ②, $\beta = 3$, $\zeta = 1$, and G_{ie} has a triple-order pole at $s = -\omega_c$. As was shown in Section 2.4.2, the maximum value of $|G_{ib}|$ is 1.30, and there is 25% overshoot of the step response. Of the three cases, this gives the best attenuation of wind gusts for a given value of ω_c. For curve ③, $\beta = \sqrt{6} = 2.45$ and $\zeta = 0.725$. As shown by the plot for loop D in Chapter 2 (Fig 2.4-16), for

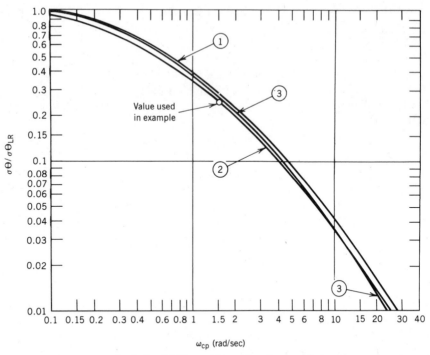

Figure 11.4-1 Plots of relative RMS tracking noise due to wind gusts on an antenna, as a function of the gain crossover frequency ω_{cp} of the tracking position loop.

$\zeta = 0.725$, the maximum value of $|G_{ib}|$ is 1.42, and there is 33% overshoot in the step response.

This analysis was first performed by Briggs [11.8], who assumed the parameters of curve ③ of Fig 11.4-1 for $\beta = \sqrt{6}$. With such a low value of β, stability is marginal. The above analysis has extended his work to include more stable loop transfer functions.

11.4.2 Tracking Error of Satellite Communication Antenna

Let us apply Fig 11.4-1 to calculate the tracking error due to wind gusts that is experienced by a large satellite communication antenna. Table 11.4-1 shows the steps for calculating the tracking error. The parameters are typical of a 32-m (105-ft) satellite communication antenna, which usually transmits at 6 GHz and receives at 4 GHz. The azimuth carriage (which weighs 500,000 lb) is carried on four railroad-car wheels, which roll on a circular track.

It is assumed that the antenna has a monopulse tracking feed which allows continuous tracking of the received signal from the satellite. The signal strength is assumed to be sufficiently high to achieve negligible tracking error due to receiver noise. The wind is assumed to have a mean velocity of 45 mph,

TABLE 11.4-1   Calculation of Wind-Gust Tracking Error of 105-ft Satellite-Communication Antenna

Parameter	Azimuth	Elevation	Units
(1) Inertia J	4.0×10^6	2.0×10^6	ft-lb-sec^2
(2) Structural resonant frequency	$f_n = 2.0$ $\omega_n = 12.6$	2.0 12.6	Hz rad/sec
(3) Stiffness $K = J\omega_n^2$	6.35×10^8	3.175×10^8	ft-lb/rad
(4) Compliance	1.57×10^{-9}	3.15×10^{-9}	rad/ft-lb
$C^m = 1/K$	0.90×10^{-7}	1.80×10^{-7}	deg/ft-lb
(5) Wind torque:			
30 mph	427,000	313,000	ft-lb
45 mph	961,000	704,000	ft-lb
(6) Gust torque: 45-mph mean, 1-sigma	480,500	352,000	ft-lb
(7) Locked-rotor gust error, 1-sigma	0.0432	0.0634	deg
(8) ω_{cp}	1.57	1.57	sec^{-1}
(9) $\sigma\Theta/\sigma\Theta_{LR}$	0.24	0.24	
(10) Actual gust error, 1-sigma	0.0104	0.0152	deg
(11) 2-axis errors:			
1-sigma		0.0184	deg
3-sigma		0.0552	deg

with gusts, where the standard deviation of the wind gust velocity is 25% of the mean wind velocity.

Item (1) of Table 11.4-1 shows the values of inertia J of the azimuth and elevation axes. As shown in (2), the servo structural resonant frequency in azimuth and elevation is 2.0 Hz, or 12.6 rad/sec. This 2-Hz servo structural resonant frequency is consistent with the data for a 105-ft antenna in Fig 10.1-1 of Chapter 10. (The mean practical structural resonant frequency for this aperture diameter is 1.5 Hz.) The stiffness, shown in item (3), is given by

$$K = J\omega_n^2 \qquad (11.4\text{-}53)$$

where J is the antenna inertia and ω_n is the servo structural resonant frequency in rad/sec. Substituting the values of J and ω_n in items (1), (2) gives the antenna stiffness values shown in (3). The reciprocal of the stiffness is the compliance, in rad/ft-lb, shown in (4). This is multiplied by 57.3 deg/rad to obtain the antenna compliance expressed in deg/ft-lb.

Item (5) shows the calculated values of the wind torque for a 30-mph wind when the antenna is at the worst-case orientation relative to the wind. These torques are scaled to apply to 45 mph, by multiplying them by [(45 mph)/(30 mph)]^2. Since the standard deviation for the wind-gust velocity is 25% of the mean wind velocity, the standard deviation of the wind-gust torque is 50% of the torque at mean wind velocity. Hence the 45-mph wind torque values in item (5) are multiplied by 0.5 to obtain the 1-sigma wind-gust torque values of item (6). These are multiplied by the antenna compliance values of (4) to obtain the 1-sigma values of the locked-rotor gust error shown in (7).

As was shown in Chapter 10, Section 10.5, with proper design the gain crossover frequency of the position loop can typically be $\frac{1}{8}$ of the structural resonant frequency. Since both axes have a structural resonant frequency of 2 Hz, the assumed value of position-loop gain crossover frequency for both axes is

$$\omega_{cp} = \omega_n/8 = \left(12.6 \ sec^{-1}\right)/8 = 1.57 \ sec^{-1} \qquad (11.4\text{-}54)$$

Curve ② of Fig 11.4-1 is assumed. For this value of ω_c, the factor $\sigma\Theta/\sigma\Theta_{LR}$ for curve ② of Fig 11.4-1 is 0.24, shown in item (9). The 1-sigma locked-rotor gust error of item (7) is multiplied by 0.24 to obtain the actual 1-sigma gust error in item (10). The total two-axis deflection Θ of the antenna beam from the direction of the satellite is

$$\Theta = \sqrt{\Theta_d^2 + \Theta_{eL}^2} \qquad (11.4\text{-}55)$$

where Θ_d, Θ_{eL} are the deflection and elevation tracking errors. Taking the square root of the sum of the squares of the deflection and elevation gust components of error of item (10) gives the two-axis wind-gust tracking error shown in item (11). This 1-sigma two-axis error (0.0184°) is multiplied by 3 to obtain the peak 3-sigma two-axis tracking error, which is 0.0552°.

The antenna aperture D is 32 m, or 3200 cm, and the wavelength at the transmit frequency, 6 GHz, is 5 cm. The expression in Section 11.3, Eq 11.3-3, gives the following approximation for the half-power beamwidth Θ_b of the antenna at the transmit frequency:

$$\Theta_b = 65\frac{\lambda}{D} \ deg = 65°\frac{5 \ cm}{3200 \ cm} = 0.102° \qquad (11.4\text{-}56)$$

Equation 11.3-6 gave the following for the transmit-signal loss that corresponds to the peak 3-sigma wind-gust tracking error, 0.0552°:

$$dB \ loss = 12\left(\frac{\Theta}{\Theta_b}\right)^2 \ dB = 12\left(\frac{0.0552°}{0.102°}\right)^2 \ dB = 3.51 \ dB \quad (11.4\text{-}57)$$

This is a rather high tracking loss. However, the assumed wind-gust specification is very stringent. A more realistic standard deviation for wind gusts is 15% of the mean wind velocity. For this assumption, the resultant tracking error would be 0.6 times the calculated error, and so the 3.51-dB loss would be reduced by the factor $(0.6)^2$, which is 0.36. The resultant peak signal loss due to wind-gust tracking error would then be 1.26 dB.

This statistical wind model assumes that the wind pressure (or wind torque) consists of a random gust component of wind pressure added to a constant mean wind pressure, where the gust component has a Gaussian distribution. Since the tracking loop has infinite gain at zero frequency, the constant average wind pressure (or wind torque) causes no tracking error. It is only the gust deviation from the mean wind torque that causes tracking error.

The concept of a "locked-rotor gust error" is somewhat fictitious. If the servo motor rotor were actually locked, there would be no tracking-loop compensation for the mean wind pressure, and so the mean wind pressure would produce error. Nevertheless, for analysis purposes it is convenient to consider a locked-rotor gust error that applies only to the random gust deviation of wind pressure, and not to the mean wind pressure.

Chapter 12

Control-System Nonlinearity

This chapter investigates the effects of nonlinearities in feedback-control systems. Saturated lock-on transients are studied with piecewise analysis and the phase plane. The describing-function method of nonlinear analysis is used to determine the effects of backlash in a servomechanism gear train. The principle of linearizing a nonlinearity about an operating point is applied to the design of a control system that controls pressure in a vacuum processing chamber. This approximate analysis is combined with an exact nonlinear simulation study, which uses the simulation methods developed in Chapter 9.

12.1 GENERAL PRINCIPLES FOR ANALYZING CONTROL-SYSTEM NONLINEARITIES

The analytical techniques presented up to this point assume that the control system is described by linear equations. Obviously, the same techniques still hold to reasonable accuracy in the presence of small nonlinearities. However, with larger nonlinearities, different approaches are required.

An important tool for studying nonlinear control systems is simulation. The serial-simulation technique developed in Chapter 9 can be readily extended to include nonlinearities. To illustrate this approach, serial simulation is applied in Section 12.4 to study a nonlinear process control system, which controls the pressure in a vacuum processing chamber.

On the other hand, brute-force simulation is not a satisfactory approach for the design of nonlinear control systems. The fundamental problem with simulation, whether the system be linear or nonlinear, is that one can easily get lost in the vagueness and complexity of the problem. Theory is needed to provide a frame of reference for understanding what is happening, and to show how changes can be made to improve performance.

The analysis techniques that are most effective for handling control-system nonlinearities are those that are extensions of linear analysis. Sections 12.3 and 12.4 discuss the powerful approach of linearizing a nonlinear response by considering small deviations from an operating point. This technique is used to analyze the pressure-control system discussed previously. For small deviations about a given pressure, the response is approximately linear. The linearized analysis provides a basis for assuring there is good dynamic performance in local regions. On the other hand, to describe the response of the pressure-control system quantitatively over large variations of pressure, simulation is needed. Thus, simulation supplements the linear analysis. The linear analysis provides a simple frame of reference for explaining the system operation, and allows the simulation to be applied in an effective manner.

Some control systems, such as a contactor (or relay) servomechanism, do not behave linearly over a small region. Nevertheless, linear analysis can still be applied to such systems by employing the describing-function (or first-harmonic) approximation. The nonlinear element is excited with a sinusoidal waveform, and the first harmonic of the output waveform is determined. The transfer function relating the input sinusoid to the first harmonic of the output is called the describing function. In many control systems, the describing function characterizes the dynamic effects of the nonlinearity quite accurately. It has been particularly useful in the analysis of contactor servomechanisms.

The reason that the describing function works is that the power-output device of a control system usually attenuates the higher harmonics of a periodic waveform much more strongly than the fundamental. Hence the effect of the higher harmonics on the loop dynamics is small. On the other hand, it is sometimes difficult to determine whether or not one can legitimately neglect the higher harmonics. This problem can be solved by supplementing the describing-function analysis with simulation, when the validity of the describing-function approach is in question.

There are some nonlinear cases where extension of linear analysis breaks down, and the system equations must be formulated in a nonlinear manner. Such a situation is discussed in Section 12.2, which considers the saturated step response of a servomechanism. This response is analyzed both with a piecewise linear analysis, and by the use of the phase plane. On the other hand, even though this saturated step response is quite nonlinear, the characteristics of this response are directly related to the major linear parameters of the control system: the gain crossover frequencies of the position and velocity loops. Thus, a linear frame of reference greatly helps in understanding the nonlinear response.

One of the most serious nonlinearities in mechanical control systems is static (or Coulomb) friction. Although this is a nonlinear effect, it has been handled in this book using linear analysis. Static friction is approximated by a step of torque or force that is applied in a random manner. This approach has provided a simple engineering estimate of the maximum error caused by static friction.

Thus, the most important tool for dealing with nonlinear control problems is a firm grasp of the linear aspects of the system. From this basis one can handle the nonlinear problems with analysis and simulation, without being overwhelmed by the multitude of parameters, variables, and configurations.

An important consequence of feedback is that it makes the process more linear. Even when the elements within a feedback-control loop are highly nonlinear, the loop is still relatively linear in response to external signals. An internal feedback loop is often closed in a control system to allow the outer feedback loop to operate in a more linear fashion.

On the other hand, a multiloop control system can sometimes exhibit serious instability when saturation of an inner loop causes the outer loop to become unstable. The results of such instability can be so dramatic that the system operation can best be described as "explosive." (For example, saturation of an AGC loop in a radar tracking system, like that discussed in Chapter 11, Section 11.3.4.1, can have this effect.) Hence, in designing multiloop control systems, one should make sure that the system degrades gracefully under saturated conditions.

12.2   EFFECT OF GAIN CROSSOVER FREQUENCY ON SATURATED STEP RESPONSE

An important nonlinear problem encountered in many feedback-control systems is a poorly damped oscillatory transient following a very large step input, which is caused by saturation in acceleration. This saturated step response occurs in many servomechanisms during lock-on transients. Although the saturated step response is a nonlinear process, its characteristics are directly related to the important linear parameters of the system. Except for the phase-plane discussion, this section is a condensation of a 1958 paper of the author [12.1], which was an extension of an earlier study by Travers [12.2].

12.2.1   Ideal Saturated Step Response

Consider the tachometer-feedback servo shown in Fig 12.2-1, which has position and velocity feedback loops. To analyze its response to a saturating step input, let us designate the absolute values of velocity and acceleration of the controlled variable as V and A:

$$V = |V_c| \qquad (12.2\text{-}1)$$

$$A = |A_c| \qquad (12.2\text{-}2)$$

The maximum absolute values of speed and acceleration are designated V_m and A_m:

$$V_m = \text{Max}[V] = \text{Max}|V_c| \qquad (12.2\text{-}3)$$

$$A_m = \text{Max}[A] = \text{Max}|A_c| \qquad (12.2\text{-}4)$$

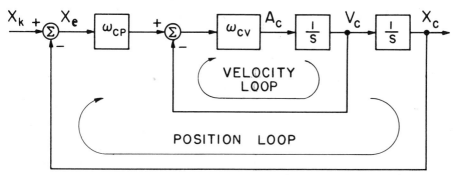

Figure 12.2-1 Velocity-feedback servo. Reprinted with permission from Ref [12.1] ©
1958 IRE (now IEEE).

The ideal saturated step response can be considered to be one in which the
controlled variable is saturated at all times either in velocity or acceleration,
and which has no overshoot. Such a transient yields the fastest possible
synchronizing response for the given saturation limits. Let us determine the
gain crossover frequencies ω_{cp} and ω_{cv} of the position and velocity loops
(shown in Fig 12.2-1) that give this ideal transient.

Figure 12.2-2 shows both the position X_c (or error X_e) and the velocity V_c
of the ideal nonlinear response of the system of Fig 12.2-1 to a large step
input. The step is applied at time t_1 (point 1), and between t_1 and t_2 (point 2)
the controlled member is under maximum acceleration A_m. At time t_2 (point
2), the controlled variable reaches maximum velocity V_m, and between t_2 and
t_3 (point 3) it moves at constant velocity. In order for the transient to end at
the shortest possible time, the system must start decelerating at a point 3,
which allows the velocity to reach zero when the error reduces to zero, at point
4. Thus, between points 3 and 4 the controlled member should be under
saturated deceleration A_m.

The time for deceleration $(t_4 - t_3)$ is equal to the time for acceleration
$(t_2 - t_1)$, which is

$$t_4 - t_3 = t_2 - t_1 = V_m/A_m \qquad (12.2\text{-}5)$$

The motion of the controlled variable during acceleration and during decelera-
tion is equal to $\frac{1}{2}A_m(\Delta t)^2$, where Δt is the time for acceleration or deceleration. By Eq 12.2-5, this motion is equal to

$$\tfrac{1}{2}A_m(\Delta t)^2 = \tfrac{1}{2}A_m\left(\frac{V_m}{A_m}\right)^2 = \tfrac{1}{2}\frac{V_m^2}{A_m} \qquad (12.2\text{-}6)$$

This distance is indicated in Fig 12.2-2. Equation 12.2-6 shows that to achieve
an ideal saturated step response, for an input large enough to produce velocity

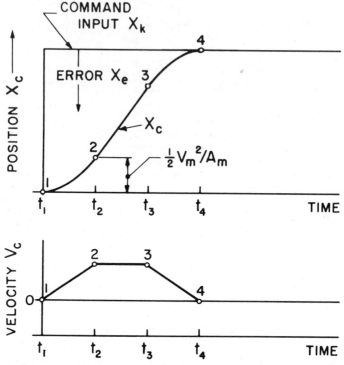

Figure 12.2-2 Ideal saturated step response. Reprinted with permission from Ref [12.1] © 1958 IRE (now IEEE).

saturation, the control system must switch to maximum deceleration at an error X_e equal to $V_m^2/2A_m$.

The manner in which the system switches to maximum deceleration can be seen from Fig 12.2-1, which shows that the acceleration A_c is equal to

$$A_c = \omega_{cv}\left(\omega_{cp}X_e - V_c\right) \qquad (12.2\text{-}7)$$

Equation 12.2-7 indicates that the acceleration is positive when

$$\omega_{cp}X_e > V_c \qquad (12.2\text{-}8)$$

The acceleration is negative when

$$\omega_{cp}X_e < V_c \qquad (12.2\text{-}9)$$

Thus, the system switches between positive and negative acceleration at an error of

$$X_e = V_c/\omega_{cp} \qquad (12.2\text{-}10)$$

If ω_{cv} is much greater than A_m/V_m, Eq 12.2-7 shows that a very small change of X_e from the value of Eq 12.2-10 produces either maximum acceleration or maximum deceleration (i.e., $A_c = +A_m$ or $A_c = -A_m$). Thus, if the velocity-loop gain crossover frequency ω_{cv} is large, the system switches almost instantaneously between maximum acceleration and maximum deceleration at the value of error given by Eq 12.2-10.

Prior to the point of switching to deceleration, the speed of the controlled variable is equal to V_m, and so the switch to maximum acceleration occurs at the error V_m/ω_{cp}. Therefore, in order that the saturated step response not overshoot, this switching must take place at an error larger than the distance the controlled member must move before it decelerates to zero speed, which is $V_m^2/2A_m$. Hence, for no overshoot,

$$\frac{V_m}{\omega_{cp}} > \frac{V_m^2}{2A_m} \qquad (12.2\text{-}11)$$

Solving for ω_{cp} gives

$$\omega_{cp} < 2A_m/V_m \qquad (12.2\text{-}12)$$

For the ideal saturated step response, ω_{cp} is equal to $2A_m/V_m$.

12.2.2   Saturated Step Response with Overshoot

When the gain crossover frequency ω_{cp} of the position loop is much greater than $2A_m/V_m$, the saturated step response is quite oscillatory. To develop a general formula characterizing the resultant transient, consider the oscillatory saturated step response shown in Fig 12.2-3, which has three nonlinear overshoots. Points 2, 4, 6, and 8 are the points where acceleration switching occurs, and so represent points of maximum speed; while points 1, 3, 5, 7, and 10 are points of zero speed. For this transient, the value of ω_{cp} is selected such that, after the last point of acceleration switching (point 2 of the third overshoot), the system decelerates with maximum deceleration, reaching zero velocity at zero error.

The following symbolism is used in the analysis:

X_n = value of X_c at point n
$X_{e(n)}$ = value of X_e at point n
V_n = speed at point n

At points 2, 4, 6, and 8, where the acceleration switching occurs, Eq 12.2-10 shows that the magnitude of error is related to the speed by

$$|X_{e(n)}| = V_n/\omega_{cp} \qquad (12.2\text{-}13)$$

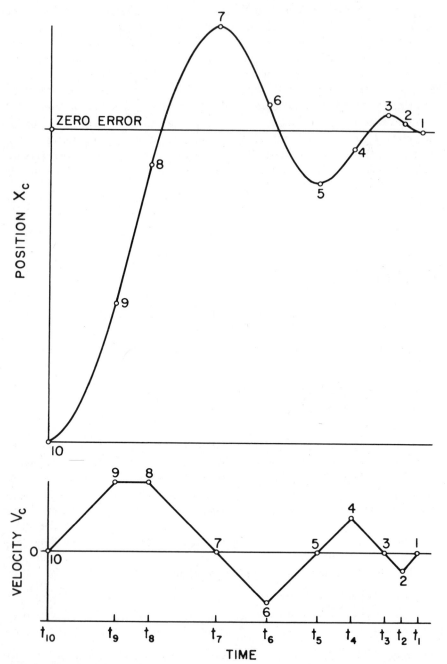

Figure 12.2-3 Saturated step response with overshoot. Reprinted with permission from Ref [12.1] © 1958 IRE (now IEEE).

This gives

$$|X_{e(2)}| = V_2/\omega_{cp} \tag{12.2-14}$$

$$|X_{e(4)}| = V_4/\omega_{cp} \tag{12.2-15}$$

$$|X_{e(6)}| = V_6/\omega_{cp} \tag{12.2-16}$$

$$|X_{e(8)}| = V_8/\omega_{cp} \tag{12.2-17}$$

Figure 12.2-3 shows that the values of error magnitude at successive switching points are related by

$$|X_{e(2)}| = |X_2 - X_1| \tag{12.2-18}$$

$$|X_{e(4)}| = |X_4 - X_3| - |X_3 - X_2| - |X_{e(2)}| \tag{12.2-19}$$

$$|X_{e(6)}| = |X_6 - X_5| - |X_5 - X_4| - |X_{e(4)}| \tag{12.2-20}$$

$$|X_{e(8)}| = |X_8 - X_7| - |X_7 - X_6| - |X_{e(6)}| \tag{12.2-21}$$

During the entire transient between point 8 and point 1, full acceleration of magnitude A_m is applied, alternately positively and negatively. Hence, the distances ΔX between successive points of zero and maximum speed are equal to

$$\Delta X = \tfrac{1}{2}A_m(\Delta t)^2 = \tfrac{1}{2}\frac{V^2}{A_m} \tag{12.2-22}$$

where V is the maximum speed for that section of the transient. Applying Eq 12.2-22 to the specific regions of the transient gives

$$|X_2 - X_1| = |X_3 - X_2| = \tfrac{1}{2}\frac{V_2^2}{A_m} \tag{12.2-23}$$

$$|X_4 - X_3| = |X_5 - X_4| = \tfrac{1}{2}\frac{V_4^2}{A_m} \tag{12.2-24}$$

$$|X_6 - X_5| = |X_7 - X_6| = \tfrac{1}{2}\frac{V_6^2}{A_m} \tag{12.2-25}$$

$$|X_8 - X_7| = |X_{10} - X_9| = \tfrac{1}{2}\frac{V_8^2}{A_m} \tag{12.2-26}$$

Substitute Eqs 17.2-18 to -21 and Eqs 12.2-23 to -26 into Eqs 12.2-14 to -17.

This gives

$$\frac{V_2}{\omega_{cp}} = \frac{1}{2A_m} V_2^2 \tag{12.2-27}$$

$$\frac{V_4}{\omega_{cp}} = \frac{1}{2A_m}\left(V_4^2 - V_2^2\right) - \frac{V_2}{\omega_{cp}} \tag{12.2-28}$$

$$\frac{V_6}{\omega_{cp}} = \frac{1}{2A_m}\left(V_6^2 - V_4^2\right) - \frac{V_4}{\omega_{cp}} \tag{12.2-29}$$

$$\frac{V_8}{\omega_{cp}} = \frac{1}{2A_m}\left(V_8^2 - V_6^2\right) - \frac{V_6}{\omega_{cp}} \tag{12.2-30}$$

The solution of Eq 12.2-27 is

$$V_2 = \frac{2A_m}{\omega_{cp}} \tag{12.2-31}$$

Equation 12.2-28 simplifies to

$$\frac{V_4 + V_2}{\omega_{cp}} = \frac{1}{2A_m}\left(V_4^2 - V_2^2\right) = \frac{1}{2A_m}(V_4 + V_2)(V_4 - V_2) \tag{12.2-32}$$

The solution of this is

$$V_4 = V_2 + \frac{2A_m}{\omega_{cp}} = \frac{2A_m}{\omega_{cp}}(2) \tag{12.2-33}$$

Similarly Eqs 12.2-29, -30 give

$$V_6 = V_4 + \frac{2A_m}{\omega_{cp}} = \frac{2A_m}{\omega_{cp}}(3) \tag{12.2-34}$$

$$V_8 = V_6 + \frac{2A_m}{\omega_{cp}} = \frac{2A_m}{\omega_{cp}}(4) \tag{12.2-35}$$

Since the speed V_8 is the maximum speed V_m, Eq 12.2-35 becomes

$$V_m = \frac{2A_m}{\omega_{cp}}(4) \tag{12.2-36}$$

This is for a saturated step response with three overshoots. When the preceding analysis is generalized to apply to a saturated step response with N

overshoots, Eq 12.2-36 becomes

$$V_m = \frac{2A_m}{\omega_{cp}}(1 + N)$$ (12.2-37)

Solving for ω_{cp} gives

$$\omega_{cp} = \frac{2A_m}{V_m}(1 + N)$$ (12.2-38)

The author [12.1] has shown that the maximum overshoot of this step response is

$$\text{Max}|X_e| = NV_m/\omega_{cp}$$ (12.2-39)

That reference also shows that the settling time of the step response is

$$T_{set} = \frac{2}{\omega_{cp}}(1 + N)(2 + N) = (2 + N)\frac{V_m}{A_m}$$ (12.2-40)

12.2.3   Effect of Velocity-Loop Gain Crossover Frequency

The preceding analysis has assumed that the gain crossover frequency ω_{cv} of the velocity loop is very much greater than the ratio A_m/V_m. The following develops a simple approximation to characterize the effect of a realistic value for the velocity-loop gain crossover frequency. Figure 12.2-1 showed that the acceleration A_c is equal to

$$A_c = \omega_{cv}(\omega_{cp}X_e - V_c)$$ (12.2-41)

Hence, the controlled member is under maximum positive acceleration $+A_m$ when

$$X_e > \frac{V_c}{\omega_{cp}} + \frac{A_m}{\omega_{cp}\omega_{cv}}$$ (12.2-42)

and it is under maximum negative acceleration when

$$X_e < \frac{V_c}{\omega_{cp}} - \frac{A_m}{\omega_{cp}\omega_{cv}}$$ (12.2-43)

Figure 12.2-4 shows the approximate step response for a realistic value of ω_{cv}, when ω_{cp} is adjusted for the maximum value with no overshoot. At the error

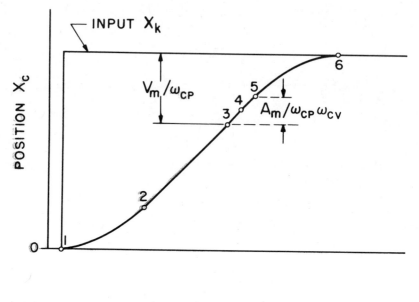

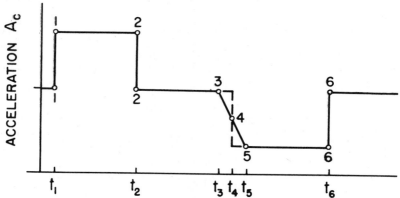

Figure 12.2-4 Effect of velocity-loop gain. Reprinted with permission from Ref [12.1] © 1958 IRE (now IEEE).

V_m/ω_{cp} (point 3) the acceleration starts to become negative, but it is not until point 5 that full saturated deceleration $-A_m$ is applied. The error at point 5 is

$$X_{e(5)} = \frac{V_5}{\omega_{cp}} - \frac{A_m}{\omega_{cp}\omega_{cv}} \qquad (12.2\text{-}44)$$

If it is assumed that there is negligible change of velocity between points 3 and 5, this error at point 5 is approximately equal to

$$X_{e(5)} \doteq \frac{V_m}{\omega_{cp}} - \frac{A_m}{\omega_{cp}\omega_{cv}} \qquad (12.2\text{-}45)$$

If the velocity is assumed to be nearly constant between points 3 and 5, the acceleration increases almost linearly with displacement in this region. The approximate acceleration plot is shown in diagram b of Fig 12.2-4. The change of velocity between points 3 and 5 is equal to the area under the acceleration curve in this region, which is

$$\Delta V = \int_{t_3}^{t_5} A \, dt = \tfrac{1}{2} A_m (t_5 - t_3) \tag{12.2-46}$$

The same change of velocity would occur in this region if the acceleration were zero up to point 4, and were equal to A_m thereafter, where point 4 is midway between points 3 and 5. The corresponding acceleration curve is shown by the dashed segment in Fig 12.2-4b.

Thus, the transient can be approximated by assuming that the acceleration switches discontinuously from zero to maximum deceleration at point 4. Since point 4 is midway between points 3 and 5, the acceleration effectively switches at an error of

$$X_{e(4)} = \frac{V_m}{\omega_{cp}} - \frac{1}{2} \frac{A_m}{\omega_{cp}\omega_{cv}} \tag{12.2-47}$$

If the controlled variable is to decelerate from point 4 to reach zero velocity when the error is zero, the error at point 4 must be equal to

$$X_{e(4)} = \frac{1}{2} \frac{V_m^2}{A_m} \tag{12.2-48}$$

Setting Eq 12.2-47 equal to Eq 12.2-48 gives

$$\omega_{cp} = \frac{2A_m}{V_m} \left[1 - \frac{2A_m/V_m}{4\omega_{cv}} \right] \tag{12.2-49}$$

The expression in the brackets is close to unity, and so $2A_m/V_m$ is approximately equal to ω_{cp}. Hence the second $(2A_m/V_m)$ term in the equation can be approximated as ω_{cp}. This gives the following approximation for Eq 12.2-49:

$$\omega_{cp} \doteq \frac{2A_m}{V_m} \left(1 - \frac{\omega_{cp}}{4\omega_{cv}} \right) \tag{12.2-50}$$

Equation 12.2-50 shows that the approximate effect of a realistic value of ω_{cv} is to require that ω_{cp} be changed by the following factor, if the number of overshoots of the saturated step response is to be the same as for infinite ω_{cv}:

$$\text{relative change of } \omega_{cp} = 1 - \frac{\omega_{cp}}{4\omega_{cv}} \tag{12.2-51}$$

The velocity-loop gain crossover frequency ω_{cv} is almost always at least twice as large as the position-loop gain crossover frequency ω_{cp}. Hence this factor of Eq 12.2-51 is almost always between $\frac{7}{8}$ and unity. This shows that the saturated-step-response equations, which assume infinite ω_{cv}, are still approximately correct for any practical value of ω_{cv}.

12.2.4 Effect of Other Dynamic Elements

Additional high-frequency lags in the system, which always occur in practical systems, do not alter the saturated step response appreciably, as long as the system has good stability under linear conditions. The primary effect of such lags on the saturated step response is to delay the switching, and hence somewhat prolong the transient.

On the other hand, a low-frequency integral network can have a very pronounced effect on the saturated step response. Generally this problem is

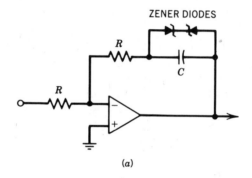

(a)

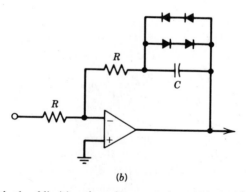

(b)

Figure 12.2-5 Methods of limiting the voltage stored on an integral-network capacitor during saturated conditions: (a) zener-diode limiter; (b) limiter using signal diodes.

avoided or minimized either by shorting the integral network during a saturated step transient, or by limiting the charge that can be stored on the integral network. Shorting an integral network does not limit the accuracy of the system, because an integral network need not operate under saturated conditions. Allowing an integral network to operate under saturated conditions leads to instability and adds an undesirable tail to the transient.

Convenient methods for limiting the charge stored on the integral-network capacitor are illustrated in Fig 12.2-5. During linear operation, the circuits provide the integral network transfer function $(1 + \omega_i/s)$, where $\omega_i = 1/RC$. In circuit a the charge on the integrator capacitor is limited by back-to-back zener diodes. To achieve a lower voltage limit, series strings of simple diodes can be used as shown in diagram b. Since each diode conducts at a voltage of about 0.7 V, this circuit limits the charge on the capacitor to ± 1.4 V.

12.2.5   Extension to Other Servos

In addition to the velocity-feedback servo described in the preceeding sections, there are two other common types of servos to which this analysis applies

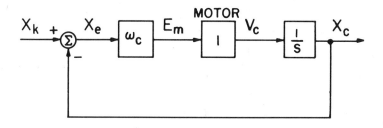

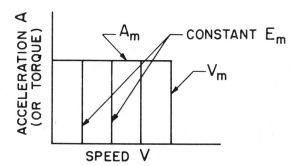

Figure 12.2-6   Velocity-source servo. Reprinted with permission from Ref [12.1] © 1958 IRE (now IEEE).

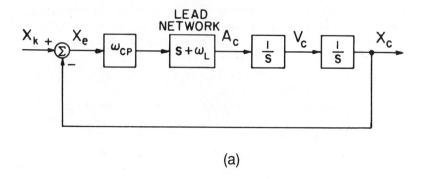

(a)

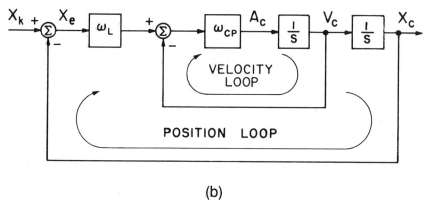

(b)

Figure 12.2-7 Lead-network-compensated servo: (*a*) lead-network servo; (*b*) equivalent velocity-feedback servo. Reprinted with permission from Ref [12.1] © 1958 IRE (now IEEE).

when suitably modified. These are the velocity-source servo illustrated in Fig 12.2-6 and the lead-network servo in Fig 12.2-7*a*.

Velocity-Source Servo. A velocity-source servo has a drive device that provides a velocity proportional to the applied signal. A common example of such a system is a hydraulic servo employing a valve that delivers an output flow proportional to valve displacement. The drive member acts like a velocity feedback loop, and so this servo does not require velocity feedback or a lead network for stabilization.

For the velocity-source servo of Fig 12.2-6, the nondimensionalized variable E_m represents the signal applied to the motor. The torque–speed curve of the motor shows that for a given value of E_m the motor supplies whatever acceleration (or torque) is required, up to the maximum acceleration capability, to drive the controlled variable at a velocity proportional to E_m. Therefore, the system acts like the velocity-feedback servo of Fig 12.2-1, with an infinite-gain crossover frequency in the velocity loop. Hence the equations that have been derived for the tachometer-feedback servo apply directly to the velocity-source servo.

Lead-Network Servo. In Fig 12.2-7a, the lead-network transfer function is shown as $(s + \omega_L)$. This approximates the following actual transfer function of a lead network, in cascade with a gain of α:

$$\frac{s + \omega_L}{s + \alpha \omega_L} = \frac{(1/\alpha)(s + \omega_L)}{1 + (s/\alpha \omega_L)} \qquad (12.2\text{-}52)$$

The diagram neglects the factor $(1 + s/\alpha \omega_L)$, but this has little effect on the saturated-step response if α is reasonably large.

An analysis of the saturated-step response of a lead-network servo is given by the author in Ref [12.1]. This shows that during the saturated-step response the lead-network servo of Fig 12.2-7a behaves essentially the same as the equivalent velocity-feedback servo of Fig 12.2-7b. This equivalent velocity-feedback servo has a position-loop gain crossover frequency equal to ω_L, and a velocity-loop gain crossover frequency equal to ω_{cp}. Applying the preceeding analyses gives

$$\omega_L = \frac{2A_m}{V_m}(1 + N) \qquad (12.2\text{-}53)$$

This holds with reasonable accuracy provided that $\omega_{cp} > 2\omega_L$, a condition that is generally satisfied.

12.2.6 Nonlinear Techniques to Improve Performance

It has been shown that the gain crossover frequency of the position loop must be limited to achieve a nonoscillatory saturated step response. However, this limitation may degrade the system performance under linear conditions. Hence, it is often desirable to employ nonlinear control techniques to achieve both a fast linear response and a well-damped saturated-step response.

The principle for achieving this is based on the equation for the ideal step response, which is

$$\omega_{cp} = 2A_m/V \qquad (12.2\text{-}54)$$

The maximum motor speed V_m has been replaced by the speed V. This shows that ideal performance can be achieved at all operating speeds if the position-

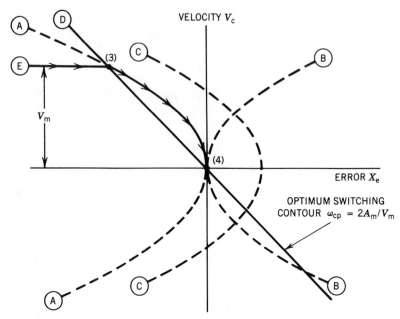

Figure 12.2-8 Phase-plane plot for optimum saturated-step response of Fig 12.2-2.

loop gain crossover frequency ω_{cp} is varied so it is inversely proportional to the speed V of the controlled variable. This is not practical at very low speeds, because ω_{cp} would be so high it would cause linear instability. Hence, at low speeds ω_{cp} is set at the maximum value commensurate with good linear stability. Equation 12.2-54 is implemented when it gives a value for ω_{cp} that is less than this linear value.

12.2.7 The Phase Plane

The nonlinear transient responses that have been analyzed can be presented conveniently on the *phase plane*, which is illustrated in Fig 12.2-8. This is a plot of velocity V_c versus position error X_e during the transient. The term "phase plane" has nothing to do with the concept of phase shift. This term was coined by Gibbs [12.3], who used the phase plane to characterize dynamic chemical reactions. The word "phase," as used by the chemist, is equivalent to the word "state" as used by the control engineer. Hence, this plot should properly be called the "state plane," and the two axes in it called "state variables." However, the term "phase plane" has such wide usage that it will remain. The two state variables plotted on the phase plane are usually the velocity and the position error, as shown in Fig 12.2-8. However, two other state variables (such as velocity and acceleration) can also be used.

During the nonlinear transients depicted in Figs 12.2-2, -3, the servo is operated continuously under saturated accelerations of $\pm A_m$, except during the initial period of saturated velocity V_m. When the acceleration is saturated, the absolute values of the controlled velocity V_c, and of the change ΔX in position error, are described by

$$|V_c| = V = A_m|\Delta t| \tag{12.2-55}$$

$$|\Delta X| = \tfrac{1}{2}A_m(\Delta t)^2 \tag{12.2-56}$$

When the speed is increasing, the quantity Δt is the time difference between the present time and the last time the speed was zero, while ΔX is the change of position since that instant of zero speed. When the speed is decreasing, the quantity Δt is the difference between the present time and the next time the speed will be zero, while ΔX is the amount the position will change before the speed is reduced to zero. Equations 12.2-55, -56 can be combined to give

$$|\Delta X| = V^2/2A_m \tag{12.2-57}$$

Equation 12.2-57 describes a family of parabolic contours on the phase plane. During saturated acceleration, the servo operates along one of these contours. Three of these parabolic contours are shown in Fig 12.2-8 as dashed curves (A), (B), (C). Curve (A) is a contour of maximum negative acceleration, while curve (B) is a contour of maximum positive acceleration. For these contours, the position error is zero when the speed is zero. The parabolic curve (C) has the same shape as curve (A), but its vertex is displaced in the direction of positive error.

In Fig 12.2-8, segment 3–4 is the contour corresponding to the optimum saturated step response that was shown in Fig 12.2-2. Line (D) is a plot of the equation

$$V_c = -\omega_{cp}X_e = -\frac{2A_m}{V_m}X_e \tag{12.2-58}$$

where ω_{cp} is set at its optimum value, $2A_m/V_m$. This is the locus of points at which the servo switches acceleration, when the gain crossover frequency is set at its optimum value. Line (E) is the contour for $V_c = V_m$, which is the portion of the transient of Fig 12.2-2 between points 2 and 3, where the speed is held constant at its saturated value V_m. Point 3 of Fig 12.2-8 is the same as point 3 in Fig 12.2-2. At this point 3, contour (E) reaches the acceleration switching contour (D), and so saturated negative acceleration $-A_m$ is applied at this point. Hence, the servo now follows the parabolic curve (A) along the solid trajectory, which brings the error to zero at point 4, at the instant that the velocity is reduced to zero. Thus, the optimum saturated-step response has been achieved.

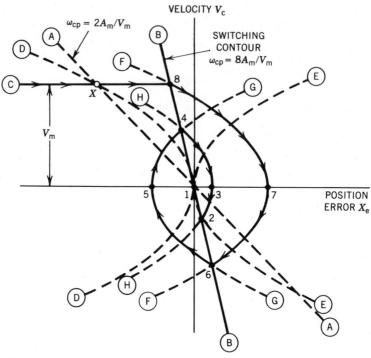

Figure 12.2-9 Phase-plane for saturated-step response of Fig 12.2-3 (three overshoots).

Figure 12.2-9 shows the phase-plane trajectory corresponding to the saturated transient of Fig 12.2-3, which has three overshoots. The gain crossover frequency ω_{cp} is increased by a factor of $(1 + N)$, which is 4, relative to the optimum value $\omega_{cp} = 2A_m/V_m$. Line Ⓐ is the switching contour for optimum ω_{cp}, which is $2A_m/V_m$, while line Ⓑ is the switching contour for this oscillatory transient, for which ω_{cp} is equal to $8A_m/V_m$. The servo is initially operating along contour Ⓒ at saturated velocity V_m. If the gain crossover frequency ω_{cp} were optimum, the acceleration would switch at point (X), where contour Ⓒ crosses the optimum acceleration switching contour Ⓐ. The servo would then operate at negative saturated acceleration, following optimum parabolic contour Ⓓ to reach zero error at the instant that the speed is reduced to zero. The parabolic contour Ⓔ is the negative of the parabolic contour Ⓓ, and is the optimum contour corresponding to saturated positive acceleration. This contour Ⓔ would be followed by an optimum saturated transient approaching zero error from the opposite direction.

However, with the increased value of ω_{cp}, the acceleration does not switch until trajectory Ⓒ reaches switch contour Ⓑ, at point 8. (Points 1 to 8 correspond to the same points indicated on the transient of Fig 12.2-3.) At point 8, maximum deceleration A_m is applied, and so the servo begins to follow the parabolic contour Ⓕ. This contour has the same shape as the optimum parabolic contour (D), but its vertex is shifted in the direction of positive error until the parabola passes through point 8 on the switching contour (B). After point 8, the servo follows contour (F), reaching zero speed at point 7, and then accelerates, reaching maximum negative velocity at point 6.

At point 6 the parabolic contour (F) again crosses the switch contour (B), and the servo switches from saturated negative acceleration to saturated positive acceleration. The servo now begins to follow the parabolic contour (G), which has the same shape as the optimum parabolic contour (E), but its vertex is shifted in the direction of negative error until the parabolic contour passes through point 6 on the switch contour (B). The servo follows contour (G) from point 6, reaching zero speed at point 5. The servo accelerates, reaching maximum positive velocity at point 4, where contour (G) again crosses the switching contour (B).

At point 4, saturated negative acceleration is applied, and the servo begins to follow the parabolic contour (H), which has the same shape as contour (D), with its vertex shifted until the parabola passes through point 4. The servo follows contour (H) from point 4, reaching zero speed at point 3. It reaches maximum negative velocity at point 2, where contour (H) crosses the switching contour (B). Again saturated positive acceleration is applied. However, point 2 lies on the optimum contour (E) for positive acceleration, which was described previously. The servo follows along this optimum contour (E) from point 2, reaching zero speed at point 1, where the error is also zero.

Thus, the phase plane shows in a direct graphical manner the same principles analyzed previously by means of equations and graphs. Some investigators have found the phase plane to be very effective for studying nonlinear transients, and an appreciable body of theory has been developed concerning its use. Extensive discussions of this theory are presented by Gibson [12.4] (Chapter 7), by Gille, Pelegrin, and Decaulne [1.5] (Chapter 25), and by Ogata [12.5] (Chapter 12).

12.3 LINEARIZATION OF NONLINEAR RESPONSE ABOUT AN OPERATING POINT

One of the primary methods of analyzing nonlinearities in feedback control systems is to consider small deviations from an operating point. For small deviations, the response is essentially linear. However, the coefficients of the linear response vary as the system operating point changes.

12.3.1 Nonlinearity of the Torque-Speed Curve of an AC Servo Motor

Let us apply this concept to analyze the response of a servo using an AC servo motor, which has very nonlinear torque–speed curves. A typical family of

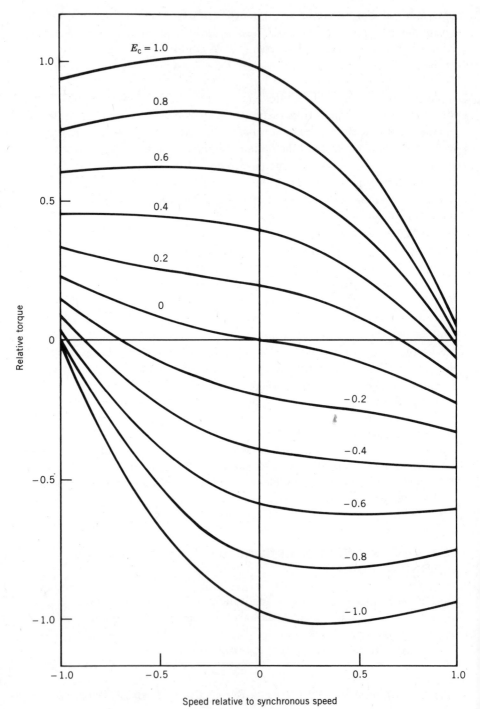

$E_c = 1.0$

0.8

0.6

0.4

0.2

0

-0.2

-0.4

-0.6

-0.8

-1.0

Relative torque

Speed relative to synchronous speed

Figure 12.3-1 Typical torque–speed curves of an AC servo motor.

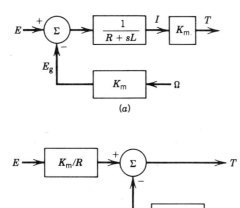

(a)

(b)

Figure 12.3-2 Signal-flow diagram of a DC servo motor: (*a*) basic diagram; (*b*) modified low-frequency form.

torque–speed curves for a two-phase AC servo motor is shown in Fig 12.3-1. This was derived from Gibson and Tuteur [8.1] (Fig 7.9). Voltage E_c is the relative amplitude of the AC voltage applied to the control winding. A negative value for E_c denotes phase reversal of the AC voltage. The AC voltage applied to the reference winding has a constant amplitude, and is phase shifted by 90 deg relative to the voltage on the control winding.

Before analyzing these nonlinear curves, let us first examine the torque–speed curves of a DC servo motor, which are linear. Diagram *a* in Fig 12.3-2 shows the signal-flow diagram for a DC servo motor, which was given in Chapter 4, Fig 4.1-3. The voltage E_m, torque T_m, angular velocity Ω_m, resistance R_m, and inductance L_m of the motor are expressed simply as E, T, Ω, R, and L. If the effect of motor inductance is ignored, the signal-flow diagram (*a*) can be simplified to diagram *b*, which is represented by the equation

$$T = \frac{K_m}{R}E - \frac{K_m^2}{R}\Omega \qquad (12.3\text{-}1)$$

This equation is expressed in graphical form by the torque–speed plots of the motor, which are shown in Fig 12.3-3. These plots show the variation of motor torque T as a function of motor angular velocity Ω, for constant values of the voltage E applied to the motor. Four plots are shown, for E equal to E_{Max}, $0.75E_{\text{Max}}$, $0.5E_{\text{Max}}$, and $0.25E_{\text{Max}}$, where E_{Max} is the rated maximum voltage of the motor. The value of the maximum torque T_{Max} is obtained by setting

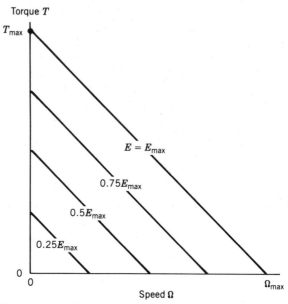

Torque T

Figure 12.3-3 Torque–speed curves of a DC servo motor.

$\Omega = 0$ and $E = E_{\text{Max}}$ in Eq 12.3-1, which gives

$$T_{\text{Max}} = \frac{K_m}{R} E_{\text{Max}} = K_m \frac{E_{\text{Max}}}{R} = K_m I_{\text{Max}} \qquad (12.3\text{-}2)$$

The current I_{Max} is the maximum current, which occurs at zero speed (when the back EMF E_g is zero), and the applied voltage E is equal to its maximum specified value E_{Max}. Thus,

$$I_{\text{Max}} = \frac{E_{\text{Max}}}{R} \qquad (12.3\text{-}3)$$

The value of the maximum speed Ω_{Max} is obtained by setting $T = 0$ and $E = E_{\text{Max}}$ in Eq 12.3-1, which gives

$$\Omega_{\text{Max}} = \frac{E_{\text{Max}}}{K_m} \qquad (12.3\text{-}4)$$

Now let us return to the nonlinear torque–speed curves of the AC servo motor. Linear analysis can be applied to these by considering small deviations from an operating point. Figure 12.3-4 shows a nonlinear torque–speed curve. The operating point is represented by point o on the torque–speed curve $\textcircled{1}$. The values of the motor voltage E, motor angular velocity Ω, and motor

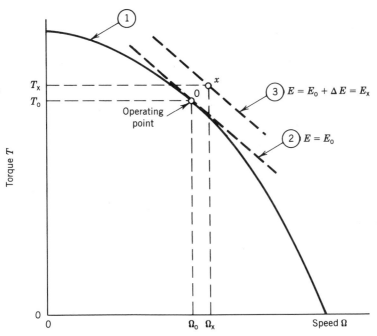

Figure 12.3-4 Torque–seed curve of an AC servo motor, showing linearization about an operating point.

torque T at this operating point are designated E_o, Ω_o, and T_o. In the vicinity of this operating point, the solid nonlinear torque–speed curve ① is approximated by the straight line tangent to the curve at the operating point, shown as the dashed line ②. The dashed line ③ is the linearized torque–speed curve for a slightly larger value of the motor voltage E. Point x is a particular point on line ③. The torque at point x can be expressed as

$$T_x = T_o + \frac{\partial T}{\partial E}(E_x - E_o) - \left(-\frac{\partial T}{\partial \Omega}\right)(\Omega_x - \Omega_o) \qquad (12.3\text{-}5)$$

The variables T_x, E_x, Ω_x are the values of T, E, Ω at the point x. The quantity $\partial T/\partial E$ is the partial derivative of the torque T with respect to voltage E, at constant angular velocity Ω, while $\partial T/\partial \Omega$ is the partial derivative of torque T with respect to angular velocity Ω, at constant voltage E. The partial derivative $\partial T/\partial \Omega$ is the slope of the torque–speed curve, which is negative. Hence, it is replaced by $-\partial T/\partial \Omega$, which is positive, in cascade with a negative sign.

Equation 12.3-5 can be expressed as

$$\Delta T = \frac{\partial T}{\partial E}\Delta E - \left(-\frac{\partial T}{\partial \Omega}\right)\Delta \Omega \qquad (11.3\text{-}6)$$

where $\Delta T = T_x - T_o$, $\Delta E = E_x - E_o$, and $\Delta \Omega = \Omega_x - \Omega_o$. The quantities ΔT, ΔE, and $\Delta \Omega$ are the deviations of the motor torque T, motor voltage E, and motor angular velocity Ω from the values T_o, E_o, Ω_o at the operating point. Equation 12.3-6 is expressed in signal-flow-diagram form in Fig 12.3-5a. Note that this has the same form as the signal-flow diagram of the DC motor shown in Fig 12.3-2b. Let us assume that the AC motor is driving a pure inertia load of inertia $J_{(m)}$. Figure 12.3-5b shows the signal-flow diagram of the AC servo motor plus inertia load. This diagram relates the applied incremental motor voltage ΔE to the resultant incremental motor angular velocity $\Delta \Omega$.

In Fig 12.3-5, diagram c shows one possible simplified form of diagram b. The feedback factor $-\partial T/\partial \Omega$ is placed in the forward path of the motor loop, while its reciprocal $(-\partial \Omega/\partial T)$ is placed in front of that loop in series with the ΔE signal. The motor loop has the loop transfer function

$$G_{(m)} = \frac{-\partial T/\partial \Omega}{s J_{(m)}} = \frac{\omega_{cm}}{s} \tag{12.3-7}$$

The gain crossover frequency of this motor loop is

$$\omega_{cm} = \frac{-\partial T/\partial \Omega}{J_{(m)}} \tag{12.3-8}$$

The feedback transfer function of the motor loop is

$$G_{ib(m)} = \frac{G_{(m)}}{1 + G_{(m)}} = \frac{1}{1 + 1/G_{(m)}} = \frac{1}{1 + s/\omega_{cm}} \tag{12.3-9}$$

Hence, the transfer function relating incremental motor voltage ΔE to incremental motor angular velocity $\Delta \Omega$ is, from Fig 12.3-5c,

$$\frac{\Delta \Omega}{\Delta E} = \frac{\partial T}{\partial E}\left(-\frac{\partial \Omega}{\partial T}\right) G_{ib(m)} = \frac{(\partial T/\partial E)(-\partial \Omega/\partial T)}{1 + s/\omega_{cm}} \tag{12.3-10}$$

The expression in the numerator is equal to the partial derivative $\partial \Omega/\partial E$ of motor speed Ω relative to motor voltage E, for constant motor torque T. Thus,

$$\frac{\partial \Omega}{\partial E} = \frac{\partial T}{\partial E}\left(-\frac{\partial \Omega}{\partial T}\right) \tag{12.3-11}$$

Substituting Eq 12.3-11 into Eq 12.3-10 gives

$$\frac{\Delta \Omega}{\Delta E} = \frac{\partial \Omega/\partial E}{1 + s/\omega_{cm}} \tag{12.3-12}$$

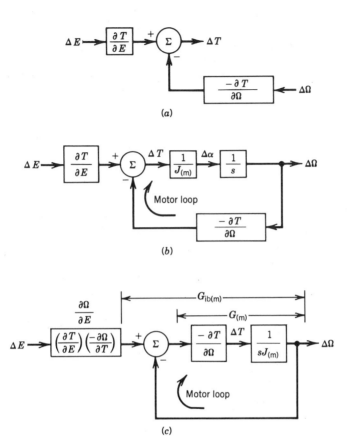

(a)

(b)

(c)

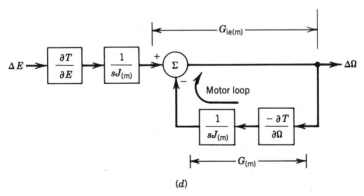

(d)

Figure 12.3-5 Signal-flow diagrams of AC servo motor: (*a*) diagram of motor for incremental variations from operating point; (*b*) motor with inertia load; (*c*) traditional simplified form of (*b*); (*d*) alternative simplified form of (*b*).

361

Diagram d of Fig 12.3-5 shows an alternative method for simplifying the signal-flow diagram b. The summation point of b is placed after the transfer function $1/sJ_{(m)}$. The resultant transfer function relating the incremental motor voltage ΔE to the incremental motor angular velocity $\Delta\Omega$ is

$$\frac{\Delta\Omega}{\Delta E} = \frac{\partial T}{\partial E}\frac{1}{sJ_{(m)}}G_{ie(m)} \qquad (12.3\text{-}13)$$

The error transfer function of the motor loop is

$$G_{ie(m)} = \frac{1}{1 + G_{(m)}} = \frac{1}{1 + \omega_{cm}/s} \qquad (12.3\text{-}14)$$

Combining Eqs 12.3-13, -14 gives the following transfer function, relating the incremental motor voltage ΔE to the incremental angular velocity $\Delta\Omega$:

$$\frac{\Delta\Omega}{\Delta E} = \frac{\partial T/\partial E}{sJ_{(m)}(1 + \omega_{cm}/s)} \qquad (12.3\text{-}15)$$

Equation 12.3-12 is the traditional form used to represent the transfer function for an AC servo motor. The constant $\partial\Omega/\partial E$ varies greatly for different operating points on the torque–speed curve. This partial derivative $\partial\Omega/\partial E$ is proportional to the difference between two adjacent curves along a horizontal (constant-torque) line. As can be seen from the AC-motor torque–speed curves of Fig 12.3-1, the horizontal difference between the curves varies drastically over the plot. Since the partial derivative $\partial\Omega/\partial E$ is the gain constant of Eq 12.3-12, this suggests that the dynamic response of a servo using an AC motor should vary greatly over its operating range, and hence the servo should have poor dynamic performance.

A much clearer picture of AC servo-motor response is obtained from the alternative form of the transfer function given in Eq 12.3-15. The gain constant of this equation is proportional to $\partial T/\partial E$, which is the partial derivative of torque with respect to motor voltage, at constant motor speed. This partial derivative is proportional to the vertical difference, at constant motor speed, between adjacent torque–speed curves. As shown in Fig 12.3-1, this vertical difference between curves is nearly constant over the torque–speed curve, and at zero velocity is exactly constant.

Figure 12.3-6a shows a magnitude asymptote plot of the transfer function $\Delta\Omega/\Delta E$. In the high-frequency region, the asymptote is proportional to the transfer function $(\partial T/\partial E)/(sJ_{(m)})$, and so is nearly constant throughout the torque–speed region. The break frequency ω_{cm} varies greatly, because ω_{cm} is proportional to $\partial T/\partial\Omega$, which is the slope of the torque–speed curve. In the negative-torque region (where the motor is being decelerated) this slope not only drops to zero; it actually changes sign. This large variation of ω_{cm}

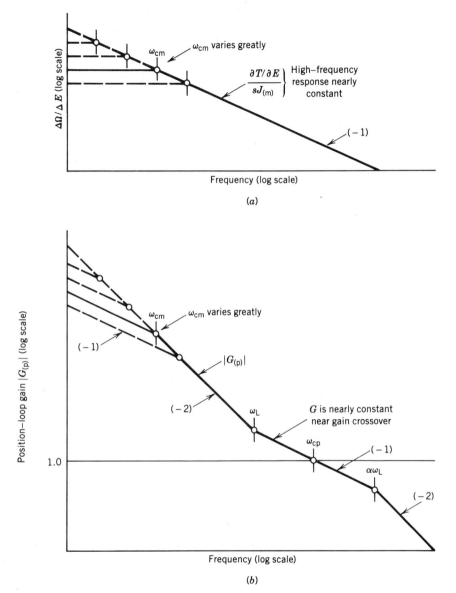

Figure 12.3-6 Magnitude asymptote plots showing effect of variation of AC servo-motor dynamics: (*a*) frequency response of AC servo motor with inertia load; (*b*) loop gain of resultant position loop with lead-network compensation.

changes the transfer function $\Delta\Omega/\Delta E$ greatly at low frequencies, but has little effect at high frequencies.

A servo using an AC motor practically always has either tachometer feedback or lead-network compensation. The slope $\partial T/\partial\Omega$ of the torque–speed curve represents a torque variation with speed, and so is equivalent to damping. Tachometer feedback provides a much stronger damping signal, and so swamps out the damping effect due to the slope of the torque–speed curve. Therefore, with tachometer feedback, the variation of the slope $\partial T/\partial\Omega$ of the torque–speed curve has very little effect on servo performance.

Figure 12.3-6b shows the position-loop frequency response of an AC motor servo with lead-network compensation. This has little variation in the high-frequency region, and so the gain crossover frequency ω_{cp} of the position loop is nearly constant. At low frequency the loop response changes greatly as the break frequency ω_{cm} varies, but this has only a secondary effect on the dynamic performance of the servo. The loop transfer function of the position loop is expressed as follows in terms of ω_{cp}:

$$G_{(P)} = \frac{\omega_{cp}(1 + \omega_L/s)}{s(1 + \omega_{cm}/s)(1 + s/\alpha\omega_L)} \qquad (12.3\text{-}16)$$

When this transfer function is written in the traditional form, it becomes

$$G_{(P)} = \frac{K_1(1 + s/\omega_L)}{s(1 + s/\omega_{cm})(1 + s/\alpha\omega_L)} \qquad (12.3\text{-}17)$$

where K_1 is the velocity constant, which is

$$K_1 = \omega_{cp}\omega_L/\omega_{cm} \qquad (12.3\text{-}18)$$

Since K_1 is inversely proportional to ω_{cm}, it is highly variable, even though the gain crossover frequency ω_{cp} is nearly constant. Thus, the traditional form of writing the loop transfer function gives the false impression that the dynamic response of the loop varies strongly.

The preceeding discussion illustrates the following general principle. When studying the effect of parameter variation in a nonlinear feedback-control loop, one should relate the parameter to the gain crossover frequency (or frequencies) of the control system. If the gain crossover frequency of a critical control loop changes greatly with the varying parameter, that parameter variation is serious. However, if the parameter variation merely affects the low-frequency characteristics of a control loop, the variation is not very important.

In the signal-flow diagram of this section, the variables were all represented as incremental values: ΔE, ΔT, $\Delta\Omega$. Whenever the feedback-control system is even slightly nonlinear, one should consider the signals of the signal-flow diagram to be incremental values. Remember that a signal-flow diagram is mathematically valid only when the system equations are linear.

As was explained in Chapter 8, Section 8.1, there appears to be a high-frequency lag in the response of an AC servo motor. When this lag is included, the transfer function of Eq 12.3-15 becomes

$$\frac{\Delta\Omega}{\Delta E} = \frac{\partial T/\partial E}{sJ_{(m)}(1 + \omega_{cm}/s)(1 + s/\omega_e)} \quad (12.3\text{-}19)$$

where ω_e is the effective electrical break frequency of the AC motor.

12.3.2 Power Dissipation in Servo Motors

Along with this discussion of torque–speed curves for DC and AC motors, let us examine a very important practical issue associated with the use of these motors: motor efficiency. It is well known that the AC servo motor has much poorer "efficiency" than the DC servo motor. However, this issue of efficiency is confusing when applied to a servo system. Many servos operate most of the time at close to zero speed. Near zero speed, the mechanical output power is very low, and so the efficiency is essentially zero for both DC and AC motors. On the other hand, when an AC motor operates at high speed, it has high efficiency. What then does it mean to say that an AC servo motor has poor "efficiency?" To understand this, let us investigate the issues of efficiency and dissipation in DC and AC servo motors.

12.3.2.1 *Power Dissipation in a DC Servo Motor.*

Although a small DC servo motor generally has a permanent magnet to generate the field, a large DC servo motor often has a separately excited field winding. Any dissipation in a separately excited field winding should properly be included as part of the motor dissipation. Nevertheless, the following discussion ignores field dissipation. It also ignores mechanical dissipation in the motor, due to windage and friction.

Figure 12.3-7a shows a particular torque–speed curve for a DC servo motor. The motor is assumed to be operating at the designated point indicated on this torque–speed curve, where it is being excited with an applied winding voltage E, and is generating a torque T, while rotating at a speed Ω. The values of maximum torque and maximum speed for this particular torque–speed curve are designated T_{Max} and Ω_{Max}. In the discussion relative to Fig 12.3-3, the parameters T_{Max} and Ω_{Max} were defined in terms of the torque–speed curve for the maximum rated winding voltage E_{Max}. However, that concept is now being generalized: T_{Max} and Ω_{Max} now apply to any torque–speed curve at constant voltage E.

The output mechanical power P_{out} from the DC servo motor is equal to the product of the torque T and the angular velocity Ω:

$$P_{out} = T\Omega \quad (12.3\text{-}20)$$

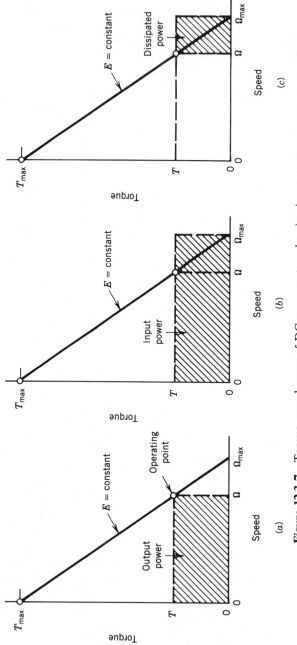

Figure 12.3-7 Torque–speed curve of DC servo motor, showing input power, output power, and dissipated power.

This output is equal to the cross-hatched area in Fig 12.3-7a. The input power to the motor is

$$P_{in} = EI \tag{12.3-21}$$

The motor torque T is related to the current I by

$$T = K_m I \tag{12.3-22}$$

By Eq 12.3-4, the maximum speed Ω_{Max} for the particular torque–speed curve is related as follows to the winding voltage E:

$$\Omega_{Max} = E/K_m \tag{12.3-23}$$

Substituting Eqs 12.3-22, -23 into Eq 12.3-21 gives the input power:

$$P_{in} = EI = K_m \Omega_{Max} \frac{T}{K_m} = T\Omega_{Max} \tag{12.3-24}$$

This input power is indicated by the cross-hatched area in Fig 12.3-7b. The difference between the input power in diagram b and the output power in diagram a is the power dissipated in the motor, which is shown as the cross-hatched area in diagram c. By Eqs 12.3-20, -24 the efficiency of the motor is

$$\eta = \frac{P_{out}}{P_{in}} = \frac{\Omega}{\Omega_{Max}} \tag{12.3-25}$$

The parameter T_{Max} shown on the torque–speed curve is a theoretical value that often cannot be reached. The maximum torque must be limited for two reasons: (1) to avoid demagnetization of the motor field, and (2) to avoid excessive power dissipation in the motor, which could cause overheating. Since torque is proportional to current, the maximum torque is restricted by limiting the maximum value of the motor current. The motor current can be limited either with a current feedback loop, or by incorporating a current-limiting circuit into the power-amplifier stage.

Since the power dissipation in the motor is proportional to the square of the current, it is proportional to the square of the torque:

$$P_{dis} = I^2 R = \left(\frac{T}{K_m}\right)^2 R = T^2 \frac{R}{K_m^2} \tag{12.3-26}$$

Many DC servo motors can provide, without demagnetization, a peak torque that is much larger than the rated continuous torque. Such motors can intermittently deliver torques that are much greater than the rated torque.

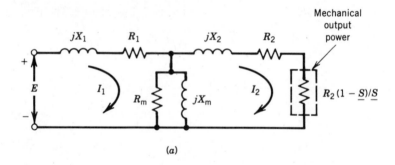

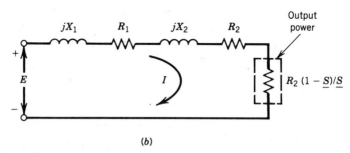

(a)

(b)

Figure 12.3-8 Equivalent circuit for one winding of a two-phase AC servo motor: (*a*) complete circuit; (*b*) approximate circuit ignoring magnetizing impedance.

However, when making use of this capability, one should remember that power dissipation varies as the square of the torque. Consequently, if the peak torque is 2 times the rated torque, the duty ratio at full torque must not exceed $\frac{1}{4}$; while a duty ratio not exceeding $\frac{1}{9}$ is required if the peak torque is 3 times the rated torque.

12.3.2.2 Power Dissipation in an AC Servo Motor.

Not let us examine the power dissipation in an AC servo motor. For simplicity it is assumed that the control and reference windings have the same number of turns, and so have the same impedance. When the AC voltages on the reference and control windings are equal, either winding can be described by the equivalent circuit of Fig 12.3-8a. (See Gibson and Tuteur [8.1], Fig 7.6.) The parameter $\underline{S}$ is the motor slip, which is defined by

$$\underline{S} = \frac{\Omega_{syn} - \Omega}{\Omega_{syn}} = 1 - \frac{\Omega}{\Omega_{syn}} \qquad (12.3\text{-}27)$$

where Ω is the motor speed and Ω_{syn} is the synchronous speed of the AC motor. (The symbol $\underline{S}$ for slip has an underline to differentiate it from the Laplace-transform variable s.) The power absorbed by the resistance

$R_2(1 - \underline{S})/\underline{S}$ represents the electrical power, per phase, that is converted to mechanical power. The power absorbed in resistors R_1 and R_2 is the power per phase actually dissipated within the motor. The effect of the magnetization impedance is relatively small, and so the equivalent circuit of diagram a can be approximated by that of diagram b. Hence, the approximate values of output and dissipated power per phase are

$$P_{out} = I^2 R_2 \frac{1 - \underline{S}}{\underline{S}} \tag{12.3-28}$$

$$P_{dis} = I^2(R_1 + R_2) \tag{12.3-29}$$

To achieve good servo performance, the torque–speed curve of a two-phase AC servo motor must have negative slope throughout the range of positive torque and positive angular velocity. When this condition is not satisfied, the motor tends to run by itself as a single-phase AC motor, being driven by the constant voltage on the reference winding, independent of the voltage on the control winding. As shown by Gibson and Tuteur [8.1] (Eq 7.53), negative slope is maintained throughout the quadrant of positive torque and positive angular velocity if the following is satisfied:

$$R_2^2 > R_1^2 + (X_1 + X_2)^2 \tag{12.3-30}$$

The reactance X_1 of the primary circuit is much greater than the primary circuit resistance R_1:

$$X_1 \gg R_1 \tag{12.3-31}$$

Combining Eqs 12.3-30, -31 shows that R_2 must be much larger than R_1:

$$R_2 \gg R_1 \tag{12.3-32}$$

Hence, R_1 in the expression for dissipated power in Eq 12.3-29 can be neglected. Dividing Eq 12.3-28 by Eq 12.3-29 gives

$$\frac{P_{out}}{P_{dis}} = \frac{(1 - \underline{S})}{\underline{S}} \tag{12.3-33}$$

Solve this for P_{out}, the output power per phase:

$$P_{out} = \frac{1 - \underline{S}}{\underline{S}} P_{dis} = \left(\frac{1}{\underline{S}} - 1\right) P_{dis} = \frac{P_{dis}}{\underline{S}} - P_{dis} \tag{12.3-34}$$

The input power per phase, P_{in}, is the sum of the output power per phase, P_{out}, plus the dissipated power per phase, P_{dis}. From Eq 12.3-34, this sum is

$$P_{in} = P_{out} + P_{dis} = \frac{P_{dis}}{\underline{S}} \tag{12.3-35}$$

Dividing Eq 12.3-34 by Eq 12.3-35 gives the ratio of output to input power, which is the efficiency η:

$$\eta = \frac{P_{out}}{P_{in}} = 1 - \underline{S} = \frac{\Omega}{\Omega_{syn}} \qquad (12.3\text{-}36)$$

The slip $\underline{S}$ has been replaced by the expression of Eq 12.3-27. Solving this for the input power gives

$$P_{in} = \frac{\Omega_{syn}}{\Omega} P_{out} \qquad (12.3\text{-}37)$$

This applies to either the power per phase or the total power. The total output power P_{out} is equal to the total output torque T multiplied by the angular velocity Ω:

$$P_{out} = T\Omega \qquad (12.3\text{-}38)$$

Combining Eqs 12.3-37, -38 gives the total input power:

$$P_{in} = T\Omega_{syn} \qquad (12.3\text{-}39)$$

Let us relate this equation to the torque–speed curve of the AC servo motor. We are considering the case where the RMS voltages applied to the control winding and the reference winding are equal. This condition corresponds to the torque–speed curve for maximum control voltage, which has the approximate form shown in Fig 12.3-9. The torque drops to zero approximately at the synchronous speed Ω_{syn}. In accordance with Eq 12.3-39, when the motor is at the specified operating point on the torque–speed curve, the input power is equal to $(T\Omega_{syn})$, which is given by the cross-hatched area in diagram a. The output power is equal to $T\Omega$, which is the cross-hatched area shown in diagram b. The difference between the cross-hatched areas of diagrams a and b is the power dissipated in the motor, which is shown by the cross-hatched area of diagram c.

Diagrams a, b, c of Fig 12.3-9 show that when the AC servo motor is operating reasonably close to synchronous speed, with full voltage on the control winding, the motor is quite efficient, and its dissipation is relatively low. Under this condition, the AC servo motor is just as efficient as a comparable DC servo motor.

Now consider what happens when the AC motor operates at zero speed. When the control winding has full voltage, the input power is equal to $T_{Max}\Omega_{syn}$. This product of synchronous speed Ω_{syn} multiplied by maximum stall torque T_{Max} is indicated by the cross-hatched area in Fig 12.3-9d. This

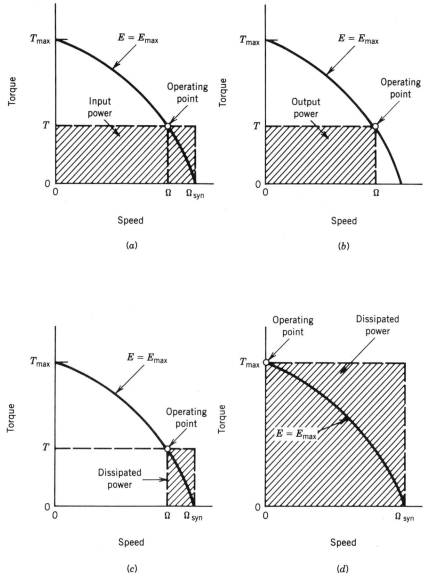

Figure 12.3-9 Torque–speed curves of an AC servo motor, illustrating motor dissipation at different speeds: (a–c) high speed; (d) zero speed.

input power is totally dissipated in the motor, because there is no output power at zero speed. When the voltage on the control winding is reduced to zero, there is no output torque and no dissipation in the control winding, but the dissipation in the reference winding is unchanged. Hence, at zero speed and zero torque, the AC servo motor has half the dissipation it has at zero speed and maximum torque. This dissipation at zero speed and zero torque is very large, and is equal to

$$P_{\text{dis}} = \tfrac{1}{2} T_{\text{Max}} \Omega_{\text{syn}} \qquad (12.3\text{-}40)$$

This dissipation at zero speed and zero torque is $\tfrac{1}{2}$ of the cross-hatched area of diagram d. This is very much greater than the dissipation shown in diagram c, which occurs at high speed with full voltage on the control winding.

Thus, the two-phase AC servo motor is efficient and has relatively low dissipation when operated at high torque and high speed. However, its dissipation is very much greater than that of a comparable DC servo motor when operated at low speed.

12.3.3   Analysis of Automatic-Gain-Control (AGC) Feedback Loop

Some control systems have nonlinearities that are much stronger than that of the torque–speed curve of an AC servo motor. For example, consider an automatic-gain-control (AGC) feedback loop, which is used to keep a nearly constant signal level at the output of an amplifier.

Figure 12.3-10a shows the elements of an AGC loop. The input signal e_i is amplified in a variable-gain amplifier, the gain of which varies in response to changes of the gain-control voltage E_{gc}. The amplitude of the input signal is designated E_i. The amplifier output signal is fed through a detector, which delivers a voltage E_o equal to the amplitude of the output signal. A reference circuit combines this voltage E_o with a negative voltage $-V_{\text{cc}}$ to form a difference signal E_x. The difference signal is fed through the DC amplifier and lowpass filter, to form the gain-control voltage E_{gc}, which controls the amplifier gain. (Many AGC loops do not have a DC amplifier, but one is included here for generality.) The difference voltage E_x is equal to

$$
\begin{aligned}
E_x &= \frac{R_2}{R_1 + R_2} E_o - \frac{R_1}{R_1 + R_2} V_{\text{cc}} \\[2mm]
&= \frac{R_2}{R_1 + R_2} \left(E_o - \frac{R_1}{R_2} V_{\text{cc}} \right)
\end{aligned}
\qquad (12.3\text{-}41)
$$

Figure 12.3-10b shows a functional block diagram of the AGC loop. (Since this is a nonlinear representation, it is not called a signal-flow diagram.) The input signal amplitude E_i is multiplied by the amplifier gain $K[E_{\text{gc}}]$ (which is a function of the gain control voltage E_{gc}) to form the output signal amplitude E_o:

$$E_o = E_i K \qquad (12.3\text{-}42)$$

The command voltage E_k is subtracted from the output voltage E_o (the controlled variable) to form the signal $-E_e$, which is the negative of the

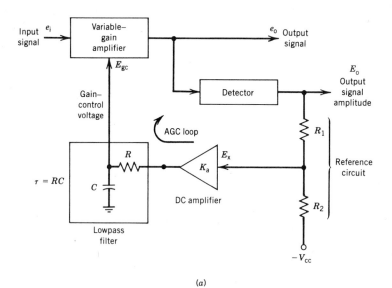

(a)

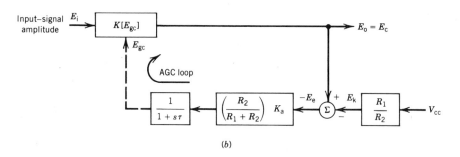

(b)

Figure 12.3-10 Representation of an AGC loop by a mathematical block diagram: (a) general circuit elements; (b) mathematical block diagram.

system error signal E_e. Equation 12.3-41 shows that the command voltage E_k is equal to $(R_1/R_2)V_{cc}$. The signal $-E_e$ is multiplied by the gain factor $[R_2/(R_1 + R_2)]K_a$ and fed through the lowpass filter $1/(1 + s\tau)$ to form the gain control voltage E_{gc}.

The nonlinear control relation of Eq 12.3-42 can be linearized by differentiating it to obtain the response at an operating point for small perturbations. This gives

$$dE_o = E_i \, dK + K \, dE_i \qquad (12.3\text{-}43)$$

Although this equation is linear for small values of the incremental variables dE_i, dE_o, and dK, it is inconvenient to use because the coefficients E_i and K vary widely. A more convenient expression can be obtained by solving Eq

12.3-42 for E_i and K:

$$E_i = E_o/K \qquad (12.3\text{-}44)$$

$$K = E_o/E_i \qquad (12.3\text{-}45)$$

Substituting these into Eq 12.3-43 gives

$$dE_o = E_o \frac{dK}{K} + E_o \frac{dE_i}{E_i} = E_o(d \ln[K] + d \ln[E_i]) \qquad (12.3\text{-}46)$$

This is expressed as a signal-flow diagram in Fig 12.3-11a. Since the AGC loop keeps the output amplitude E_o nearly constant, the parameter E_o in the gain block is a nearly constant gain factor. This diagram applies for appreciable deviations from the operating point, and so the derivative signals dE_o, dE_i, etc. are replaced by ΔE_o, ΔE_i, etc. Equation 12.3-46 can be derived more simply by taking the natural logarithm of the basic control expression of Eq 12.3-42, which gives

$$\ln[E_o] = \ln[E_i K] = \ln[E_i] + \ln[K] \qquad (12.3\text{-}47)$$

Differentiating this gives

$$d \ln[E_o] = \frac{dE_o}{E_o} = d \ln[E_i] + d \ln[K] \qquad (12.3\text{-}48)$$

It is usually convenient to express the logarithmic values of the input signal amplitude and amplifier gain in terms of decibels (dB). It can be shown that

$$\log[x] = \log[e]\ln[x] = 0.4343 \ln[x] \qquad (12.3\text{-}49)$$

Hence,

$$\ln[x] = \frac{20 \log[x]}{20(0.4343)} = 0.1151 \,(20 \log[x]) \qquad (12.3\text{-}50)$$

Applying Eq 12.3-50 to the signal-flow diagram a of Fig 12.3-11 gives diagram b, which is then simplified to diagram c.

Apply the incremental signal-flow diagram of Fig 12.3-11c to the functional block diagram of the AGC loop in Fig 12.3-10b. This gives the incremental signal-flow diagram for the AGC loop shown in Fig 12.3-12a. The amplifier gain characteristic is described by a plot such as that of Fig 12.3-13, which shows the variation of amplifier gain in decibels versus the gain-control voltage E_{gc}. At any operating point, this characteristic is approximated by the tangent to the curve at that point, as indicated by the dashed line. The tangent has a slope of $d(20 \log[K])/dE_{gc}$, which has the units of dB/volt. This slope is the gain factor shown in the signal-flow diagram of Fig 12.3-12a.

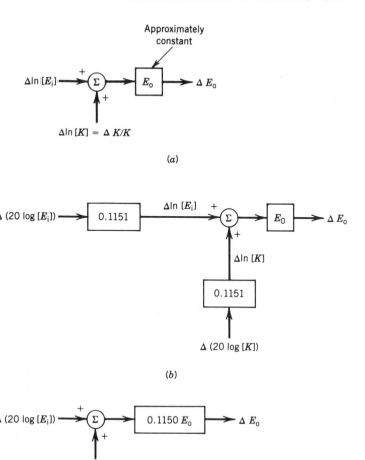

Figure 12.3-11 Linearized signal-flow diagram of variable-gain amplifier: (*a*) diagram expressed in terms of natural logarithms; (*b*) diagram expressed in terms of decibel values of gain; (*c*) simplified form of (*b*).

The slope of gain in decibels versus gain-control voltage E_{gc} in volts is negative. Hence, in the signal-flow diagram, it is convenient to use the negative of this slope, which is positive. When this is done, diagram *a* of Fig 12.3-12 changes to diagram *b*. The signs of the variables ΔE_{gc}, $-\Delta E_e$, $-\Delta E_r$ are reversed, and the loop takes on the form of the basic feedback loop.

The loop transfer function of the AGC loop in Fig 12.3-12*b* is

$$G = \frac{(0.1151 E_o)\left(-d(20\log[K])/dE_{gc}\right)K_a}{(s\tau + 1)(1 + R_2/R_1)} \qquad (12.3\text{-}51)$$

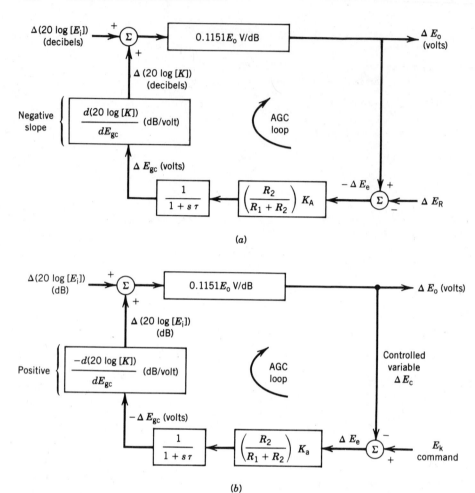

(a)

(b)

Figure 12.3-12 Signal-flow diagram of AGC loop for small perturbations from operating point: (*a*) signal-flow diagram; (*b*) sign change to convert AGC loop to form of the basic feedback loop.

A magnitude asymptote plot of this loop transfer function is shown in Fig 12.3-14. The loop gain at zero frequency is

$$K_0 = \frac{0.1151 E_o K_a}{1 + R_2/R_1} \left(\frac{-d(20\log[K])}{dE_{gc}} \right) \qquad (12.3\text{-}52)$$

The gain crossover frequency is

$$\omega_c = K_0/\tau \qquad (12.3\text{-}53)$$

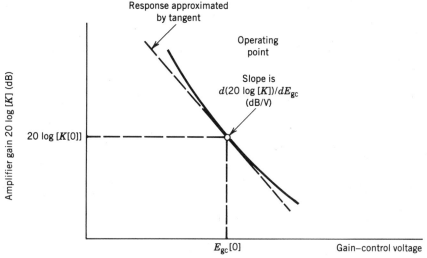

Figure 12.3-13   General plot of amplifier gain in decibels versus gain-control voltage.

Since the variation of the output voltage E_o is small, the DC gain K_0 and the gain crossover frequency ω_c are essentially constant when the slope of the logarithmic gain characteristic of Fig 12.3-13 is constant. It is desirable to keep ω_c fixed, and so the amplifier gain in decibels should ideally be a linear function of the gain control voltage E_{gc}. This condition is approximated in well-designed AGC loops. An equivalent statement of this condition is that the amplifier gain K should be an exponential function of the control voltage E_{gc}:

$$K = ae^{-bE_{gc}} \qquad (12.3\text{-}54)$$

where a and b are constants.

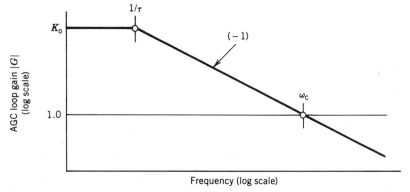

Figure 12.3-14   Frequency response of loop transfer function of an AGC loop, for small perturbations from an operating point.

At GTE, the author designed an AGC loop with a fast response and a very wide dynamic range (1000 : 1 in voltage gain). Its purpose was to counteract rapid signal fading in a laser tracking system. This was achieved by cascading several gain-control stages, which resulted in a logarithmic gain in decibles that was closely proportional to the gain-control voltage over a 60-dB range. This characteristic allowed the gain crossover frequency to be held fixed at a high value.

12.4 ANALYSIS AND SIMULATION OF A NONLINEAR PRESSURE-CONTROL SYSTEM

Strong nonlinearities, similar to those of the AGC loop discussed in Section 12.3.3, occur frequently in process control systems. The same linearization approach described in Section 12.3 can generally be used in such systems to analyze the response about different operating points. The linearization provides a simple frame of reference for understanding the dynamics of the process and for guiding the system design. An accurate representation of system response, including the effects of nonlinearities, can be obtained by applying the serial computer simulation approach described in Chapter 9.

These principles for designing nonlinear process-control systems are illustrated in this section by studying a system that controls the pressure in a vacuum chamber, which is used in a materials processing operation. The pressure in such a vacuum chamber is normally held between 0.001 and 10 torr, where on torr is the pressure of one millimeter of mercury. Standard atmospheric pressure is 760 mm of mercury (760 torr).

12.4.1 System Equations

Figure 12.4-1 is a block diagram of the pressure control system. The selected gas is fed into the vacuum chamber at a constant input mass-flow rate. The pressure chamber is exhausted by a vacuum pump, which generates a much lower pressure. A throttle valve between the vacuum chamber and the vacuum pump varies the rate of gas flow from the vacuum chamber, and thereby controls the pressure in the chamber.

The pressure-control loop works as follows. A pressure sensor measures the pressure in the chamber, and its pressure signal is subtracted from the command-pressure setting to form the pressure error signal. The pressure error signal is amplified and fed through a lead (or "derivative") network. The resultant signal drives a motor at a velocity proportional to the amplifier output voltage. The motor is coupled through a gear train to the control shaft of the throttle valve, and varies the valve opening to control the exit flow of gas from the chamber.

Since this pressure-control system operates rather slowly, a stepper motor is generally adequate for the valve drive mechanism. Hence, a stepper motor is assumed in the following example, because it is simple and reliable.

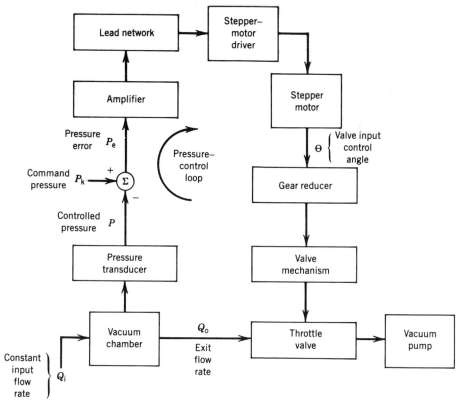

Figure 12.4-1   Block diagram of system for controlling pressure in a vacuum chamber.

For a given mass, the product of the pressure P and volume V of a gas is proportional to the absolute temperature. Hence, if the temperature is constant, the PV product is an indirect measure of mass. Since pressure is measured in torr, and volume is measured in liters, the equivalent PV mass of a gas is measured in torr-liters. Thus, the equivalent mass $M_{(tL)}$ in torr-liters is designated as follows:

$$M_{(tL)} = P_{(t)}V_{(L)} \qquad (12.4\text{-}1)$$

where $P_{(t)}$ is the pressure in torr, and $V_{(L)}$ is the volume in liters. The value of the pressure P in absolute units is related to the pressure $P_{(t)}$ in torr by

$$P = P_{(t)}\frac{P_a}{760} \qquad (12.4\text{-}2)$$

where P_a is atmospheric pressure. The actual mass M of the gas is related as follows to the mass $M_{(tL)}$ in torr-liters:

$$M = M_{(tL)}\frac{\rho_a}{760} \qquad (12.4\text{-}3)$$

where ρ_a is the density of the gas at atmospheric pressure. As shown in Ref [12.13] (p. 2-133), standard atmospheric pressure at sea level is exactly 760 mm of mercury, which corresponds to

$$P_a = 1{,}013.250 \text{ millbar} = 1.013250 \times 10^6 \text{ dyne/cm}^2$$
$$= 1.013250 \times 10^5 \text{ N/m}^2 \tag{12.4-4}$$

Let us assume that the gas has the density of air. From p. 2-137 (Table 21-5) of Ref [12.13], the density and temperature of the standard atmosphere at sea level are

$$\rho_a = 1.2250 \text{ g/liter} = 1.2250 \text{ kg/m}^3 \tag{12.4-5}$$
$$T = 15.00°C = 59.00°F = 288.15°K \tag{12.4-6}$$

The equivalent mass flow rate is measured in torr-liter/sec, and so is designated $Q_{(tL)}$. The true mass flow rate Q_m is related to equivalent mass flow rate $Q_{(tL)}$ as follows:

$$Q_m = Q_{(tL)} \frac{\rho_a}{760} \tag{12.4-7}$$

The net equivalent mass flow rate of gas into a chamber of volume $V_{(L)}$ is equal to

$$Q_{c(tL)} = \frac{dM_{(tL)}}{dt} = \frac{d(P_{(t)}V_{(L)})}{dt} = V_{(L)} \frac{dP_{(t)}}{dt} \tag{12.4-8}$$

where $Q_{c(tL)}$ is the mass flow rate into the chamber in torr-liter/sec, and $P_{(t)}$ is the pressure in the chamber, measured in torr.

Figure 12.4-2 is a mathematical block diagram of the pressure-control system, which describes the linear and nonlinear dynamic equations of the process. The input mass-flow rate, which is normally held constant, is designated Q_i, and the varying output mass flow rate is designated Q_o. In accordance with Eq 12.4-8, the rate of change of pressure in the vacuum chamber is

$$\dot{P} = \frac{dP}{dt} = \frac{Q_c}{V} = \frac{Q_i - Q_o}{V} \tag{12.4-9}$$

All of these variables are expressed in equivalent torr-liter units. The pressure P in the chamber is the integral of the rate of change of pressure. In terms of the Laplace transforms, this is

$$P = \frac{1}{s} \dot{P} \tag{12.4-10}$$

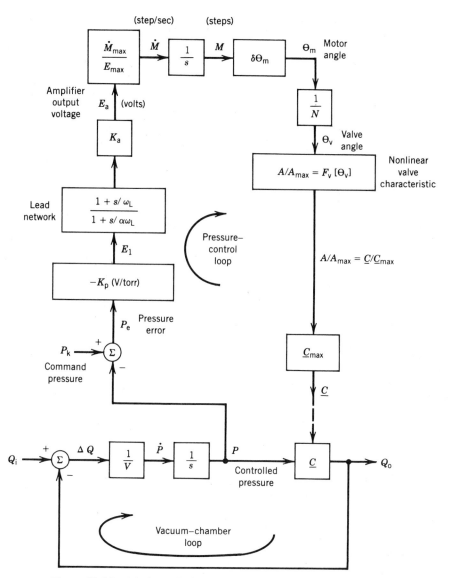

Figure 12.4-2 Mathematical block diagram of pressure-control system.

The mass flow rate Q_o flowing through the valve is proportional to the product of the cross-section area A of the valve orifice, multiplied by the pressure drop across the orifice. The pressure at the output of the throttle valve is much smaller than the pressure P in the chamber, and so can be considered to be zero. Hence, the output flow rate from the chamber can be expressed as

$$Q_o = k_v AP \qquad (12.4\text{-}11)$$

where k_v is a constant that characterizes the valve. The quantity $k_v A$ is called the conductance of the valve; which we designate $\underline{C}$. (The underline distinguishes this parameter from electrical capacitance.) Thus, Eq 12.4-11 is expressed as

$$Q_o = \underline{C} P \qquad (12.4\text{-}12)$$

Since the valve conductance $\underline{C}$ is proportional to the area A of the valve orifice, it can be related to area A by

$$\frac{\underline{C}}{\underline{C}_{\text{Max}}} = \frac{A}{A_{\text{Max}}} \qquad (12.4\text{-}13)$$

where $\underline{C}_{\text{Max}}$ is the maximum conductance of the valve, which occurs when the valve is fully open, and A_{Max} is the orifice area of the fully open valve. The quantitative relation between valve conductance $\underline{C}$ and valve orifice area A will be derived in Section 12.4.2.

The controlled pressure P in the vacuum chamber is measured by the pressure transducer, which has a sensitivity in volt/torr designated as K_p. The resultant voltage, proportional to pressure, is subtracted from the pressure command voltage, to form the pressure error voltage. However, it is convenient to express this subtraction directly in terms of pressure signals, and so the pressure-transducer sensitivity K_p is placed after the summation point in Fig 12.4-2. In order to achieve negative feedback, there must be positive gain between chamber pressure P and valve conductance $\underline{C}$, because increasing the chamber pressure P produces increasing valve conductance $\underline{C}$, which, in turn, decreases the chamber pressure P. Hence the negative sign at the summing point, for the pressure command signal P_k, must be offset by a subsequent negative sign in a gain element. This negative sign is included by expressing the gain block for the pressure transducer as $-K_p$.

The pressure error voltage is fed to the lead network, which has the transfer function

$$H_L = \frac{1 + s/\omega_L}{1 + s/\alpha \omega_L} \qquad (12.4\text{-}14)$$

A magnitude asymptote plot of this lead-network transfer function was given in Chapter 4, Fig 4.3-1. Networks for implementing this transfer function were given in Figs 4.3-2, -3.

The output from the lead network is fed through the amplifier gain K_a to form the amplifier output voltage E_a. A pulse-generator circuit produces pulses that drive the stepper motor, forward and backward, at a stepping rate that is proportional to the amplifier output voltage E_a. The stepping rate $\dot{M}$

can thus be expressed as

$$\dot{M} = \frac{\dot{M}_{\text{Max}}}{E_{\text{Max}}} E_{\text{a}} \qquad (12.4\text{-}15)$$

where E_{Max} is the maximum absolute value of the amplifier output voltage E_{a}, and $\dot{M}_{\text{Max}}$ is the maximum absolute value of the stepping rate in steps/sec, which occurs when $E_{\text{a}} = \pm E_{\text{Max}}$. The net number of steps taken by the motor, which is designated M, is the integral of $\dot{M}$. Hence, these variables are related by

$$M = \frac{1}{s}\dot{M} \qquad (12.4\text{-}16)$$

Note that the variable $\dot{M}$ can be either positive or negative. The step increment of the stepper motor is designated $\delta\Theta_{\text{m}}$. The net angular displacement of the motor shaft, which is designated Θ_{m}, is equal to the step increment $\delta\Theta_{\text{m}}$ multiplied by the number of motor steps M:

$$\Theta_{\text{m}} = M\,\delta\Theta_{\text{m}} \qquad (12.4\text{-}17)$$

The angular displacement of the throttle valve is designated Θ_{v}. This is equal to the motor angular displacement Θ_{m} divided by the gear ratio N of the gear train:

$$\Theta_{\text{v}} = \Theta_{\text{m}}/N \qquad (12.4\text{-}18)$$

The orifice area A of the throttle valve is a nonlinear function of the valve angle Θ_{v}. The actual function is considered below in Section 12.4.4. Figure 12.4-2 represents this as the general function $F_{\text{v}}[\Theta_{\text{v}}]$, which is the ratio of the orifice area A divided by the maximum orifice area A_{Max}. Thus

$$A/A_{\text{Max}} = F_{\text{v}}[\Theta_{\text{v}}] \qquad (12.4\text{-}19)$$

By Eq 2.4-13, the relative valve conductance $\underline{C}/\underline{C}_{\text{Max}}$ is equal to the relative orifice area A/A_{Max}. Hence, the valve conductance $\underline{C}$ is equal to

$$\underline{C} = \underline{C}_{\text{Max}}\frac{A}{A_{\text{Max}}} = C_{\text{Max}}F_{\text{v}}[\Theta_{\text{v}}] \qquad (12.4\text{-}20)$$

By Eq 2.4-12, the output flow Q_{o} from the chamber is equal to the product of the valve conductance $\underline{C}$ multiplied by the pressure P in the chamber.

12.4.2 Conductance of Throttle Valve

To derive the equation relating valve conductance $\underline{C}$ to valve orifice area A_{v}, consider the simple physical model of Fig 12.4-3. The constant force F_{p} on the

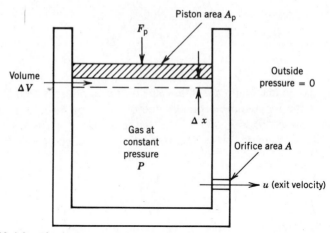

Figure 12.4-3   Physical model for deriving pressure-flow equation for throttle valve.

piston compresses the air in the chamber to a constant pressure P equal to

$$P = F_p / A_p \qquad (12.4\text{-}21)$$

where A_p is the area of the piston. The air if forced out of the chamber through an orifice of area A, and emerges at a velocity u. The pressure outside the chamber is considered to be zero.

In a period of time Δt, the piston moves downward a distance Δx. Hence, the potential energy is decreased during this time increment by

$$\text{potential-energy change} = F_p\,\Delta x = \frac{F_p}{A_p} A_p\,\Delta x$$

$$= P\,\Delta V \qquad (12.4\text{-}22)$$

where ΔV is the decrease in air volume in the chamber during the time Δt, which is equal to

$$\Delta V = A_p\,\Delta x \qquad (12.4\text{-}23)$$

Since the chamber pressure is constant, the density ρ of the air is constant. Hence, the mass Δm forced from the chamber during this time interval Δt is equal to

$$\Delta m = \rho\,\Delta V = \rho A_p\,\Delta x \qquad (12.4\text{-}24)$$

The potential-energy loss is converted to kinetic energy in the air forced through the orifice in the time interval Δt, which is equal to

$$\text{kinetic energy} = \tfrac{1}{2}\Delta m\, u^2 = \tfrac{1}{2}\rho\, \Delta V u^2 \qquad (12.4\text{-}25)$$

Equating the expressions for potential and kinetic energy in Eqs 12.4-22, -25 gives

$$P\Delta V = \tfrac{1}{2}\rho\, \Delta V u^2 \qquad (12.4\text{-}26)$$

Dividing both sides of Eq 12.4-26 by ΔV gives

$$P = \tfrac{1}{2}\rho u^2 \qquad (12.4\text{-}27)$$

The volumetric flow rate of air through the orifice is equal to the product of the ejection velocity u multiplied by the orifice area A:

$$Q_v = Au \qquad (12.4\text{-}28)$$

The mass flow rate Q_m is equal to the volumetric flow rate Q_v multiplied by the density ρ:

$$Q_m = \rho Q_v = \rho Au \qquad (12.4\text{-}29)$$

Combining Eqs 12.4-27, -29 gives

$$P = \frac{1}{2\rho}(\rho u)^2 = \frac{1}{2\rho}\left(\frac{Q_m}{A}\right)^2 \qquad (12.4\text{-}30)$$

Solving for the mass flow rate Q_m gives

$$Q_m = A\sqrt{2\rho P} \qquad (12.4\text{-}31)$$

The density ρ of the air is proportional to the pressure P. Assuming the air is at standard atmospheric temperature (15°C), the density ρ is equal to

$$\rho = \frac{\rho_a}{P_a}P \qquad (12.4\text{-}32)$$

where P_a and ρ_a are the standard values of atmospheric pressure and density given in Eqs 12.4-4, -5. Combining Eqs 12.4-31, -32 gives

$$Q_m = A\sqrt{2\frac{\rho_a}{P_a}P^2} = AP\sqrt{\frac{2\rho_a}{P_a}} \qquad (12.4\text{-}33)$$

The preceeding analysis is approximate, because it assumes that the air molecules are all forced through the orifice at a constant velocity u. However, because of viscosity in the air, the velocity is low near the edges of the orifice. This effect can be accounted for by reducing the mass flow rate by an orifice discharge coefficient, designated k_d. Thus, Eq 12.4-33 becomes

$$Q_m = k_d A P \sqrt{\frac{2\rho_a}{P_a}} \qquad (12.4\text{-}34)$$

The discharge coefficient k_d is usually between 0.8 and unity.

Equation 12.4-34 applies to the throttle valve, where A is the valve orifice area A_v. The conductance of the throttle valve is defined as

$$\underline{C} = Q_m/P \qquad (12.4\text{-}35)$$

Combining Eqs 12.4-34, -35 gives for the valve conductance

$$\underline{C} = \frac{Q_m}{P} = k_d A_v \sqrt{\frac{2\rho_a}{P_a}} \qquad (12.4\text{-}36)$$

By Eqs 12.4-2, -7, this ratio Q_m/P can be expressed as follows in terms of the corresponding ratio in torr-liter units:

$$\underline{C} = \frac{Q_m}{P} = \frac{Q_{(tL)}(\rho_a/760)}{P_{(t)}(P_a/760)}$$

$$= \frac{\rho_a}{P_a} \frac{Q_{(tL)}}{P_{(t)}} = \frac{\rho_a}{P_a} \underline{C}_{(L)} \qquad (12.4\text{-}37)$$

where $Q_{(tL)}$ is the mass flow rate in torr-liter/sec, $P_{(t)}$ is the pressure in torr, and $\underline{C}_{(L)}$ is the value conductance in liter/sec. Combining Eqs 12.4-36, -37 gives the following for the valve conductance expressed in liter/sec:

$$\underline{C}_{(L)} = \frac{P_a}{\rho_a} \underline{C} = k_d A_v \sqrt{\frac{2P_a}{\rho_a}} \qquad (12.4\text{-}38)$$

Assume that the valve orifice area A_v is 1.0 cm², and that the discharge coefficient k_d is unity. Applying the values for P_a and ρ_a in Eqs 12.4-4, -5 to Eq 12.4-38 gives the following for the valve conductance:

$$\underline{C}_{(L)} = (1.0 \text{ cm}^2) \sqrt{2 \frac{1.01325 \times 10^5 \text{ N/m}^2}{1.2250 \text{ kg/m}^3}}$$

$$= (1.0 \text{ cm}^2)\sqrt{1.6543 \times 10^3 \text{ m}^2/\text{sec}^2}$$

$$= 406.7 \text{ cm}^2(\text{m/sec}) = 40.67 \times 10^3 \text{ cm}^3/\text{sec}$$

$$= 40.67 \text{ liter/sec} \qquad (12.4\text{-}39)$$

In converting units, note that the newton (a unit of force) is equivalent to the units of mass-times-acceleration. Hence

$$1\,\text{N} = 1\,\text{kg-m}/\text{sec}^2 \tag{12.4-40}$$

Combining Eqs 12.4-38, -39 gives the following general expression for valve conductance in liter/sec (where A_v is expressed in cm^2):

$$\underline{C}_{(\text{L})} = (40.67\,\text{liter}/\text{sec})\,k_\text{d}\big(A_v/\text{cm}^2\big) \tag{12.4-41}$$

12.4.3 Linearized Response at Operating Point

Now consider the linear response of the control system in terms of small deviations from an operating point. Take the natural logarithm of Eq 12.4-12 $(Q_\text{o} = P\underline{C})$:

$$\ln[Q_\text{o}] = \ln[P\underline{C}] = \ln[P] + \ln[\underline{C}] \tag{12.4-42}$$

Differentiating this gives

$$d\,\ln[Q_\text{o}] = d\,\ln[P] + d\,\ln[\underline{C}] \tag{12.4-43}$$

This can be expressed as

$$\frac{dQ_\text{o}}{Q_\text{o}} = \frac{dP}{P} + \frac{d\underline{C}}{\underline{C}} \tag{12.4-44}$$

By Eq 12.4-13, the conductance $\underline{C}$ is proportional to the throttle-valve orifice area A. Hence,

$$\frac{d\underline{C}}{\underline{C}} = \frac{dA}{A} = d\,\ln\!\left[\frac{A}{A_\text{Max}}\right] \tag{12.4-45}$$

This equation indicates that for linear analysis one needs the nonlinear function relating the valve angle Θ_v to the logarithm of the relative valve area A/A_Max.

The resultant linearized signal-flow diagram of the control system is shown in Fig 12.4-4. The loop transfer function of the vacuum-chamber loop is

$$G_{(\text{c})} = \frac{Q_\text{o}}{sVP} = \frac{\omega_{\text{cc}}}{s} \tag{12.4-46}$$

The parameter ω_{cc} is the asymptotic gain crossover frequency of the vacuum-

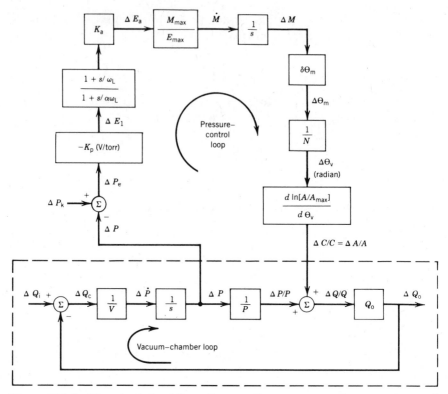

Figure 12.4-4 Linearized signal-flow diagram of pressure-control system, for small deviations from an operating pressure.

chamber loop, which is equal to

$$\omega_{cc} = \frac{Q_o}{VP} = \frac{C}{V} \qquad (12.4\text{-}47)$$

The feedback transfer function corresponding to the loop transfer function $G_{(c)}$ of Eq 12.4-46 is

$$G_{ib(c)} = \frac{G_{(c)}}{1 + G_{(c)}} = \frac{Q_o/sVP}{1 + Q_o/sVP} = \frac{Q_o/VP}{s + Q_o/VP} \qquad (12.4\text{-}48)$$

Figure 12.4-4 shows that the signal $\Delta P/P$ is the negative of the feedback signal of the vacuum-chamber loop for the loop input $\Delta C/C$. Hence,

$$\frac{\Delta P}{P} = -G_{ib(c)}\frac{\Delta C}{C} \qquad (12.4\text{-}49)$$

Solve Eq 12.4-49 for the transfer function $\Delta P/(\Delta \underline{C}/\underline{C})$, using the expression for G_{ib} of the vacuum-chamber loop in Eq 12.4-48. This gives

$$\frac{\Delta P}{\Delta \underline{C}/\underline{C}} = -PG_{ib(c)} = \frac{-Q_o/V}{s + Q_o/VP} \tag{12.4-50}$$

This represents the transfer function of the dashed block in Fig 12.4-4, and relates the input $\Delta \underline{C}/\underline{C}$ of that block to the output ΔP. Using Eq 12.4-50, the loop transfer function of the pressure loop can be obtained by inspection from Fig 12.4-4 as follows:

$$\begin{aligned}
G_{(pr)} &= \frac{\Delta P}{\Delta \underline{C}/\underline{C}} \frac{-K_p(1 + s/\omega_L)K_a(\dot{M}_{Max}/E_{Max})\,\delta\Theta_m}{(1 + s/\alpha\omega_L)Ns} \frac{d\ln[A/A_{Max}]}{d\Theta_v} \\
&= \frac{-Q_o/V}{s + Q_o/VP} \frac{-K_p(1 + s/\omega_L)K_a(\dot{M}_{Max}/E_{Max})\,\delta\Theta_m}{(1 + s/\alpha\omega_L)Ns} \frac{d\ln[A/A_{Max}]}{d\Theta_v}
\end{aligned} \tag{12.4-51}$$

Make the following changes in Eq 12.4-51:

$$1 + \frac{s}{\omega_L} = \frac{s}{\omega_L}\left(1 + \frac{\omega_L}{s}\right) \tag{12.4-52}$$

$$s + \frac{Q_o}{VP} = s\left(1 + \frac{Q_o}{VPs}\right) \tag{12.4-53}$$

Equation 12.4-51 becomes

$$\begin{aligned}
G_{(pr)} &= \frac{\omega_{cpr}(1 + \omega_L/s)}{s(1 + Q_o/sVP)(1 + s/\alpha\omega_L)} \\
&= \frac{\omega_{cpr}(1 + \omega_L/s)}{s(1 + \omega_{cc}/s)(1 + s/\alpha\omega_L)}
\end{aligned} \tag{12.4-54}$$

The parameter ω_{cpr} is the asymptotic gain crossover frequency of the pressure loop, which is equal to

$$\omega_{cpr} = K_p K_a \frac{\dot{M}_{Max}}{E_{Max}} \frac{\delta\Theta_m}{N} \frac{Q_o}{V} \frac{1}{\omega_L} \frac{d\ln[A/A_{Max}]}{d\Theta_v} \tag{12.4-55}$$

The lead-network time constant τ_L is defined as the reciprocal of the lead-network break frequency ω_L:

$$\tau_L = 1/\omega_L \tag{12.4-56}$$

The resolution of the valve angle Θ_v, corresponding to a step of the stepper motor, is designated $\delta\Theta_v$. This valve-angle resolution is equal to the size of the stepper-motor step $\delta\Theta_m$ divided by the gear ratio N:

$$\delta\Theta_v = \delta\Theta_m / N \tag{12.4-57}$$

Substituting Eqs 12.4-56, -57 into Eq 12.4-55 simplifies the expression

$$\omega_{cpr} = K_p K_a \frac{\dot{M}_{Max}}{E_{Max}} \delta\Theta_{vTL} \frac{Q_o}{V} \frac{d \ln[A/A_{Max}]}{d\Theta_v} \tag{12.4-58}$$

A magnitude asymptote plot of the pressure-loop transfer function of Eq 12.4-54 is shown in Fig 12.4-5. The break frequency ω_{cc}, which is the asymptotic gain crossover frequency of the vacuum-chamber loop, is equal to Q_o/VP, and so varies with pressure P. However, this usually affects the pressure-loop transfer function only at low frequency, and so does not critically alter the response. On the other hand, the gain crossover frequency ω_{cpr} of the pressure loop, given by Eq 12.4-78, can cause instability if it is too large. This gain crossover frequency is proportional to the slope of the logarithmic valve characteristic: $\ln[A/A_{Max}]$ versus Θ_v.

12.4.4   Nonlinear Valve Characteristic

To obtain a representative function relating valve orifice area A to valve control angle Θ_v, consider first the simple flapper valve illustrated in Fig 12.4-6. The height of the valve opening is equal to

$$h = R_v(1 - \cos[\Theta_v]) \tag{12.4-59}$$

where R_v is the radial distance from the vane axis to the vane tip. The cross section of the valve opening is rectangular, and has a width designated as w. Hence the total valve orifice area for the two sides of the vane is

$$A = 2wh = 2wR_v(1 - \cos[\Theta_v]) \tag{12.4-60}$$

The maximum valve area, when the valve is fully open ($\Theta_v = 90°$), is

$$A_{Max} = 2wR_v \tag{12.4-61}$$

Dividing Eq 12.4-60 by Eq 12.4-61 gives

$$A/A_{Max} = 1 - \cos[\Theta_v] \tag{12.4-62}$$

This expression assumes that the valve is completely closed when Θ_v is zero. However, because of physical tolerances, a small leakage occurs when the

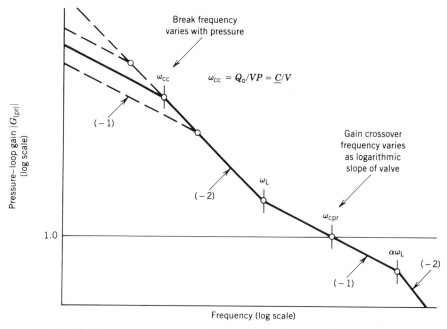

Figure 12.4-5   Frequency response of pressure loop, for small deviations of pressure.

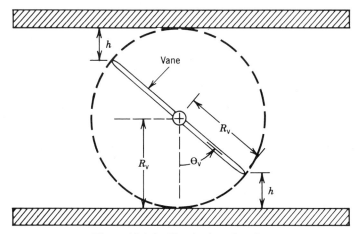

Figure 12.4-6   Simple throttle valve.

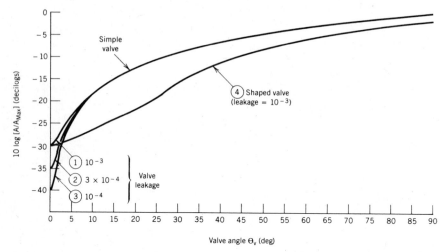

Figure 12.4-7   Logarithm of relative valve area A/A_{Max} in decilogs, versus valve angle Θ_v, for simple and shaped valves.

valve is closed. The ratio of this leakage area to the maximum valve area A_{Max} is designated δL_v:

$$\delta L_v = \frac{\text{leakage orifice area}}{A_{Max}} \qquad (12.4\text{-}63)$$

When the effect of leakage is included, Eq 12.4-62 becomes

$$A/A_{Max} = 1 - \cos[\Theta_v] + \delta L_v \qquad (12.4\text{-}64)$$

As was shown in Fig 12.4-4, for linear analysis, one should consider the function $\ln[A/A_{Max}]$. However, rather than plot the natural logarithm of A/A_{Max}, it is more convenient to express the ratio A/A_{Max} in decilogs. Decilog plots of this ratio for the valve characteristic of Eq 12.4-64 are shown as curves ① to ③ in Fig 12.4-7. These curves are for relative-leakage values δL_v of 10^{-3}, 3×10^{-4}, and 10^{-4}.

These decilog plots of A/A_{Max} can be related to the natural logarithm of A/A_{Max} in the following manner. The base-10 logarithm is related to the natural logarithm by

$$\log[x] = \log[e]\ln[x] = 0.4343 \ln[x] \qquad (12.4\text{-}65)$$

Hence, the natural logarithm of A/A_{Max} is related as follows to the decilog value for A/A_{Max}:

$$\ln[A/A_{\text{Max}}] = \frac{1}{0.4343}\log[A/A_{\text{Max}}]$$

$$= 0.2303(10\log[A/A_{\text{Max}}]) \qquad (12.4\text{-}66)$$

Equation 12.4-58 showed that the gain crossover frequency of the pressure loop, ω_{cpr}, is proportional to the slope of the natural logarithm of A/A_{Max} versus valve angle Θ_v. By Eq 12.4-66, this is equal to 0.2303 times the slope of the characteristic curve in Fig 12.4-7, with the valve angle expressed in radians. Curves ① to ③ show that this slope varies strongly with valve angle Θ_v. Therefore, one would like a valve characteristic curve that more nearly approximates a straight line on the decilog plot of A/A_{Max} versus valve angle Θ_v.

A better valve characteristic can be obtained by shaping the valve seat, as shown in Fig 12.4-8. The valve is fully closed when the vane is at point (1). Between points (1) and (2), the valve-seat contour is circular. The center of rotation for the circular valve contour is point (6), and the radius is $(1 + a_0)R_v$. The vane rotates about point (5), and the radial distance to the vane edge is R_v. Between point (2) and point (3), the valve-seat contour is a straight line that is tangent to the circle at point (2). Point (3) corresponds to maximum valve opening.

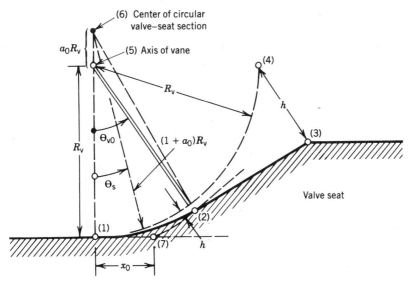

Figure 12.4-8   Geometry of valve seat for shaped throttle valve.

The angle of the vane corresponding to point (2) is designated Θ_{v0}. For $\Theta_v < \Theta_{v0}$, the height of the valve opening is

For $\Theta_v \leq \Theta_{v0}$: $\qquad h = 1 + a_0 - \sqrt{1 + a_0^2 + 2a_0\cos[\Theta_v]}$ $\qquad$ (12.4-67)

The angle of the radial vector from point (6) to the valve seat is designated Θ_s. The particular angle of this vector for point (2) is designated Θ_{s0}, and is equal to

$$\Theta_{s0} = \arctan\left[\frac{\sin[\Theta_{v0}]}{\cos[\Theta_{v0}] + a_0}\right] \qquad (12.4\text{-}68)$$

The linear valve-seat contour between points (2) and (3), if extended, would intersect the horizontal line passing through point (1) at point (7). The distance between points (1) and (7) is designated x_0, and is equal to

$$x_0 = \frac{(1 + a_0)(1 - \cos[\Theta_{s0}])}{\sin[\Theta_{s0}]} \qquad (12.4\text{-}69)$$

When the vane angle Θ_v exceeds Θ_{v0}, the valve-opening height is equal to

For $\Theta_v \geq \Theta_{v0}$:

$$\begin{aligned} h &= \cos[\Theta_{s0}] + x_0\sin[\Theta_{s0}] - \cos[\Theta_{s0}]\cos[\Theta_v] - \sin[\Theta_{s0}]\sin[\Theta_v] \\ &= \cos[\Theta_{s0}] + (1 + a_0)(1 - \cos[\Theta_{s0}]) - \cos[\Theta_v - \Theta_{s0}] \\ &= 1 + a_0(1 - \cos[\Theta_{s0}]) - \cos[\Theta_v - \Theta_{s0}] \qquad (12.4\text{-}70) \end{aligned}$$

The following specific parameters are assumed for the shaped valve:

$$a_0 = 0.1 \qquad (12.4\text{-}71)$$

$$\Theta_{v0} = 20° \qquad (12.4\text{-}72)$$

From Eq 12.4-68, the corresponding parameter Θ_{s0} is

$$\Theta_{s0} = 18.2093° \qquad (12.4\text{-}73)$$

From Eqs 12.4-67, -70, the resultant equations for the valve-opening height are

For $\Theta_v < 20°$: $\qquad h = 1.1 - \sqrt{1.01 + 0.2\cos[\Theta_v]}$ $\qquad$ (12.4-74)

For $\Theta_v \geq 20°$: $\qquad h = 1.005\,008 - \cos[\Theta_v - 18.2093°]$

$$\qquad\qquad\qquad = 1.005\,008 - \cos[\Theta_v - 0.317\,812 \text{ rad}] \qquad (12.4\text{-}75)$$

When valve leakage is included, the relative valve area is

$$A/A_{\text{Max}} = h + \delta L_{\text{v}} \qquad (12.4\text{-}76)$$

The height h is obtained from Eq 12.4-74 or -75, depending on whether Θ_{v} is less than or greater than 20°.

In Fig 12.4-7, curve ④ is the decilog plot of the relative area A/A_{Max} for the shaped valve, where the relative valve leakage δL_{v} is assumed to be 10^{-3}. As the figure shows, the characteristic of the shaped valve is much better than that of the simple valve. At ($\Theta_{\text{v}} = 90°$), this characteristic does not reach zero decilogs, and so the orifice area A_{Max} is not reached. Hence, for the shaped valve, the actual maximum valve orifice area is somewhat less than A_{Max}.

The slope of the logarithmic valve characteristic needed for linear analysis could be measured from the corresponding curve in Fig 12.4-7. However, it is simpler to obtain this slope by differentiating the equation for the valve orifice area. Differentiating Eq 12.4-64 gives the following for the simple valve:

$$d[A/A_{\text{Max}}] = \sin[\Theta_{\text{v}}]\, d\Theta_{\text{v}} \qquad (12.4\text{-}77)$$

Dividing this by A/A_{Max} in Eq 12.4-64 gives

$$\frac{dA}{A} = \frac{d[A/A_{\text{Max}}]}{A/A_{\text{Max}}} = \frac{\sin[\Theta_{\text{v}}]\, d\Theta_{\text{v}}}{1 - \cos[\Theta_{\text{v}}] + \delta L_{\text{v}}} \qquad (12.4\text{-}78)$$

This is equal to $d\ln[A/A_{\text{Max}}]$. Thus the slope of the natural logarithm of A/A_{Max} is

Simple valve:

$$\frac{d\ln[A/A_{\text{Max}}]}{d\Theta_{\text{v}}} = \frac{dA/A}{d\Theta_{\text{v}}} = \frac{\sin[\Theta_{\text{v}}]}{1 - \cos[\Theta_{\text{v}}] + \delta L_{\text{v}}} \qquad (12.4\text{-}79)$$

For the shaped valve (with $\Theta_{\text{v}} < 20°$), dA/A_{Max} is obtained by differentiating Eq 12.4-74:

$$\frac{dA}{A_{\text{Max}}} = \frac{0.1 \sin[\Theta_{\text{v}}]\, d\Theta_{\text{v}}}{\sqrt{1.01 + 0.2 \cos[\Theta_{\text{v}}]}} \qquad (12.4\text{-}80)$$

From Eqs 12.4-74, -76, the expression for A/A_{Max} is as follows for δL_{v} equal to 10^{-3}:

$$A/A_{\text{Max}} = 1.101 - \sqrt{1.01 + 0.2 \cos[\Theta_{\text{v}}]} \qquad (12.4\text{-}81)$$

Dividing Eq 12.4-80 by Eq 12.4-81 gives for the slope of the natural logarithm

Shaped valve, $\Theta_v \leq 20°$:

$$\frac{dA/A}{d\Theta_v} = \frac{0.01 \sin[\Theta_v]}{\left(1.101 - \sqrt{1.01 + 0.2\cos[\Theta_v]}\right)\sqrt{1.01 + 0.2\cos[\Theta_v]}} \quad (12.4\text{-}82)$$

For $\Theta_v > 20°$, dA/A_{Max} for the shaped value is obtained by differentiating Eq 12.4-75:

$$dA/A_{\text{Max}} = \sin[\Theta_v - 0.317\,812\ \text{rad}]\,d\Theta_v \quad (12.4\text{-}83)$$

From Eqs 12.4-75, -76, the expression for A/A_{Max} is as follows for δL_v equal to 10^{-3}:

$$A/A_{\text{Max}} = 1.006\,008 - \cos[\Theta_v - 0.317\,812\ \text{rad}] \quad (12.4\text{-}84)$$

Dividing Eq 12.4-83 gives for the slope of the natural logarithm

Shaped valve, $\Theta_v \geq 20°$:

$$\frac{dA/A}{d\Theta_v} = \frac{\sin[\Theta_v - 0.317\,812\ \text{rad}]}{1.006\,008 - \cos[\Theta_v - 0.317\,812\ \text{rad}]} \quad (12.4\text{-}85)$$

Figure 12.4-9 shows plots of the slope $(dA/A)/d\Theta_v$, expressed in radian^{-1}, versus valve angle Θ_v in degrees. These are plots of the expressions of Eq 12.4-79 (for the simple valve), and of Eqs 12.4-82, -85 (for the shaped valve). Plots are shown for the simple valve with relative leakage values of 10^{-3}, 3×10^{-4}, and 10^{-4}. The relative leakage for the shaped valve is 10^{-3}.

The plot for the shaped valve shows that a better shaping function should be used to avoid the sharp slope variation from 15° to 30°, and to increase the slope at large angles. With a more complex surface for the valve seat, a better valve characteristic could be obtained. On the other hand, this example is quite adequate for our purpose. It is meant to illustrate the principles for optimizing the valve, and not to provide a final design.

12.4.5   Example of Pressure-Control System

Let us apply these valve equations to a specific pressure-control system design. The valve conductance for a fully open valve is assumed to be

$$\underline{C}_{\text{Max}} = 100\ \text{liter/sec}\ (\text{L/sec}) \quad (12.4\text{-}86)$$

Assume that the discharge coefficient k_d for the valve orifice is 0.9. By Eq

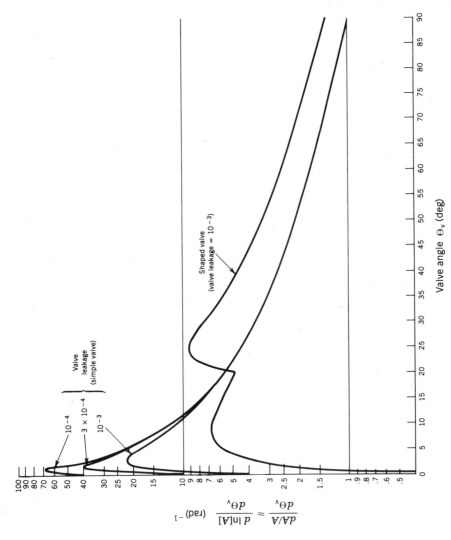

Figure 12.4-9 Slope of logarithmic valve characteristic versus valve angle Θ_v.

TABLE 12.4-1   Valve Parameters and Vacuum-Chamber Gain Crossover Frequency ω_{cc} for Steady-State Operation at Pressures of 0.3 and 3 torr

Pressure (torr)	Valve Conductance (liter/sec)	Logarithmic Area A/A_{Max} (decilog)	Valve Angle (deg)		ω_{cc} Q_o/VP (sec^{-1})
			Simple	Shaped	
0.030	33.3	-4.776	48.5	66.0	1.67
3.0	0.333	-24.778	4.0	13.0	0.00167

12.4-41, the maximum valve orifice area is

$$A_{\text{Max}} = \frac{1}{k_d} \frac{C_{\text{Max}}}{(40.67 \text{ L/sec})} \text{ cm}^2$$

$$= \frac{100}{(40.67)(0.9)} \text{ cm}^2 = 2.73 \text{ cm}^2 \qquad (12.4\text{-}87)$$

For a square valve orifice, this corresponds to a valve width of 1.65 cm, or 0.65 in. The vacuum chamber is assumed to have a volume of

$$V = 20 \text{ liter (L)} \qquad (12.4\text{-}88)$$

This is equivalent to a cube, 27 cm (or 11 in.) on a side. The input mass flow rate is assumed to be

$$Q_i = 1.0 \text{ torr-liter/sec (T-L/sec)} \qquad (12.4\text{-}89)$$

The chamber is to be operated over a pressure range of 0.03 to 3 torr (T). The steady-state valve conductance at a pressure of 0.03 torr (T) is

$$\underline{C} = \frac{Q_i}{P} = \frac{1.0 \text{ T-L/sec}}{0.03 \text{ T}} = 33.3 \text{ L/sec} \qquad (12.4\text{-}90)$$

The ratio $\underline{C}/C_{\text{Max}}$ in decilogs for 0.03-torr pressure is

$$10 \log \left[\frac{\underline{C}}{C_{\text{Max}}} \right] = 10 \log \left[\frac{22.4 \text{ L/sec}}{100 \text{ L/sec}} \right] = -4.776 \text{ dg} \qquad (12.4\text{-}91)$$

This is equal to the relative valve orifice area A/A_{Max}, expressed in decilogs, which is plotted in Fig 12.4-7. The corresponding valve angles read from the curves are $\Theta_v = 48.5°$ for the simple valve and $\Theta_v = 66°$ for the shaped valve.

Table 12.4-1 lists the parameters computed from Eqs 12.4-90, -91 for the pressures: $P = 0.03$ and 3 torr. The last column gives the values for the vacuum-chamber gain crossover frequency ω_{cc}. In accordance with Eq 12.4-58, the steady-state value of ω_{cc} at 0.03-torr pressure is

$$\omega_{cc} = \frac{Q_o}{VP} = \frac{1.0 \text{ T-L/sec}}{(20 \text{ L})(0.03 \text{ T})} = 1.67 \text{ sec}^{-1} \qquad (12.4\text{-}92)$$

TABLE 12.4-2   Characteristics of Vernitech No. 11MRA-AD3-D0 Stepping Motor

Stepping angle:	15°
Max. unloaded stepping rate:	800 pulse/sec
Max. running torque at 10 pulse/sec:	1.2 oz-in.
Input power:	15 W
Dimensions:	1.0 in. diam. by 1.3 in. long

As was shown in the loop gain plot of the pressure loop in Fig 12.4-5, the parameter ω_{cc} is a break frequency of $G_{(pr)}$.

The following characteristics are assumed for the elements of the pressure loop. The sensitivity of the pressure transducer is

$$K_p = 2.0 \text{ volt/torr (V/T)} \tag{12.4-93}$$

The lead network is assumed to have an attenuation factor α of

$$\alpha = 15 \tag{12.4-94}$$

The maximum (saturated) output voltage from the amplifier is

$$E_{Max} = 10 \text{ V} \tag{12.4-95}$$

The stepper motor is assumed to be a Vernitech size-11, No. 11MRA-AD3-D0, which has the characteristics shown in Table 12.4-2. Thus its step angle is

$$\delta\Theta_m = 15° \tag{12.4-96}$$

It is assumed that the resolution of the valve angle Θ_v is

$$\delta\Theta_v = 0.05° = 8.727 \times 10^{-4} \text{ rad} \tag{12.4-97}$$

To achieve this resolution, the gear reduction between the stepper motor and the valve should be

$$N = \frac{\delta\Theta_m}{\delta\Theta_v} = \frac{15°}{0.05°} = 300 \tag{12.4-98}$$

With such a large gear ratio, the motor is essentially unloaded, and so the maximum unloaded stepping rate (800 step/sec) given in Table 12.4-2 should apply. However, we conservatively assume half of this rate, and so the maximum stepping rate is

$$\dot{M}_{Max} = 400 \text{ pulse/sec (p/sec)} \tag{12.4-99}$$

Hence, the ratio of maximum stepping rate to maximum motor voltage is

$$\frac{\dot{M}_{Max}}{E_{Max}} = \frac{400 \text{ p/sec}}{10 \text{ V}} = 40 \frac{\text{p/sec}}{\text{V}} \tag{12.4-100}$$

A reasonable criterion for adjusting the lead-network break frequency ω_L is to set ω_L equal to the maximum value of ω_{cc} experienced under operational conditions, which occurs at 0.030-torr pressure. This yields

$$\omega_L = \text{Max}[\omega_{cc}] = 1.67 \text{ sec}^{-1} \tag{12.4-101}$$

The lead-network time constant τ_L is the reciprocal of ω_L, which is

$$\tau_L = \frac{1}{\omega_L} = \frac{1}{1.67 \text{ sec}^{-1}} = 0.60 \text{ sec} \tag{12.4-102}$$

The upper break frequency of the lead network is

$$\alpha\omega_L = 15(1.67 \text{ sec}^{-1}) = 25 \text{ sec}^{-1} \tag{12.4-103}$$

There must also be a lowpass filter in the pressure loop to attenuate noise from the pressure transducer. A single-order lowpass filter is assumed, having the transfer function

$$H_f = \frac{1}{1 + s/\omega_f} \tag{12.4-104}$$

The filter break frequency ω_f is assumed to be 3 times the upper break frequency of the lead network. Thus,

$$\omega_f = 3\alpha\omega_L = 3(25 \text{ sec}^{-1}) = 75 \text{ sec}^{-1} \tag{12.4-105}$$

This lowpass filter simulates the total effect of all high-frequency phase lags in the pressure-control loop. Thus, the time constant (13.3 msec) of this lowpass filter is assumed to include the time delay in the stepper-motor operation. Combining the transfer function H_f of Eq 12.4-104 with the loop transfer function of the pressure loop in Eq 12.4-54 gives:

$$G_{(pr)} = \frac{\omega_{cpr}(1 + \omega_L/s)}{s(1 + \omega_{cc}/s)(1 + s/\alpha\omega_L)(1 + s/\omega_f)} \tag{12.4-106}$$

Magnitude asymptote plots of $G_{(pr)}$ are shown in Fig 12.4-10 for different values of pressure, but the variation of ω_{cpr} with pressure is not indicated. The extreme pressures under operational conditions are 0.03 and 3 torr. However, when the valve is fully open, the pressure P is $Q_i/\underline{C}_{Max}$, which is 0.01 torr. The plot for this fully open case is shown by curve ④. For the fully open valve, the break frequency ω_{cc} is C_{Max}/V, which is 5 sec^{-1}.

Applying the system parameters given above to Eq 12.4-58 yields the following expression for the asymptotic gain crossover frequency of the

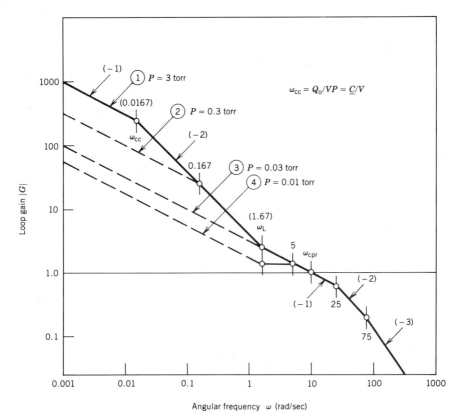

Figure 12.4-10 Asymptote plots of loop gain for pressure feedback loop at various values of steady-state pressure.

pressure loop:

$$\omega_{cpr} = K_p K_a \frac{\dot{M}_{Max}}{E_{Max}} \delta\Theta_v \tau_L \frac{Q_o}{V} \frac{d \ln[A/A_{Max}]}{d\Theta_v}$$

$$= K_x K_a \frac{d \ln[A/A_{Max}]}{d\Theta_v} \tag{12.4-107}$$

where K_x is a constant, which is equal to

$$K_x = K_p \frac{\dot{M}_{Max}}{E_{Max}} \delta\Theta_v \tau_L \frac{Q_o}{V}$$

$$= \frac{\left(2 \frac{V}{T}\right)\left(40 \frac{p/sec}{V}\right)(8.727 \times 10^{-3}\ rad)(0.6\ sec)\left(1.0 \frac{T\text{-}L}{sec}\right)}{20\ L}$$

$$= 2.094 \times 10^{-3}\ rad \tag{12.4-108}$$

TABLE 12.4-3   Calculation of Pressure-Loop Asymptotic Gain Crossover Frequency ω_{cpr} for Steady-State Operation at Pressures of 0.03 and 3 Torr

	Amplifier Gain K_a	Pressure (torr)	Θ_v (deg)	$(dA/A)/d\Theta_v$ (rad^{-1})	ω_{cpr} (sec^{-1})
Simple valve	240	0.030	48.5	2.20	1.11
	240	3.0	4.0	20.5	10.3
Shaped valve	776	0.030	66	2.20	3.57
	776	3.0	13	6.15	10.0

Combining Eqs 12.4-107, -108 gives

$$\omega_{cpr} = \left(2.094 \times 10^{-3} \text{ rad}\right) K_a \frac{d \ln[A/A_{Max}]}{d\Theta_v}$$

$$= \left(2.094 \times 10^{-3} \text{ rad}\right) K_a \frac{dA/A}{d\Theta_v} \qquad (12.4\text{-}109)$$

The following values of amplifier gain are assumed for the two pressure control systems, having simple and shaped valves:

Simple valve: $K_a = 240$ (12.4-110a)

Shaped valve: $K_a = 776$ (12.4-110b)

By applying these values of K_a to Eq 12.4-109, the values for the pressure-loop asymptotic gain crossover frequency are computed in Table 12.4-3, for pressures of 0.03 and 3 torr. The logarithmic slope $d \ln[A/A_{Max}]/d\Theta_v$ is equal to $(dA/A)/d\Theta_v$, and is read from Fig 12.4-9, for the corresponding valve angle Θ_v. Table 12.4-3 shows that ω_{cpr} varies with increasing pressure from 3.57 to 10.0 sec^{-1} for the shaped valve, and from 1.11 to 10.3 sec^{-1} for the simple valve.

12.4.6   Computer Simulation of Pressure-Control System

The linearized analysis allows one to evaluate the stability of the control system, and to establish basic requirements for system parameters. However, to obtain precise data concerning dynamic behavior, the system response should be simulated. The following shows how the serial simulation approach described in Chapter 9 can be applied to simulate a nonlinear system.

The simulation equations are derived from the nonlinear block diagram given in Fig 12.4-2. The only one-cycle time delay of the simulation occurs in the feedback path of the vacuum-chamber loop. Figure 12.4-11 shows how the compensation factor $(1 + sT)$ is inserted to correct for this time delay. In

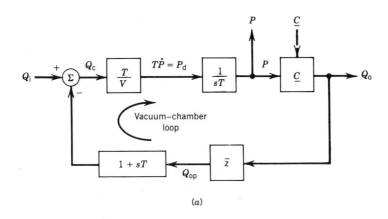

(a)

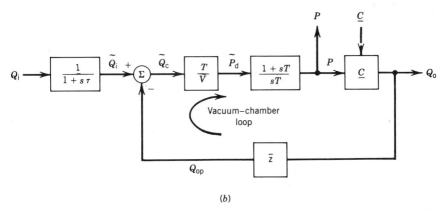

(b)

Figure 12.4-11 Steps for expressing the dynamics of a vacuum-chamber loop in the form for serial simulation: (a) compensation of time delay and normalization of integrations; (b) combination of $(1 + sT)$ factor with integration.

diagram a the integration $1/s$ is normalized to $1/sT$, which requires that the rate of change of pressure, $\dot{P}$, be multiplied by T. This normalized pressure derivative has the units of pressure, and so is designated P_d:

$$P_d = T\dot{P} \qquad (12.4\text{-}111)$$

The subscript d indicates that this is a derivative. The factor $(1 + sT)$ in the feedback path is moved to the forward path and combined with the normalized integration $1/sT$. The variables Q_i, Q_c, and P_d are delayed by one sample period at low frequencies, and so are represented as $\tilde{Q}_i$, $\tilde{Q}_c$, and $\tilde{P}_d$. In the calculations, the input flow-rate variable $\tilde{Q}_i$ can be approximated as the past value of Q_i.

Figure 12.4-11 is applied to Fig 12.4-2, and the effect of the lowpass filter of Eq 12.4-104 is included, along with several nonlinearity constraints. This yields the mathematical block diagram of Fig 12.4-12, from which one can derive the equations for serial computer simulation of the pressure-control system.

The following is a summary of the computation steps to simulate the pressure-control loop. The chamber pressure P is subtracted from the pressure command signal P_k to form the pressure error signal P_e. This is multiplied by the pressure transducer sensitivity $-K_p$ to form the voltage E_1. The voltage E_1 is fed through the lowpass filter to form the voltage E_2, which is fed through the lead network to form the voltage E_3. The voltage E_3 is amplified by the amplifier gain K_a to form the amplifier output voltage E_a. Saturation limits are placed on E_a, to keep this amplifier output voltage within the range ± 10 V. The voltage E_o is the ideal amplifier output voltage, prior to saturation.

The steeper-motor pulse rate (or step rate) $\dot{M}$ is equal to the amplifier output voltage E_a multiplied by the sensitivity of the pulse generator, $\dot{M}_{Max}/E_{Max}$, which is 40 (pulse/sec)/V. The steeper pulse rate $\dot{M}$ is integrated to obtain the number of motor steps, M. It is convenient not to normalize this integration, and so the integration is represented as $1/s$ rather than $1/sT$.

To model the quantization of the stepper-motor operation, the variable M, which is a real number, is converted to an integer, designated M_{int}, which discards the fractional part of M. The variable M_{int} is the actual number of steps that are taken by the stepper motor. This variable M_{int} is multiplied by the step increment of the motor, $\delta\Theta_m$ (which is $15°$), to obtain the angular displacement Θ_m of the motor. The motor angular displacement Θ_m is divided by the gear ratio N (which is 300) to obtain the angular displacement of the valve, Θ_v. Limit stops in the valve mechanism restrict the range of the valve angle Θ_v, to keep it between zero and $90°$. These limits are implemented in the simulation by limiting the range of the variable M_{int} by an equivalent amount.

The nonlinear equations for the valve (obtained from Eqs 12.4-74, -75, -76, or from Eq 12.4-64) are applied to compute A/A_{Max}, which is equal to $\underline{C}/\underline{C}_{Max}$. The ratio A/A_{Max} is multiplied by $\underline{C}_{Max}$ to obtain the valve conductance $\underline{C}$. The chamber pressure P is multiplied by valve conductance $\underline{C}$ to obtain the output flow rate Q_o.

The computer difference equations that are used to simulate the various dynamic transfer functions are as follows:

Normalized integration with lead (step 402):

$$\frac{Y}{X} = \frac{1 + sT}{sT} \qquad (12.4\text{-}112)$$

$$y = y_p + (3/2) * x - (1/2) * x_p \qquad (12.4\text{-}113)$$

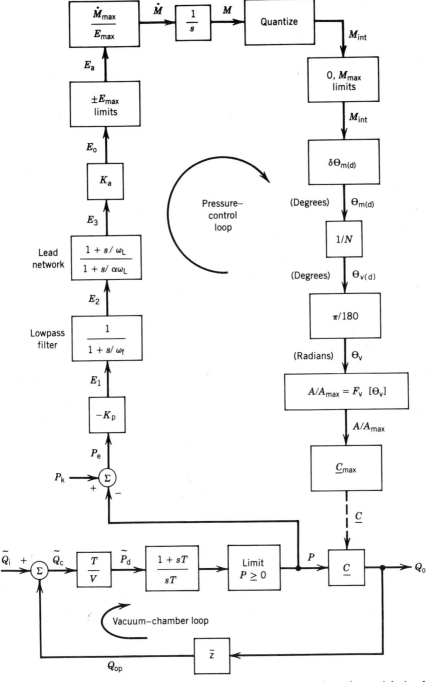

Figure 12.4-12 Mathematical block diagram describing equations for serial simulation.

Integration (step 416):

$$\frac{Y}{X} = \frac{1}{s} \tag{12.4-114}$$

$$y = y_p + (T/2) * (x + x_p) \tag{12.4-115}$$

Lowpass filter (step 406):

$$\frac{Y}{X} = \frac{1}{1 + \tau_f s} \tag{12.4-116}$$

$$y = \left(y_p(2 * \tau_f/T - 1) + x + x_p\right)/(2 * \tau_f/T + 1) \tag{12.4-117}$$

Lead network (steps 407, 408):

$$\frac{Y}{X} = \frac{1 + s/\omega_L}{1 + s/\alpha\omega_L} = \frac{1 + s\tau_L}{1 + s(\tau_L/\alpha)} \tag{12.4-118}$$

$$B = 2 * \tau_L/T \tag{12.4-119}$$

$$y = \left(y_p * (B/\alpha - 1) + x * (B + 1) - x_p * (B - 1)\right)/(B/\alpha + 1) \tag{12.4-120}$$

The steps of the computer program are shown below. Steps 204 and 209 set the initial values of the pressure and valve angle. The initial valve angle is set in terms of the motor step count M, which is equal to

$$M = \frac{\Theta_{m(d)}}{\delta\Theta_{m(d)}} = \frac{N\Theta_{v(d)}}{\delta\Theta_{m(d)}} \tag{12.4-121}$$

where N is the gear ratio, $\delta\Theta_{m(d)}$ is the step increment of the stepping motor in degrees (15°), $\Theta_{m(d)}$ is the angular displacement of the motor in degrees, and $\Theta_{v(d)}$ is the angular displacement of the valve angle in degrees. The valve is fully open when $\Theta_{v(d)} = 90°$, and so the maximum value of M for a fully open valve is

$$M_{Max} = \frac{90N}{\delta\Theta_{m(d)}} = \frac{90(300)}{15} = 1800 \tag{12.4-122}$$

The pressure-control system is normally turned on with the valve fully open. Hence, the normal startup values for P and M in steps 204, 209 are $P_p = 0$, $M_p = M_{Max}$.

It is often desirable to measure the transient response when the command pressure jumps from one pressure level to another. To minimize the time to reach steady state at the first pressure setting, the value for P_p in step 204 should be set equal to the first pressure, and M_p in step 209 should be set equal to the step number than corresponds to the steady-state valve angle for that pressure. In the steady state, the valve conductance is controlled to the value

$$\underline{C} = Q_i/P \tag{12.4-123}$$

where Q_i is the input flow rate. Since $\underline{C}$ is equal to $(A/A_{Max})\underline{C}_{Max}$, the corresponding value for A/A_{Max} is

$$\frac{A}{A_{Max}} = \frac{Q_i}{P\underline{C}_{Max}} = \frac{1 \text{ T-L/sec}}{(100 \text{ L/sec})P} = \frac{0.01 \text{ T}}{P} \tag{12.4-124}$$

For this value of A/A_{Max}, one can determine the corresponding value of valve angle Θ_v from Fig 12.4-7. The past value of the variable M in step 209 should be set equal to

$$M_p = \frac{N\Theta_{v(d)}}{\delta\Theta_{m(d)}} = \frac{300}{14}\Theta_{v(d)} = 20\Theta_{v(d)} \tag{12.4-125}$$

where $\Theta_{v(d)}$ is the valve angle in degrees obtained from Fig 12.4-7.

Steps 701–704 in the program allows the operator to command a sequence of step responses from one command pressure to another. After each time interval, the operator can command a new pressure for a new time interval. The program is terminated by commanding a zero value for the time interval ($\text{TME}_{Max} = 0$).

SPECIFY PROGRAM PARAMETERS:

100   $T = 0.005$ (sec) (Sample period)
101   $\text{TME}_{Max} = ?$ (Maximum time for simulation)
102   $L_{Max} = 5$ (Number of computation cycles per print cycle)

SPECIFY SYSTEM PARAMETERS:

103   $V = 20$ (liter) (Chamber volume)
104   $\underline{C}_{Max} = 100$ (liter/sec) (Maximum valve conductance)
105   $\delta L_v = 0.001$ (Relative valve leakage area)
106   $K_p = 2$ (V/torr) (Pressure-transducer sensitivity)
107   $\tau_f = 1/75$ (sec) (Filter time constant)
108   $\tau_L = 0.6$ (sec) (Lead-network time constant)
109   $\alpha = 15$ (Lead-network attenuation factor)
110   $E_{Max} = 10$ (V) (Amplifier-output saturation voltage)
111   $M_{Max} = 400$ (pulse/sec) (Maximum stepping rate of motor)

112 $\delta\Theta_{m(d)} = 15$ (deg) (Motor stepping increment)
113 $N = 300$ (Gear ratio)
114 $M_{Max} = 90 * N/\delta\Theta_{m(d)} = 1800$ (Limit on M for fully open valve)
115 $K_{a1} = 240$ (Amplifier gain for simple valve)
116 $K_{a2} = 776$ (Amplifier gain for shaped valve)

CHOOSE VALVE TYPE:

112 $VT = ?$ (-1 for simple valve, $+1$ for shaped valve)

INITIALIZE PROGRAM PARAMETERS:

200 TME $= 0$ (Initial time)
201 $L = 0$ (Initiate print counter)

INITIAL VARIABLES:

202 $Q_{op} = 0$
203 $\tilde{P}_{dp} = 0$
204 $P_p = ?$ ($P_p = 0$ at startup)
205 $E_{1p} = 0$
206 $E_{2p} = 0$
207 $E_{3p} = 0$
208 $\dot{M}_p = 0$
209 $M_p = ?$ ($M_p = M_{Max}$ at startup)

INPUT SIGNALS:

300 TME $=$ TME $+ T$
301 $\tilde{Q}_i = 1.0$ (torr-liter/sec) (Input mass flow rate)
302 $P_k = ?$ (torr) (Command pressure)

SYSTEM EQUATIONS:

400 $\tilde{Q}_c = \tilde{Q}_i - Q_{op}$
401 $\tilde{P}_d = (T/V) * \tilde{Q}_c$
402 $P = P_p + (3/2) * \tilde{P}_d - (1/2) * \tilde{P}_{dp}$
403 IF $P < 0$ THEN $P = 0$ (Pressure cannot be negative)
404 $P_e = P_k - P$
405 $E_1 = -K_p * P_e$
406 $E_2 = (E_{2p} * (2 * \tau_f/T - 1) + E_1 + E_{1p})/(2 * \tau_f/T + 1)$
407 $B = 2 * \tau_L/T$
408 $E_3 = (E_{3p} * (B/\alpha - 1) + E_2 * (B + 1) - E_{2p} * (B - 1))/(B/\alpha + 1)$
409 IF $VT < 0$ THEN $K_a = K_{a1}$ (Simple valve)

410   IF VT > 0 THEN $K_a = K_{a2}$ (Shaped valve)
411   $E_o = K_a * E_3$
412   $E_a = E_o$
413   IF $E_a > E_{Max}$ THEN $E_a = E_{Max}$ (Amplifier saturation limit)
414   IF $E_a < -E_{Max}$ THEN $E_a = -E_{Max}$ (Amplifier saturation limit)
415   $\dot{M} = (\dot{M}_{Max}/E_{Max}) * E_a$
416   $M = M_p + (T/2)*(\dot{M} + \dot{M}_p)$
417   $M_{int} = $ INT(M) (Integer part of M) (Stepper-motor quantization)
418   IF $M_{int} < 0$ THEN $M_{int} = 0$ (Closed-valve limit)
419   IF $M_{int} > M_{Max}$ THEN $M_{int} = M_{Max}$ (Wide-open-valve limit)
420   $\Theta_{m(d)} = M_{int} * \delta\Theta_{m(d)}$ (Motor angle in degrees)
421   $\Theta_{v(d)} = \Theta_{m(d)}/N$ (Valve angle in degrees)
422   $\Theta_v = (\pi/180)*\Theta_{v(d)}$ (Valve angle in radians)
423   IF VT > 0 GOTO 426

SIMPLE VALVE:

424   $A/A_{Max} = 1 - \cos(\Theta_v) + \delta L_v$
425   GOTO 430

SHAPED VALVE:

426   IF $\Theta_{v(d)} > 20$ GOTO 429
427   $A/A_{Max} = 1.101 - $ SQR$(1.01 + 0.2*\cos(\Theta_v))$
428   GOTO 430
429   $A/A_{Max} = 1.006008 - \cos(\Theta_v - 0.317812)$

COMPUTE VALVE CONDUCTANCE AND FLOW:

430   $\underline{C} = (A/A_{Max}) * \underline{C}_{Max}$
431   $Q_o = \underline{C} * P$

CHECK PRINT COUNT, PRINT DATA:

500   $L = L + 1$
501   IF $L < L_{Max}$ GOTO 400
502   $L = 0$
503   PRINT DATA   (TME, P, Q_o, $\underline{C}$, E_a, etc.)

SET PAST VALUES EQUAL TO PRESENT VALUES:

600   $Q_{op} = Q_o$
601   $\tilde{P}_{dp} = \tilde{P}_d$
602   $P_p = P$

603   $E_{1p} = E_1$
604   $E_{2p} = E_2$
605   $E_{3p} = E_3$
606   $\dot{M}_p = \dot{M}$
607   $M_p = M$

CHECK TIME LIMIT, REPEAT, START NEW TRANSIENT:

700   IF TME $<$ TME$_{\text{Max}}$ GOTO 300
701   INPUT TME$_{\text{Max}}$ (For next transient response)
702   INPUT P_k (For next transient response)
703   TME $= 0$
704   IF TME$_{\text{Max}} > 0$ GOTO 300
705   END

To obtain a stable computation, the following upper limits must be placed on the sample period T:

$$T < \frac{0.4}{\omega_{cc}} = \frac{0.4VP}{Q_o} = \frac{0.4V}{\underline{C}} \qquad (12.4\text{-}126)$$

$$T < \frac{0.4}{\omega_{cpr}} \qquad (12.4\text{-}127)$$

Equation 12.4-126 should be computed for a fully closed valve, where $\underline{C} = \underline{C}_{\text{Max}}$. This gives

$$T < \frac{0.4V}{\underline{C}_{\text{Max}}} = \frac{0.4(20 \text{ L})}{100 \text{ L/sec}} = 0.08 \text{ sec} \qquad (12.4\text{-}128)$$

Equation 12.4-127 should be computed at the maximum possible value of the pressure-loop gain crossover frequency ω_{cpr}, which is proportional to the logarithmic slope of the valve characteristic. From Fig 12.4-9, the maximum logarithmic slope (which occurs for the simple valve) is

$$\text{Max}\left[\frac{dA/A}{d\Theta_v}\right] = 24 \text{ sec}^{-1} \qquad (12.4\text{-}129)$$

Applying Eq 12.4-129 to Eq 12.4-58 gives $\omega_{cpr} = 12.0 \text{ sec}^{-1}$ as the maximum gain crossover frequency of the pressure loop. Hence, Eq 12.4-127 gives the following limit on the sample period T

$$T < \frac{0.4}{\omega_{cpr}} = \frac{0.4}{12.0 \text{ sec}^{-1}} = 0.033 \text{ sec} \qquad (12.4\text{-}130)$$

To obtain accurate dynamic simulation, the following additional limits should be placed on T:

$$T < \frac{0.4}{\alpha \omega_L} = \frac{0.4 \tau_L}{\alpha} = \frac{0.4(0.6 \text{ sec})}{15} = 0.016 \text{ sec} \qquad (12.4\text{-}31)$$

$$T < 0.4/\omega_f = 0.4\tau_f = 0.4(\tfrac{1}{75}) \text{ sec} = 0.0053 \text{ sec} \qquad (12.4\text{-}132)$$

In accordance with Eq 12.4-132, the sample period T was set at 0.005 sec in step 100 of the program.

12.4.7 Results of Computer Simulation

The startup transients of the pressure-control system, as the pressure is raised from zero to a fixed level, are not very interesting, because the system operates under saturated conditions during most of this transient. When the pressure is rising, its rate of change is limited by the input flow rate. The maximum rate of increase of pressure is

$$\text{Max}[\dot{P}] = \frac{Q_i}{N} = \frac{1.0 \text{ T-L/sec}}{20 \text{ L}} = 0.5 \text{ torr/sec} \qquad (12.4\text{-}133)$$

At this rate, it takes 60 sec for the pressure to reach 3 torr, the upper limit of the operational pressure range considered in the example. At high pressures, the rate of pressure increase is lower than the value of Eq 12.4-133, because of valve leakage. For a fully closed valve, the valve conductance is proportional to the relative leakage δL_v, and is equal to

$$\text{Min}[C] = \delta L_v C_{\text{Max}} = (0.001)(100 \text{ L/sec}) = 0.1 \text{ liter/sec} \quad (12.4\text{-}134)$$

The flow through the closed valve is

$$Q_o = \text{Min}[C]\, P = (0.1 \text{ L/sec})\, P \qquad (12.4\text{-}135)$$

Hence, the actual rate of increase of pressure is reduced from the value of Eq 12.4-133 to

$$\text{Max}[\dot{P}] = \frac{Q_i}{V} - \frac{Q_o}{V} = 0.05 \text{ T/sec} - \frac{0.1 \text{ L/sec}}{20 \text{ L}} P$$

$$= 0.05 \left(1 - \frac{P}{10 \text{ torr}} \right) \text{ torr/sec} \qquad (12.4\text{-}136)$$

The control system also experiences saturation caused by the velocity limit of the stepper motor. The following is the time for the valve to change from fully open to fully closed:

$$\text{slew time} = \frac{M_{\text{Max}}}{\dot{M}_{\text{Max}}} = \frac{1800}{400 \text{ p/sec}} = 4.5 \text{ sec} \qquad (12.4\text{-}137)$$

The computer program was implemented to record step responses in the region of minimum operating pressure (0.03 torr) and in the region of maximum operating pressure (3 torr). Figure 12.4-13 shows the pressure responses for the simple and shaped valves, for step changes of command pressure from 0.03 to 0.04 torr. Figure 12.4-14 shows the pressure responses for the two valves, for step changes of command pressure from 3 to 2.95 torr and from 2.95 to 3 torr. For these responses, steady-state conditions were reached at the initial pressure, before the step change of command pressure was made.

Let us compared these transient responses with the frequency responses derived from the linearized analysis. At 0.03 torr, the pressure-loop gain crossover frequencies in Table 12.4-3 are 1.11 sec^{-1} for the simple valve, and 3.57 sec^{-1} for the shaped valve. For linear responses, the approximate values

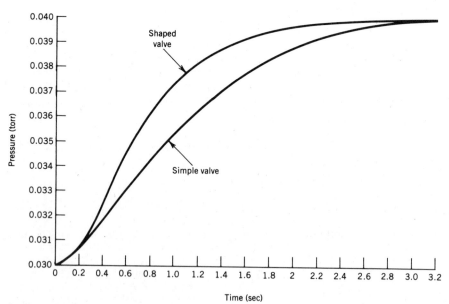

Figure 12.4-13 Pressure-step responses with simple and shaped valves for command pressure changes from 0.03 to 0.04 torr.

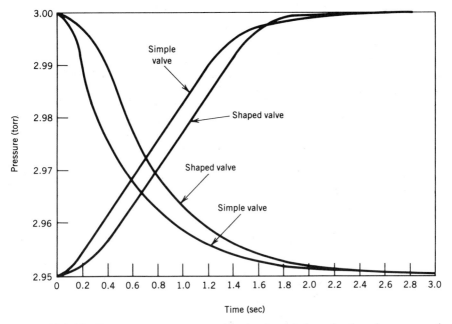

Figure 12.4-14 Pressure-step responses with simple and shaped valves for command pressure changes from 3 to 2.95 torr, and from 2.95 to 3 torr.

of delay time for the step responses should be

Simple valve:
$$T_d \doteq \frac{1}{\omega_{cpr}} = \frac{1}{1.11 \; sec^{-1}} = 0.90 \text{ sec} \qquad (12.4\text{-}138)$$

Shaped valve:
$$T_d \doteq \frac{1}{\omega_{cpr}} = \frac{1}{3.57 \; sec^{-1}} = 0.28 \text{ sec} \qquad (12.4\text{-}139)$$

Remember that the delay time T_d is the time for the step response to reach 50% of the final value. The actual values of T_d measured from the step responses of Fig 12.4-13 are

Simple valve: $\qquad\qquad\qquad T_d = 0.957$ sec $\qquad\qquad\qquad$ (12.4-140)

Shaped valve: $\qquad\qquad\qquad T_d = 0.686$ sec $\qquad\qquad\qquad$ (12.4-141)

The measured value of T_d for the simple valve (0.957 sec) is quite close to the theoretical value (0.90 sec) of Eq 12.4-138. However, the measured T_d for the shaped valve (0.686 sec) is 2.5 times greater than the theoretical value (0.28 sec) in Eq 12.4-139 for linear response. This indicates that the transient for the shaped valve is quite nonlinear. A significant amount of this nonlinearity is due to saturation of the motor speed. The amplifier is strongly saturated for nearly 0.3 sec after the step change of command pressure.

For the transient responses from 3 torr to 2.95 torr in Fig 12.4-14, the measured values of delay time are

Simple valve: $T_d = 0.428$ sec   (12.4-142)

Shaped valve: $T_d = 0.650$ sec   (12.4-143)

The gain crossover frequencies in Table 12.4-3 at this pressure are both approximately 10 sec^{-1}, which corresponds to a delay time T_d of 0.1 sec. Since this theoretical linear delay time is much smaller than the measured values of Eqs 12.4-142, -143, the responses are strongly nonlinear. Also, the magnitude asymptote plot of G at 3-torr pressure (curve ① in Fig 12.4-10) shows that the pressure loop (at 3 torr) has a double integration at low frequency. This indicates that the linear step response should have appreciable overshoot; whereas the simulated step responses in Fig 12.4-14 have essentially no overshoot.

The relative change of pressure in Fig 12.4-14 is very small: 2.95 to 3 torr. Hence, one might expect that a linearized analysis, based on small variations from an operating point, should apply. However, even though the relative change of pressure is quite small, the relative change of output flow rate is very large.

Figure 12.4-15 shows the flow rate Q_o responses for the shaped valve that correspond to the pressure responses of Fig 12.4-14. Curve ① is the valve-flow response for the shaped valve as the pressure drops from 3 to 2.95 torr, while curve ② is the valve-flow response as the pressure rises from 2.95 to 3 torr. Between 0.6 and 1.25 sec, curve ② indicates that the valve is completely closed. (The flow in this interval is 0.30 torr-liter/sec, which is due to valve leakage.) This indicates that the pressure response of Fig 12.4-14, for a pressure increase from 2.95 to 3 torr, should be strongly nonlinear. The valve flow rate is not saturated for curve ①. Nevertheless, the relative variation of flow rate is 2.5 : 1, which is a strong variation. Hence, the linear assumption of small variations from an operating point does not apply. Also, the stepper-motor velocity is strongly saturated during this transient for the first 0.5 sec.

In Fig 12.4-15, the curve-① response of the valve flow Q_o has a peculiar one-cycle oscillation between 0.6 and 0.8 sec. This corresponds approximately to a valve angle of 20°. The oscillatory transient component is caused by the irregular shape of the slope of the logarithmic valve characteristic in this region, shown in Fig 12.4-9.

For the system to exhibit a linear step response of pressure in the region of 3 torr, the step change of pressure must be much smaller than that shown in Fig 12.4-14. In Fig 12.4-16, curve ② shows the pressure response to a command pressure step from 3.000 to 2.999 torr. Even this response is slightly

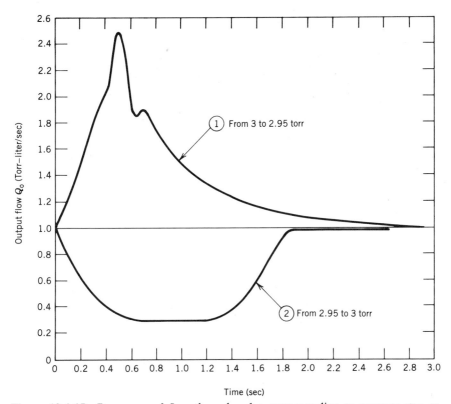

Figure 12.4-15 Responses of flow through valve corresponding to pressure-step responses of Fig 12.4-14 for shaped valve.

nonlinear, because the motor velocity is weakly saturated between 0.15 and 0.45 sec. The linear output E_o from the amplifier exceeds the amplifier saturation limit (± 10 V) during this interval, reaching a maximum value of 13.8 V. The valve flow rate Q_o during this transient varies over the range 0.988–1.145 torr-liter/sec, which is a small relative change.

Linear operation with a larger step change of pressure can be achieved by increasing the stepper-motor velocity. For response curve ④ in Fig 12.4-16, the saturated stepper-motor velocity $\dot{M}_{Max}$ was increased from 400 to 800 step/sec. To keep the gain crossover frequency of the pressure loop constant, the amplifier gain K_a was decreased by a factor of 2, from 776 to 388. Curve ④ is the response to the step change of command pressure from 3.000 to 2.998 torr, shown by curve ③. During this response, the valve flow rate Q_o varies over the range 0.977–1.302 torr-liter/sec. Again the motor velocity is weakly saturated in the range 0.15–0.45 sec. The linear-amplifier output E_o reaches a maximum of 13.8 V in this range.

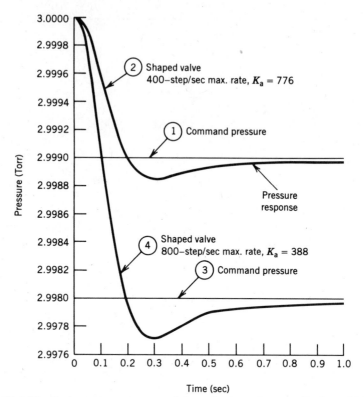

Figure 12.4-16 Pressure-step responses of shaped valve at low amplitudes, which show linear responses: $\omega_{c(pr)} = 10 \ sec^{-1}$.

For the shaped valve, the gain crossover frequency at 3-torr pressure from Table 12.4-3 is 10.0 sec^{-1}. Hence, the step-response delay time for a linear response is approximately

$$T_d \doteq \frac{1}{\omega_{c(pr)}} = \frac{1}{10.0 \ sec^{-1}} = 0.10 \ sec \qquad (12.4\text{-}144)$$

This is very close to the actual values of delay time of the nearly linear responses in Fig 12.4-16, which are $T_d = 0.108$ sec for curve ②, and $T_d = 0.101$ sec for curve ④. These step responses have a delay time and overshoot that are consistent with the frequency-response plot of $|G|$ in Fig 12.4-10 (curve ①).

Although linear step responses are experienced only at very small step changes of pressure, in the region of 3-torr pressure, they are still important. It is essential that the linear response of the pressure loop have good stability; otherwise, the loop would tend to oscillate.

12.4.8 General Application of Concepts in Process Control

This example has illustrated a number of important issues concerning nonlinear process control. Linearized analysis is an important step for describing the response of the system to incremental perturbations. Even a control system that is highly nonlinear (in its normal transient response) must have good incremental stability to operate properly. The nonlinear behavior of the control system can be studied by simulation. Serial simulation is particularly useful in such a study, because it is closely related to the engineering design steps, and can be easily modified to include additional dynamic elements and nonlinear constraints.

Much of process control is based on the PID ("proportional, integral, and derivative") controller. These control units provide proportional gain control, along with an integral network and a lead (or derivative) network. The controller can vary the attenuation factors α_L and α_i of the lead and integral networks, and the break frequencies ω_L and ω_i of those networks. The general forms of these compensation transfer functions are as follows:

Lead (or derivative) network:

$$H_L = \frac{1 + s/\omega_L}{1 + s/\alpha_L\omega_L} \qquad (12.4\text{-}145)$$

Integral network:

$$H_i = \frac{1 + \omega_i/s}{1 + \omega_i/\alpha_i s} = \frac{\alpha_i(1 + s/\omega_i)}{1 + s\alpha_i/\omega_i} \qquad (12.4\text{-}146)$$

The attenuation parameters α_L and α_i can be set to unity, to eliminate the derivative or integral compensation. In Chapter 4, Sections 4.2 and 4.3 give general discussions of the use of the integral network and the lead (or derivative) network for dynamic compensation.

Modern PID controllers often incorporate microprocessors, which implement the derivative and integral network functions with digital-computer algorithms. These algorithms can be readily derived by applying the principles presented in Chapter 9. The algorithms corresponding to the transfer functions of Eqs 12.4-145, -146 are as follows:

Lead (or derivative) network:

$$B_L = 2/(\omega_L * T) = 2 * \tau_L/T \qquad (12.4\text{-}147)$$

$$y = \left(y_p * (B_L/\alpha_L - 1) + x * (B_L + 1) - x_p * (B_L - 1)\right)/(B_L/\alpha_L + 1) \qquad (12.4\text{-}148)$$

Integral network:

$$B_i = 2/(\omega_i * T) = 2 * \tau_i / T \qquad (12.4\text{-}149)$$

$$y = \left(y_p * (\alpha_i * B_i - 1) + x * \alpha_i (\beta_i + 1) - x_p * \alpha_i (B_i - 1)\right) / (B_i * \alpha_i + 1)$$
$$(13.4\text{-}150)$$

The attenuation parameter α_i of the integral network can be made infinite. For this case, the transfer function of the integral network (which was given in Eq 12.4-146) becomes

$$H_i = 1 + \frac{\omega_i}{s} \qquad (12.4\text{-}151)$$

The digital-computer algorithm to implement this transfer function is

$$y = y_p + (\omega_i * T/2) * (x + x_p) + x \qquad (12.4\text{-}152)$$

In the pressure-control system, the stepper motor provides a pure integration between the amplifier voltage and the valve angle. A valve actuator could also be implemented by spring-loading the valve vane, and driving the vane with a torque motor. The vane displacement would then be proportional to the motor torque, which would be proportional to the motor current supplied by the amplifier. For this implementation, the amplifier would require integral compensation, rather than lead compensation, to achieve the same transfer function in the pressure loop.

It is common practice in the process-control field to apply PID controllers in an empirical manner, in which the desired controller parameters are determined by trial-and-error adjustment of the PID controls. Clearly, there is much to be gained from quantitative studies of PID process-control applications, using linear analysis and computer simulation.

12.5   DESCRIBING-FUNCTION ANALYSIS

12.5.1   Background of the Describing Function

The describing function has proven to be quite effective in the analysis of nonlinearities in feedback-control systems. It is determined by feeding a sinusoidal signal into a nonlinear element, and examining the fundamental component of the output waveform. The describing function is the transfer relating the sinusoidal input waveform to the fundamental of the output waveform. It varies with the amplitude of the input, but often does not vary with frequency.

Feedback-control loops generally provide an open-loop frequency-response characteristic that attenuates with increasing frequency. Because of this, the harmonics at the output of the nonlinearity usually have small effect on the dynamic response of the loop, relative to the fundamental. When this condition is satisfied, the describing function, which describes only the response of the fundamental, can characterize the dynamic effect of the nonlinearity quite effectively.

The describing-function approach was first presented in the United States by Kochenburger [12.7], who developed the technique in his Ph.D. research, applying it to a contactor servomechanism. However, as explained by Gibson [12.4] (Chapter 9), other authors (Goldfarb in Russia, and Tustin and Oppelt in England) independently presented essentially the same approach somewhat earlier. On the other hand, it was Kochenburger who coined the term "describing function", and that term is generally accepted.

This section introduces the describing function, by applying it to analyze the effect of backlash in a gear train. For more detailed discussions of the describing function, the reader is referred to books by Gibson [12.4], by Gille, Pelegrin, and Decaulne [1.5], and by Grabbe, Ramo, and Wooldridge [12.8] (Chapter 25 written by Gaines [12.9]). Of particular merit is a paper by Sridhar [12.10], who developed general formulas for the describing functions of a wide class of nonlinearities. These formulas by Sridhar are reprinted in Gibson's book [12.4] (Chapter 9).

12.5.2 Derivation of the Describing Function for Gearing Backlash

An early study of the dynamic effects of gearing backlash using the describing function was presented by Johnson [12.11]. However, the following presentation more closely follows the subsequent paper by Nichols [12.12], who plotted the describing function on the Nichols chart, rather than on the polar locus.

Figure 12.5-1 shows the input–output response characteristic for gear-train backlash, which is an example of a pure hysteresis nonlinearity. To simplify the analysis, the hysteresis characteristic is defined as having a linear slope of unity between the input variable x and the output variable y. The input variable x represents the shaft angle at the input to the gear train (the motor shaft). The output variable y is the shaft angle at the output of the gear train, which is divided by the gear ratio N to normalize it relative to the gear-train input shaft. The gear train backlash B, normalized relative to the gear-train input, can be measured by locking the output shaft and measuring the peak-to-peak amount the gear-train input shaft can be rotated.

Figure 12.5-2 shows the input and output waveforms versus normalized time for a particular signal amplitude. The input signal x is a sinusoidal waveform with a peak-to-peak angular amplitude (designated $2|X|$) that is equal to $3B$, where B is the peak-to-peak gear-train backlash. The solid contour in Fig 12.5-1 corresponds to this particular amplitude of input signal x. The points $1, 2, 3, 4$ on this contour of Fig 12.5-1 are equivalent to the corresponding points on the waveforms of x and y in Fig 12.5-2.

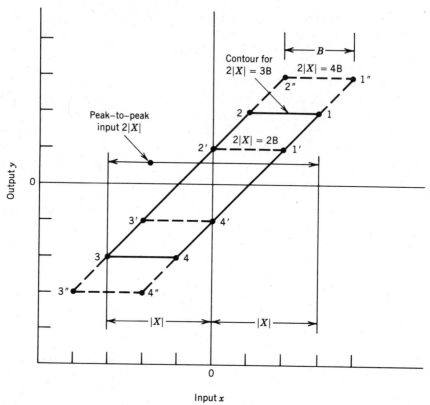

Figure 12.5-1 Input–output characteristic of normalized gear train with backlash, where x and y are the normalized input and output angles.

The size of the hysteresis-loop contour of Fig 12.5-1 depends on the amplitude of the input sinusoid x. The solid loop bounded by points 1, 2, 3, 4 is the contour for $2|X| = 3B$. The dashed loop bounded by points $1', 2', 3', 4'$ is the contour for $2|X| = 2|B|$; while the dashed loop bounded by points $1'', 2'', 3'', 4''$ is the contour for $2|X| = 4B$.

The time scale of Fig 12.5-2 is normalized in terms of the angular frequency ω of the input sinusoid, and is equal to ωt, where t is time. This normalized time ωt is defined as Θ, and is expressed in radians. The gear-train input x, shown as solid curve ①, is equal to

$$x = |X|\cos[\Theta] \tag{12.5-1}$$

The gear-train output y is shown by solid curve ②. The variable portions of the y response follow curves ③ and ④, which are equal to $(x + B/2)$ and $(x - B/2)$. Between points 1 and 2, the output y is a constant equal

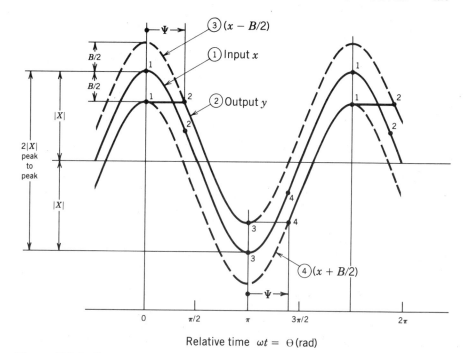

Figure 12.5-2 Response to sinusoidal input waveform of normalized gear train with backlash.

to $(|X| - B/2)$; while between points 3 and 4, y is a constant equal to $-(|X| - B/2)$. The normalized time difference ωt between points 1 and 2, or between points 3 and 4, is defined as Ψ, which is equal to

$$\Psi = \arccos\left[1 - \frac{B}{|X|}\right] \tag{12.5-2}$$

The output waveform y is described by

$$y = |X| - \frac{B}{2} \qquad\qquad \text{for } 0 < \Theta < \Psi \tag{12.5-3}$$

$$y = x + \frac{B}{2} = |X|\cos[\Theta] + \frac{B}{2} \qquad \text{for } \Psi < \Theta < \pi \tag{12.5-4}$$

$$y = -\left(|X| - \frac{B}{2}\right) \qquad\qquad \text{for } \pi < \Theta < \pi + \Psi \tag{12.5-5}$$

$$y = x - \frac{B}{2} = |X|\cos[\Theta] - \frac{B}{2} \qquad \text{for } \pi + \Psi < \Theta < 2\pi \tag{12.5-6}$$

In Chapter 2, Eqs 2.2-1, -3, and -4 gave the Fourier series expansion and its coefficients. If the DC term is assumed to be zero, these equations can be expressed as

$$f[t] = \frac{2}{T}(a_1\cos[\omega t] + b_1\sin[\omega t]) + \text{harmonics} \qquad (12.5\text{-}7)$$

$$a_1 = \int_{(T)} f[t]\cos[\omega t]\, dt \qquad (12.5\text{-}8)$$

$$b_1 = \int_{(T)} f[t]\sin[\omega t]\, dt \qquad (12.5\text{-}9)$$

The expression (T) indicates that the integrations are performed over a period T of the waveform. We are concerned only with the fundamental components. The integration can be performed over any part of a complete period T of the fundamental. The parameters α, β are defined as follows: $\alpha = 2a_1/T$, $\beta = 2b_1/T$. The function $f[t]$ is replaced by $y[t]$, the output from the nonlinearity, and the time t is expressed in terms of the normalized variable $\Theta = \omega t$. Equations 12.5-7 to -9 become

$$y[\Theta] = \alpha\cos[\Theta] + \beta\sin[\Theta] + \text{harmonics} \qquad (12.5\text{-}10)$$

$$\alpha = \frac{2}{\omega T}\int_{(T)} y[\omega t]\cos[\omega t]\, d\omega\, t = \frac{1}{\pi}\int_{(2\pi)} y[\Theta]\cos[\Theta]\, d\Theta \quad (12.5\text{-}11)$$

$$\beta = \frac{2}{\omega T}\int_{(T)} y[\omega t]\sin[\omega t]\, d\omega\, t = \frac{1}{\pi}\int_{(2\pi)} y[\Theta]\sin[\Theta]\, d\Theta \quad (12.5\text{-}12)$$

The expression $2/\omega T$ is replaced by

$$\frac{2}{\omega T} = \frac{2}{(2\pi F)T} = \frac{1}{\pi} \qquad (12.5\text{-}13)$$

where F is the frequency of the fundamental in hertz, which is equal to $1/T$. The expression (2π) indicates that the integration is performed over the angle 2π for the variable Θ.

By Eq 12.5-1, the nonlinearity input x has an amplitude $|X|$, and zero phase relative to $\cos[\Theta]$. By Eq 12.5-10, the output y of the nonlinearity has a fundamental component of the amplitude α for the cosine term, the amplitude β for the sine term. The describing function of the nonlinearity is designated H_{nL}, and represents the transfer function relating the sinusoidal input x to the fundamental component of the nonlinearity output y. Hence, the real and imaginary parts of the nonlinear describing function H_{nL} are equal to

$$\text{Re}[H_{nL}] = \alpha/|X| \tag{12.5-14}$$

$$\text{Im}[H_{nL}] = -\beta/|X| \tag{12.5-15}$$

Note that when β is positive, the sine term of y lags the x sinusoidal input, and so the imaginary part of H_{nL} should be negative. Hence, Eq 12.5-15 has a negative sign. It is convenient to define a normalized backlash ratio b as

$$b = \frac{B}{2|X|} \tag{12.5-16}$$

This parameter b is the ratio of peak-to-peak backlash B to peak-to-peak sinusoidal motion $2|X|$, at the input to the gear train. Let us also define a normalized nonlinearity output y' that is equal to $y/|X|$. Hence, Eqs 12.5-3 to -6 for y can be normalized as follows:

$$y' = y/|X| = (1 - b) \qquad \text{for} \quad 0 < \Theta < \Psi \tag{12.5-17}$$

$$y' = \cos[\Theta] + b \qquad \text{for} \quad \Psi < \Theta < \pi \tag{12.5-18}$$

$$y' = -(1 - b) \qquad \text{for} \quad \pi < \Theta < \pi + \Psi \tag{12.5-19}$$

$$y' = \cos[\Theta] - b \qquad \text{for} \quad \pi + \Psi < \Theta < 2\pi \tag{12.5-20}$$

By Eq 12.5-2, Ψ is equal to

$$\Psi = \arccos[1 - 2b] \tag{12.5-21}$$

The real part of H_{nL} is obtained as follows by combining Eqs 12.5-17 to -20 with Eqs 12.5-11, -14:

$$\pi \, \text{Re}[H_{nL}] = \int_{(2\pi)} y'[\Theta]\cos[\Theta] \, d\Theta$$

$$= \int_0^\Psi (1 - b)\cos[\Theta] \, d\Theta + \int_\Psi^\pi (\cos[\Theta] + b)\cos[\Theta] \, d\Theta$$

$$- \int_\pi^{\pi + \Psi} (1 - b)\cos[\Theta] \, d\Theta + \int_{\pi + \Psi}^{2\pi} (1 - b)\cos[\Theta] \, d\Theta \tag{12.5-22}$$

The imaginary part of H_{nL} is obtained as follows by combining Eqs 12.5-17 to

-20 with Eqs 12.5-12, -15:

$$-\pi \operatorname{Im}[H_{nL}] = \int_{(2\pi)} y'[\Theta]\sin[\Theta]\, d\Theta$$

$$= \int_0^{\Psi}(1-b)\sin[\Theta]\, d\Theta + \int_{\Psi}^{\pi}(\cos[\Theta]+b)\sin[\Theta]\, d\Theta$$

$$- \int_{\pi}^{\pi+\Psi}(1-b)\sin[\Theta]\, d\Theta + \int_{\pi+\Psi}^{2\pi}(1-b)\sin[\Theta]\, d\Theta \quad (12.5\text{-}23)$$

Solving Eqs 12.5-22, -23 gives

$$\operatorname{Re}[H_{nL}] = \frac{1}{\pi}\{\pi - \Psi + (1-2b)\sin[\Psi]\} \qquad (12.5\text{-}24)$$

$$\operatorname{Im}[H_{nL}] = \frac{4}{\pi}b(b-1) \qquad (12.5\text{-}25)$$

where Ψ is given in Eq 12.5-21.

12.5.3 Application of Backlash Describing Function

Figure 12.5-3 is a plot of the magnitude and phase lag of the describing function H_{nL} versus the normalized backlash b, which is equal to the ratio of peak-to-peak backlash divided by the peak-to-peak sinusoidal displacement of the gear-train input. Note that when b is greater than unity, the backlash region is greater than the peak-to-peak displacement of the gear-train input shaft, and so there is no motion of the gear-train output.

Figure 12.5-4 is a plot of the describing function H_{nL} on a polar locus. The parameter indicated along the locus is the relative backlash parameter b. Figure 12.5-5 shows the same locus expressed in the rectangular coordinates used on the Nichols chart: logarithmic magnitude in decilogs versus phase angle in degrees. These are derived from figures given by Johnson [12.11] and by Nichols [12.12]. The vectors drawn in Figs 12.5-4, -5 show the complex values of the transfer function H_{nL} for a normalized backlash b of 0.5. (For this value of b, the peak-to-peak displacement of the gear train input shaft is twice the peak-to-peak backlash range.) From Eqs 12.5-24, -25 the describing function at this relative signal amplitude ($b = 0.5$) is

$$H_{nL} = 0.500 - j0.318 = 0.592 \underline{/-32.5^{\circ}} = (-2.27\text{ dg})\underline{/-32.5^{\circ}} \quad (12.5\text{-}26)$$

The transfer function H_{nL} is independent of frequency. Hence, at the amplitude corresponding to $b = 0.5$, all points on the linear G-locus of the servo position loop experiences a gain change of -2.27 dg, and a phase shift of -32.5°, because of the backlash nonlinearity.

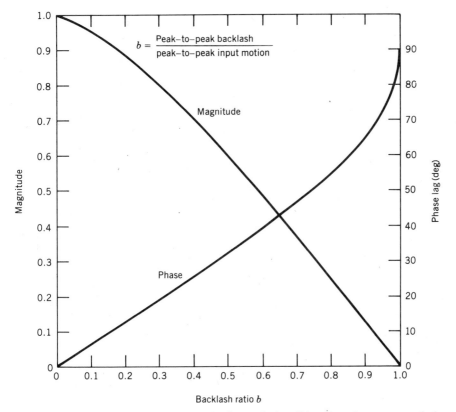

Figure 12.5-3 Plot of magnitude and phase of describing function versus relative backlash factor b.

Let us assume the servo position loop has the ideal double-integration transfer function that was plotted on the Nichols chart of Fig 3.4-4 in Chapter 3. Its transfer function is

$$G = \frac{\omega_c(1 + \omega_{LF}/s)}{s(1 + s/\omega_{HF})} \qquad (12.5\text{-}27)$$

The asymptotic gain crossover frequency ω_c is set equal to 1.0 rad/sec, and the frequency parameters are optimized for $\text{Max}|G_{ib}| = 2$ dB $= 1$ dg $= 1.26$. This locus is shown in the Nichols-chart plot of Fig 12.5-6 as curve ①. This is the linear G-locus of the servo position loop, which applies when the gear-train amplitude is so large that the relative-backlash parameter b is much less than unity. Curve ② is the nonlinear G-locus, which applies for $b = 0.5$, when the peak-to-peak gear-train input motion is twice the gear-train backlash. Locus (2) is formed by shifting each point on locus ① by -2.27 dg in amplitude

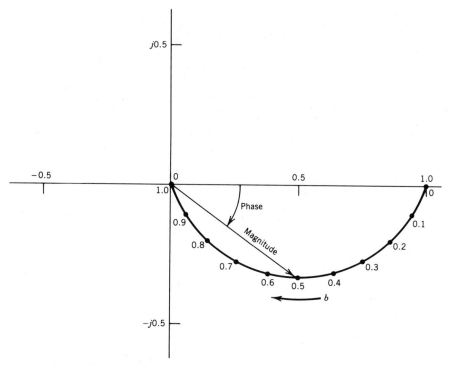

Figure 12.5-4   Polar describing function for backlash.

and by $-32.5°$ in phase. The feedback loop is still stable at this low amplitude, but $\text{Max}|G_{ib}|$ has increased from its linear value of 1.0 dg = 1.26 to a value of 4.5 dg = 2.8.

Since the describing function H_{nL} varies with signal amplitude, a number of nonlinear G-loci similar to curve ② of Fig 12.5-6 are needed to show the nonlinear responses at different amplitudes. However, the effect of the nonlinear response can be characterized adequately for most purposes merely by observing the relation between the nonlinear G-locus and the $G = -1$ point. This can be achieved by keeping the G-locus fixed, and moving the $G = -1$ point in accordance with the reciprocal of the describing function H_{nL}.

This approach is illustrated in Fig 12.5-7. Curve ① is the linear G-locus assumed for the position loop. Curve ② is the effective motion of the $G = -1$ point corresponding to the backlash nonlinearity, and so is a plot of the function

$$G = 1/H_{nL} \qquad (12.5\text{-}28)$$

The parameter indicated along curve ② is the relative backlash parameter b. Curve ② is formed by shifting the phase of the $G = -1$ point by the

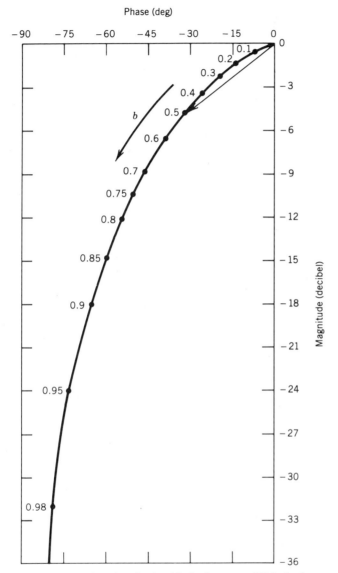

Figure 12.5-5 Describing function for backlash, drawn on Nichols-chart coordinates.

negative of the phase of H_{nL} plotted in Fig 12.5-5, and by shifting the logarithmic magnitude of that point by the negative of the decilog magnitude value of H_{nL}. Thus, in accordance with Eq 12.5-26, at the point on curve ② for $b = 0.5$, the effective $G = -1$ point is shifted in magnitude by $+2.27$ dg, and in phase by $+35.0°$, relative to $G = -1$, which occurs where $|G| = 0$ dg and $\text{Ang}[G] = -180°$.

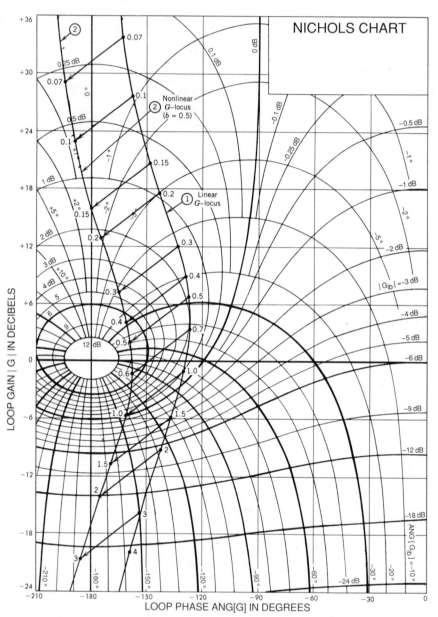

Figure 12.5-6 Shift on Nichols chart of *G*-locus caused by backlash nonlinearity for *b* = 0.5 (backlash is 0.5 times peak-to-peak displacement).

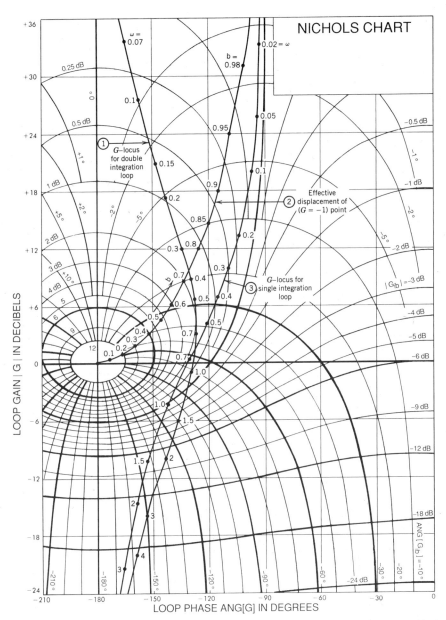

Figure 12.5-7 Effective shift of $(G = -1)$ point on Nichols chart caused by backlash nonlinearity, presented with two typical G-loci.

The effective $G = -1$ contour, shown as curve ② in Fig 12.5-7, intersects the linear G-locus of the position loop, shown by curve ①, at a frequency of 0.37 rad/sec (shown on curve ①), and at a relative backlash value of $b = 0.73$ (shown on curve ②). This indicates that the position loop oscillates at 0.37 rad/sec (i.e., at 37% of its asymptotic gain crossover frequency) with a relative amplitude equal to $1/b = 1/0.73 = 1.37$. Thus, the peak-to-peak oscillation is 1.37 times the gearing backlash.

The reason that the position loop oscillates is that the integral network in the position loop continually tries to correct for the error due to backlash. If this integral network is removed, the oscillation stops. Curve ③ shows the linear G-locus of the following alternative position-loop transfer function, which has no integral network:

$$G = \frac{\omega_c}{1 + s/\omega_x} \qquad (12.5\text{-}29)$$

Again, the asymptotic gain crossover frequency ω_c is set equal to 1 rad/sec. The break frequency ω_x is adjusted to give a value for $\text{Max}|G_{ib}|$ of 2 dB = 1 dg = 1.26, just as for the G-locus of curve ①. This G-locus of curve ③ does not intersect the effective $G = -1$ contour given by curve ②. Hence, this alternative position-loop transfer function does not oscillate.

On the other hand, since this position-loop oscillation occurs at low frequency, it may not be harmful if the backlash is small. Although the loop without the integral network does not experience an oscillation, it is not necessarily better. This loop can have inferior performance because its error is appreciably larger.

The preceding analysis has described the effect of gear-train backlash on the position feedback loop of the servo. Backlash also has another, quite different effect on the tachometer velocity feedback loop. When the gearing is within the backlash region, the reflected inertia of the controlled member is decoupled from the motor, and so the effective inertia experienced by the motor is reduced. The inertias experienced by the motor are as follows:

Gearing in backlash region:

$$J_{(m)} = J_m \qquad (12.5\text{-}30)$$

Gearing in solid contact:

$$J_{(m)} = J_m + J_{c(m)} = J_m\left(1 + \frac{J_{c(m)}}{J_m}\right) \qquad (12.5\text{-}31)$$

In these equations, J_m is the motor inertia, $J_{(m)}$ is the total effective inertia experienced by the motor, and $J_{c(m)}$ is the inertia of the controlled member reflected to the motor shaft. The inertia $J_{c(m)}$ is equal to J_c/N^2, where J_c is the inertia of the controlled member, and N is the gear ratio.

The gain crossover frequency ω_{cv} of the velocity loop is proportional to $1/J_{(m)}$. Hence, ω_{cv} is increased by the factor $(1 + J_{c(m)}/J_m)$ when the gearing is operating in the backlash region. This increased gain tends to make the velocity loop oscillate within the backlash region. If such an oscillation occurs, it can be quite harmful because the frequency is high. Acceleration, which is proportional to the square of the frequency, is appreciable even for a very small backlash angle. This problem can be minimized by coupling a flywheel to the motor shaft. This makes J_m much larger then $J_{c(m)}$, and so the gain variation factor $(1 + J_{c(m)}/J_m)$ remains close to unity.

Because of the problems associated with backlash, high-performance control systems often use two motors, with separate gear trains, to drive the controlled member. The motors operate in opposition, one providing the torque for clockwise motion, and the other the torque for counterclockwise motion. A constant preload torque is exerted between the motors, to eliminate the effect of backlash in the gear trains.

Chapter 13

Advanced Analysis of Stability and Transient Response

This chapter presents advanced analysis techniques for evaluating stability, and for calculating closed-loop poles and the corresponding transient coefficients. Section 13.1 describes two criteria for determining the absolute stability of a feedback loop: the Nyquist criterion and the Routh criterion.

Section 13.2 describes two methods for computing the closed-loop poles of a feedback loop. The first is a generalized frequency-response approach developed by Kusters and Moore and extended by the author, and the second is an iterative procedure developed by the author. Based on the methods of Section 13.2, simple techniques are developed in Section 13.3 for approximating the closed-loop poles and the corresponding transient coefficients, by inspecting magnitude asymptote frequency-response plots.

Section 13.4 summarizes the root-locus method of Evans. It also shows how root-locus plots can be developed using the frequency-response technique described in Section 13.2.

13.1 CRITERIA FOR ABSOLUTE STABILITY

There are two techniques that are useful for determining whether a loop transfer function is stable in an absolute sense: the Nyquist criterion and the Routh criterion. The Routh criterion is simpler to use, but, unlike the Nyquist criterion, gives little insight into why the loop is unstable, or what one should do to eliminate the instability.

13.1.1 Nyquist Stability Criterion

The Nyquist stability criterion provides a general method for determining from the frequency response of a loop transfer function whether or not the loop is stable in an absolute sense. It is implemented by examining a polar plot of the G-locus for positive and negative frequencies. However, the magnitude scale of the G-locus must be distorted to show the behavior of the locus at very low and very high frequencies. More extensive explanations of the Nyquist stability theory are presented by Gille, Pelegrin, and Decaulne [1.5] (Chapter 16), and by Sollecito and Reque [13.1], who wrote Chapter 21 in Grabbe, Ramo, and Wooldrige [12.8].

13.1.1.1 *Conformal-Mapping Principle.*

The Nyquist criterion is based on the principle of conformal mapping, which is illustrated in Fig 13.1-1. In diagram a, the variable s varies along this contour in the clockwise direction, as shown by the arrows. A point s_1 is shown which lies inside the contour, and a second point s_2 is shown which lies outside the contour. Vectors are drawn from points s_1 and s_2 to the moving point s on the contour, and so represent the functions $(s - s_1)$ and $(s - s_2)$. Diagram b shows the plots of these functions $(s - s_1)$ and $(s - s_2)$ as the variable s moves along the contour. The contour of $(s - s_1)$ encircles the origin in the clockwise direction, while the contour of $(s - s_2)$ does not encircle the origin.

Diagram c shows the contour of the function $1/(s - s_1)$. This also encircles the origin, but in the counterclockwise direction. The reason for this counter-clockwise motion of the contour is that the phase of the vector $1/(s - s_1)$ is the negative of the phase of the vector $(s - s_1)$:

$$\text{Ang}\left[\frac{1}{s - s_1}\right] = -\text{Ang}[s - s_1] \qquad (13.1\text{-}1)$$

Hence the vector representing $1/(s - s_1)$ must move in the opposite direction to the vector representing $(s - s_1)$. The contour for $1/(s - s_2)$ is not shown. However, like the contour of $(s - s_2)$, it does not encircle the origin.

This principle can be extended to show that the contour of any function of s encircles the origin in the clockwise direction once for every zero lying inside the s-plane contour, and encircles the origin in the counterclockwise direction once for every pole lying inside the counter. Hence, the net number of clockwise encirclements of the origin is equal to the number of zeros lying inside the s-plane contour minus the number of poles.

The Nyquist theory applies this conformal-mapping principle by drawing a closed contour on the s-plane that encircles the right half of the s-plane, as shown in Fig 13.1-2. This contour follows the imaginary axis from $s = -j\infty$ at point A to $s = +j\infty$ at point D. Then it circles back in the clockwise direction to point A, around the right half of the s-plane, at $|s| = \infty$. The contour takes a circular detour around the origin, as shown in the segment between points B and C, to avoid the poles of G at the origin.

The feedback loop is stable if G_{ib} or G_{ie} has no poles that lie within this contour. It is simpler to consider G_{ie}, which is equal to

$$G_{ie} = \frac{1}{1 + G} \qquad (13.1\text{-}2)$$

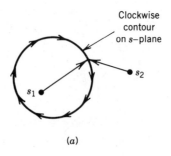

(a)

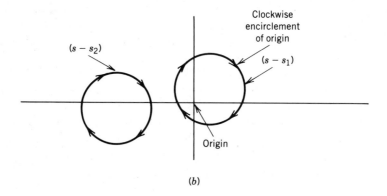

(b)

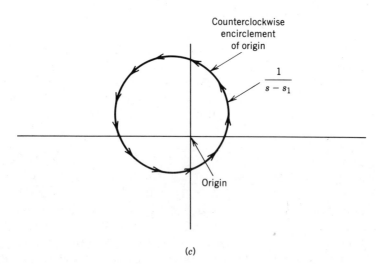

(c)

Figure 13.1-1 Conformal-mapping principle on which Nyquist stability criterion is based: (*a*) clockwise contour of *s* on *s*-plane; (*b*) corresponding contours of $(s - s_1)$ and $(s - s_2)$; (*c*) corresponding contour of $1/(s - s_1)$.

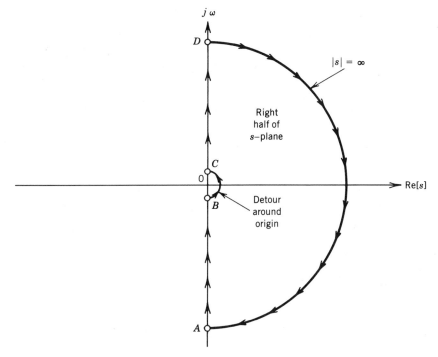

Figure 13.1-2 Contour around right half of s-plane (distorted scale).

The requirement that G_{ie} should have no poles within this contour is equivalent to requiring that $(1 + G)$ have no zeros within the contour.

13.1.1.2 Feedback Loops That Are Open-Loop Stable. Let us first consider the usual case where G is open-loop stable, and so has no poles in the right half plane. For this case, the function $(1 + G)$ should have neither poles nor zeros within the s-plane contour of Fig 13.1-2. Hence, the loop is stable if the plot of $(1 + G)$ does not encircle the origin as s varies along the contour.

Instead of plotting the function $(1 + G)$, it is simpler to plot the function G. Since the origin of $(1 + G)$ corresponds to the point $(G = -1)$, the loop is stable if the plot of G does not encircle the $(G = -1)$ point, as s varies along the contour of Fig 13.1-2.

An important part of the s-plane contour of Fig 13.1-2 is the detour around the origin, which is taken to avoid poles of G at the origin. A magnified plot of this detour is shown in Fig 13.1-3. At very low frequencies, the loop transfer function can be accurately approximated by

$$G = K_n/s^n \qquad (13.1\text{-}3)$$

where n is the number of poles of G at the origin. In the detour region,

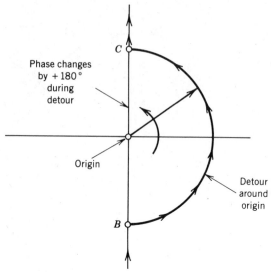

Figure 13.1-3   Enlarged view of detour around origin, taken by s-plane contour when there is a pole at the origin.

between points B and C, the phase of s changes by $+180°$. Since G is proportional to $1/s^n$ along this detour, the phase of G varies as follows in the detour from point B to point C:

$$\text{phase change of } G = -180n \text{ deg} \qquad (13.1\text{-}4)$$

We are not concerned with the case where G has one or more zeros at the origin, because in that case the detour need not be taken.

The Nyquist theory can be explained by applying it to some examples. The first is the following loop transfer function G_A:

$$G_A = \frac{\omega_c}{s(1 + s/\omega_1)(1 + s/\omega_2)} \qquad (13.1\text{-}5)$$

Consider the two cases G_{A1} and G_{A2} illustrated by the magnitude asymptote plots shown in Fig 13.1-4, where loop G_{A2} has a much higher value of ω_c than loop G_{A1}. For loop G_{A1}, the phase lag at gain crossover is less than $180°$, whereas for loop G_{A2} it is greater than $180°$.

Diagrams a and b of Fig 13.1-5 give sketches, with distorted magnitude scales, of the G-loci for loops G_{A1} and G_{A2}. These show the variation of G corresponding to the s-plane contour of Fig 13.1-2. The solid curve is the portion of the G-locus for positive frequency. The dashed curve is the portion

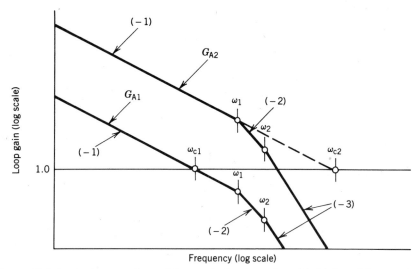

Figure 13.1-4 Asymptote plots of loop gain for loop G_A, with high and low settings for ω_c.

for negative frequency, and is the conjugate of the solid curve for positive frequency. Point C occurs at very low frequency and at very high loop gain $|G|$, while point D occurs at very high frequency, where the loop gain $|G|$ is essentially zero.

The dashed portion from point A to B is the mirror image about the real axis of the solid curve, and so represents the conjugate of the solid curve. Point A corresponds to a large negative frequency ($s \doteq -j\infty$), while point B corresponds to a very small negative frequency ($s = -j\epsilon$), where ϵ is very small.

As shown in Eq 13.1-5, the G-locus has a single pole at the origin. Hence, by Eq 13.1-4, the G-locus should change by $-180°$ as the s-plane contour detours around the origin from point B to point C. Therefore, in diagrams a and b of Fig 13.1-5, the G-locus at large values of $|G|$ changes by $-180°$, to close the contours between points B and C.

In diagram a of Fig 13.1-5, the G-locus does not encircle the ($G = -1$) point, and so loop G_{A1} is stable. However, in diagram b, when ω_c is increased to the value of loop G_{A2}, the G-locus encircles the ($G = -1$) point, which indicates that the loop is unstable. For the stable loop G_{A1}, the phase lag is less than $180°$ when the G-locus passes through the ($|G| = 1$) contour (shown by the dashed curve). However, in the unstable loop G_{A2} the phase lag at $|G| = 1$ is greater than $180°$.

Note that the G-locus for loop G_{A2} in diagram b encircles the ($G = -1$) point twice in the clockwise direction. This indicates that there are two poles in the right half plane.

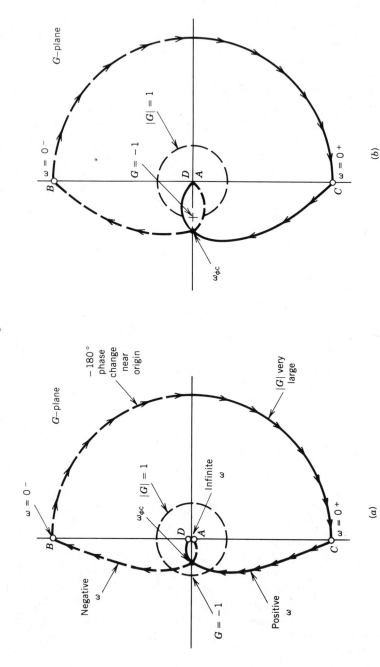

Figure 13.1-5 Nyquist plots of loop transfer functions G_{A1} and G_{A2}: (a) loop G_{A1} (stable); (b) loop G_{A2} (unstable).

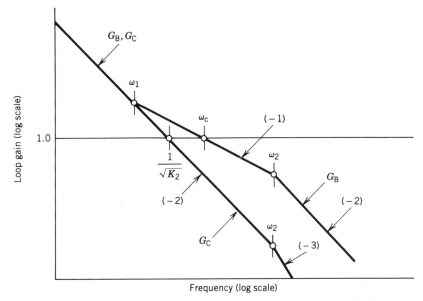

Figure 13.1-6 Asymptote plots of loop gain for loops G_B and G_C.

Let us now consider the following loop transfer functions G_B and G_C, which are illustrated by the frequency response asymptote plots of Fig 13.1-6:

$$G_B = \frac{\omega_c(1 + \omega_1/s)}{s(1 + s/\omega_2)} = \frac{\omega_c\omega_1(1 + s/\omega_1)}{s^2(1 + s/\omega_2)} = \frac{K_2(1 + s/\omega_1)}{s^2(1 + s/\omega_2)} \quad (13.1\text{-}6)$$

$$G_C = \frac{K_2}{s^2(1 + s/\omega_2)} \quad (13.1\text{-}7)$$

Both of these loops have a double-order pole at zero frequency, and a low-frequency asymptote slope of -2. Loop G_B is a stable loop, because it has a zero of break frequency ω_1, which changes the asymptote slope to -1 at gain crossover. However, loop G_C does not have a region of -1 slope near gain crossover, and so is unstable.

The Nyquist plots for these loop transfer functions G_B and G_C are shown in Fig 13.1-7. Since there are two poles at the origin, the phase of the locus changes by $-2(180°) = -360°$ as the s-contour takes its detour around the origin between points B and C. Hence, the two G-loci progress in the clockwise direction, with a phase change of $-360°$, between points B and C.

The locus of G_B in Fig 13.1-7 does not encircle the ($G = -1$) point, and so that loop is stable. However, the locus for G_C encircles the ($G = -1$) point twice in the clockwise direction. This indicates that G_C is unstable, and has two closed-loop poles in the right half of the S-plane.

13.1.1.3 *Feedback Loops That Are Open-Loop Unstable.* When a feedback loop is open-loop unstable, it is very rare that the instability is due to

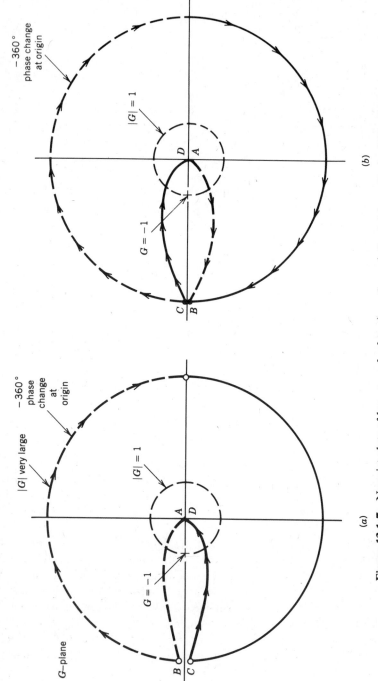

Figure 13.1-7 Nyquist plots of loop transfer functions G_B and G_C: (a) loop G_B (stable); (b) loop G_C (unstable).

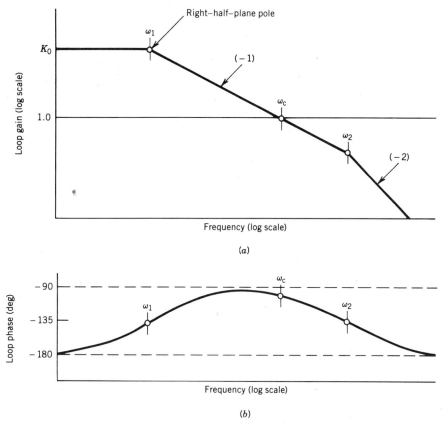

Figure 13.1-8 Magnitude and phase plots for loop G_D, which is open-loop unstable: (*a*) asymptote plot of loop gain $|G|$; (*b*) loop phase.

open-loop poles that are exactly on the $j\omega$ axis. The unstable poles are practically always in the right half plane. If one is faced with the rare case where the open-loop poles are on the $j\omega$ axis, rather than in the right half plane, the s-plane contour must detour around these poles as it does the poles at the origin. The method for doing this is described by Gille, Pellegrin, and Decaulne [1.5] (Chapter 16) and by Sollecito and Reque [13.1].

Each right-half-plane zero of $(1 + G)$ (which is an unstable closed-loop pole) causes the G-locus to encircle the $(G = -1)$ point once in the *clockwise* direction. Since a pole of G is a pole of $(1 + G)$, each unstable open-loop right-half-plane pole of G causes the G-locus to encircle the $(G = -1)$ point once in the *counterclockwise* direction. Hence, if an unstable open-loop transfer function is to be stable when the loop is closed, the G-locus must encircle the $(G = -1)$ point in the counterclockwise direction once for each right-half-plane open-loop pole of G.

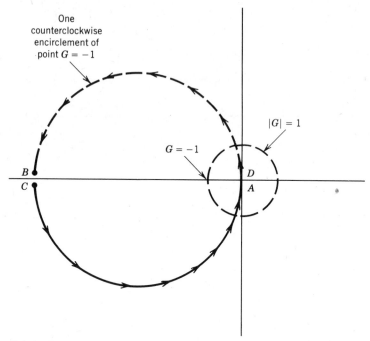

Figure 13.1-9 Nyquist plot for loop G_D, which is open-loop unstable, but closed-loop stable.

To illustrate the use of the Nyquist theory when the loop is unstable open loop, consider the following loop transfer function:

$$G_D = \frac{\omega_c}{s(1 - \omega_1/s)(1 + s/\omega_2)} = \frac{-K_0}{(1 - s/\omega_1)(1 + s/\omega_2)} \quad (13.1\text{-}8)$$

where K_0 is equal to ω_c/ω_1. This has a right-half-plane pole at $s = +\omega_1$, and so is open-loop unstable. Sketches of the magnitude and phase of this loop transfer function are shown in Fig 13.1-8.

Figure 13.1-9 shows the Nyquist sketches of this G_D-locus. Since this loop has no poles at the origin, the detour around the origin is unnecessary, and points B and C coincide. This locus encircles the $(G = -1)$ point once in the counterclockwise direction. Hence the loop is stable.

The fact that this loop is stable can also be seen by applying the following practical stability rules, which will be developed in Section 13.2. A loop is generally stable if the following conditions are satisfied:

a. If the magnitude slope is negative in any region near the frequency where $|G| = 1$, then G should approximate the function ω_c/s in that region.

b. If the magnitude slope is positive in any region near the frequency where $|G| = 1$, then G should approximate the function s/ω_r in that region.

c. At the break frequency of a right-half-plane pole of G, the asymptotic value of $|G|$ should be greater than unity.

d. At the break frequency of a right-half-plane zero of G, the asymptotic value of $|G|$ should be less than unity.

For this loop G_D there is only one frequency where $|G|$ passes through unity, and the slope there is negative. Hence, requirement (b) does not apply. By requirement (a), G should approximate ω_c/s at gain crossover. The fact that G_D can be written in terms of ω_c in the form of the first expression of Eq 13.1-8 indicates that this requirement is satisfied.

This loop has no right-half-plane zeros, and so requirement (d) does not apply. There is a right-half-plane pole of break frequency ω_1. Figure 13.1-8 shows that the asymptote of $|G|$ at this break frequency ω_1 is greater than unity, and so requirement (c) is satisfied. Thus, the practical criteria indicate that the loop should be stable. As explained in Section 13.2, additional stability requirements must be added to these when the loop transfer function G has poorly damped left-half-plane poles and zeros.

13.1.2 Routh Stability Criterion

The Routh stability criterion allows one to determine from the coefficients of a transfer function whether or not a system is stable. The loop transfer function G and the feedback transfer function G_{ib} can be expressed as

$$G = \frac{N[s]}{D[s]} \tag{13.1-9}$$

$$G_{ib} = \frac{G}{1+G} = \frac{N[s]}{N[s]+D[s]}. \tag{13.1-10}$$

where $N[s]$, $D[s]$ are the numerator and denominator polynomials of G. The closed-loop poles are calculated by setting the sum polynomial ($N[s] + D[s]$) equal to zero:

$$N[s] + D[s] = 0 \tag{13.1-11}$$

For the loop to be stable, the roots of this characteristic equation must be in the left half plane. Let us designate the coefficients of the characteristic equation as follows:

$$N[s] + D[s] = a_0 s^n + a_1 s^{n-1} + \cdots + a_{n-1}s + a_n = 0 \tag{13.1-12}$$

An essential requirement for stability is that all of the coefficients a_n of the polynomials of Eq 13.1-11 must be nonzero and of the same sign, which normally means they must all be positive.

Satisfying the preceding requirement does not assure stability, but this can be achieved by supplementing it with the Routh stability criterion. To apply the Routh criterion, construct the following Routh array, consisting of the a_m coefficients of the characteristic equation defined in Eq 13.1-12, plus a series of coefficients designated b_m, c_m, d_m, etc., which are derived from the a_m coefficients:

$$
\begin{matrix}
a_0 & a_2 & a_4 & a_6 & \cdot & \cdot & \cdot \\
a_1 & a_3 & a_5 & a_7 & \cdot & \cdot & \cdot \\
b_1 & b_3 & b_5 & & \cdot & \cdot & \cdot \\
c_1 & c_3 & c_5 & & \cdot & \cdot & \cdot \\
d_1 & d_3 & & \cdot & \cdot & \cdot & \cdot \\
e_1 & e_3 & & \cdot & \cdot & \cdot & \cdot \\
f_1 & & \cdot & \cdot & \cdot & \cdot & \cdot \\
g_1 & & \cdot & \cdot & \cdot & \cdot & \cdot
\end{matrix}
\qquad (13.1\text{-}13)
$$

The first row gives the even a_m coefficients and the second row gives the odd a_m coefficients. The other coefficients are derived using the determinant formulas given below. (The rules for computing determinant values are summarized in Appendix F.)

$$
b_1 = -\frac{1}{a_1}\det\begin{bmatrix} a_0 & a_2 \\ a_1 & a_3 \end{bmatrix} = \frac{a_1 a_2 - a_0 a_3}{a_1} \qquad (13.1\text{-}14)
$$

$$
b_3 = -\frac{1}{a_1}\det\begin{bmatrix} a_0 & a_4 \\ a_1 & a_5 \end{bmatrix} = \frac{a_1 a_4 - a_0 a_5}{a_1} \qquad (13.1\text{-}15)
$$

$$
b_5 = -\frac{1}{a_1}\det\begin{bmatrix} a_0 & a_6 \\ a_1 & a_7 \end{bmatrix} = \frac{a_1 a_6 - a_0 a_7}{a_1} \qquad (13.1\text{-}16)
$$

$$
c_1 = -\frac{1}{b_1}\det\begin{bmatrix} a_1 & a_3 \\ b_1 & b_3 \end{bmatrix} = \frac{b_1 a_3 - a_1 b_3}{b_1} \qquad (13.1\text{-}17)
$$

$$
c_3 = -\frac{1}{b_1}\det\begin{bmatrix} a_1 & a_5 \\ b_1 & b_5 \end{bmatrix} = \frac{b_1 a_5 - a_1 b_5}{b_1} \qquad (13.1\text{-}18)
$$

$$
c_5 = -\frac{1}{b_1}\det\begin{bmatrix} a_1 & a_7 \\ b_1 & b_7 \end{bmatrix} = \frac{b_1 a_7 - a_1 b_7}{b_1} \qquad (13.1\text{-}19)
$$

$$
d_1 = -\frac{1}{c_1}\det\begin{bmatrix} b_1 & b_3 \\ c_1 & c_3 \end{bmatrix} = \frac{c_1 b_3 - b_1 c_3}{c_1} \qquad (13.1\text{-}20)
$$

$$
d_3 = -\frac{1}{c_1}\det\begin{bmatrix} b_1 & b_5 \\ c_1 & c_5 \end{bmatrix} = \frac{c_1 b_5 - b_1 c_5}{c_1} \qquad (13.1\text{-}21)
$$

$$
e_1 = -\frac{1}{d_1}\det\begin{bmatrix} c_1 & c_3 \\ d_1 & d_3 \end{bmatrix} = \frac{d_1 c_3 - c_1 d_3}{d_1} \qquad (13.1\text{-}22)
$$

These calculations are continued until the coefficients are all zeros. For example, if there are seven coefficients in the characteristic equation (i.e., if a_0 to a_7 are nonzero), then there are three nonzero b_m coefficients (b_1, b_3, b_5), three nonzero c_m coefficients (c_1, c_3, c_5), two nonzero d_m coefficients (d_1, d_3), two nonzero e_m coefficients (e_1, e_3), one nonzero f_m coefficient (f_1), and one nonzero g_m coefficient (g_1).

The necessary and sufficient requirement for stability is that all of the coefficients in the first column of the Routh array of Eq 13.1-13 ($a_0, a_1, b_1, c_1, d_1, e_1, f_1,$ and g_1) must be positive. When this requirement is not satisfied, the system is unstable, and the number of changes of sign in this column is equal to the number of right-half-plane poles.

Examples illustrating the application of the Routh criterion are given by Chestnut and Mayer [1.4] (pp. 134–137). The Routh criterion has been extended by Hurwitz, who showed that the Routh criterion is equivalent to requiring that a series of Hurwitz determinants be positive. The Hurwitz extension of the Routh criterion is described by Kuo [13.2] (pp. 285–293).

13.2   CALCULATION OF CLOSED-LOOP POLES

13.2.1   Summary of Methods

This section presents two methods for calculating the closed-loop poles of a feedback-control loop, which are much simpler to implement than the root-locus method, commonly used for that purpose. (See Section 13.4.) On the other hand, it is rare that the exact values of the closed-loop poles are actually needed in a design; approximate values are generally adequate. In Section 13.3, the approaches of this section are extended to allow one to estimate the closed-loop poles by inspecting the open-loop magnitude asymptote plot. Most of the material in this section is a condensation of a 1953 paper by the author [1.11].

The discussion in Sections 13.2 and 13.3 justifies the approximations of G_{ib} and G_{ie} developed in Section 3.3, which are based on the following approximations of the closed-loop poles:

1. The zeros of G at which $\text{Asm}|G| > 1$.
2. The poles of G at which $\text{Asm}|G| < 1$.
3. The pole $s = -\omega_c$.

These approximate poles were first presented by Harris, Kirby, and Von Arx [1.10]. If the loop has a low-frequency region where the loop gain is less than unity, there is also an approximate pole given by:

4. The pole $s = -\omega_r$, where ω_r is the gain riseover frequency.

The closed-loop poles occur where $G[s] = -1$, which corresponds to the conditions $|G| = 1$, $\text{Ang}[G] = \pm 180°$. The closed-loop poles are determined by searching over the s-plane to find those values of the complex variable s at which these conditions are satisfied. Two approaches for doing this are presented. The first is a generalized frequency-response method, in which the magnitude and phase of G are plotted along different radial axes of the s-plane. The second approach is an analytical method using an iteration procedure, which starts at an approximate value of a closed-loop pole, and calculates the shift of the exact closed-loop pole from the approximate value.

13.2.2   Generalized Frequency-Response Method of Kusters and Moore

13.2.2.1   Loop Transfer Functions with Real Poles and Zeros.
As shown in the s-plane plot of Fig 13.2-1a, the frequency response of a transfer function $G[s]$ is a plot of $G[s]$ for values of s along the $j\omega$ axis. This concept can be generalized to develop plots of $G[s]$ along various radial axes in the s-plane, such as axes Ⓐ and Ⓑ. In these generalized frequency-response plots, the frequency variable ω is replaced by $|s|$, which is the distance from the origin of the s-plane to a particular point on the radial axis. The following is an extension of an approach first developed by Kusters and Moore [13.3], who derived frequency-response plots along the negative real axis of the s-plane.

Let us first consider the magnitude plot along one of these radial axes. The magnitude asymptote plots (versus $|s|$) are the same for all radial axes of the s-plane, but the magnitude deviations differ. Figure 13.2-2 gives a family of general plots of the magnitude deviation for a zero at $s = s_k$, as a function of the relative frequency $|s|/|s_k|$. The parameter ζ_x is the cosine of the angle between the radial s-plane axis of the plot and the vector from the origin to the zero at $s = s_k$. This is illustrated in Fig 13.2-1b. The deviation for a pole at $s = s_k$ is obtained by inverting the plot for a zero. These magnitude deviations of Fig 13.2-2 are plots of the following, which were derived by the author in Ref [1.11]:

$$\text{mag. dev.} = \sqrt{1 + u^2 + 2\zeta_x u} \qquad \text{for} \quad u < 1 \qquad (13.2\text{-}1)$$

$$\text{mag. dev.} = \sqrt{1 + \frac{1}{u^2} + \frac{2\zeta_x}{u}} \qquad \text{for} \quad u > 1 \qquad (13.2\text{-}2)$$

$$u = \frac{|s|}{|s_k|} \qquad (13.2\text{-}3)$$

Let us apply this to the following loop transfer function:

$$G = \frac{\omega_c(1 + \omega_2/s)(1 + \omega_3/s)}{s(1 + \omega_1/s)^2(1 + s/\omega_4)^2} \qquad (13.2\text{-}4)$$

where $\omega_1 = 0.04$, $\omega_2 = 0.2$, $\omega_3 = 1.0$, $\omega_c = 4$, and $\omega_4 = 16$, all with units of $\sec^{-1}$. A magnitude asymptote plot of this is shown in Fig 13.2-3a. Diagram b

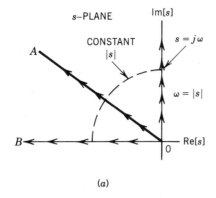

(a)

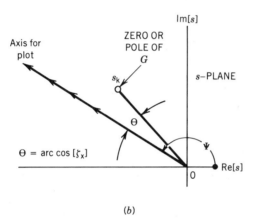

(b)

Figure 13.2-1   Definition of coordinates for generalized frequency-response plot: (a) radial s-plane axes; (b) definition of the angles Θ and Ψ.

is an s-plane plot of the open-loop poles and zeros, which all lie on the negative real s-axis.

Figure 13.2-4 shows the plot of $|G|$ along the negative real axis of the s-plane. Since all of the poles and zeros of G lie on the negative real axis, the value for ζ_x is unity for all of them. Hence the deviation plot in Fig 13.2-2 for $\zeta_x = 1$ is used for the zeros of G, and is inverted for the poles. These deviation plots are shown as dashed curves in Fig 13.3-4. Since there are double-order poles at $|s| = \omega_1, \omega_4$, the dashed curves for these poles are multiplied by 2. The dashed curves (in decilogs) of Fig 13.2-4 are added to the broken-line magnitude asymptote plot, to obtain the exact magnitude plot shown by the solid curve.

Now consider the phase plots. It is convenient to express the phase as the sum of a phase asymptote plus a phase deviation. When all poles and zeros of G are in the left half of the s-plane, the phase asymptote is directly related to the slope of the magnitude asymptote. In a frequency region where the

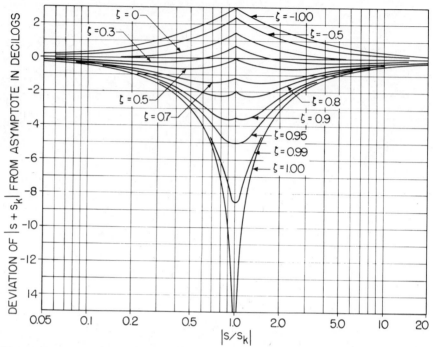

Figure 13.2-2 Deviation of magnitude of $(s - s_k)$ from asymptote, along s-plane axis at angle $\arccos[\zeta]$ relative to s_k. Reprinted with permission from Ref [13.4] © 1956 AIEE (now IEEE).

magnitude asymptote slope is $+n$, the phase asymptote is

$$\text{phase asymptote} = +n\Psi \qquad (13.2\text{-}5)$$

The angle Ψ, which is defined in Fig 13.2-1b, is the angle of the s-plane axis along with the function is plotted.

Figure 13.2-5 is a family of phase deviation curves for a zero at $s = s_k$. Just as for the magnitude deviation curves, these are shown for various values of ζ_x, which is the cosine of the angle between the s-plane axis of the plot and the radial vector to the zero at $s = s_k$. The phase deviation is either added to, or subtracted from, the phase asymptote. If the s-plane axis along which G is being plotted is clockwise relative to a zero at s_k, then the deviation is positive for $|s| < |s_k|$, and is negative for $|s| > |s_k|$. The reverse holds if the s-plane axis is counterclockwise from the zero at s_k. The phase deviation for a pole at s_k is the negative of that for a zero. The plot of the total phase (phase deviation plus phase asymptote) is a continuous curve, except at points that fall exactly on a pole or zero of G. The equations for the phase deviation plots, which

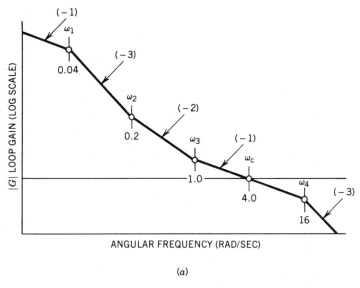

(a)

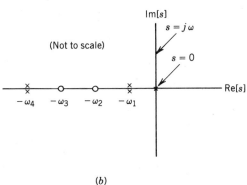

(b)

Figure 13.2-3   Graphical representation of G for feedback-loop example (Eq 13.2-4): (a) magnitude asymptote plot; (b) s-plane plot of open-loop poles and zeros (distorted scale).

were derived by the author in Ref [1.11], are

$$\text{phase dev.} = \arctan\left[\frac{u\sqrt{1 - \zeta_x^2}}{1 - \zeta_x u}\right] \qquad \text{for} \quad u < 1 \qquad (13.2\text{-}6)$$

$$\text{phase dev.} = \arctan\left[\frac{(1/u)\sqrt{1 - \zeta_x^2}}{1 - (\zeta_x/u)}\right] \qquad \text{for} \quad u > 1 \qquad (13.2\text{-}7)$$

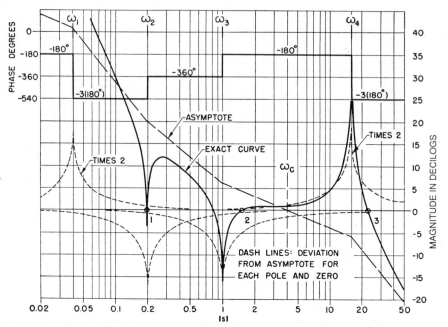

Figure 13.2-4 Magnitude plot of G for feedback-loop example (Eq 13.2-4) along negative real axis of s-plane. Reprinted with permission from Ref [1.11] © 1953 AIEE (now IEEE).

where $u = |s|/|s_k|$, as was shown in Eq 13.2-3. There is no phase deviation for $\zeta_x = 1$.

For a plot along the negative real axis, the phase deviation is always zero, and so the phase is equal to the phase asymptote. This phase plot is shown in Fig 13.2-4 for the example. By Eq 13.2-5, the phase is equal to $+180n$ deg, where n is the magnitude asymptote slope. Below ω_1, the magnitude-asymptote slope is -1, and so the phase is $-180°$; between ω_1 and ω_2, the slope is -3, and so the phase is $-3(180°) = -540°$; between ω_2 and ω_3, the slope is -2, and so the phase is $-2(180°) = -360°$; etc.

A closed-loop pole occurs where $G = -1$, which corresponds to $|G| = 1$ and Ang$[G] = \pm180°$. For the example, the phase is $-180°$ or odd multiples of $-180°$, for all frequencies except those in the ω_1–ω_2 range. Hence, any point where $|G| = 1$ is a closed-loop pole, except in the ω_1–ω_2 range. This condition is satisfied at three points, which are labeled 1, 2, 3. Note that point 1 is slightly less than ω_2. The values of $|s|$ at these three points are as follows, expressed to the accuracy readable from the plot:

$$|s_k| = 0.199, 1.5, 23 \qquad (13.2\text{-}8)$$

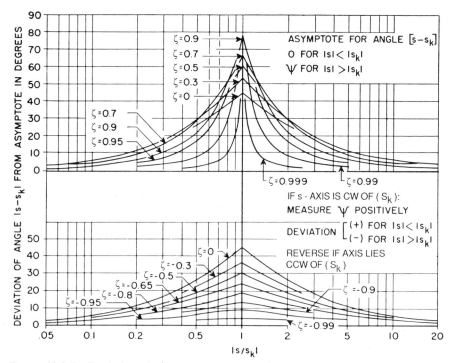

Figure 13.2-5   Deviation of phase $(s - s_k)$ from phase asymptote, along s-plane axis at angle $\arccos[\zeta]$ relative to s_k. Reprinted with permission from Ref [1.11] © 1953 AIEE (now IEEE).

The corresponding closed-loop poles are

$$s_k = -0.199, -1.5, -23 \qquad (13.2\text{-}9)$$

The approximate closed-loop poles (described in Section 13.2.1) are as follows for the example:

1. Zeros of $G[s]$ where $\text{Asm}|G| > 1$:

$$s = -\omega_2 = -0.2 \qquad (13.2\text{-}10)$$

$$s = -\omega_3 = -1.0 \qquad (13.2\text{-}11)$$

2. Poles of $G[s]$ where $\text{Asm}|G| < 1$:

$$s = -\omega_4 = -16 \quad \text{(a double-order pole)} \qquad (13.2\text{-}12)$$

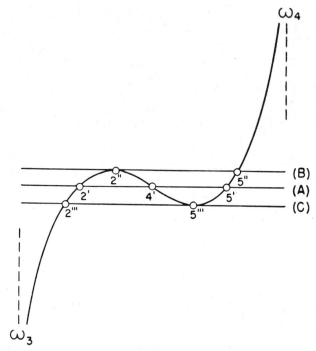

Figure 13.2-6 Magnitude plot of G along negative real axis of s-plane for the feedback-loop example. Reprinted with permission from Ref [1.11] © 1953 AIEE (now IEEE).

3. The pole $-\omega_c$:

$$s = -\omega_c = -4 \qquad (13.2\text{-}13)$$

The approximate pole at $s = -0.2$ is shifted slightly downward in frequency to $s = -0.199$; the approximate pole at $s = -1.0$ is shifted upward in frequency to -1.5; and one pole of the approximate pole pair at $s = -16$ is shifted upward in frequency to -23. The other approximate pole at $s = -16$ is shifted downward in frequency, and is combined with the approximate pole at $s = -4$ to form a complex pair of closed-loop poles.

If the value of ω_4 in the loop transfer function G is increased somewhat, with the other parameters kept constant, the magnitude plot of Fig 13.2-4 between ω_3 and ω_4 takes on the form shown in Fig 13.2-6. If the gain is set so that 0 dg occurs at axis Ⓐ, the magnitude plot intersects the 0-dg axis at three points (2', 4', and 5') between ω_3 and ω_4, rather than one point (2), as it does in Fig 13.3-4. Hence all five closed-loop poles lie on the negative real axis.

If the gain is then decreased, so that the 0-dg axis in Fig 13.2-5 moves from Ⓐ to Ⓑ, the closed loop pole at 5' moves to 5'', and the closed poles at 2'

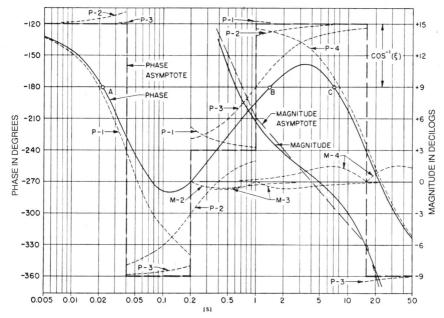

Figure 13.2-7 Magnitude and phase plots of G along upper-half-plane axis corresponding to $\zeta = 0.5$ for feedback-loop example. Reprinted with permission from Ref [1.11] © 1953 AIEE (now IEEE).

and 4′ move together to form a double-order pole at 2″. If the gain is decreased further, this double-order pole at 2″ forms a complex pole pair. Similarly, if the gain is increased, so that the 0-dg axis moves from Ⓐ to Ⓒ, the real closed-loop poles at 4′ and 5′ move together to form a double-order pole at 5‴. If the gain is increased further, this double-order pole becomes a complex pole pair.

Thus, in a region on the negative real axis where the phase is ±180°, a simple intersection of the 0-dg axis by $|G|$ indicates a single-order pole, while a tangency at a maximum or minimum point indicates a double-order pole. An intersection at a point of inflection of G indicates a triple-order pole. Such an intersection would occur in Fig 13.2-4 if the gain were increased by 1.0 dg.

When the poles are complex, they can be calculated graphically by plotting G along other s-plane axes. Figure 13.2-7 shows the magnitude and phase plots of G along the s-plane axes defined by $\Psi = 120°$. This axis is 60° clockwise from the negative real axis, and corresponds to the following effective damping ratio:

$$\zeta = \cos[60°] = 0.5 \qquad (13.2\text{-}14)$$

The phase asymptote along this axis is given by

$$\text{phase asymptote} = 120n \text{ deg} \qquad (13.2\text{-}15)$$

where n is the slope of the magnitude asymptote. The phase asymptote values are as follows: below ω_1, $-1(120°) = -120°$; $\omega_1-\omega_2$, $-3(120°) = -360°$; $\omega_2-\omega_3$, $-2(120°) = -240°$; $\omega_3-\omega_4$, $-1(120°) = -120°$; and above ω_4, $-3(120°) = -360°$. These phase asymptotes are shown by the broken lines in the figure.

The magnitude asymptote in Fig 13.2-7 is also shown by a broken line. This is the same as the magnitude-asymptote plot of Fig 13.2-4, but the magnitude scale is expanded. This shows the magnitude plot in the frequency region of interest.

The dashed curves in Fig 13.2-7 labeled P-1, P-2, P-3, and P-4 are the deviations from the phase asymptotes corresponding to the poles and zeros at $-\omega_1$, $-\omega_2$, $-\omega_3$, and $-\omega_4$ respectively. Similarly, the dashed curves labeled M-2, M-3, and M-4 are the deviations from the magnitude asymptotes corresponding to the poles and zeros at $-\omega_2$, $-\omega_3$, and $-\omega_4$ respectively. Curves P-2, P-3, M-2, and M-3 were obtained directly from Figs 13.2-5, -2 for the curves corresponding to $\zeta = 0.5$. Curves P-1, P-4, and M-4 were obtained by doubling these curves of Figs 13.2-2, -5, because the open-loop poles at $-\omega_1$ and $-\omega_4$ are of double order. The solid curves, which are the exact magnitude and phase curves, were obtained by graphically adding the dashed curves to the asymptotes.

The phase deviations in Fig 13.2-7 are derived from the phase-deviation plot for $\zeta = 0.5$ in Fig 13.2-5. The axis is clockwise from the poles and zeros. Hence for a zero, the deviation is positive for $|s| < |s_k|$ and negative for $|s| > |s_k|$; while the reverse holds for a pole. Phase deviation P-1 is for a pole of break frequency 0.04, and so is negative for $|s| < 0.04$ and is positive for $|s| > 0.04$. Since this is a double order pole, P-1 is twice the value of the plot in Fig 13.2-5. Phase deviation P-2 is for a zero of break frequency $\omega_2 = 0.2$, and so is positive for $|s| < 0.2$ and negative for $|s| > 0.2$. Phase deviation P-3 is for a zero of break frequency $\omega_3 = 1$, and so is negative for $|s| < 1$ and positive for $|s| > 1$. Phase deviation P-4 is for a double-order pole of break frequency $\omega_4 = 16$, and so is negative for $|s| < 16$ and positive for $|s| > 16$, with an amplitude twice that of the plot in Fig 13.2-5.

The phase curve intersects the 180° line at three points, labeled A, B, and C. The value of the magnitude at frequency B is 4 dg, and at frequency C it is -1.25 dg. Hence, if the gain were decreased by 4 dg, a closed-loop pole would occur at B with $\zeta = 0.5$ and (since $|s|$ at B is 1.4) $\omega_n = 1.4$. If the gain were increased by 1.25 dg, a closed-loop pole would occur at frequency C with $\zeta = 0.5$ and $\omega_n = 7.2$. The point A corresponds to an extremely low value of gain, and so is of no interest.

In Fig 13.2-8, curves of the magnitude and phase of G are drawn along s-plane axes corresponding to the following values of ζ: 1.0, 0.9, 0.8, 0.7, 0.5, and 0. An intersection of a phase curve for a given ζ with the 180° axis determines a possible closed-loop pole (for the proper value of gain). The reciprocal of $|G|$ at this frequency is the increase of gain needed to produce this closed-loop pole. The natural frequency ω_n of the pole is equal to the

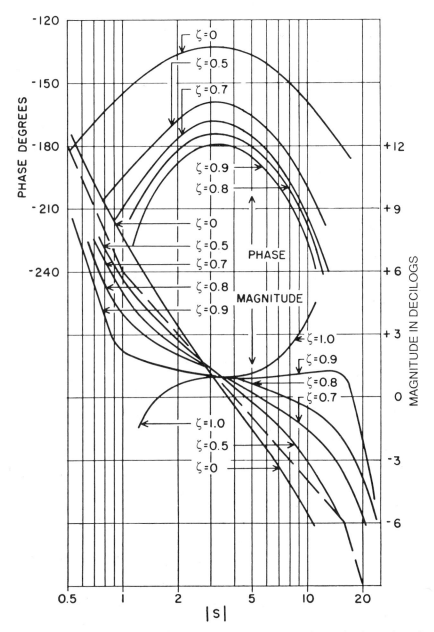

Figure 13.2-8 Magnitude and phase plots of G along several axes in the s-plane for the feedback-loop example. Reprinted with permission from Ref [1.11] © 1953 AIEE (now IEEE).

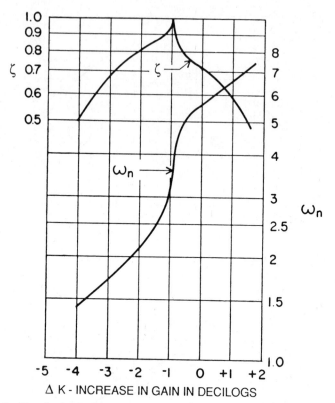

Figure 13.2-9 For the feedback-loop example, the effect of variation of loop gain on the values of ζ and ω_n for the complex closed-loop poles. Reprinted with permission from Ref [1.11] © 1953 AIEE (now IEEE).

value of $|s|$ at the intersection, and the damping ratio of the pole is equal to the value of ζ for the s-plane axis. These values of ζ and ω_n are plotted in Fig 13.2-9 as functions of the required increase in gain, ΔK. The values of these curves for $\Delta K = 0$ dg (the present gain setting) are $\zeta = 0.72$ and $\omega_n = 5.6$. These are the actual parameters of the complex pole pair for the example.

13.2.2.2 Underdamped Transfer Functions. The loop transfer function of the example has only real poles and zeros. To illustrate generalized s-plane plots of underdamped transfer functions, Fig 13.2-10 shows magnitude and phase plots of an underdamped factor along the s-plane axis for $\zeta_x = 0.5$. The factor is

$$H[s] = \frac{1}{s^2 + 2\zeta\omega_n s + \omega_n^2} \tag{13.2-16}$$

$$\zeta = 0.3, \qquad \omega_n = 2.0 \tag{13.2-17}$$

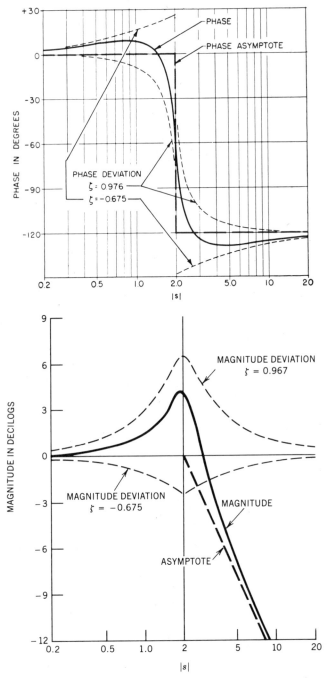

Figure 13.2-10 Magnitude and phase plots of $1/(s^2 + 2\zeta\omega_n s + \omega_n^2)$ for $\omega_n = 2$, $\zeta = 0.3$, along upper-half-plane axis corresponding to $\zeta_x = 0.5$. Reprinted with permission from Ref [1.11] © 1953 AIEE (now IEEE).

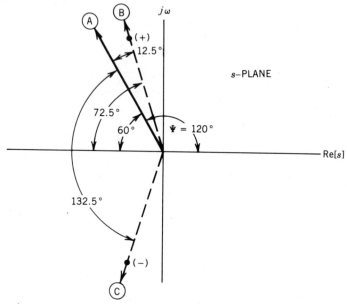

Figure 13.2-11   The axis (A) along which transfer function is to be plotted, and the location of the poles.

In Figure 13.2-11 Ⓐ is the axis along which H is to be plotted, and Ⓑ and Ⓒ are the lines on which the complex open-loop poles lie. Since arccos[0.3] = 72.5°, lines Ⓑ, Ⓒ are 72.5° from the negative real axis, while Ⓐ is located at the angle arccos[0.5] = 60° from the negative real axis. The angle Ψ of this axis is 120°.

The magnitude asymptote of Fig 13.2-10 has a slope of zero for $|s| < 2.0$, and a slope of -2 for $|s| > 2.0$. The phase asymptote is therefore 0° for $|s| < 2.0$, and $-2(120°) = -240°$ for $|s| > 2.0$.

In Fig 13.2-11, the angle between axes Ⓐ and Ⓒ is 12.5°, which corresponds to the damping ratio

$$\zeta = \cos[12.5°] = 0.976 \qquad (13.2\text{-}18)$$

The angle between axes Ⓐ and Ⓒ is 132.5°, which corresponds to the damping ratio

$$\zeta = \cos[132.5°] = -0.676 \qquad (13.2\text{-}19)$$

Therefore, the magnitude and phase deviations for $\zeta = 0.976$ and $\zeta = -0.676$ are desired. These are obtained from Figs 13.2-2 and 13.2-5 by interpolating between the curves. The magnitude deviations obtained from Fig 13.2-2 are inverted to obtain the deviations for the poles. Since the axis Ⓐ is clockwise from the upper-half-plane *pole* (for $\zeta = 0.976$), the phase deviation for $\zeta = 0.976$ is negative for $|s| < 2$, and is positive for $|s| > 2$. Since the axis is counterclockwise from the lower-half-plane *pole* (for $\zeta = -0.675$), the phase deviation for $\zeta = -0.675$ is positive for $|z| < 2$, and is negative for $|s| > 2$.

In Fig 13.2-10, these magnitude and phase deviations are shown as dashed lines. The magnitude and phase deviations are added to the magnitude and phase asymptote plots to obtain the total magnitude and phase plots, shown by the solid curves.

13.2.2.3 Loop Transfer Functions With Right-Half-Plane Poles and Zeros.
When the transfer function G has a pole or zero in the right half plane, the phase asymptote plot cannot be determined from the slope of the magnitude plot, and so an alternative method is needed to obtain the phase asymptote plot. To explain this, consider the following loop transfer function:

$$G = \frac{\omega_c(1 + \omega_2/s)(1 - s/\omega_4)}{s(1 - \omega_1/s)(1 + s/\omega_3)} \qquad (13.2\text{-}20)$$

The closed loop is generally stable because: (1) it approximates ω_c/s at gain crossover; (2) the right-half-plane pole ($s = \omega_1$) occurs at low frequencies where $\text{Asm}|G| > 1$; and (3) the right-half-plane zero ($s = \omega_4$) occurs at high frequencies where $\text{Asm}|G| < 1$. To plot G, express all factors in the form $(s + s_k)$ as follows:

$$G = -\frac{(\omega_c\omega_3/\omega_4)(s + \omega_2)(s - \omega_4)}{s(s - \omega_1)(s + \omega_3)} \qquad (13.2\text{-}21)$$

Figure 13.3-12 is an s-plane plot which shows the vectors corresponding to the factors $(s + \omega_1)$ and $(s - \omega_2)$ at two values of s (at s_1, s_2), along a radial s-plane axis Ⓐ at the angle Ψ. At s_1, which is close to the origin, the vector for the left-half-plane zero factor $(s + \omega_1)$ is at an angle that is close to zero, while the vector for the right-half-plane zero factor $(s - \omega_2)$ is at an angle that is close to 180°. As the point s moves along axis Ⓐ to s_2 and beyond, the angles of the vectors $(s + \omega_1)$ and $(s - \omega_2)$ both approach the angle Ψ of the s-plane axis Ⓐ.

This shows that the phase for a left-half-plane zero $(s + \omega_1)$ varies from 0 to Ψ as $|s|$ varies from zero to infinity, while the phase for a right-half-plane zero $(s - \omega_2)$ varies from 180° to Ψ as $|s|$ varies from zero to infinity. Therefore, the phase asymptotes for a zero factor of the form $(s + s_k)$ are as follows:

Left-half-plane zero:

$$\text{phase asymptote} = 0 \qquad \text{for} \quad |s| < |s_k| \qquad (13.2\text{-}22)$$

$$\text{phase asymptote} = \Psi \qquad \text{for} \quad |s| > |s_k| \qquad (13.2\text{-}23)$$

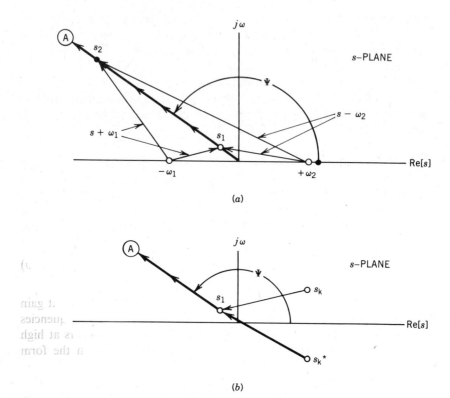

(a)

(b)

Figure 13.2-12 s-plane plots illustrating the phase asymptotes.

Right-half-plane zero:

$$\text{phase asymptote} = 180° \qquad \text{for} \quad |s| < |s_k| \qquad (13.2\text{-}24)$$

$$\text{phase asymptote} = \Psi \qquad \text{for} \quad |s| > |s_k| \qquad (13.2\text{-}25)$$

For a pole, the phase asymptote is the negative of that for a zero.

Figure 13.2-12b shows the vectors for two complex right-half-plane zeros at s_k and at s_k^*, for the value s_1 on axis Ⓐ. The average phase of these two vectors is close to 180° at s_1, and it varies from 180° to Ψ as $|s|$ varies from zero to infinity along axis Ⓐ. Hence, the phase asymptotes for each of these zeros is the same as is given in Eqs 13.2-24, -25. The total phase asymptote for the two zeros is 360° for $|s| < |s_k|$ and 2Ψ for $|s| > |s_k|$. Thus the rules for phase asymptotes in Eqs 13.2-24, -25 apply for each zero (or pole) in the right half plane, regardless of whether it is real or complex.

The phase asymptotes for the individual factors of G are plotted in Fig 13.2-13a for an s-plane axis at the angle Ψ. Curve ⓪ is the phase asymptote for the factor $1/s$, a pole at the origin. This has a phase of $-\Psi$ at all

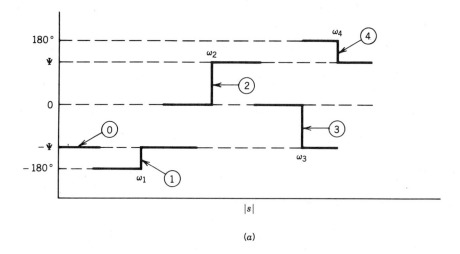

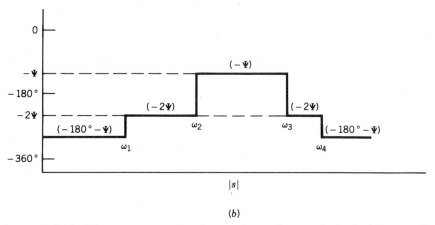

Figure 13.2-13 Phase-asymptote plot along s-plane axis at angle Ψ, for loop transfer function with right-half-plane pole and zero (Eq 13.2-21): (*a*) phase asymptotes for individual factors; (*b*) total phase-asymptote plot.

frequencies. Curves ① is the phase asymptote for the factor $1/(s - \omega_1)$, a right-half-plane pole at $s = \omega_1$. This has a phase of $-180°$ for $|s| < \omega_1$, and a phase of $-\Psi$ for $|s| > \omega_1$. Curve ② is the phase asymptote for the factor $(s + \omega_2)$, a left-half-plane zero at $s = -\omega_2$. This has zero phase for $|s| < \omega_2$ and a phase of $+\Psi$ for $|s| > \omega_2$. Curve ③ is the phase asymptote for the factor $1/(s + \omega_3)$, a left-half-plane pole at $s = -\omega_3$. This has zero phase for $|s| < \omega_3$ and a phase of $-\Psi$ for $|s| > \omega_3$. Curve ④ is the phase asymptote for the factor $(s - \omega_4)$, a right-half-plane zero at $s = \omega_4$. This has a phase of $+180°$ for $|s| < \omega_4$, and a phase of $+\Psi$ for $|s| > \omega_4$.

At zero frequency, the sum of the phase plots of diagram a of Fig 13.2-13 is equal to $-\Psi$. However, there is an additional $\pm 180°$ of phase shift due to the $(-)$ sign of the loop transfer function of Eq 13.2-21. This additional phase shift is assumed to be $-180°$, and so the total zero-frequency phase shift is $-180° - \Psi$, as shown in diagram b. The total phase plot of diagram b is obtained by adding $-180°$ to the sum of the individual phase plots of diagram a.

Figure 13.2-13b gives the phase asymptote plot for the loop transfer function of Eq 13.2-21 along any radial axis of the s-plane, where Ψ is the angle of that s-plane axis. The phase deviations for this transfer function with right-half-plane poles and zeros are constructed in the same manner described previously.

13.2.3   Iterative Method for Computing Closed-Loop Poles

13.2.3.1   *General Approximation of Closed-Loop Poles.* The generalized frequency-response plots show that the magnitude of G is approximately equal to the magnitude asymptote of G throughout the s-plane, except near the poles and zeros of G. Close to a zero, the value of $|G|$ is much less than the asymptote; close to a pole, the value of $|G|$ is much greater than the asymptote.

This indicates that the condition $|G| = 1$ occurs close to any zero of G for which Asm$|G| \gg 1$, and it occurs close to any pole of G for which Asm$|G| \ll 1$. Along each of the contours of $|G| = 1$, close to such a pole or zero, there is one point where the phase of G is $-180°$ or an odd multiple of $-180°$, and so that point is the location of the closed-loop pole. Hence, there is a closed-loop pole close to each zero of G where Asm$|G| \gg 1$, and close to each pole of G at which Asm$|G| \ll 1$.

This concept can be extended as follows. There is a closed-loop pole associated with each zero of G at which Asm$|G| > 1$, and one associated with each pole of G at which Asm$|G| < 1$. When Asm$|G|$ is close to unity at the corresponding pole of zero of G, the shift of the closed-loop pole from the open-loop zero or pole can be quite large, and may shift the closed-loop pole away from the negative real axis. Nevertheless, as long as the loop has good stability, it is still reasonable to associate the resultant closed-loop pole with the corresponding open-loop pole or zero.

Thus, for a loop with good stability, there are closed-loop poles approximately equal to each zero of G where Asm$|G| > 1$, and approximately equal to each pole of G where Asm$|G| < 1$. Besides these poles, there is an additional pole associated with the magnitude asymptote itself. In the frequency region near gain crossover, the loop transfer function for a loop with good stability can be approximated roughly as

$$G \doteq \omega_c/s \qquad\qquad (13.2\text{-}26)$$

This transfer function has a closed-loop pole where

$$s = -\omega_c \tag{13.3-27}$$

For a loop with loop gain less than unity at low frequency, there is a second frequency region where the asymptote of G is close to unity, in the vicinity of the gain riseover. In this region, G can be approximated as

$$G = s/\omega_r \tag{13.3-28}$$

This transfer function has the following closed-loop pole:

$$s = -\omega_r \tag{13.2-29}$$

Hence, the approximate closed loop poles of a feedback loop having good stability are:

1. The zeros of G for which $\mathrm{Asm}|G| > 1$.
2. The poles of G for which $\mathrm{Asm}|G| < 1$.
3. The pole $s = -\omega_c$.
4. The pole $s = -\omega_r$, if the loop has less than unity loop gain at low frequencies.

13.2.3.2 Iterative Method. Based on this approximation of the closed-loop poles, the values of most of the closed-loop poles can be readily calculated by an iterative method that computes the shifts of the exact closed-loop poles from the approximate poles. This approach is explained by applying it to the example that was given in Eq 13.2-4.

The magnitude asymptote of G is greater than unity at the break frequencies ω_2, ω_3 of the zeros of G at $s = -\omega_2$, $s = -\omega_3$. Hence these are approximate closed-loop poles. The magnitude asymptote of G is less than unity at the break frequency ω_4 of the double-order pole of G at $s = -\omega_4$, and so there is an approximate double-order closed-loop pole at $s = -\omega_4$. There is also an approximate closed-loop pole at $s = -\omega_c$. Hence, the factors for the approximate closed-loop poles are

$$(s + \omega_2)(s + \omega_3)(s + \omega_c)(s + \omega_4)^2 = 0 \tag{13.2-30}$$

Numerically, this expression is

$$(s + 0.2)(s + 1)(s + 4)(s + 16)^2 = 0 \tag{13.2-31}$$

The loop transfer function of Eq 13.2-4 is expressed as follows when all factors are put in the form $(s + \omega_x)$:

$$
\begin{aligned}
G &= \frac{\left(\omega_c \omega_4^2/\omega_3\right)(s + \omega_2)(s + \omega_3)}{s(s + \omega_1)^2(s + \omega_4)^2} \\[2mm]
&= \frac{1024(s + 0.2)(s + 1)}{s(s + 0.04)^2(s + 16)^2} = \frac{N}{D}
\end{aligned}
\qquad (13.2\text{-}32)
$$

The closed-loop transfer function G_{ib} is

$$
\begin{aligned}
G_{ib} &= \frac{N}{N + D} \\[2mm]
&= \frac{1024(s + 0.2)(s + 1)}{s(s + 0.04)^2(s + 16)^2 + 1024(s + 0.2)(s + 1)}
\end{aligned}
\qquad (13.2\text{-}33)
$$

This closed-loop denominator expression $(N + D)$ is a fifth-order polynomial in s, and so there are five closed-loop poles. This is consistent with the approximate closed-loop pole factors of Eq 13.2-30, which yields five closed-loop poles.

To find the closed-loop poles, the loop transfer function of Eq 13.2-32 is set equal to -1 as follows:

$$
G = \frac{1024(s + 0.2)(s + 1)}{s(s + 0.04)^2(s + 16)^2} = -1
\qquad (13.2\text{-}34)
$$

This expression is solved for each of the zero and pole factors of G that correspond to closed-loop poles. Let us start with the factor $(s + \omega_2) = (s + 0.2)$ for the open-loop zero of break frequency ω_2. Factoring $(s + 0.2)$ from Eq 13.3-34 gives

$$
(s + 0.2) = -\frac{s(s + 0.04)^2(s + 16)^2}{1024(s + 1)}
\qquad (13.2\text{-}35)
$$

The exact pole corresponding to the approximate pole $s = -\omega_2$ is defined as

$$
s = -\omega_2 + \delta_2 = -0.2 + \delta_2
\qquad (13.2\text{-}36)
$$

where δ_2 is the shift of the exact closed-loop pole s_2 from the approximate pole $s = -\omega_2$. Equation 13.2-35 holds at a closed-loop pole, and so holds at

$s = s_2$. Substituting Eq 13.2-36 into Eq 13.3-35 gives

$$\delta_2 = -\left. \frac{s(s + 0.04)^2(s + 16)^2}{1024(s + 1)} \right|_{s=s_2=-0.2+\delta_2} \qquad (13.2\text{-}37)$$

The right-hand expression varies slowly with s in the region near $s = s_2$, and so can be approximated by its value at $s = -0.2$. Setting $s = -0.2$ in the right-hand expression gives the following approximation for δ_2:

$$\delta_2 \doteq -\frac{(-0.2)(-0.2 + 0.04)^2(-0.2 + 16)^2}{1024(-0.2 + 1)} = +0.0016 \quad (13.2\text{-}38)$$

This is substituted into Eq 13.2-36 to obtain the following more accurate approximation for the closed-loop pole:

$$s_2 = -0.2 + \delta_2 \doteq -0.2 + 0.0016 = -0.1984 \qquad (13.2\text{-}39)$$

When this value for s_2 is substituted into the right-hand side of Eq 13.2-37, it yields the more accurate value $\delta_2 = 0.00151$. Substituting this into Eq 13.2-36 gives the following very accurate value for s_2:

$$s_2 = -0.2 + 0.0015 = -0.1985 \qquad (13.2\text{-}40)$$

Now let us calculate the closed-loop pole corresponding to the zero factor of G of break frequency ω_3: $(s + \omega_3) = (s + 1)$. Solving Eq 13.2-34 for this factor gives

$$(s + 1) = -\frac{s(s + 0.04)^2(s + 16)^2}{1024(s + 0.2)} \qquad (13.2\text{-}41)$$

The exact closed-loop pole is designated

$$s_3 = -\omega_3 + \delta_3 = -1 + \delta_3 \qquad (13.2\text{-}42)$$

Substituting Eq 13.2-42 into Eq 13.2-41 gives

$$\delta_3 = -\left. \frac{s(s + 0.04)^2(s + 16)^2}{1024(s + 0.2)} \right|_{s=s_3=-1+\delta_3} \qquad (13.2\text{-}43)$$

To approximate δ_3, compute the right-hand expression at $s = -1$, which gives

$$\delta_3 \doteq -\frac{(-1)(-1 + 0.04)^2(-1 + 16)^2}{1024(-1 + 0.2)} = -0.253 \qquad (13.2\text{-}44)$$

Substituting this into Eq 13.2-42 gives the following closer approximation for the closed-loop pole:

$$s_3 = -\omega_3 + \delta_3 = -1 + \delta_3 \doteq -1 - 0.253 = -1.253 \qquad (13.2\text{-}45)$$

This is substituted for s_3 in Eq 13.2-43 to give a more accurate value for δ_3: $\delta_3 \doteq = -0.372$. By Eq 13.2-42 the more accurate value for s_3 is

$$s_3 = -1 + \delta_3 \doteq -1 - 0.372 = -1.372 \qquad (13.2\text{-}46)$$

This is again substituted for s_3 in Eq 13.2-43 to give $\delta_3 \doteq -0.434$. Substituting this into Eq 13.2-46 gives $s_3 \doteq -1.434$. Repeating the process yields the following successive values for s_3: -1.485, -1.496, -1.503, -1.507, -1.509, and finally -1.511.

For this pole at s_3, the process converges slowly because the closed-loop pole is close to the gain crossover frequency. The reason for this is obvious from the frequency-response plot along the negative real axis of the s-plane given in Fig 13.2-4. This is the pole corresponding to point 2, and the magnitude slope at point 2 is very gradual. If the loop gain were decreased by 1 dg, the closed-loop pole would move to the inflection point at $|s| = 3$, and so could not be calculated by this method.

Now let us consider the closed-loop poles corresponding to the double-order open-loop pole of break frequency $\omega_4 = 16$. Solving Eq 13.2-34 for $(s + \omega_4)^2 = (s + 16)^2$ gives

$$(s + 16)^2 = -\frac{1024(s + 0.2)(s + 1)}{s(s + 0.04)^2} \qquad (13.2\text{-}47)$$

The exact poles corresponding to the approximate pole at $s = -\omega_4 = -16$ are defined by

$$s_4 = -\omega_4 + \delta_4 = -16 + \delta_4 \qquad (13.2\text{-}48)$$

Substituting this into Eq 13.2-47 gives

$$\delta_4^2 = -\frac{1024(s + 0.2)(s + 1)}{s(s + 0.04)^2}\Bigg|_{s=s_4=-\omega_4+\delta_4} \qquad (13.2\text{-}49)$$

Solving this at $s = -\omega_4$ gives the approximation

$$\delta_4^2 \doteq -\frac{1024(-16 + 0.2)(-16 + 1)}{(-16)(-16 + 0.04)^2} = 59.55 \qquad (13.2\text{-}50)$$

Solving for δ_4 gives

$$\delta_4 \doteq \pm\sqrt{59.55} = \pm 7.72 \qquad (13.2\text{-}51)$$

This shows that the approximate double-order closed-loop pole at $s = -\omega_4$ splits in two, with one pole moving downward in frequency and the other moving upward. The pole moving downward in frequency combines with the approximate pole at $s = -\omega_c$ at gain crossover to form a pair of complex poles; hence, it cannot be computed by this method. However, the pole moving away from gain crossover can be calculated easily. Substitute into Eq 13.2-48 the negative value for δ_4 in Eq 13.2-51. The gives the following approximation for the pole moving upward in frequency:

$$s_4 = -\omega_4 + \delta_4 \doteq -16 - 7.72 = -23.72 \qquad (13.2\text{-}52)$$

Substituting this for s_4 in Eq 13.2-49 gives $\delta_4^2 \doteq 41.14$, which yields $\delta_4 \doteq \pm 6.41$. Again, the negative value for δ_4 is used to compute s_4:

$$s_4 \doteq -\omega_4 + \delta_4 = -16 - 6.41 = -22.41 \qquad (13.2\text{-}53)$$

Continuing this process yields the following successive values: $s_4 \doteq -22.6$, -22.56.

The exact values for three of the five closed-loop poles have been computed. The remaining two cannot be obtained by this method because they are too close to gain crossover. However, they can be calculated by dividing the denominator of G_{ib} by the expression containing the poles. The denominator of G_{ib} in Eq 13.2-33 is as follows when expanded:

$$N + D = s(s + 0.04)^2(s + 16)^2 + 1024(s + 0.2)(s + 1)$$

$$= s^5 + 32.08s^4 + 258.6s^3 + 1044.5s^2 + 1229.2s + 204.8 \quad (13.2\text{-}54)$$

The factored expression for the three closed-loop poles that have been calculated is

$$(s + 0.1985)(s + 1.511)(s + 22.56)$$

$$= s^3 + 24.27s^2 + 38.87s + 6.766 \quad (13.2\text{-}55)$$

Dividing Eq 13.2-55 into Eq 13.2-54 by long division yields the quotient

$$(s^2 + 7.81s + 30.2) = (s - 3.91 + j3.87)(s - 3.91 - j3.87) \quad (13.2\text{-}56)$$

with a remainder of $(1.1s^2 + 3.5s + 0.5)$, which can be neglected. Comparing this quadratic with the standard form $(s^2 + 2\zeta\omega_n s + \omega_n^2)$ gives the natural frequency and damping ratio of the complex poles:

$$\omega_n = \sqrt{30.2} = 5.50 \quad (13.2\text{-}57)$$

$$\zeta = \frac{7.81}{2\omega_n} = \frac{7.81}{2(5.50)} = 0.710 \quad (13.2\text{-}58)$$

13.2.3.3 Calculation of Shifts from Complex Approximate Poles.

In the preceeding examples, the approximate closed-loop poles were real. Let us now consider the shifts when the approximate closed-loop poles are complex. The approximate closed-loop poles are the low-frequency zeros of G and the high-frequency poles of G. In practical systems it is rare that low-frequency zeros are complex, but complex high-frequency poles occur frequently.

As an example of a feedback loop with complex high-frequency poles, consider the loop transfer function discussed in Chapter 3, Section 3.1. From Eq 3.1-5, G is

$$
\begin{aligned}
G &= \frac{\omega_c(1 + \omega_2/s)}{s(1 + \omega_1/s)\left[1 + 2\zeta(s/\omega_n) + (s/\omega_n)^2\right]} \\[2mm]
&= \frac{\omega_c\omega_n^2(s + \omega_2)}{s(s + \omega_1)(s^2 + 2\zeta\omega_n s + \omega_n^2)} \\[2mm]
&= \frac{\omega_c\omega_n^2(s + \omega_2)}{s(s + \omega_1)(s + \zeta\omega_n - j\omega_o)(s + \zeta\omega_n + j\omega_o)} \quad (13.2\text{-}59)
\end{aligned}
$$

where $\omega_1 = 4$, $\omega_2 = 30$, $\omega_c = 100$, $\omega_n = 500$ sec^{-1}, and $\zeta = 0.2$. The step response, $|G_{ib}|$ frequency response, and Nichols-chart plot for this loop transfer function were shown in Figs 3.1-4, -5, -6. The parameters $\zeta\omega_n$ and ω_o are

$$\zeta\omega_n = 100 \quad (13.2\text{-}60)$$

$$\omega_o = \omega_n\sqrt{1 - \zeta^2} = 489.9 \quad (13.2\text{-}61)$$

Let us calculate the shift δ of the complex closed-loop pole from the complex open-loop upper-half-plane pole $(s = -\zeta\omega_n + j\omega_o)$. Set G of Eq 13.2-59 equal to -1, and solve for $(s + \zeta\omega_n - j\omega_o)$:

$$(s + \zeta\omega_n - j\omega_o) = -\frac{\omega_c\omega_n^2(s + \omega_2)}{s(s + \omega_1)(s + \zeta\omega_n + j\omega_o)} \qquad (13.2\text{-}62)$$

In Eq 13.2-62, set s equal to s_k, given by

$$s_k = -\zeta\omega_n + j\omega_o + \delta = -100 + j489.9 + \delta \qquad (13.2\text{-}63)$$

This gives

$$\delta = -\frac{\omega_c\omega_n^2(s + \omega_2)}{s(s + \omega_1)(s + \zeta\omega_n + j\omega_o)}\bigg|_{s=s_k}$$

$$= -\frac{2.5 \times 10^7(s + 30)}{s(s + 4)(s + 100 + j489.9)}\bigg|_{s=s_k} \qquad (13.2\text{-}64)$$

In Eq 13.2-63, assume that $\delta = 0$, and substitute the resultant s_k into Eq 13.2-64:

$$\delta \doteq -\frac{2.5 \times 10^7(30 - 100 + j489.9)}{(-100 + j489.9)(4 - 100 + j489.9)(100 + j489.9 - 100 + j489.9)}$$

$$\doteq -\frac{(2.5 \times 10^7)494.9\big/98.1°}{500\big/101.5° \; 499.2\big/101.1° \; 979.8\big/90°}$$

$$\doteq -50.6\big/194.5° = 50.6\big/-14.5° = 49.0 - j12.7 \qquad (13.2\text{-}65)$$

Substitute this into Eq 13.2-63 to obtain the following more accurate approximation for s_k:

$$s_k = -100 + j489.9 + \delta \doteq -51.0 + j477.2 \qquad (13.2\text{-}66)$$

This is substituted into Eq 13.2-64 to obtain $\delta = 53.3 - j5.9$. Equation 13.2-63 in turn yields $s_k = -46.7 + j484.0$. Another cycle is performed to yield the following final value of s_k:

$$s_k = -47.2 + j484.9 = 487.2\big/95.8° \qquad (13.2\text{-}67)$$

The natural frequency and damping ratio of the complex closed-loop pole pair are

$$\omega_n' = 487.2 \qquad (13.2\text{-}68)$$

$$\zeta' = \frac{47.2}{\omega_n} = \frac{47.2}{487.2} = 0.097 \qquad (13.2\text{-}69)$$

Thus, the open-loop underdamped poles of natural frequency $\omega_n = 500$ and damping ratio $\zeta = 0.2$ have been shifted to form underdamped closed-loop poles of natural frequency $\omega'_n = 487.2$ and damping ratio $\zeta' = 0.097$.

An advantage of this method for calculating the closed-loop poles is that it can yield simple approximations. For example, in the frequency region near these closed-loop poles, the loop transfer function in Eq 13.2-59 can be approximated as follows in terms of the asymptote value of $|G|$ at the break frequency ω_n, which is designated $\mathrm{Asm}|G[\omega_n]|$:

$$G \doteq \frac{\mathrm{Asm}|G[\omega_n]|}{(s/\omega_n)\left[(s^2 + 2\zeta\omega_n s + \omega_n^2)/\omega_n^2\right]}$$

$$\doteq \frac{\omega_n^3 \mathrm{Asm}|G[\omega_n]|}{s(s + \zeta\omega_n - j\omega_o)(s + \zeta\omega_n + j\omega_o)} \qquad (13.2\text{-}70)$$

Set $G = -1$, solve for $(s + \zeta\omega_n - j\omega_o)$, and set s equal to

$$s_k = -\zeta\omega_n + j\omega_o + \delta \qquad (13.2\text{-}71)$$

This gives

$$\delta \doteq - \left.\frac{\omega_n^3 \mathrm{Asm}|G[\omega_n]|}{s(s + \zeta\omega_n + j\omega_o)}\right|_{s = -\zeta\omega_n + j\omega_o + \delta} \qquad (13.2\text{-}72)$$

Setting $\delta = 0$ in the right-hand expression gives

$$\delta \doteq - \frac{\omega_n^3 \mathrm{Asm}|G[\omega_n]|}{(-\zeta\omega_n + j\omega_o)(2 j\omega_o)} = \frac{\omega_n^3 \mathrm{Asm}|G[\omega_n]|}{2 j\omega_o(\zeta\omega_n - j\omega_o)} \qquad (13.2\text{-}73)$$

Multiply numerator and denominator by $(\zeta\omega_n + j\omega_o)$:

$$\delta \doteq \frac{\omega_n^3 \mathrm{Asm}|G[\omega_n]|(\zeta\omega_n + j\omega_o)}{2 j\omega_o(\zeta\omega_n - j\omega_o)(\zeta\omega_n + j\omega_o)} = \frac{\omega_n^3 \mathrm{Asm}|G[\omega_n]|(\zeta\omega_n + j\omega_o)}{2 j\omega_o(\omega_n^2)}$$

$$\doteq \frac{\omega_n}{2} \mathrm{Asm}|G[\omega_n]|\left(-j\zeta\frac{\omega_n}{\omega_o} + 1\right)$$

$$\doteq \frac{\omega_n}{2} \mathrm{Asm}|G[\omega_n]|\left(1 - \frac{j\zeta}{\sqrt{1 - \zeta^2}}\right) \qquad (13.2\text{-}74)$$

If the damping ratio ζ is reasonably small, the peaking of $|G|$ above the asymptote at the natural frequency ω_n is approximately equal to $1/2\zeta$. Hence, the magnitude of G at the natural frequency ω_n can be approximated by

$$|G[\omega_n]| \doteq \frac{1}{2\zeta} \text{Asm}|G[\omega_n]| \qquad (13.2\text{-}75)$$

Combining Eqs 13.2-74, -75 gives

$$\delta \doteq \zeta\omega_n\left(1 - j\zeta/\sqrt{1 - \zeta^2}\right)|G[\omega_n]| \qquad (13.2\text{-}76)$$

Substitute this into Eq 13.3-71 to obtain the closed-loop pole s_k:

$$s_k \doteq -\zeta\omega_n\left(1 - |G[\omega_n]|\right) + j\omega_o\left(1 - \frac{\zeta^2|G[\omega_n]|}{1 - \zeta^2}\right) \qquad (13.2\text{-}77)$$

Unless ζ is close to unity, this can be closely approximated by

$$s_k \doteq -\zeta\omega_n\left(1 - |G[\omega_n]|\right) + j\omega_o \qquad (13.2\text{-}78)$$

The natural frequency ω_n' of these closed-loop poles is approximately equal to the natural frequency ω_n of the open-loop poles of G. The damping ratio ζ' of the closed-loop poles is approximately

$$\zeta' \doteq \zeta\left(1 - |G[\omega_n]|\right) \qquad (13.2\text{-}79)$$

For the example, $|G[\omega_n]|$ is equal to 0.5 (as was shown in Fig 3.9-3). The open-loop poles have a damping ratio ζ of 0.2 and a natural frequency ω_n of 500. Hence, by Eq 13.2-19, the approximate damping ratio for the closed-loop poles is

$$\zeta' \doteq \zeta(1 - 0.5) = 0.5\zeta = 0.1. \qquad (13.2\text{-}80)$$

The approximate natural frequency is $\omega_n' = 500 \text{ sec}^{-1}$. By Eqs 13.2-68, -69, the exact values are $\zeta' = 0.097$, $\omega_n' = 487 \text{ sec}^{-1}$, which are both within 3% of the approximate values.

13.2.4   Practical Rules for Stability

The preceeding discussion will now be applied to develop simple practical rules for assuring that a feedback loop is stable. The s-plane can be separated into the following regions of $|s|$:

1. Values of $|s|$ at which $\text{Asm}|G| \ll 1$.
2. Values of $|s|$ at which $\text{Asm}|G| \gg 1$.
3. Values of $|s|$ at which $\text{Asm}|G|$ is near unity.

In a frequency region described by (1), where $Asm|G| \ll 1$, the closed loop poles are near the poles of G. Hence, there should be no right-half-plane poles of G in this frequency region. Thus, the first stability requirement is

(A) *There should be no right-half-plane pole of G with a break frequency in a frequency region where $Asm|G| < 1$.*

On the other hand, this condition by itself is not sufficient to assure that there are no unstable closed-loop poles in this frequency region. Poorly damped left-hand-plane poles of G can be so close to the $j\omega$ axis that the corresponding closed-loop poles fall in the right half plane. This possibility can be avoided by adding the requirement:

(B) *For an underdamped pair of poles of G with a break frequency in a frequency region where $Asm|G| \ll 1$, the magnitude of G for real frequencies should be less than unity for frequencies near this break frequency.*

The loop can still be stable when condition (B) is violated. However, when it is violated, the stability of the loop should be carefully examined. Requirement (B) is a good general principle to follow whenever possible.

In a frequency region described by (2), where $Asm|G| \gg 1$, the closed-loop poles are near the zeros of G. Hence, there should be no right-half-plane zeros in this region:

(C) *There should be no right-half-plane zero of G with a break frequency in a frequency region where $Asm|G| > 1$.*

To this should be added the following requirement, which is similar to requirement (B);

(D) *For an underdamped pair of zeros of G with a break frequency in a frequency region where $Asm|G| \gg 1$, the magnitude of G for real frequencies should be greater than unity for frequencies near this break frequency.*

In a frequency region described by (3), where $Asm|G|$ is close to unity, the following general requirements apply:

(E) *In a frequency region where $Asm|G|$ is near to unity, and the average magnitude asymptote slope is negative, G should approximate the transfer function ω_c/s, and the locus of G should pass below the $(G = -1)$ point.*

(F) *In a frequency region where $Asm|G|$ is near to unity and the average magnitude asymptote slope is positive, G should approximate the transfer function s/ω_r, and the locus of G should pass above the $(G = -1)$ point.*

13.3   APPROXIMATION OF CLOSED-LOOP POLES AND TRANSIENT COEFFICIENTS BY INSPECTING MAGNITUDE ASYMPTOTE PLOTS

This section shows how the values of the closed-loop poles and the transient coefficients for these poles can be approximated by examining the magnitude asymptote plots of the loop transfer function G. The material of this section is a condensation of a 1956 paper by the author [13.4]. This section provides a mathematical justification for the simple approximations for G_{ib} and G_{ie} presented in Chapter 3.

13.3.1   Approximation of Closed-Loop Poles

As was shown in Section 13.2, the closed-loop poles of a feedback loop that has high gain at low frequency are approximately given by

1. The low-frequency zeros of G (where $\text{Asm}|G| < 1$).
2. The high-frequency poles of G (where $\text{Asm}|G| > 1$).
3. The closed-loop pole $s = -\omega_c$.

A loop with gain less than unity at low frequencies also has a closed-loop pole approximately equal to $s = -\omega_r$, where ω_r is the gain riseover frequency. However, this special case will not be considered here. In the following, it is assumed that the poles and zeros of G are all of single order. (In Ref [13.4], the author considered the more general case where this restriction does not apply.) It is also assumed that all poles and zeros of G are in the left half plane.

The iteration method of Section 13.2.3 uses the following to calculate the shift δ of an actual closed-loop pole from an approximate pole:

1. Shift from low-frequency zero of G at $s = s_a$:

$$\delta_a = -\left.\frac{s - s_a}{G}\right|_{s = s_a'} \tag{13.3-1}$$

2. Shift from high-frequency pole of G at $s = s_b$:

$$\delta_b = -(s - s_b)G\big|_{s = s_b'} \tag{13.3-2}$$

Here s_a', s_b' are the actual closed-loop poles that are shifted from the approximate closed-loop poles s_a, s_b. Although G has a zero at $s = s_a$, the quantity $(s - s_a)/G$ does not. If there are no other poles or zeros of G close to s_a or s_a', the magnitude of $(s - s_a)/G$ evaluated at s_a' can be approximated by the magnitude asymptote:

$$\left| \frac{s - s_a}{G} \right|_{s = s_a'} \doteq \text{Asm} \left| \frac{s - s_a}{G} \right|_{s = s_a'}$$

$$\doteq \text{Asm} |s_a' - s_a| \text{Asm} |1/G[s_a']| \qquad (13.3\text{-}3)$$

The magnitude asymptote of $|s_a' - s_a|$ is equal to either $|s_a'|$ or $|s_a|$, whichever is greater. Combining Eqs 13.3-1, -3 gives the following for the magnitude of the shift of the closed-loop pole from the low-frequency zero of G:

$$|\delta_a| \doteq \text{Greater}[|s_a|, |s_a'|] \text{Asm} |1/G[s_a']| \qquad (13.3\text{-}4)$$

The function Greater$[|s_a|, |s_a'|]$ is defined as either $|s_a|$ or $|s_a'|$, whichever is greater. Since the quantities $|s_a|$, $|s_a'|$ are the break frequencies of the open-loop zero and the corresponding closed-loop pole, it is convenient to designate them as ω_a and ω_a'. Therefore, Eq 13.4-4 can be expressed as

$$|\delta_a| \doteq \text{Greater}[\omega_a, \omega_a'] \text{Asm} |1/G[\omega_a']| \qquad (13.3\text{-}5)$$

If the shift of the closed-loop pole from the open-loop zero is downward in frequency, then $\omega_a > \omega_a'$; whereas $\omega_a' > \omega_a$ if the shift is upward in frequency. Applying the same concepts to Eq 13.4-2 gives the following for the magnitude of the shift of a closed-loop pole from a high-frequency pole of G:

$$|\delta_b| \doteq \text{Greater}[\omega_b, \omega_b'] \text{Asm} |G[\omega_b']| \qquad (13.3\text{-}6)$$

Let us apply these approximations to calculate the closed-loop poles of the following loop transfer function:

$$G = \frac{\omega_c(1 + \omega_3/s)(1 + \omega_4/s)}{s(1 + \omega_1/s)(1 + \omega_2/s)(1 + s/\omega_5)(1 + s/\omega_6)} \qquad (13.3\text{-}7)$$

where $\omega_1 = 0.01$, $\omega_2 = 0.02$, $\omega_3 = 0.08$, $\omega_4 = 0.20$, $\omega_c = 1.0$, $\omega_5 = 2.5$, and $\omega_6 = 8 \text{ sec}^{-1}$. An asymptote plot of $|G|$ is shown in Fig 13.3-1. To perform the analysis, it is convenient to consider the magnitude asymptote plot of Fig 13.3-2, which is a plot of $|1/G|$ for $\omega < \omega_c$, and a plot of $|G|$ for $\omega > \omega_c$. This plot is called the magnitude asymptote of the fictitious function F defined by

$$F = 1/G \qquad \text{for} \quad \text{Asm} |G| > 1 \qquad (13.3\text{-}8)$$

$$F = G \qquad \text{for} \quad \text{Asm} |G| < 1 \qquad (13.3\text{-}9)$$

The approximate closed-loop poles are the poles of F. In Fig 13.3-2, the break frequencies of these poles are designated with the symbol ($\bigcirc$), and occur at ω_3, ω_4, ω_5, and ω_6. There is also an approximate closed-loop pole at ω_c, but

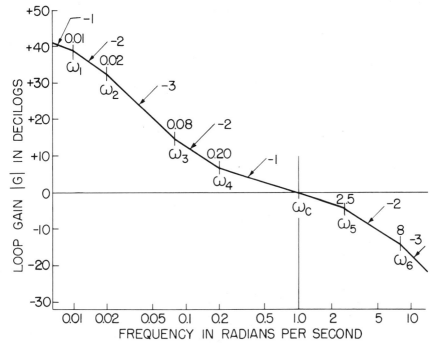

Figure 13.3-1 Magnitude and phase plots of $G[s]$ for illustrative feedback loop. Reprinted with permission from Ref [13.4] © 1956 AIEE (now IEEE).

this point is a discontinuity of F, not a break frequency of F. In terms of the fictitious function F, the magnitude of the shift of the actual closed-loop pole from the approximate pole given in Eqs 13.3-5, -6 can be replaced by the single equation

$$|\delta_x| \doteq \text{Greater}[\omega_x, \omega_x']\,\text{Asm}|F[\omega_x']| \qquad (13.3\text{-}10)$$

The parameter ω_x represents a break frequency of a pole of F, ω_x' is the corresponding closed-loop pole, and the function $\text{Greater}[\omega_x, \omega_x']$ is equal to either ω_x or ω_x', whichever is greater.

If the loop transfer function G has no right-half-plane poles or zeros, the direction of shift of a closed-loop pole from a pole of F can be determined from the asymptote slope of F. The actual closed-loop poles are shifted from the approximate poles (the poles of F) in the direction where the asymptote slope of G or F is odd. Figure 13.3-2 shows that the approximate pole at ω_3 shifts downward in frequency into a region of $+3$ slope for F; the approximate pole at ω_4 shifts upward in frequency into a region of slope $+1$ for F; the approximate pole at ω_5 shifts downward in frequency into a region of -1 slope for F; and the approximate pole at ω_6 shifts upward in

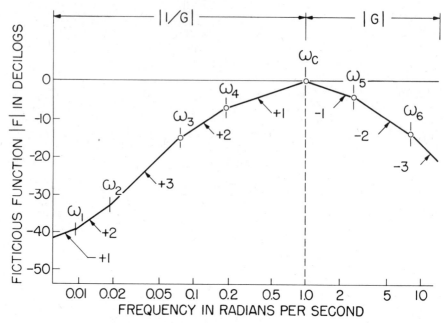

Figure 13.3-2 Magnitude and phase plots of fictitious function $F[s]$ for illustrative loop. Reprinted with permission from Ref [13.4] © 1956 AIEE (now IEEE).

frequency into a region of -3 slope for F. Hence the actual closed-loop poles can be calculated as follows, using Eq 13.3-10 for the magnitude of the shift:

$$\omega_3' = \omega_3 - |\delta_3| \doteq \omega_3 - \omega_3 \mathrm{Asm}\,|F[\omega_3']| \qquad (13.3\text{-}11)$$

$$\omega_4' = \omega_4 + |\delta_4| \doteq \omega_4 - \omega_4' \mathrm{Asm}\,|F[\omega_4']| \qquad (13.3\text{-}12)$$

$$\omega_5' = \omega_5 - |\delta_5| \doteq \omega_5 - \omega_5 \mathrm{Asm}\,|F[\omega_5']| \qquad (13.3\text{-}13)$$

$$\omega_6' = \omega_6 + |\delta_6| \doteq \omega_6 + \omega_6' \mathrm{Asm}\,|F[\omega_3']| \qquad (13.3\text{-}14)$$

Table 13.3-1 shows the first step for calculating the closed-loop poles from the asymptotes. Since the values of the closed-loop poles are not known, Eq 13.3-10 is initially approximate as

$$|\delta_x| \doteq \omega_x \mathrm{Asm}\,|F[\omega_x]| \qquad (13.3\text{-}15)$$

This is used for $|\delta|$ in Eqs 13.3-11 to -14. Column (1) shows the break frequencies ω_x of the poles of F. Column (2) shows the asymptote values of $|F|$ at these break frequencies. In accordance with Eq 13.3-15, the values of columns (1) and (2) are multiplied together to obtain the magnitude of the shift, $|\delta|$, given in column (3). In accordance with Eqs 13.3-11 to 14, this shift

TABLE 13.3-1   First Step for Calculating Break Frequencies of Closed-Loop Poles

Break Frequency of Approx. Pole, ω_x	Asymptote of $\lvert F \rvert$ at Break Frequency, $\mathrm{Asm}\lvert F[\omega_x] \rvert$	Magnitude of Shift, $\lvert \delta_x \rvert$	Approximate Break Frequency of Actual Pole, ω_x'
$\omega_3 = 0.08$	$0.2(\omega_3/\omega_4)^2 = 0.0320$	0.0026	$\omega_3 - \lvert \delta_3 \rvert = 0.0774$
$\omega_4 = 0.2$	$\omega_4/\omega_c = 0.20$	0.04	$\omega_4 + \lvert \delta_4 \rvert = 0.24$
$\omega_5 = 2.5$	$\omega_c/\omega_5 = 0.40$	1.0	$\omega_5 - \lvert \delta_5 \rvert = 1.5$
$\omega_6 = 8$	$0.4(\omega_5/\omega_6)^2 = 0.0391$	0.31	$\omega_6 + \lvert \delta_6 \rvert = 8.31$
(1)	(2)	(3)	(4)

magnitude is either added to, or subtracted from, the break frequency of F in column (1), to obtain the approximate break frequency of the closed-loop pole in column (4).

The per-unit shift of the closed-loop pole is equal to the magnitude of the shift $\lvert \delta_x \rvert$ divided by the break frequency ω_x of the pole of F. The per-unit shift $\lvert \delta_x \rvert/\omega_x$ is approximately equal to the value of $\mathrm{Asm}\lvert F[\omega_x] \rvert$. For the poles at ω_3', ω_6', the per-unit shift is very small, and so the values calculated for these poles can be considered to be exact. However, the per-unit shifts for the poles at ω_4', ω_5' are so great that a more accurate calculation is needed. This is performed using Eqs 13.3-12, -13, where the values for ω_4', ω_5' are obtained from column (4) of Table 13.3-1. The values of $\mathrm{Asm}\lvert F[\omega_x'] \rvert$ at the approximate break frequencies of the closed-loop poles given in column (4) are

$$\mathrm{Asm}\lvert F[\omega_4'] \rvert = \frac{\omega_4'}{\omega_c} = \frac{0.24}{1.0} = 0.24 \tag{13.3-16}$$

$$\mathrm{Asm}\lvert F[\omega_5'] \rvert = \frac{\omega_c}{\omega_5'} = \frac{1.0}{1.5} = 0.67 \tag{13.3-17}$$

By Eq 13.3-10, the values of $\lvert \delta_x \rvert$ are

$$\lvert \delta_4 \rvert = \omega_4' \mathrm{Asm}\lvert F[\omega_4'] \rvert = 0.24(0.24) = 0.0576 \tag{13.3-18}$$

$$\lvert \delta_5 \rvert = \omega_5 \mathrm{Asm}\lvert F[\omega_5'] \rvert = 2.5(0.67) = 1.675 \tag{13.3-19}$$

Applying these to Eqs 13.3-12, -13 gives the following for ω_4', ω_5':

$$\omega_4' = \omega_4 + \lvert \delta_4 \rvert = 0.2 + 0.0576 = 0.2576 \tag{13.3-20}$$

$$\omega_5' = \omega_5 - \lvert \delta_5 \rvert = 2.5 - 1.675 = (0.825) \tag{13.3-21}$$

This new value for ω_5' is below the gain crossover frequency, which indicates that the per-unit shift is so great that the poles cannot be calculated by this method. The approximate pole at $s = -\omega_5$ is combined with the approximate pole at $s = -\omega_c$ to form a complex pair of poles.

A more accurate value for ω_4' can be obtained by solving Eqs 13.3-16, -18, -20 simultaneously, which gives

$$\omega_c \omega_4' = \omega_c \omega_4 + \left(\omega_4'\right)^2 \tag{13.3-22}$$

Solving for ω_4' gives

$$\omega_4' = \left(\omega_c/2\right)\left(1 - \sqrt{1 - 4(\omega_4/\omega_c)}\right) = 0.276 \tag{13.3-23}$$

A similar equation for ω_5' can be obtained from Eqs 13.3-17, -19, -21, and is

$$\omega_5' = \frac{\omega_5}{2}\left(1 + \sqrt{1 - 4(\omega_c/\omega_5)}\right) \tag{13.3-24}$$

This yields a complex value for ω_5', which indicates that the resultant pole is probably complex.

13.3.2   Developing Accurate Magnitude Asymptote Plots of G_{ib} and G_{ie}

The simple approximations yield reasonably accurate values for the break frequencies ω_3', ω_4', and ω_6', but do not give the break frequency ω_5' for the complex pole pair. However, as will now be shown, if the values for ω_3', ω_4', and ω_6' are accurately known, accurate asymptotic plots of $|G_{ib}|$ and $|G_{ie}|$ can be derived from them, without knowing the value for ω_5'. The equations for G, G_{ib}, and G_{ie} can be expressed as

$$G = \frac{N}{D} \tag{13.3-25}$$

$$G_{ib} = \frac{G}{1 + G} = \frac{N}{N + D} \tag{13.3-26}$$

$$G_{ie} = \frac{1}{1 + G} = \frac{D}{N + D} \tag{13.3-27}$$

The roots of N are the zeros of G, the roots of D are the poles of G, and the roots of $(N + D)$ are the closed-loop poles. The zeros of G are the zeros of G_{ib}, and the poles of G are the zeros of G_{ie}. Table 13.3-2 lists the break frequencies of the transfer functions G, $1/G$, G_{ib}, and G_{ie}. Note that G and G_{ib} have the same zeros, while $1/G$ and G_{ie} have the same zeros.

In Fig 13.3-3 are the approximations for the asymptote plots of $|G_{ib}|$ and $|G_{ie}|$ that were presented in Chapter 3. Figure 13.3-4 shows how these are modified to form accurate magnitude asymptote plots. These accurate plots are constructed by using the break-frequency lists of Table 13.3-2, along with the break-frequency values given in column (4) of Table 13.3-2 for ω_3', ω_6' and

TABLE 13.3-2   List of Break Frequencies of Transfer Functions G, $1/G$, G_{ib}, and G_{ie}

Function	Arrangement of Break Frequencies							
						ω_c		
(a) G				ω_3	ω_4			
	0	ω_1	ω_2				ω_5	ω_6
(b) $1/G$	0	ω_1	ω_2				ω_5	ω_6
				ω_3	ω_4			
(c) G_{ib}				ω_3	ω_4			
			ω_3'		ω_4'	$(\omega_5')^2$		ω_6'
(d) G_{ie}	0	ω_1	ω_2				ω_5	ω_6
			ω_3'		ω_4'	$(\omega_5')^2$		ω_6'

the value in Eq 13.3-23 for ω_4'. These values are

$$\omega_3' = 0.0774, \qquad \omega_4' = 0.276, \qquad \omega_6' = 8.31 \qquad (13.3\text{-}28)$$

Diagram a of Fig 13.3-4 shows the plots for $\omega < \omega_c$. The solid curve is the asymptote plot of $|1/G|$. One of the dashed curves is the asymptote plot of $|G_{ie}|$. At zero frequency, $|G|$ is infinite, and so G_{ie} is exactly equal to $1/G$. Hence, G_{ie} and $1/G$ have the same low-frequency magnitude asymptote. Also, as shown in rows (b) and (d) of Table 13.3-2, they have the same break frequencies up to the pole $-\omega_3'$ of G_{ie}. Hence, as shown in Fig 13.3-4a, at frequencies below ω_3' the asymptote plot of $|G_{ie}|$ exactly follows that of $|1/G|$. At $\omega_3' = 0.0774$, the asymptote plot of $|G_{ie}|$ breaks downward from a slope of $+3$ to a slope of $+2$. It continues along this slope until frequency $\omega_4' = 0.24$, where the slope decreases to $+1$.

The other dashed plot in Fig 13.3-4a is the asymptote plot of $|G_{ib}|$. At zero frequency, G_{ib} is exactly unity, because $|G|$ is infinite. As shown in row (c) of Table 13.3-2, the lowest break frequency of G_{ib} is ω_3'. Hence at frequencies below ω_3' the asymptote of $|G_{ib}|$ is unity. The table shows that G_{ib} has a denominator break frequency at ω_3', followed by a numerator break frequency at ω_3. Hence, at $\omega_3' = 0.0774$, the asymptote of $|G_{ib}|$ breaks downward to a slope of -1, and follows that slope until $\omega_3 = 0.08$, where it breaks upward to a slope of zero. Then at $\omega_4 = 0.2$, the plot breaks upward to a slope of $+1$,

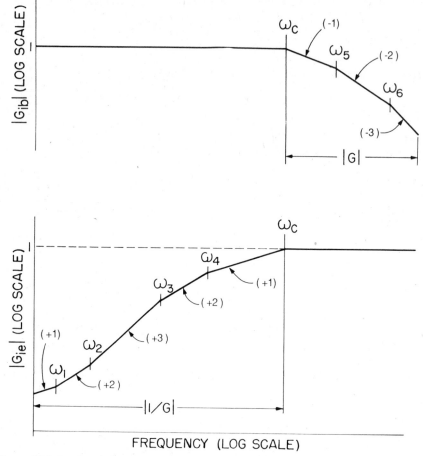

Figure 13.3-3 Approximate magnitude asymptote plots of G_{ib} and G_{ie}. Reprinted with permission from Ref [13.4] © 1956 AIEE (now IEEE).

which it follows until ω_4', where it breaks downward again to a slope of zero. These break frequencies ω_3', ω_3, ω_4, ω_4' can be read from row (c) of Table 13.3-2.

Notice in Fig 13.3-4a that the dashed asymptote plot of $|G_{ib}|$ departs from the unity-gain axis axis by the same amount (on this logarithmic scale) as the asymptote plot of $|G_{ie}|$ departs from $|1/G|$. The reason for this is that the ratio $G_{ib}/1$ is equal to the ratio $G_{ie}/(1/G)$, as shown by

$$\frac{G_{ib}}{1} = \frac{G/(1+G)}{1} = \frac{G}{1+G} \qquad (13.3\text{-}29)$$

$$\frac{G_{ie}}{1/G} = \frac{1/(1+G)}{1/G} = \frac{G}{1+G} \qquad (13.3\text{-}30)$$

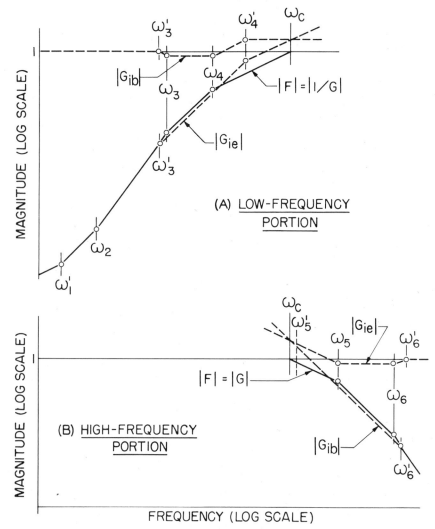

Figure 13.3-4 Construction of magnitude asymptote plots of G_{ib} and G_{ie}: (*a*) low-frequency portion; (*b*) high-frequency portion. Reprinted with permission from Ref [13.4] © 1956 AIEE (now IEEE).

A consequence of this is that the asymptote plots of $|G_{ib}|$ and $|G_{ie}|$ must be exactly equal at the asymptotic gain crossover frequency ω_c, as shown in the figure.

Diagram *b* of Fig 13.3-4 shows the asymptote plots for $\omega > \omega_c$. The solid curve is a plot of $|G|$, and one of the dashed curves is a plot of $|G_{ib}|$. At infinite frequency, $G = 0$, and so G_{ib} is exactly equal to G. Table 13.3-2 shows that neither G nor G_{ib} has a break frequency above ω_6'. Therefore the asymptote plot of $|G_{ib}|$ follows that of $|G|$ at frequencies above ω_6', where the asymptote slope is -3. Since G_{ib} has a denominator break frequency at $\omega_6' = 8.31$, the asymptote slope of $|G_{ib}|$ changes from -3 above ω_6' to -2 below ω_6'. Below ω_6', the asymptote follows this -2 slope until reaching the frequency ω_5'. However, the value for ω_5' is not known, and so the -2 asymptote of $|G_{ib}|$ is projected from ω_6' downward in frequency until it drops below ω_c.

The other dashed curve of Fig 13.3-4b is the asymptote plot of $|G_{ie}|$. At infinite frequency, G_{ie} is exactly unity, because G is zero. As shown in row (d) of Table 13.3-2, the highest break frequency of G_{ie} is ω_6', and so the asymptote of $|G_{ie}|$ is unity at frequencies greater than ω_6'. The break-frequency list for G_{ie} in the table shows that the asymptote plot of $|G_{ie}|$ is constructed as follows, progressing downward in frequency. Above frequency $\omega_6' = 8.31$, the slope is zero. Below ω_6', the slope changes to $+1$, and stays at that slope down to the frequency $\omega_6 = 8$. At frequencies below ω_6, the slope changes back to zero, and the asymptote follows zero slope until the frequency $\omega_5 = 2.5$. At frequencies below ω_5, the asymptote slope changes to -1, and follows this slope until the frequency ω_5' (which is unknown).

In diagram b of Fig 13.3-4, the dashed asymptote plot of $|G_{ie}|$ departs from the unity-gain axis by the same amount that the asymptote plot of $|G_{ib}|$ departs from the plot of $|G|$. The reason for this is that the ratio $G_{ie}/1$ is equal to the ratio G_{ib}/G, as shown by

$$\frac{G_{ie}}{1} = \frac{1/(1 + G)}{1} = \frac{1}{1 + G} \qquad (13.3\text{-}31)$$

$$\frac{G_{ib}}{G} = \frac{G/(1 + G)}{G} = \frac{1}{1 + G} \qquad (13.3\text{-}32)$$

The total asymptote plots of $|G_{ie}|$ and $|G_{ib}|$ are formed as shown in Fig 13.3-5 by superimposing the low-frequency and high-frequency plots of $|G_{ie}|$ and $|G_{ib}|$ developed in Fig 13.3-4. When the low-frequency and high-frequency plots of G_{ie} are superimposed, the frequency at which they cross is the unknown break frequency ω_5' of the complex poles. The asymptote plots $|G_{ib}|$ also cross at exactly this same frequency. By this means, the value for ω_5' is determined, and accurate magnitude asymptote plots of $|G_{ie}|$ and $|G_{ib}|$ are obtained.

The unknown value of ω_5' can be calculated as follows. The value of $\text{Asm}|G_{ib}|$ at ω_5' is the same as that at ω_4', which is

$$\text{Asm}|G_{ib}[\omega_5']| = \text{Asm}|G_{ib}[\omega_4']| = \frac{\omega_3' \, \omega_4'}{\omega_3 \, \omega_4}$$

$$= \frac{0.0774}{0.08} \frac{0.276}{0.2} = 1.3352 \qquad (13.3\text{-}33)$$

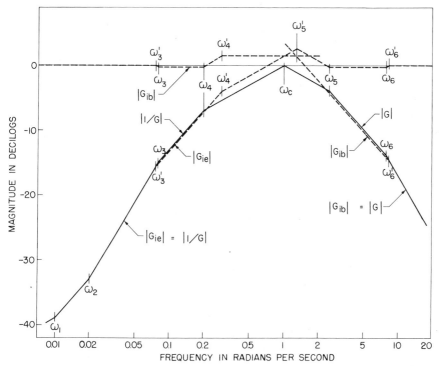

Figure 13.3-5 Total magnitude asymptote plots of $G_{\rm ib}$ and $G_{\rm ie}$. Reprinted with permission from Ref [13.4] © 1956 AIEE (now IEEE).

The value of $\mathrm{Asm}|G_{\rm ib}|$ at ω_6' is equal to the value of $\mathrm{Asm}|G|$ at ω_6', which is

$$\mathrm{Asm}\big|G_{\rm ib}[\omega_6']\big| = \mathrm{Asm}\big|G[\omega_6']\big| = \frac{\omega_c}{\omega_5}\left(\frac{\omega_5}{\omega_6}\right)^2\left(\frac{\omega_6}{\omega_6'}\right)^3$$

$$= \frac{\omega_2\omega_5\omega_6}{\left(\omega_6'\right)^3}$$

$$= \frac{(1)(2.5)(8)}{(8.31)^3} = 0.03485 \qquad (13.3\text{-}34)$$

The value of $\mathrm{Asm}|G_{\rm ib}|$ at ω_5' is related to this as follows:

$$\mathrm{Asm}\big|G_{\rm ib}[\omega_5']\big| = \mathrm{Asm}\big|G_{\rm ib}[\omega_6']\big|\left(\frac{\omega_6'}{\omega_5'}\right)^2 = 0.03485\left(\frac{\omega_6'}{\omega_5'}\right)^2 \qquad (13.3\text{-}35)$$

TABLE 13.3-3   Comparison of Approximate and Exact Values of Closed-Loop Poles and Their Break Frequencies

Break Frequencies		Poles	
Approximate	Exact	Approximate	Exact
$\omega_3' = 0.0774$	0.0776	-0.0774	-0.0776
$\omega_4' = 0.276$	0.2885	-0.276	-0.2885
$\omega_5' = 1.34$	1.308	$\begin{cases} \omega_n = 1.34 \\ \zeta = ? \end{cases}$	$\begin{aligned} \omega_n &= 1.308 \\ \zeta &= 0.67 \end{aligned}$
$\omega_6' = 8.31$	8.392	-8.31	-8.392
(1)	(2)	(3)	(4)

Equating the expressions of Eqs 13.3-33, -35 gives

$$\omega_5' = \omega_6' \sqrt{\frac{0.03485}{1.3352}} = 8.61(0.1616) = 1.34 \qquad (13.3\text{-}36)$$

The approximate break frequencies of the closed-loop poles that have been computed from the magnitude asymptote plots are compared with the exact values in Table 13.3-3. Column (1) shows the approximate break frequencies, and column (2) shows the exact break frequencies. The values of the closed-loop poles are compared in columns (3) and (4). There is very good agreement for the break frequencies, even for the complex poles of break frequency $\omega_5' = \omega_n$. However, the approximate method is unable to calculate reliably the damping ratio ζ of these complex poles.

13.3.3   Approximation of Transient Coefficients for Step Response

The coefficients of the transient response for the individual closed-loop poles can be approximated by inspecting the asymptote plots of $|G_{ib}|$ and $|G_{ie}|$. The transform of the error response to a unit step is

$$X_e = G_{ie} X_i = G_{ie}(1/s)$$

$$= \frac{c_0}{s} + \frac{K_3}{s + \omega_3'} + \frac{K_4}{s + \omega_4'} + \frac{K_6}{s + \omega_6'} + \frac{K_5}{s - s_5} + \frac{K_5^*}{s - s_5^*} \qquad (13.3\text{-}37)$$

where s_5 is one of the complex poles, shifted from the approximate pole $s = -\omega_5$. It can be expressed as

$$s_5 = -\zeta\omega_n + j\omega_o \qquad (13.3\text{-}38)$$

The quantity s_5^* is the conjugate of s_5. The coefficient c_0 is the position error coefficient, which is equal to

$$c_0 = s \frac{G_{ie}}{s} \Big|_{s=0} = G_{ie}[0] = 0 \qquad (13.3\text{-}39)$$

Similarly, the transform of the feedback response to a unit step is

$$X_b = G_{ib} X_i = G_{ib}(1/s)$$

$$= \frac{c_0'}{s} + \frac{K_3'}{s + \omega_3'} + \frac{K_4'}{s + \omega_4'} + \frac{K_6'}{s + \omega_6'} + \frac{K_5'}{s - s_5} + \frac{K_5'^*}{s - s_5^*} \qquad (13.3\text{-}40)$$

The coefficient c_0' is equal to

$$c_0' = s \frac{G_{ib}}{s} \Big|_{s=0} = G_{ib}[0] = 1 \qquad (13.3\text{-}41)$$

For a unit step input, the feedback and error responses are related by

$$x_b = x_i - x_e = 1 - x_e \qquad (13.3\text{-}42)$$

Hence,

$$c_0' = 1 - c_0 \qquad (13.3\text{-}43)$$

$$K_x' = -K_x \qquad (13.3\text{-}44)$$

where K_x is any one of the coefficients $(K_3, K_4, K_6, K_5, K_5^*)$ of the error response, and K_x' is the corresponding coefficient of the feedback response.

Approximations of the error response coefficients K_3, K_4 for the low-frequency real poles can be derived in the following manner from the asymptote plot of $|G_{ie}|$. Figure 13.3-6a shows a general magnitude asymptote plot of a G_{ie} transfer function at the break frequency ω_x' of a low-frequency real pole $s = -\omega_x'$. If there are no other zeros or poles of G_{ie} close to this pole, the G_{ie} transfer function in the vicinity of this pole can be approximated as

$$G_{ie} \doteq \frac{\text{Asm}|G_{ie}[\omega_x']|(s/\omega_x')^{p+1}}{(s + \omega_x')/\omega_x'} \qquad (13.3\text{-}45)$$

where p is the magnitude-asymptote slope at frequencies above the break frequency ω_x' of the closed-loop pole. (It is assumed that there are no poles or zeros of G in the right half plane.) The error-response transient coefficient of

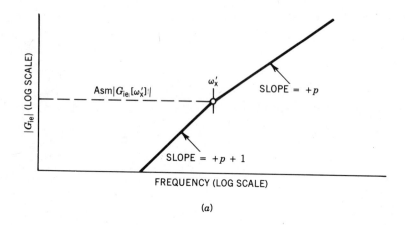

(a)

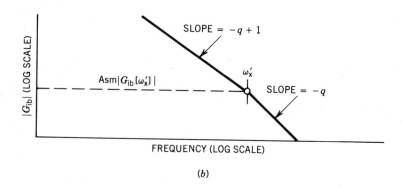

(b)

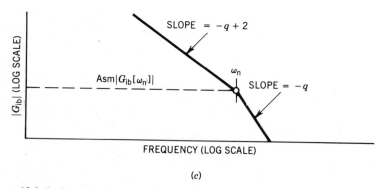

(c)

Figure 13.3-6 Magnitude asymptote plots of G_{ib} and G_{ie} in the vicinity of the closed-loop poles: (a) $|G_{ie}|$ near single-order low-frequency closed-loop pole; (b) $|G_{ib}|$ near single-order high-frequency closed-loop pole; (c) $|G_{ib}|$ near underdamped high-frequency closed-loop pole pair.

this pole, for a unit-step input, is

$$K_x = (s + \omega_x') \frac{G_{ie}}{s} \Bigg|_{s = -\omega_x'} \tag{13.3-46}$$

Substituting Eq 13.3-45 into this gives the following approximation for the transient coefficient:

$$K_x \doteq \mathrm{Asm} |G_{ie}[\omega_x']| \left| \left(\frac{s}{\omega_x'}\right)^P \right|_{s = -\omega_x'} = (-1)^P \mathrm{Asm} |G_{ie}[\omega_x']| \tag{13.3-47}$$

This shows that the magnitude of the step-response coefficient is approximately equal to the asymptote value of $|G_{ie}|$ at the break frequency of the low-frequency closed-loop pole. The sign of the coefficient is positive if the $|G_{ie}|$ asymptote slope p is even at frequencies above the break frequency, and is negative if that slope is odd.

For the $|G_{ie}|$ plot of the example, the slope p above the break frequency ω_3' is $+2$ (even), and the slope above ω_4' is $+1$ (odd). Hence, the error-response transient coefficients of these poles for a unit step input are

$$K_3 \doteq \mathrm{Asm} |G_{ie}[\omega_3']| \tag{13.3-48}$$

$$K_4 \doteq - \mathrm{Asm} |G_{ie}[\omega_4']| \tag{13.3-49}$$

It can be seen from Fig 13.3-4 that the asymptote value of $|1/G|$ at ω_3' is

$$\mathrm{Asm} |1/G[\omega_3']| = \frac{\omega_4}{\omega_c} \left(\frac{\omega_3}{\omega_4}\right)^2 \left(\frac{\omega_3'}{\omega_3}\right)^3 = \frac{(\omega_3')^3}{\omega_3 \omega_4 \omega_c} \tag{13.3-50}$$

This is equal to the value of $\mathrm{Asm} |G_{ie}|$ at ω_3', which is

$$\mathrm{Asm} |G_{ie}[\omega_3']| = \mathrm{Asm} |1/G[\omega_3']| = \frac{(\omega_3')^3}{\omega_3 \omega_4 \omega_c}$$

$$= \frac{(0.0774)^3}{(0.08)(0.2)(1)} = 0.0290 \tag{13.3-51}$$

The value for ω_3' is obtained from column (1) of Table 13.3-3. The asymptote value of $|G_{ie}|$ at ω_4' is related to this by

$$\mathrm{Asm} |G_{ie}[\omega_4']| = \left(\frac{\omega_4'}{\omega_3'}\right)^2 \mathrm{Asm} |G_{ie}[\omega_3']| = \frac{\omega_3'(\omega_4')^2}{\omega_3 \omega_4 \omega_c}$$

$$= \frac{(0.0774)(0.0276)^2}{(0.08)(0.2)(1)} = 0.369 \tag{13.3-52}$$

The value for ω_4' is also obtained from column (1) of Table 13.3-3. Substituting Eqs 13.3-51, -52 into Eqs 13.3-48, -49 gives the following approximations for the transient coefficients:

$$K_3 \doteq \text{Asm}|G_{ie}[\omega_3']| = 0.0290 \tag{13.3-53}$$

$$K_4 \doteq -\text{Asm}|G_{ie}[\omega_4']| = -0.369 \tag{13.3-54}$$

The coefficient K_6 for the high-frequency pole of break frequency ω_6' cannot be approximated by Eq 13.3-47, because the G_{ie} transfer function has an open-loop zero close to each high-frequency closed-loop pole. Instead, the transient coefficient for a high-frequency pole should be derived from the G_{ib} response. Figure 13.3-6b shows a general magnitude asymptote plot of a G_{ib} transfer function, in the vicinity of the break frequency ω_x' of a high-frequency closed-loop pole at $s = -\omega_x'$. If there are no other zeros or poles of G_{ib} close to this pole, G_{ib} in the vicinity of the pole can be approximated as

$$G_{ib} \doteq \frac{\text{Asm}|G_{ib}[\omega_x']|}{[(s + \omega_x')/\omega_x'](s/\omega_x')^{q-1}} \tag{13.3-55}$$

The magnitude asymptote slope above the break frequency ω_x' is defined as $-q$. Therefore, the approximate feedback-response transient coefficient of this pole, for a unity-step input, is

$$K_x' = (s + \omega_x') \left. \frac{G_{ib}}{s} \right|_{s=-\omega_x'} = \left. \frac{\text{Asm}|G_{ib}[\omega_x']|}{(s/\omega_x')^q} \right|_{s=-\omega_x'}$$

$$\doteq (-1)^q \text{Asm}|G_{ib}[\omega_x']| \tag{13.3-56}$$

Hence, the magnitude of the feedback-response transient coefficient K_x' of a real high-frequency pole, for a unit-step input, is approximately equal to the asymptote of $|G_{ib}|$ at the break frequency of the pole. The sign of the coefficient is positive if the asymptote slope of $|G_{ib}|$ above that pole is even, and is negative if that slope is odd. The corresponding transient coefficient K_x for the error response is the negative of this coefficient K_x' for the feedback response.

The asymptote value of $|G_{ib}|$ at the break frequency ω_6' is the same as that of $|G|$, which is

$$\text{Asm}|G_{ib}[\omega_6']| = \text{Asm}|G[\omega_6']| = \frac{\omega_c}{\omega_5}\left(\frac{\omega_5}{\omega_6}\right)^2\left(\frac{\omega_6}{\omega_6'}\right)^3$$

$$= \frac{\omega_c \omega_5 \omega_6}{(\omega_6')^3} = \frac{(1)(2.5)(8)}{(8.31)^3} = 0.0349 \tag{13.3-57}$$

The value for ω_6' is obtained from column (1) of Table 13.3-3. The asymptote slope of $|G_{ib}|$ above ω_6' is -3 (odd). Hence, by Eq 13.3-56, the approximate feedback transient coefficient of the pole at ω_6', for a unit-step input, is

$$K_6' \doteq -\text{Asm}|G_{ib}[\omega_6']| = -0.0349 \qquad (13.3\text{-}58)$$

The error-response transient coefficient is the negative of this, which is

$$K_6 = -K_6' = 0.0349 \qquad (13.3\text{-}59)$$

The complex pole pair of break frequency ω_5' is greater than ω_c, and so should be derived from the G_{ib} response. Figure 13.3-6c shows a general magnitude asymptote plot of a G_{ib} transfer function at the break frequency ω_n of a high-frequency complex pole pair. If there are no other zeros or poles of G_{ib} close to this pole pair, the G_{ib} transfer function in the vicinity of these poles can be approximated by

$$G_{ib} \doteq \frac{\text{Asm}|G_{ib}[\omega_n]|}{\left[(s^2 + 2\zeta\omega_n s + \omega_n^2)/\omega_n^2\right](s/\omega_n)^{q-2}} \qquad (13.3\text{-}60)$$

The asymptote slope of $|G_{ib}|$ above the break frequency ω_n is defined as $-q$. By Eq 2.1-46 of Chapter 2, the transient coefficient of a complex pole pair for the feedback response to a unit-step input is

$$K_n' = \frac{1}{j\omega_o}(s^2 + 2\zeta\omega_n s + \omega_n^2)\frac{G_{ib}}{s}\bigg|_{s=-\zeta\omega_n+j\omega_o} \qquad (13.3\text{-}61)$$

where $F[s]$ was replaced by X_b, which by Eq 13.3-40 is equal to G_{ib}/s. Substitute into this the approximation for G_{ib} in Eq 13.3-60. This gives

$$K_n' \doteq \text{Asm}|G_{ib}[\omega_n]|\frac{\omega_n/j\omega_o}{(s/\omega_n)^{q-1}}\bigg|_{s=-\zeta\omega_n+j\omega_o=\omega_n e^{j\phi}} \qquad (13.3\text{-}62)$$

The s-plane plot of the pole $(-\zeta\omega_n + j\omega_o)$ is shown in Fig 13.3-7. The angle of the vector to this pole is defined as ϕ, and is equal to

$$\phi = \pi - \arccos[\zeta] \qquad (13.3\text{-}63)$$

Equation 13.3-61 reduces to

$$K_n' \doteq \frac{\text{Asm}|G_{ib}[\omega_n]|(\omega_n/\omega_o)}{j(e^{j\phi})^{q-1}} \qquad (13.3\text{-}64)$$

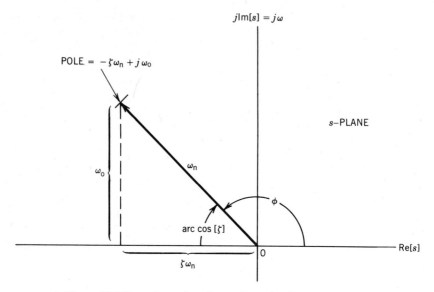

Figure 13.3-7   *s*-plane plot of complex pole, showing angle ϕ.

The magnitude and phase of this are

$$|K_n'| \doteq \text{Asm}|G_{ib}[\omega_n]| \frac{\omega_n}{\omega_o} = \frac{\text{Asm}|G_{ib}[\omega_n]|}{\sqrt{1 - \zeta^2}} \qquad (13.3\text{-}65)$$

$$\text{Ang}[K_n'] \doteq -(q - 1)\phi - \pi/2 \qquad (13.3\text{-}66)$$

From Eq 2.1-47, the feedback-signal transient term is

$$\Delta x_b[t] = |K_n'|e^{-\zeta\omega_n t}\cos[\omega_o t + \text{Ang}[K_n']] \qquad (13.3\text{-}67)$$

For the complex pole pair of break frequency ω_ζ', the slope of the asymptote of $|G_{ib}|$ above the pole is $-q = -2$. Set $q = 2$ in Eq 13.3-66 and combine this with Eq 13.3-63. This gives

$$\text{Ang}[K_\zeta'] \doteq -\phi - \frac{\pi}{2} = \arccos[\zeta] - \frac{3\pi}{2} = \arccos[\zeta] + \frac{\pi}{2} \qquad (13.3\text{-}68)$$

To simplify the expression, an angle of 2π radians was added. By Eqs 13.3-67, -68, the value of the transient term at time $t = 0$ is

$$\Delta x_b[0] = |K_n'|\cos[\arccos[\zeta] + \pi/2] = -|K_n'|\sin[\arccos[\zeta]]$$
$$= -|K_n'|\sqrt{1 - \zeta^2} \qquad (13.3\text{-}69)$$

Substitute into this the expression for $|K'_n|$ in Eq 13.3-65. This gives

$$\Delta x_b[0] \doteq -\text{Asm}|G_{ib}[\omega'_5]| \qquad (13.3\text{-}70)$$

Thus, for the example, the asymptote value of $|G_{ib}|$ at the break frequency of the complex pole pair is approximately equal to the magnitude of the initial value of the transient term for that pole pair.

The asymptote of $|G_{ib}|$ at ω'_5 is related as follows to the asymptote value at ω'_6 given in Eq 13.3-57:

$$\text{Asm}|G_{ib}[\omega'_5]| = \left(\frac{\omega'_6}{\omega'_5}\right)^2 \text{Asm}|G_{ib}[\omega'_6]|$$

$$= \left(\frac{\omega'_6}{\omega'_5}\right)^2 \frac{\omega_c \omega_5 \omega_6}{(\omega'_6)^3} = \frac{\omega_c \omega_5 \omega_6}{(\omega'_5)^2 \omega'_6}$$

$$= \frac{(1)(2.5)(8)}{(1.34)^2(8.31)} = 1.34 \qquad (13.3\text{-}71)$$

where ω'_5 is obtained from column (1) of Table 13.3-3. Substituting this into Eq 13.3-70 gives for the initial value of the transient term

$$\Delta x_b[0] \doteq -\text{Asm}|G_{ib}[\omega'_5]| = -1.34 \qquad (13.3\text{-}72)$$

The results of the preceeding approximations are summarized in Table 13.3-4 and compared with the exact values. The transient coefficients for the real poles are also the initial values of the transient terms. This table gives the approximate and exact initial values of the transient terms for the feedback response to a unit step. The approximate values for the poles at the break frequencies ω'_3, ω'_4 are the negative of the values given in Eqs 13.3-53, -54. The approximate values for the poles of break frequencies ω'_6, ω'_5 are obtained from Eqs 13.3-58, -72. The agreement between the exact and approximate values is good.

TABLE 13.3-4   Comparison of Approximate and Exact Initial Values of Feedback-Response Transient Terms for a Unit-Step Input

Break Frequency	Initial Values of Transient Terms	
	Approximate	Exact
ω'_3	-0.0290	-0.0271
ω'_4	0.369	0.584
ω'_5	-1.34	-1.506
ω'_6	-0.0338	-0.0421

In calculating the approximate transient-response coefficients, the $|G_{ib}|$ plot was not used at low frequencies, nor the $|G_{ie}|$ plot at high frequencies, because in those regions a zero is close to the particular closed-loop pole. However, this problem can be corrected by taking into account the deviation produced by the zero. In the vicinity of of the closed-loop pole of break frequency ω_3', the G_{ib} transfer function can be approximated as

$$G_{ib} \doteq \frac{\text{Asm}|G_{ib}[\omega_3']|(1 + s/\omega_3)}{1 + s/\omega_3'} = \frac{\text{Asm}|G_{ib}[\omega_3']|(s + \omega_3)/\omega_3}{(s + \omega_3')/\omega_3'} \qquad (13.3\text{-}73)$$

The feedback-response transient coefficient of this pole, for a unit-step input, is

$$K_3' = (s + \omega_3') \left. \frac{G_{ib}}{s} \right|_{s=-\omega_3'}$$

$$\doteq \left. \frac{\text{Asm}|G_{ib}[\omega_3']|(s + \omega_3)/\omega_3}{s/\omega_3'} \right|_{s=-\omega_3'}$$

$$\doteq -\frac{\text{Asm}|G_{ib}[\omega_3']|(\omega_3 - \omega_3')}{\omega_3} \doteq -\frac{|\delta_3|}{\omega_3}\text{Asm}|G_{ib}[\omega_3']| \qquad (13.3\text{-}74)$$

The quantity $\omega_3' - \omega_3$ is equal to the shift δ_3 of ω_3' from ω_3. Since $\omega_3 > \omega_3'$, the magnitude $|\delta_3|$ of this shift is equal to $\omega_3 - \omega_3'$. By Eq 13.3-10, $|\delta_3|$ is approximately equal to

$$|\delta_3| \doteq \text{Greater}[\omega_3, \omega_3']\text{Asm}|1/G[\omega_3']| = \omega_3\text{Asm}|1/G[\omega_3']| \qquad (13.3\text{-}75)$$

Substituting Eq 13.3-75 into Eq 13.3-74 gives

$$K_3' \doteq -\text{Asm}|1/G[\omega_3']|\text{Asm}|G_{ib}[\omega_3']| = -\text{Asm}|G_{ie}[\omega_3']| \qquad (13.3\text{-}76)$$

This applies the relation $G_{ib}/G = G_{ie}$. Since K_3' is equal to $-K_3$, this is equivalent of the expression for K_3 derived in Eq 13.3-48.

By applying this approach, one can calculate the transient coefficient for the pole at ω_4' from the G_{ib} response, and that for the pole at ω_6' from the G_{ie} response. This indicates that the asymptote approximations of G_{ib} and G_{ie} are mathematically consistent, at least for the real closed-loop poles, because the same approximate transient coefficients are derived from the two asymptote approximations.

13.3.4 Approximation of Transient Coefficients for Response to a Derivative or Integral Step

It has been shown that the magnitude of the initial value of a transient term is approximately equal to $\text{Asm}|G_{ie}|$ at the break frequency of a low-frequency closed-loop pole, and is approximately equal to $\text{Asm}|G_{ib}|$ at the break frequency of a high-frequency closed-loop pole. This applies also for the complex pole pair.

This principle can be generalized to provide estimates of the transient coefficients of the response to a step of any derivative or integral of the input. A ramp is a step of the first derivative. The transform of the error response to a unit ramp can be expressed as

$$X_e = G_{ie}\frac{1}{s^2} = \frac{G_{ie}}{s}\frac{1}{s} \tag{13.3-77}$$

This response is equal to the unity-step response of the transfer function G_{ie}/s. Hence, to obtain the transient coefficient for a unit ramp, magnitude asymptote plots are constructed for the transfer functions G_{ie}/s and G_{ib}/s. These are obtained by multiplying the asymptote plots of $|G_{ie}|$ and $|G_{ib}|$ by $1/\omega$.

In Fig 13.3-8, diagram a shows the asymptote plots of $|G_{ie}|$ and $|G_{ib}|$. The circles shows the asymptote values at the break frequencies of the closed-loop poles. These asymptote plots of diagram a are multiplied by ω_c/ω to form the normalized plots of G_{ie}/s and G_{ib}/s in diagram b. The actual asymptote values of G_{ie}/s and G_{ib}/s at the break frequencies are obtained by multiplying the values shown by $1/\omega_c$, which is equal to 1.0 sec. The coefficients of the unit-ramp-response transient terms are derived from the values of the plot of G_{ie}/s for the low-frequency poles, and from the values of the plot of G_{ib}/s for the high-frequency poles.

In diagram c the plot of $|G_{ie}/s|$ in diagram b is multiplied by ω_c/ω to obtain the normalized plot of $|G_{ie}/s^2|$. The coefficients of the response to a unit step of the second derivative are derived from this plot. This plot is multiplied by ω_c/ω to obtain the normalized plot of $|G_{ie}/s^3|$, from which one can derive the transient coefficients for a step of the third derivative. The transient coefficients for steps of the second and third derivative for high-frequency poles are very small relative to those for low-frequency poles. Since these transient terms can usually be ignored, $|G_{ib}/s^2|$ and $|G_{ib}/s^3|$ have not been plotted.

In diagram d, the plots of $|G_{ib}|$ and $|G_{ie}|$ of diagram a are multiplied by ω/ω_c to obtain normalized plots of $|sG_{ie}|$ and $|sG_{ib}|$. The asymptote values at the appropriate break frequencies of these plots are the approximate transient coefficients for the unit-impulse response (a unit step of the first integral of the input).

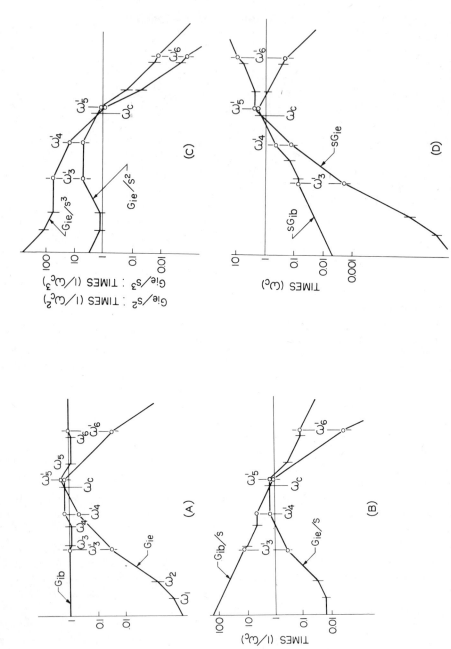

Figure 13.3-8 Magnitude asymptote plots for estimating the transient coefficients for steps of derivatives and steps of integrals of the input: (*a*) G_{ie} and G_{ib}, for estimating step-response coefficients; (*b*) G_{ie}/s and G_{ib}/s, for estimating ramp-response coefficients; (*c*) G_{ie}/s^2 and G_{ie}/s^3, for estimating coefficients of responses to steps of acceleration and rate of acceleration; (*d*) sG_{ie} and sG_{ib}, for estimating impulse-response coefficients. Reprinted with permission from Ref [13.4]

13.4 THE ROOT-LOCUS METHOD

13.4.1 Summary of the Root-Locus Method

The root-locus method developed by Evans [1.13] provides plots of the loci of the closed-loop poles of a loop transfer function as the loop gain is varied. The loop transfer function can be expressed in the form

$$G = K_n G_n \tag{13.4-1}$$

where K_n is a normalized gain parameter, which is varied, and G_n is the normalized loop transfer function. The root loci are plots of the closed-loop poles (the poles of G_{ib}) as the gain parameter K_n is varied from zero to infinity.

Figure 13.4-1 is the root-locus plot of the following normalized loop transfer function:

$$G_n = \frac{1}{s(s+1)(s+2)} = \frac{\omega_{cn}}{s(1+s/1)(1+s/2)} \tag{13.4-2}$$

where ω_{cn} is the normalized asymptotic gain crossover frequency, which is $\omega_{cn} = 0.5$. The parameters along the root loci are the values of the gain parameter K_n. The values of the complex variable s at those points are the closed-loop poles (the poles of G_{ib}) corresponding to the specified values of K_n. This case was analyzed in detail by Truxal [11.3] (p. 231).

The poles of G_{ib}, which are the closed-loop poles, are the values of the complex variable s for which $G[s] = -1$, which correspond to the conditions

$$|G[s]| = K_n|G_n[s]| = 1 \tag{13.4-3}$$

$$\text{Ang}[G[s]] = \text{Ang}[G_n[s]] = 180 \pm 360N \text{ deg} \tag{13.4-4}$$

where N is any integer. Any value of s which satisfies Eq 13.4-4 lies on a root locus, because Eq 13.4-3 can always be satisfied by selecting the appropriate value for K_n.

To illustrate the application of Eq 13.4-4, consider the following normalized loop transfer function:

$$G_n[s] = \frac{(s+z_1)(s+z_2)}{s(s+p_2)(s+p_3)(s+p_4)} \tag{13.4-5}$$

This loop transfer function has zeros at $s = -z_1, -z_2$ and poles at $s = 0, -p_2, -p_3, -p_4$. An s-plane plot showing these poles and zeros is shown in Fig 13.4-2. The poles at $s = -p_3, -p_4$ are a complex pair, and so $-p_4 = -p_3^*$ (the conjugate of $-p_3$).

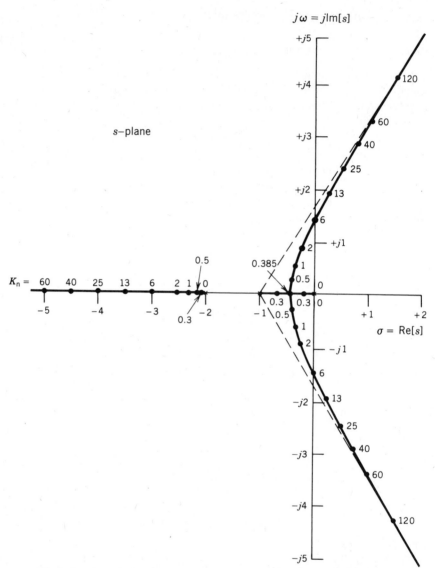

Figure 13.4-1   Root-locus for loop transfer function $1/[s(s + 1)(s + 2)]$.

Consider the value of $G_n[s]$ at the point $s = s_1$ on the complex s-plane, as shown in Fig 13.4-2. The value of $G_n[s]$ at this point $s = s_1$ is

$$G_n[s_1] = \frac{(s_1 + z_1)(s_1 + z_2)}{s_1(s_1 + p_2)(s_1 + p_3)(s_1 + p_4)} \qquad (13.4\text{-}6)$$

The phase of $G_n[s]$ at $s = s_1$ is

$$\text{Ang}[G_n[s_1]] = \text{Ang}[V_{z1}] + \text{Ang}[V_{z2}]$$
$$- \text{Ang}[V_{p1}] - \text{Ang}[V_{p2}] - \text{Ang}[V_{p3}] - \text{Ang}[V_{p4}] \qquad (13.4\text{-}7)$$

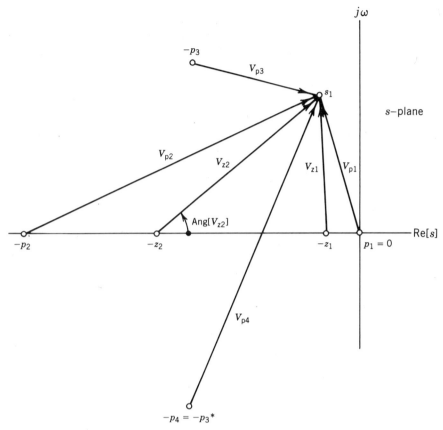

Figure 13.4-2   *s*-plane plot of poles and zeros of general loop transfer function.

As shown in Fig 13.4-2, V_{z1}, V_{z2} are the vectors drawn from the zeros at $s = -z_1, -z_2$ to the point $s = s_1$; and $V_{p1}, V_{p2}, V_{p3}, V_{p4}$ are the vectors drawn from the poles at $s = 0, -p_2, -p_3, -p_4$ to the point $s = s_1$. Combining Eqs 13.4-4, -7 shows that a point on a root locus occurs at those points on the *s*-plane where the following is satisfied:

$$180 \pm 360N \text{ deg} = \text{Ang}[V_{z1}] + \text{Ang}[V_{z2}] - \text{Ang}[V_{p1}]$$

$$- \text{Ang}[V_{p2}] - \text{Ang}[V_{p3}] - \text{Ang}[V_{p4}] \quad (13.4\text{-}8)$$

Figure 13.4-2 shows how the angle for the vector V_{z2} is defined. The angles for the other vectors are defined in a similar manner.

The root loci are plotted by a trial-and-error procedure which finds those points on the *s*-plane for which Eq 13.4-8 is satisfied. This procedure is simplified by applying a set of 10 rules, which place constraints on the root loci. These rules are as follows:

1. *Roots at zero gain.* For $K_n = 0$, the roots (closed-loop poles) are the poles of $G_n[s]$.

2. *Roots at infinite gain.* For $K_n = \infty$, the roots are the zeros of $G_n[s]$, including the zeros of $G_n[s]$ at $|s| = \infty$. (If the high-frequency magnitude asymptote of $G_n[j\omega]$ has a logarithmic slope of $-n$, there are n zeros of $G_n[j\omega]$ at $|s| = \infty$.)

3. *Number of separate root loci.* Define the numerator and denominator polynomials of $G_n[s]$ as $N[s]$, $D[s]$:

$$G_n[s] = \frac{N[s]}{D[s]} \tag{13.4-9}$$

The number of separate root loci is equal to the order of the sum polynomial $(N[s] + D[s])$.

4. *Symmetry.* The root-locus plot is symmetric about the real axis of the s-plane. In other words, the root-locus plot in the lower half plane is the mirror image of the plot in the upper half plane.

5. *Asymptotes for large $|s|$.* For large values of $|s|$, the root loci are asymptotic to straight lines at the following angles:

$$\Theta_k = \frac{(2k + 1)180°}{p - q} \tag{13.4-10}$$

where p is the order of the numerator polynomial $N[s]$ of $G_n[s]$, and q is the order of the denominator polynomial $D[s]$. The parameter k is an integer which takes on the values

$$k = 0, 1, \dots, (|p - q| - 1) \tag{13.4-11}$$

These asymptotes intersect on the real s-axis, at the following value of s:

$$s_{int} = \frac{\Sigma \, \mathrm{Re}[\text{poles}] - \Sigma \, \mathrm{Re}[\text{zeros}]}{q - p} \tag{13.4-12}$$

where $\mathrm{Re}[\text{poles}]$, $\mathrm{Re}[\text{zeros}]$ represent the real parts of the poles and zeros. For example, consider $G_n[s]$ in Eq 13.4-5. Since $p = 2$ and $q = 4$, the parameter k takes on the values $k = 0, 1$. There are two high-frequency asymptotes, at the angles

$$\Theta_0 = \frac{[2(0) + 1]180°}{2 - 4} = -90° \tag{13.4-13}$$

$$\Theta_1 = \frac{[2(1) + 1]180°}{2 - 4} = -270° \tag{13.4-14}$$

These asymptotes intersect on the real s-axis at the following value of s:

$$s_{int} = \frac{(0 - p_2 - 2\omega_n\cos[\zeta]) - (-z_1 - z_2)}{4 - 2} \qquad (13.4\text{-}15)$$

where ω_n, ζ are the natural frequency and damping ratio of the complex pole pair at $s = -p_3, -p_3^*$.

6. *Root loci along real s-axis.* The regions of root loci along the real s-axis depend only on the real open-loop poles and zeros, because complex poles are in pairs. Any point on the real s-axis lies on a root locus if the number of real poles and zeros to the right of that point is odd.

7. *Intersections with imaginary axis.* The root-locus points on the imaginary s-axis can be derived from the frequency-response phase plot of the loop transfer function $G_n[j\omega]$. Any frequency where the phase of $G_n[j\omega]$ is $180 \pm 360N$ deg characterizes an intersection of the root locus with the $s = j\omega$ axis. Since the root-locus plot is symmetric about the horizontal real s-axis, any such frequency corresponds to two separate root loci, one for the upper half plane and one for the lower half plane.

8. *Locus near open-loop pole or zero.* As was shown in rules (1),(2), the root loci in the region near $K_n = 0$ are near the open-loop poles, and the root loci in the region near $K_n = \infty$ are near the open-loop zeros. The angle of a root locus in the region near an open-loop pole or zero is described by

$$\text{Ang}[V_{1y}] = \overset{\text{other}}{\underset{\text{zeros}}{\sum}} \text{Ang}[V_{xy}] - \overset{\text{other}}{\underset{\text{poles}}{\sum}} \text{Ang}[V_{xy}] \qquad (13.4\text{-}16)$$

As shown in Fig 13.4-3, we are concerned with the locus near an open-loop pole or zero at $s = s_y$. Point s_1 lies close to s_y on the root locus. The vector drawn from s_1 to s_y is designated V_{1y}, and Eq 13.4-16 gives the angle of this vector, designated $\text{Ang}[V_{iy}]$. Point s_x represents any other open-loop pole or zero. The quantity V_{xy} is the vector drawn from s_x to s_y.

9. *Points of breakaway from real s-axis.* The points at which the root loci break away from the real s-axis are real values of s where the derivative $dG_n[s]/ds$ is zero. In terms of the numerator and denominator polynomials $N[s]$, $D[s]$, as defined in Eq 13.4-9, this becomes

$$N[s]\frac{dD[s]}{ds} = D[s]\frac{dN[s]}{ds} \qquad (13.4\text{-}17)$$

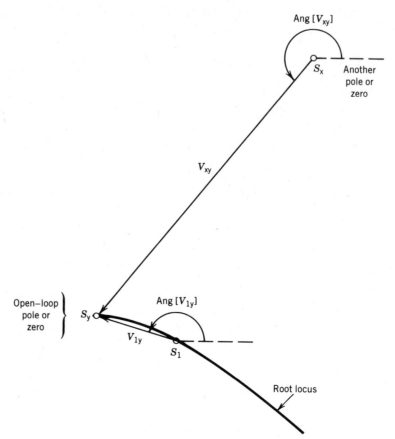

Figure 13.4-3 Root-locus plot near open-loop pole or zero.

The real values of s that satisfy this equation are breakaway points on the real s-axis, provided that these points also satisfy the requirements of rule (6) for root loci on the real s-axis.

10. *Values of gain along root loci.* In accordance with Eq 13.4-3, the value of K_n corresponding to any point $s = s_1$ along a root locus is calculated from

$$K_n = \frac{1}{|G_n[s_1]|} \qquad (13.4\text{-}18)$$

Rules (1) to (9) are equivalent to rules (1) to (9), respectively, given by Truxal [11.3] (pp. 227–231), and rule (10) is equivalent to rule (11) of Truxal. Rule (10) of Truxal is equivalent to Eq 13.4-4.

Let us apply these rules to the loop transfer function of Eq 13.4-2 to determine the major characteristics of the root-locus plot, which is given in Fig 13.4-1. The loci start (for $K_n = 0$) at the open-loop poles at $s = 0, -1, -2$, in accordance with rule (1). Since there are no finite zeros, the root loci end at the zeros of G at infinite frequency, in accordance with rule (2). The polynomial $(N[s] + D[s])$ is

$$N[s] + D[s] = 1 + D[s] = 1 + s(s + 1)(s + 2)$$
$$= 1 + 2s + 3s^2 + s^3 \qquad (13.4\text{-}19)$$

Since this is third order, by rule (3) there are three separate root loci.

The order of the polynomial $N[s]$ is $p = 0$, and the order of the polynomial $D[s]$ is $q = 3$. Hence, by rule (5) (Eq 13.4-10) the angles Θ_k of the high-frequency asymptotes are given by

$$\Theta_k = \frac{(2k + 1)180°}{0 - 3} = -(2k - 1)60° \qquad (13.4\text{-}20)$$

By Eq 13.4-11, the integer k takes on the values

$$k = 0, 1, 2 \qquad (13.4\text{-}21)$$

Comparing Eqs 13.4-20, -21 gives the following three angles Θ_k:

$$\Theta_0 = +60°, \qquad \Theta_1 = -60°, \qquad \Theta_2 = -180° \qquad (13.4\text{-}22)$$

By Eq 13.4-12, the three asymptotes intersect at the following value of s on the real s-axis:

$$s_{\text{int}} = \frac{\Sigma \, \text{Re(poles)}}{q - p} = \frac{(0) + (-1) + (-2)}{3 - 0} = -1 \qquad (13.4\text{-}23)$$

Note that the loop has three poles at $s = 0, -1, -2$, and no zeros. Two of the asymptotes are shown as dashed lines in Fig 13.4-1. The third asymptote (corresponding to $\Theta_1 = -180°$) lies along the real s-axis, to the left of the intercept point at $s_{\text{int}} = -1$.

To apply rule (6), separate the real s-axis into the following regions:

(A)	$s = +\infty$ to $s = 0$	(none)	(13.4-24)
(B)	$s = 0$ to $s = -1$	(one)	(13.4-25)
(C)	$s = -1$ to $s = -2$	(two)	(13.4-26)
(D)	$s = -2$ to $s = -\infty$	(three)	(13.4-27)

The words in parentheses give the number of real open-loop poles and zeros to the right of each region. (Remember that the loop has poles at $s = 0, -1, -2$

and no zeros.) Since there are an odd number of real open-loop poles and zeros to the right of regions (B) and (D), these are the regions of root loci.

To apply rule (7), note that the phase of G_n for real frequencies is

$$\text{Ang}[G_n[j\omega]] = -90° - \arctan[1 + j\omega] - \arctan[2 + j\omega] \quad (13.4\text{-}28)$$

This is equal to $-180°$ at $\omega = 1.414$. Hence, the root loci intersect the imaginary s-axis at $s = \pm j1.414$.

To apply rule (8), let us designate the three open-loop poles as s_a, s_b, s_c:

$$s_a = 0, \qquad s_b = -1, \qquad s_c = -2 \quad (13.4\text{-}29)$$

In the region of the root loci near these poles, the angles of the vectors drawn from the loci to the poles are given by

$$\text{Ang}[V_{1a}] = -\overset{\text{other}}{\underset{\text{poles}}{\sum}} \text{Ang}[V_{xa}] = -\text{Ang}[V_{ba}] - \text{Ang}[V_{ca}] \quad (13.4\text{-}30)$$

$$\text{Ang}[V_{1b}] = -\text{Ang}[V_{ab}] - \text{Ang}[V_{cb}] \quad (13.4\text{-}31)$$

$$\text{Ang}[V_{1c}] = -\text{Ang}[V_{ac}] - \text{Ang}[V_{bc}] \quad (13.4\text{-}32)$$

The vectors V_{1a}, V_{1b}, V_{1c} are drawn from a point s_1 on a locus to points s_a, s_b, s_c, where s_1 is close to points s_a, s_b, s_c. Vectors V_{ba}, V_{ca} are drawn from s_b, s_c to s_a; vectors V_{ab}, V_{cb} are drawn from s_a, s_c to s_b, and vectors V_{ac}, V_{bc} are drawn from s_a, s_b to s_c. The angles of the vectors drawn between the points s_a, s_b, s_c are

$$\text{Ang}[V_{ab}] = 180° \qquad \text{Ang}[V_{ba}] = 0 \quad (13.4\text{-}33)$$

$$\text{Ang}[V_{ac}] = 180° \qquad \text{Ang}[V_{ca}] = 0 \quad (13.4\text{-}34)$$

$$\text{Ang}[V_{bc}] = 180° \qquad \text{Ang}[V_{cb}] = 0 \quad (13.4\text{-}35)$$

Combine Eqs 13.4-33 to -35 with Eqs 13.4-30 to -32:

$$\text{Ang}[V_{1a}] = -0 - 0 = 0 \quad (13.4\text{-}36)$$

$$\text{Ang}[V_{1b}] = -180° - 0 = -180° = 180° \quad (13.4\text{-}37)$$

$$\text{Ang}[V_{1c}] = -180° - 180° = -360° = 0 \quad (13.4\text{-}38)$$

Equations 3.4-36 to -38 describe the directions of the loci in regions close to the closed-loop poles at $s_a = 0$, $s_b = -1$, $s_c = -2$.

To apply rule (9), note that

$$N[s] = 1 \quad (13.4\text{-}39)$$

$$D[s] = s(s + 1)(s + 2) \quad (13.4\text{-}40)$$

Hence, Eq 13.4-17 becomes

$$(1)\frac{d[s(s+1)(s+2)]}{ds} = s(s+1)(s+2)\frac{d[1]}{ds} \qquad (13.4\text{-}41)$$

The right-hand expression is zero. To calculate the left-hand derivative, note that

$$d[xyz] = yz\,dx + xz\,dy + xy\,dz \qquad (13.4\text{-}42)$$

Hence, Eq 13.4-41 becomes

$$s(s+1)\frac{d[s+2]}{ds} + s(s+2)\frac{d[s+1]}{ds} + (s+1)(s+2)\frac{ds}{ds} = 0 \quad (13.4\text{-}43)$$

This reduces to

$$s^2 + 2s + \tfrac{2}{3} = 0 \qquad (13.4\text{-}44)$$

Using the quadratic formula gives

$$s = -1 \pm \sqrt{1/3} = -1.5774, -0.4226 \qquad (13.4\text{-}45)$$

The point $s = -1.5774$ does not lie on the region of a root locus, and so is discarded. Hence the breakaway point is $s = -0.4226$, which agrees with the root-locus plot of Fig 13.4-1.

Let us apply rule (11) to calculate the values of the gain K_n corresponding to particular points on the root loci. At the breakaway point calculated from rule (9), $s = -0.4226$. By Eq 13.4-2, the value of G_n at this point is

$$G_n = \frac{1}{s(1+s)(2+s)}$$

$$= \frac{1}{(-0.4226)(1-0.4226)(2-0.4226)} = -2.598 \quad (13.4\text{-}46)$$

By Eq 13.4-18, the value of K_n at the breakaway point is

$$K_n = \frac{1}{|G_n|} = \frac{1}{|-2.598|} = 0.3849 \qquad (13.4\text{-}47)$$

Rule (7) showed that the loci intersect the $j\omega$ axis at $s = \pm j1.414$. The value of $1/G_n$ at $s = +j1.414$ is

$$1/G_n = j1.414(1+j1.414)(2+j1.414)$$

$$= 1.414\underline{/90°}\ 1.732\underline{/54.73°}\ 2.449\underline{/35.26°}$$

$$= 5.998\underline{/179.99°} \qquad (13.4\text{-}48)$$

Hence, by Eq 13.4-18, the value of K_n at this point is

$$K_n = \frac{1}{|G_n|} = 5.998 \tag{13.4-49}$$

13.4.2 Root-Locus Plot Using Generalized Frequency Response

13.4.2.1 Kusters and Moore's Discussion. The generalized frequency-response method described in Section 13.2 can also be used to derive a root-locus plot. This approach is described in the following (slightly edited) discussion by N. L. Kusters and W. J. M. Moore, applied to a paper given by the author [1.11] (pp. 69–70). The author's paper extended the work of Kusters and Moore [13.3].

"The root locus can be obtained by considering a plot of the log-magnitude and the phase curves of the open-loop transfer function against $\log[|s|]$ along both the negative real and positive imaginary axes of the s-plane. The curves are quite easily drawn, since:"

1. Along the negative real axis, the phase curves are straight lines.

2. Along the positive imaginary axis, the log-magnitude curves can be approximated by straight-line asymptotes.

From these open-loop curves, a root-locus plot of the closed loop can easily be drawn on the same diagram.

A plot of these curves is shown in Fig 13.4-4 for the following open-loop transfer function:

$$G = \frac{(s + 0.2)(s + 1)}{s(s + 0.04)^2(s + 40)^2} \tag{13.4-50}$$

The corresponding root-locus plot (not to scale) is shown in Fig 13.4-5.

"At very low values of gain, such as K_1, there are five negative real roots r_1, r_2, r_3, r_4, and r_5, because at these points on the plot of the open-loop function along the negative real axis, the magnitude is unity and the phase is $-180°$, or an odd multiple thereof. The locations of these roots at this low gain are largely determined by the poles of the open-loop transfer function, which occur at $|s| = 0$, $|s| = 0.04$, and $|s| = 40$."

"As the gain is increased, the roots r_1 and r_2 approach one another on the plot of the function along the negative real axis ($s = -\omega$) and eventually coincide to become a double-order root. As the gain is increased further, they form a conjugate complex pair. Their loci proceed toward the imaginary axis plot ($s = j\omega$), reaching it at point a where the corresponding phase plot indicates a phase lag of 180° (a'). The loci then cross the $s = j\omega$ curve, indicating instability (an excursion into the right half plane)."

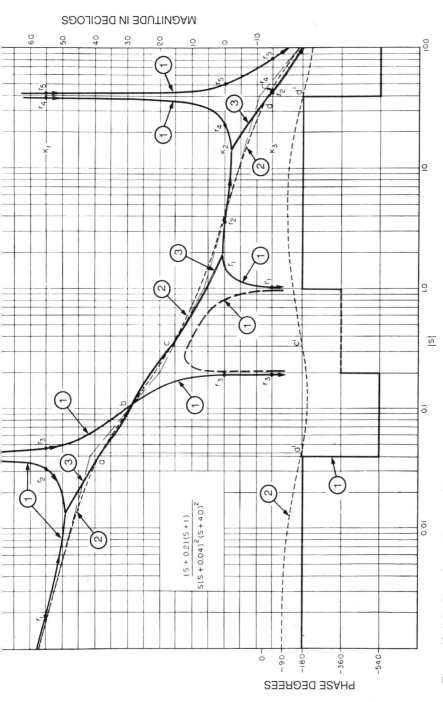

$$\frac{(S+0.2)(S+1)}{S(S+0.04)^2(S+40)^2}$$

Figure 13.4-4 Plot of open-loop transfer function and closed-loop locus: curve ① open-loop transfer function along negative real axis; ② open-loop transfer function along positive imaginary axis; ③ closed-loop locus. Reprinted with permission from Ref [1.11] © 1953 AIEE (now IEEE).

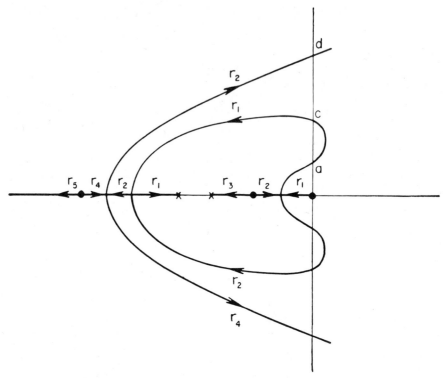

Figure 13.4-5   Root-locus plot on s-plane (not to scale). Reprinted with permission from Ref [1.11] © 1953 AIEE (now IEEE).

"Note that the region between the open-loop transfer-function plot along the negative real axis and the plot along the $s = j\omega$ axis is a stable region for the root-locus plot. Thus, as the gain is increased from point a, the root-locus plot has to pass through point b. It has to stay in the unstable region until it reaches point c, where the corresponding phase curve of the $s = j\omega$ plot is again 180°. At a gain slightly smaller than K_2, these complex roots coincide again on the negative real axis. At a gain of K_2, the system has five separate real roots, r_3, r_1, r_2, r_4, r_5. At this gain region, the roots below ω_c are controlled by the open-loop zeros at $|s| = 0.2$ and $|s| = 1$, while those above ω_c are controlled by the pole $|s| = 40$."

"This situation of five negative real roots continues, with r_1 approaching the zero at $|s| = 1$, and r_2 approaching root r_4 along the $s = -\omega$ plot, until r_2 equals r_4. At that point, r_2 and r_4 form another conjugate complex pair. At a gain of K_3, the system has two stable complex roots r_2 and r_4, and three negative real roots r_3, r_1, and r_5. The system finally becomes unstable at a gain corresponding to point d, at which the phase lag of the $s = j\omega$ plot is again 180° (d'), and the complex roots r_2 and r_4 assume positive real parts."

"The foregoing could also have been explained by using the root-locus method of Evans, which would give a plot similar to Fig 14.3-5. However, any attempt to plot this figure accurately for the function described is practically impossible, because of the large separation of its poles and zeros. This inherent disadvantage of Evans' root-locus method can be overcome by making several plots of different scales."

13.4.2.2 Setting the Damping Ratio of Dominant Closed-Loop Poles.

The usual goal of root-locus design is to set the loop gain so as to satisfy a given damping ratio for the dominant closed-loop poles. This can be achieved by plotting the phase of the loop transfer function G along the radial s-plane axis corresponding to the desired damping ratio.

Consider, for example, the following loop transfer function, obtained from Eq 13.2-4, which was analyzed in Section 13.2:

$$G = \frac{\omega_c(1 + \omega_2/s)(1 + \omega_3/s)}{s(1 + \omega_1/s)^2(1 + s/\omega_4)^2} = \frac{\omega_c\omega_4^2(s + \omega_2)(s + \omega_3)}{s(s + \omega_1)^2(s + \omega_4)^2} \quad (13.4\text{-}57)$$

where $\omega_1 = 0.04$, $\omega_2 = 0.2$, $\omega_3 = 1$, $\omega_4 = 16$. The normalized loop transfer function G_n and the gain parameter K_n are given by

$$G_n = \frac{(s + \omega_2)(s + \omega_3)}{s(s + \omega_1)^2(s + \omega_4)^2} = \frac{(s + 0.2)(s + 1)}{s(s + 0.04)^2(s + 16)^2} \quad (13.4\text{-}52)$$

$$K_n = \omega_4^2\omega_c = 256\omega_c \quad (13.4\text{-}53)$$

This is the same as the loop transfer function considered above by Kusters and Moore in Figs 13.4-3, -4, except that the upper break frequency ω_4 is 16 rather than 40. Hence, the general forms of the root loci are similar.

Figure 13.2-8 showed plots of the magnitude and phase of this loop transfer function of Eq 13.4-52 along radial s-plane axes corresponding to the following damping ratios of the closed-loop poles: $\zeta = 0, 0.5, 0.7, 0.8, 0.9$. These are shown in the frequency region near the dominant closed-loop poles. In a root-locus design, only the phase plot of Fig 13.2-8 need be constructed, and this only for the desired value of damping ratio of the dominant closed-loop poles.

Assume, for example, that the desired damping ratio of the dominant closed-loop poles is $\zeta = 0.7$. The phase plot for $\zeta = 0.7$ shown in Fig 13.2-8 would be constructed. This phase plot intersects the $-180°$ phase axis at the following values of $|s|$:

$$|s| = 1.75, 5.7 \quad (13.4\text{-}54)$$

Assuming that one wants to maximize the natural frequency of the dominant closed-loop poles, the case $|s| = 5.7$ is selected. The parameters for the dominant closed-loop poles are then

$$\zeta = 0.7 \tag{13.4-55}$$
$$\omega_n = |s| = 5.7 \tag{13.4-56}$$
$$\zeta\omega_n = 3.99 \tag{13.4-57}$$
$$\omega_o = \omega_n\sqrt{1 - \zeta^2} = 4.07 \tag{13.4-58}$$

Hence, the dominant closed-loop poles are

$$s = -\zeta\omega_n \pm j\omega_o = -3.99 \pm j4.07 \tag{13.4-59}$$

In accordance with Eq 13.4-18, the corresponding value of the normalized gain K_n is found by calculating $1/G_n$ at the upper half-plane pole:

$$\frac{1}{G_n} = \left.\frac{s(s + 0.04)^2(s + 16)^2}{(s + 0.2)(s + 1)}\right|_{s = -3.99 + j4.02}$$

$$= \frac{(-3.99 + j4.02)(0.04 - 3.99 + j4.02)^2(16 - 3.99 + j4.02)^2}{(0.2 - 3.99 + j4.02)(1 - 3.99 + j4.02)}$$

$$= \frac{(5.664\underline{/134.79°})(5.636\underline{/134.50°})^2(12.66\underline{/18.51°})^2}{(5.525\underline{/133.31°})(5.010\underline{/126.64°})}$$

$$= 1041.8\underline{/180.86°} \tag{13.4-60}$$

By Eq 13.4-18, the value of K_n is

$$K_n = \frac{1}{|G_n|} = 1041.8 \tag{13.4-61}$$

By Eq 13.4-53, the corresponding value for ω_c is

$$\omega_c = \frac{K_n}{256} = \frac{1041.8}{256} = 4.07 \tag{13.4-62}$$

The phase of $1/G_n$ in Eq 13.4-60 should ideally be 180°, but because of inaccuracy in the plot of Fig 13.2-8, and in reading that plot, this phase is in error by 0.86°. Correcting these errors would yield an exact value of ω_c of 4.053, and exact dominant closed-loop poles of

$$s = -3.892 \pm j3.971 \tag{13.4-63}$$

The values of the other closed-loop poles (which are real) can be computed by using the methods described in Section 13.2.

Chapter 14

Response to an Arbitrary Input

This chapter presents a method for analyzing the response of a system to an arbitrary input signal. This provides background for the concepts developed in Chapter 5 for characterizing the general time behavior of feedback-control loops.

14.1   SUMMARY OF APPROACH

An error-coefficient method was presented in Chapter 5 for calculating the response to an arbitrary input that was based on approximations of the G_{ie} frequency response. This chapter gives a more general approach, which addresses high-frequency inputs as well as low-frequency inputs. It provides a basic philosophy for the approximations of system response given earlier. Most of the material of this section is a condensation of a 1955 paper by the author [14.1].

If one has an input signal that is prescribed explicitly either by a set of equations or by tabulated data, the simplest way to calculate the response of a feedback system to that input is to feed the data into a simulation of the system. However, this situation rarely occurs in practice. One is usually faced with vague descriptions of tasks the system must perform, and disturbances it will experience. The response of the system must be characterized in a general fashion so that the effects of these vaguely defined commands and disturbances can be evaluated. Therefore, one needs insight and general quantitative means of characterizing the response to various inputs. The approach of this chapter is directed to that end.

Most methods for calculating the response of a system to an arbitrary input require that the input be approximated by a series of impulses, or steps, or straight lines, or parabolas. These may be considered to represent steps of derivatives or integrals of the input: a straight line is a step of the first derivative, a parabola is a step of the second derivative, and an impulse is a step of the first integral. By using a sufficient number of any one of these elements, any of these methods can approximate the response to the accuracy desired.

This section shows how to break an input down optimally into a sum of not just one type of element, but a number of different types, using for each portion of the input curve the order of derivative step that best approximates that portion. For example, an impulse is used to approximate a pulselike portion of the input, while a parabola is used to approximate a slowly-varying portion. Approximating the input in this manner requires much fewer approximating elements than when one type is used. Even more important, it gives insight into why the system responds the way it does, and how the system transfer function can be changed to improve the response.

Thus, any curve can be represented as a sum of steps of various derivatives and integrals of the input. The steps of derivatives approximate the low-frequency portions of the input, and the steps of integrals approximate the high-frequency portions. To calculate the response to an arbitrary input, general expressions are needed for the responses to steps of derivatives and integrals of the input. In feedback-control applications, low-frequency inputs are usually of more interest, and so the response to a step of a derivative is considered first.

14.2 RESPONSES TO STEPS OF DERIVATIVES AND INTEGRALS OF THE INPUT

14.2.1 Response to a Step of a Derivative

The following analysis yields an expression for the response to a step of a derivative, which contains a series of steady-state coefficients and a transient term proportional to the size of the step. To simplify the symbolism, the nth time derivative is designated D^n:

$$D^n x_i[t] = \frac{d^n x_i[t]}{dt^n} \tag{14.2-1}$$

Similarly, the mth time integral of $x_i[t]$ is designated $D^{-m} x_i[t]$. Assume that the nth derivative of the input experiences a step of amplitude A_n:

$$D^n x_i[t] = A_n \quad \text{for} \quad t > 0 \tag{14.2-2}$$

Since A_n is a constant, the Laplace transform of this derivative is

$$\mathscr{L}[D^n x_i[t]] = A_n/s \tag{14.2-3}$$

The Laplace transform of the input is

$$X_i = \frac{1}{s^n} \mathscr{L}[D^n x_i[t]] = \frac{A_n}{s^{n+1}} \tag{14.2-4}$$

If this signal is fed into a feedback loop, the Laplace transform of the resultant feedback signal is

$$X_b = G_{ib} X_i = \frac{A_n G_{ib}}{s^{n+1}} \tag{14.2-5}$$

This can be expanded in partial fractions as follows:

$$X_b = A_n \left(\frac{c_0'}{s^{n+1}} + \frac{c_1'}{s^n} + \cdots + \frac{c_n'}{s} + T_n'[s] \right) \tag{14.2-6}$$

The constants c_0', c_1', c_n' are called the steady-state coefficients of the feedback response, and $T_n'[s]$ is the transform of the nth-derivative transient term. The steady-state coefficients are designated with primes to distinguish them from the error coefficients, which are derived from an equivalent expansion of the error response. Gardner and Barnes [2.1] (Chapter VI, p. 162, Eq 27) gives a general expression for the constants of the partial-fraction expansion for multiple-order poles. This yields the following equation for the steady-state coefficients:

$$c_n' = \frac{1}{n!} \frac{d^n G_{ib}}{ds^n} \bigg|_{s=0} \tag{14.2-7}$$

The equation for the transform of the nth-derivative transient term is

$$T_n'[s] = \sum_{k=1}^{K} \frac{K_{kn}'}{s - s_k} \tag{14.2-8}$$

Each value s_k represents one of the K poles of G_{ib}. To simplify the presentation it is assumed that G_{ib} has only single-order poles, but the same conclusions would apply without this restriction. The coefficients K_{kn}' are given by

$$K_{kn}' = \frac{(s - s_k) G_{ib}}{s^{n+1}} \bigg|_{s=s_k} \tag{14.2-9}$$

Factor A_n / s^{n+1} from the steady-state coefficient part of the expansion of Eq. 14.2-6:

$$X_b = \frac{A_n}{s^{n+1}} \left(c_0' + c_1' s + c_2' s^2 + \cdots + c_n' s^n \right) + A_n T_n'[s] \tag{14.2-10}$$

In accordance with Eq 14.2-4, substitute X_i for A_n / s^{n+1}:

$$\begin{aligned} X_b &= X_i \left(c_0' + c_1' s + c_2' s^2 + \cdots + c_n' s^n \right) + A_n T_n'[s] \\ &= c_0' X_i + c_1' s X_i + c_2' s^2 X_i + \cdots + c_n' s^n X_i + A_n T_n'[s] \end{aligned} \tag{14.2-11}$$

The inverse Laplace transform of this is

$$x_b = c_0' x_i + c_1' D x_i + c_2' D^2 x_i + \cdots + c_n' D^n x_i + A_n T_n'[t] \quad (14.2\text{-}12)$$

where $T_n'[t]$ is the inverse Laplace transform of $T_n'[s]$, and is equal to

$$T_n'[t] = \sum_{k=1}^{K} K_{kn}' e^{s_k t} \quad (14.2\text{-}13)$$

This is the nth-derivative transient term. By Eq 14.2-2, the constant A_n is the value of $D^n x_i$. Replacing A_n by $D^n x_i$ gives

$$x_b = c_0' x_i + c_1' D x_i + c_2' D^2 x_i + \cdots + (c_n' + T_n'[t]) D^n x_i \quad (14.2\text{-}14)$$

The quantity $(c_n' + T_n'[t])$ is called the nth-derivative composite transient term. It can be shown that $c_n' = - T_n'[t]$ at $t = 0$, and so the composite transient term is zero at time $t = 0$.

As will be shown, by successively differentiating an input signal and applying straight-line approximations to the result, an input signal can be broken down into a sum of steps of derivatives. Equation 14.2-14 shows that the resultant response can be calculated by: (1) including a composite transient term for each step of derivative, proportional to the size of the step; and (2) adding to the sum of these transients a steady-state portion, consisting of the lower-order derivatives, multiplied by the corresponding steady-state coefficients.

For a low-frequency input signal, it is usually more convenient to consider the error response rather than the feedback response, because the feedback signal follows the input very closely. The expansion can be expressed in terms of the error response by noting that

$$x_e = x_i - x_b \quad (14.2\text{-}15)$$

Substituting Eq 14.2-11 into Eq 14.2-15 gives

$$x_e = (1 - c_0') x_i - c_1' D x_i - c_2' D^2 x_i - \cdots - (c_n' + T_n'[t]) D^n x_i \quad (14.2\text{-}16)$$

This can be expressed as

$$x_e = c_0 x_i + c_1 D x_i + c_2 D^2 x_i + \cdots + (c_n + T_n[t]) D^n x_i \quad (14.2\text{-}17)$$

where c_0, c_1, etc. are the steady-state error coefficients, and $T_n[t]$ is the nth-derivative transient term for the error response. These are equal to

$$c_0 = 1 - c_0' \quad (14.2\text{-}18)$$

$$c_n = -c_n' \quad \text{for} \quad n > 0 \quad (14.2\text{-}19)$$

$$T_n[t] = -T_n'[t] \quad (14.2\text{-}20)$$

14.2.2   Calculation of Steady-State Coefficients and Transient Terms

The error coefficients c_n can be calculated by applying Eqs 14.2-18, -19 to Eq 14.2-7, but can also be calculated as follows, directly from the G_{ie} transfer function:

$$c_n = \frac{1}{n!}\frac{d^n G_{ie}}{ds^n}\bigg|_{s=0} \qquad (14.2\text{-}21)$$

On the other hand, it is much simpler to calculate the steady-state coefficients c_n and c_n' by long-division expansions of the transfer functions G_{ie} and G_{ib}, as described in Appendix J.

Since a transient term $T_n[t]$ for the error response is the negative of the transient term $T_n'[t]$ for the feedback response, the coefficient K_{kn} for the error-response transient term is equal to the negative of the coefficient K_{kn}' for the feedback response:

$$K_{kn} = -K_{kn}' \qquad (14.2\text{-}22)$$

The simplest way to obtain the coefficients K_{kn} or K_{kn}' is to calculate the coefficients for a step of the input (the zero derivative) and derive the others from these. An error-response coefficient for a step of x_i is given by

$$K_{k0} = \frac{(s - s_k)G_{ie}}{s}\bigg|_{s=s_k} \qquad (14.2\text{-}23)$$

Comparing this with Eq 14.2-9 shows that the coefficient for a step of the nth derivative is related to this by

$$K_{kn} = \frac{K_{k0}}{s_k^n} \qquad (14.2\text{-}24)$$

Equation 14.2-24 shows that as n is increased the coefficients for high-frequency poles (i.e., the poles with large values of $|s_k|$) are decreased at a much faster rate than the coefficients for low-frequency poles. Consequently, if n is sufficiently large, the transient coefficient for the lowest-frequency pole is much greater than all the other coefficients. Thus, for high-order derivatives, the transient terms are determined almost entirely by the lowest-frequency closed-loop pole (or poles). If this pole is real, which is often the case, the transient terms for the higher derivatives are simply real exponentials.

Combining Eqs 14.2-13, -20, -22, and -24 gives the following for the nth-derivative transient term:

$$T_n[t] = \sum \frac{K_{k0}}{s_k^n}e^{s_k t} \qquad (14.2\text{-}25)$$

Comparing the transient terms for successive derivatives shows that

$$T_{n-1}[t] = \frac{dT_n[t]}{dt} \qquad (14.2\text{-}26)$$

Although Eq 14.2-25 does not apply to transfer functions with multiple-order poles, it can be shown that Eq 14.2-26 applies to all transfer functions.

Practical feedback loops have zero gain at infinite frequency. For such loops, the feedback signal x_b cannot change instantaneously for a step of the input, or for a step of a derivative of the input. Hence for $n \geq 0$, the expansion of Eq 14.2-14 can be set equal to zero at $t = 0+$. This requires that

$$c_n' + T_n'[0] = 0 \qquad \text{for} \quad n \geq 0 \qquad (14.2\text{-}27)$$

For the error response the same relation holds for $n > 0$:

$$c_n + T_n[0] = 0 \qquad \text{for} \quad n > 0 \qquad (14.2\text{-}28)$$

Thus, for the derivatives, the composite transient terms for both the error and feedback responses start at zero at time $t = 0$, and for the zeroth derivative ($n = 0$), the composite transient term for the feedback signal starts at zero. Integrating Eq 14.2-26 gives

$$T_n[t] - T_n[0] = \int_0^t T_{n-1}[t]\, dt \qquad (14.2\text{-}29)$$

Applying Eqs 14.2-27, -28 to this gives

$$c_n' + T_n'[t] = \int_0^t T_{n-1}'[t]\, dt \qquad \text{for} \quad n \geq 0 \qquad (14.2\text{-}30)$$

$$c_n + T_n[t] = \int_0^t T_{n-1}[t]\, dt \qquad \text{for} \quad n > 0 \qquad (14.2\text{-}31)$$

14.2.3   Response to a Step of an Integral

If the input signal contains a high-frequency component, which rises and falls like a pulse in a period significantly shorter than the rise time of the step response, this component should be approximated by an impulse, a doublet, etc., which can be regarded as a step of an integral of the input. Hence, for high-frequency signals one should consider the expansion for a step of an integral of the input. Assume that the mth integral of the input is a step of magnitude A_{-m}:

$$D^{-m}x_i = A_{-m} \qquad \text{for} \quad t > 0 \qquad (14.2\text{-}32)$$

The Laplace transform of the mth integral is

$$\mathscr{L}\left[D^{-m}x_i\right] = A_{-m}/s \qquad (14.2\text{-}33)$$

The input x_i is the mth derivative of this step, and so its transform is

$$X_i = s^m\mathscr{L}\left[D^{-m}x_i\right] = A_{-m}s^{m-1} \qquad (14.2\text{-}34)$$

The feedback signal response for this input is

$$X_b = G_{ib}X_i = A_{-m}\left(s^{m-1}G_{ib}\right) \qquad (14.2\text{-}35)$$

Equation 14.2-35 can be expanded as follows:

$$X_b = A_{-m}\left(a_0 s^{m-1} + a_1 s^{m-2} + \cdots + a_{m-1} + T'_{-m}[s]\right) \qquad (14.2\text{-}36)$$

The transient term $T'_{-m}[s]$ is obtained by replacing n with $-m$ in the expression for $T'_n[s]$. The coefficients $a_0, a_1, \ldots, a_m$ are called the initial-value coefficients. The terms of Eq 14.2-36 which contain these initial-value coefficients produce impulses, doublets, triplets, etc. in the x_b response. Substitute Eq 14.2-34 into Eq 14.2-36:

$$X_b = a_0 X_i + a_1 \frac{X_i}{s} + \cdots + a_{m-1}\frac{X_i}{s^{m-1}} + A_{-m}T'_{-m}[t] \qquad (14.2\text{-}37)$$

Take the inverse transform, and replace A_{-m} by $D^{-m}x_i$:

$$x_b = a_0 x_i + a_1 D^{-1}x_i + \cdots + a_{m-1}D^{-m+1}x_i + \left(D^{-m}x_i\right)T'_{-m}[t] \qquad (14.2\text{-}38)$$

Equation 14.2-38 shows that the feedback response to high-frequency portions of the input can be calculated as follows. Integrate each high-frequency portion successively, and, at an appropriate integral, approximate the curve by steps. Multiply each input curve, except the last, by the corresponding initial-value coefficient a_m, and add the resultant curves to the calculated response. For each step, add a transient to the response proportional to the size of the step.

The simplest general way to calculate the initial-value coefficients is to expand the G_{ib} transfer function by long division, as shown in Appendix J, with the terms arranged in the reverse order used for calculating the steady-state coefficients. The first coefficient a_0 is zero if G_{ib} has zero gain at infinite frequency. The initial-value coefficients with zero values, and the first nonzero coefficient, can be calculated from

$$a_m = s^m G_{ib}|_{s=\infty} = s^m G|_{s=\infty} \qquad (14.2\text{-}39)$$

The a_m coefficients are called initial-value coefficients because they are the initial values of the corresponding transient terms:

$$a_m = T'_{-m}[0] \tag{14.2-40}$$

The initial-value coefficients are defined in terms of the feedback response, rather than the error response, because the feedback response is generally used to characterize the response to high-frequency inputs. Nevertheless, it is sometimes desirable to consider the equivalent expansion in terms of the error response. Use the expression for x_b in Eq 14.2-38, and set x_e equal to $(x_i - x_b)$. This gives

$$x_e = (1 - a_0)x_i - a_1 D^{-1}x_i - \cdots - a_{m-1}D^{m-1}x_i - D^{-m}x_i T'_{-m}[t] \tag{14.2-41}$$

This can be expressed as

$$x_e = a_0' x_i + a_1' D^{-1}x_i + \cdots + a_{m-1}' D^{m-1}x_i + D^{-m}x_i T_{-m}[t] \tag{14.2-42}$$

where

$$a_0' = 1 - a_0 \tag{14.2-43}$$

$$a_m' = -a_m \qquad \text{for} \quad m > 0 \tag{14.2-44}$$

The a_m' parameters are the initial-value coefficients for the error response.

14.3 FEEDBACK-LOOP EXAMPLE

14.3.1   System Parameters

The following loop transfer function is used as an example:

$$G = \frac{\omega_c(1 + \omega_2/s)}{s(1 + \omega_1/s)(1 + s/\omega_3)} = \frac{\omega_c \omega_3(s + \omega_2)}{s(s + \omega_1)(s + \omega_3)} = \frac{N}{D} \tag{14.3-1}$$

An asymptotic magnitude plot of this is shown in Fig 14.3-1. The assumed values of the system parameters are shown in the left-hand part of Table 14.3-1. The G_{ib} transfer function is

$$G_{ib} = \frac{G}{1 + G} = \frac{N}{N + D}$$

$$= \frac{\omega_c \omega_3(s + \omega_2)}{s^3 + s^2(\omega_1 + \omega_3) + s\omega_3(\omega_1 + \omega_c) + \omega_c \omega_2 \omega_3} \tag{14.3-2}$$

The closed-loop poles can be calculated using the techniques of Chapter 13,

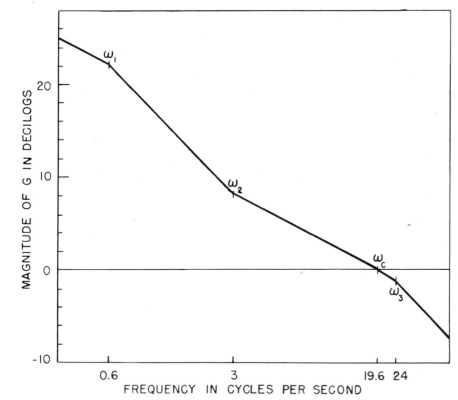

Figure 14.3-1 Magnitude asymptotes for illustrative feedback loop. Reprinted with permission from Ref [14.1]© 1955 AIEE (now IEEE).

Section 13.2. This allows the denominator to be factored, to give

$$G_{ib} = \frac{\omega_c \omega_3 (s + \omega_2)}{(s + \omega_a)(s^2 + 2\zeta\omega_n s + \omega_n^2)} \qquad (14.3\text{-}3)$$

The values for ω_a, ω_n, and ζ are given in the right-hand part of Table 14.3-1. The equation for G_{ie} is

$$
\begin{aligned}
G_{ie} &= \frac{1}{1 + G} = \frac{D}{N + D} \\
&= \frac{s(s + \omega_1)(s + \omega_3)}{s^3 + s^2(\omega_1 + \omega_3) + s\omega_3(\omega_1 + \omega_c) + \omega_c\omega_2\omega_3} \\
&= \frac{s(s + \omega_1)(s + \omega_3)}{(s + \omega_a)(s^2 + 2\zeta\omega_n s + \omega_n^2)} \qquad (14.3\text{-}4)
\end{aligned}
$$

TABLE 14.3-1   Parameters of Feedback-Loop Transfer Function

Open-Loop Parameters	Closed-Loop Parameters
$\omega_c = 123.0 \ \text{sec}^{-1}$ (19.6 Hz)	$\omega_a = 21.53 \ \text{sec}^{-1}$
$\omega_1 = 3.77 \ \text{sec}^{-1}$ (0.30 Hz)	$\omega_n = 127.0 \ \text{sec}^{-1}$
$\omega_2 = 18.55 \ \text{sec}^{-1}$ (6.00 Hz)	$\zeta = 0.522$
$\omega_3 = 150.8 \ \text{sec}^{-1}$ (24.0 Hz)	$\omega_o = 108.3 \ \text{sec}^{-1}$
	$\zeta\omega_n = 66.3 \ \text{sec}^{-1}$

The error-response nth-derivative transient term has the form

$$T_n[t] = K_{an}e^{-\omega_a t} + |K_{bn}|e^{-\zeta\omega_n t}\cos[\omega_o t + \text{Ang}[K_{bn}]] \qquad (14.3\text{-}5)$$

The coefficients for a unit step of the input are

$$K_{a0} = (s + \omega_a)G_{ie}|_{s=-\omega_a} = \frac{(s + \omega_1)(s + \omega_3)}{(s^2 + 2\zeta\omega_n s + \omega_n^2)}\Bigg|_{s=-\omega_a} \qquad (14.3\text{-}6)$$

$$K_{b0} = \frac{1}{j\omega_0}(s^2 + 2\zeta\omega_n s + \omega_n^2)\frac{G_{ie}}{s}\Bigg|_{s=-\zeta\omega_n+j\omega_o}$$

$$= \frac{(s + \omega_1)(s + \omega_3)}{(j\omega_o)(s + \omega_a)}\Bigg|_{s=-\zeta\omega_n+j\omega_o} \qquad (14.3\text{-}7)$$

The coefficients for derivatives of the input are related to these by

$$K_{an} = \frac{K_{a0}}{(-\omega_a)^n} \qquad (14.3\text{-}8)$$

$$K_{bn} = \frac{K_{b0}}{(-\zeta\omega_n + j\omega_o)^n} \qquad (14.3\text{-}9)$$

The magnitude and phase of the expression $(-\zeta\omega_n + j\omega_o)$ are

$$|-\zeta\omega_n + j\omega_o| = \omega_n \qquad (14.3\text{-}10)$$

$$\text{Ang}[-\zeta\omega_n + j\omega_o] = \pi - \arccos[\zeta] \ \text{radian} \qquad (14.3\text{-}11)$$

Hence, by Eq 14.3-9 the magnitude and phase of K_{bn} are related as follows to

TABLE 14.3-2   Coefficients for Error-Response Transient Terms $T_n[t]$ for Steps of Derivatives and Integrals of the Input

n	K_{an}	$\|K_{bn}\|$	Ang[K_{bn}] (rad)
-2	-76.9 sec^{-2}	$2.17 \times 10^4 \text{ sec}^{-2}$	-2.571
-1	3.57 sec^{-1}	171.5 sec^{-1}	1.593
0	-0.166	1.35	-0.529
1	$7.71 \times 10^{-3} \text{ sec}$	$1.06 \times 10^{-2} \text{ sec}$	-2.649
2	$-3.58 \times 10^{-4} \text{ sec}^2$	$8.37 \times 10^{-5} \text{ sec}^2$	1.513
3	$1.66 \times 10^{-5} \text{ sec}^3$	$6.59 \times 10^{-7} \text{ sec}^3$	-0.607

the magnitude and phase of K_{b0}:

$$|K_{bn}| = |K_{b0}|/\omega_n^n \tag{14.3-12}$$

$$\text{Ang}[K_{bn}] = \text{Ang}[K_{b0}] - n(\pi - \arccos[\zeta]) \text{ radian} \tag{14.3-13}$$

Table 14.3-2 shows the values of K_{an}, $|K_{bn}|$, and Ang[K_{bn}] for values of n from -2 to $+3$. The values for $n = 0$ are calculated from Eqs 14.3-6, -7, and the values for other values of n are obtained by applying Eqs 14.3-8, -12, and -13.

The simplest way to calculate the error coefficients is to expand the transfer function G_{ie} in long division as described in Appendix J. By Eq 14.3-4 the expression for G_{ie} using the values of Table 14.3-1 is

$$G_{ie} = \frac{s^3 + (\omega_1 + \omega_3)s^2 + \omega_1\omega_3 s}{s^3 + (\omega_1 + \omega_3)s^2 + \omega_3(\omega_1 + \omega_c)s + \omega_c\omega_2\omega_3}$$

$$= \frac{s^3 + 154.57s^2 + 568.52s}{s^3 + 154.57s^2 + 19{,}117s + 349{,}640} \tag{14.3-14}$$

The expressions in the numerator and the denominator are arranged in ascending powers of s, and long division is used to divide the numerator by the denominator:

$$
\begin{array}{r}
1.6260 \times 10^{-3}s + 3.5319 \times 10^{-4}s^2 - 1.717 \times 10^{-5}s^3 \\
\hline
a_0 + a_1 s + a_2 s^2 + s^3 \,\big)\, 568.52s + \quad 154.57s^2 + \quad s^3 \\
568.52s + \quad 31.08s^2 + \quad 0.2513s^3 \\
\hline
123.49s^2 + \quad 0.7487s^3 \\
123.49s^2 + \quad 6.7519s^3 \\
\hline
- \quad 6.0032s^3
\end{array}
$$

$$\tag{14.3-15}$$

TABLE 14.3-3   Accurate Values of Error Coefficients for Example Compared with Approximate Values

	Accurate Value	Approximate Value
c_1	1.626×10^{-3} sec	$1/\alpha\omega_c = 1.626 \times 10^{-3}$ sec
c_2	3.53×10^{-4} sec^2	$1/\omega_c\omega_i = 4.31 \times 10^{-4}$ sec^2
c_3	-1.717×10^{-5} sec^3	$-\tau_{ic}c_2 = -2.64 \times 10^{-5}$ sec^3

where $a_0 = 349{,}640$, $a_1 = 19{,}117$, and $a_2 = 154.57$. This yields the following error-coefficient expansion:

$$G_{ie} = c_1 s + c_2 s^2 + c_3 s^3 + \cdots$$
$$= 1.6260 \times 10^{-3}s + 3.5319 \times 10^{-4}s^2 - 1.717 \times 10^{-5}s^3 + \cdots \quad (14.3\text{-}16)$$

These error coefficient values c_1, c_2, and c_3 are listed in the first column of Table 14.3-3. The second column gives the approximate error-coefficient values based on the equations derived in Section 5.4 of Chapter 5. The equations for the approximate velocity and acceleration error coefficients, as given in Chapter 5, are

$$c_1 = \frac{1}{\alpha\omega_c} = \frac{1}{(\omega_2/\omega_1)\omega_c} = \frac{\omega_1}{\omega_2\omega_c} \quad (14.3\text{-}17)$$

$$c_2 = \frac{1}{\omega_c\omega_i} = \frac{1}{\omega_c\omega_2} \quad (14.3\text{-}18)$$

Chapter 5 did not include an error term for the rate-of-acceleration error coefficient c_3 for this class of loop transfer function. However, the time delay applied to the error-coefficient terms has an effect that is equivalent to a rate-of-acceleration error component. If the third derivative D^3x_i is constant, the second derivative (acceleration) D^2x_i is a ramp equal to $(D^3x_i)t$, and the acceleration error is

$$\text{acceleration error} = c_2\big(D^2x_i[t]\big) = c_2\big(D^3x_i\big)t \quad (14.3\text{-}19)$$

This error is shown in Fig 14.3-2 as curve ①. Curve ② is the corrected acceleration error, which is obtained by delaying curve ① by the time delay τ_{ic}. As shown, curve ② could be formed by adding a correction to curve ① equal to

$$\text{Correction} = -\tau_{ic}\big(c_2 D^3x_i\big) = c_3 D^3x_i \quad (14.3\text{-}20)$$

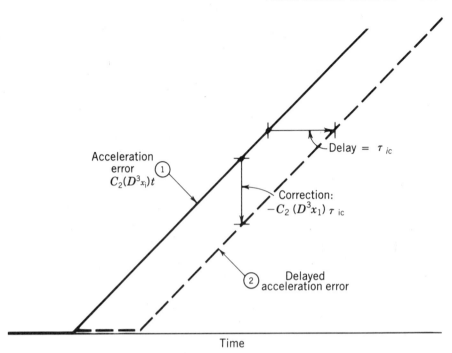

Figure 14.3-2 Calculation of a rate-of-acceleration error that is equivalent to the time delay applied to the acceleration error.

This is an error component proportional to the third derivative, and so can be described by a rate-of-acceleration error coefficient equal to

$$c_3 = -\tau_{ic} c_2 \qquad (14.3\text{-}21)$$

For this example, the value for τ_{ic} is

$$\tau_{ic} = \frac{1}{\omega_i} + \frac{1}{\omega_c} = \frac{1}{\omega_2} + \frac{1}{\omega_c} = 0.0612 \text{ sec} \qquad (14.3\text{-}22)$$

The approximate value for c_3 shown in Table 14.3-3 is obtained by multiplying the approximate value of c_2 by the negative of Eq 14.3-22.

The approximate and exact values for c_1 in Table 14.3-3 are the same, but those for c_2 differ by about 20%, and those for c_3 differ by about 50%. This loop has a value for α of 5, which is quite low. If α were higher, the approximate and exact values for these error coefficients would be much closer.

In accordance with Eq 14.2-39, the initial-value coefficients can be calculated by examining the loop transfer function G as s approaches infinity, which by Eq 14.3-1 is

$$G \doteq \frac{\omega_c \omega_3}{s^2} \qquad \text{as} \quad s \to \infty \qquad (14.3\text{-}23)$$

Hence, by Eq 14.2-39, the first three initial-value coefficients are

$$a_0 = G[s]\big|_{s=\infty} = \frac{\omega_c \omega_3}{s^2}\bigg|_{s=\infty} = 0 \qquad (14.3\text{-}24)$$

$$a_1 = sG[s]\big|_{s=\infty} = \frac{\omega_c \omega_3}{s}\bigg|_{s=\infty} = 0 \qquad (14.3\text{-}25)$$

$$a_2 = s^2 G[s]\big|_{s=\infty} = \omega_c \omega_3\big|_{s=\infty} = \omega_c \omega_3 = 1.85 \times 10^4 \text{ sec}^{-2} \quad (14.3\text{-}26)$$

If the higher-order initial-value coefficients are desired, they can be calculated by long-division expansion of G_{ib}, as described in Appendix J.

14.3.2   Transient Terms for Example

The composite error-response transient terms for the example are shown in Figs 14.3-3 to -6 for the input (position), the first derivative (velocity), the second derivative (acceleration), and the third derivative (rate-of-acceleration).

Position Transient. The solid curve of Fig 14.3-3 is the error response to a unit step and is the position transient term $T_0[t]$. Note that the position error coefficient c_0 is zero, and so the final value of the response is zero. The dashed curve ① shows the component due to the underdamped pole pair, and the dashed curve ② shows the component due to the low-frequency real pole.

Velocity Transient. The solid curve of Fig 14.3-4 is the error response to a unit ramp, and is the composite velocity transient term $(c_1 + T_1[t])$. This has a final value equal to the velocity error coefficient c_1. The dashed curves ① and ② show the components due to the underdamped poles and the low-frequency real pole. For this transient, the amplitudes of the transient components for the two sets of poles are approximately equal, whereas for the position transient the component for the high-frequency pole pair is much larger than that for the low-frequency pole.

Acceleration Transient. The solid curve of Fig 14.3-5 is the composite acceleration transient term $(c_2 + T_2[t])$. The final value of this is the acceleration error coefficient c_2. A unit step of acceleration produces a unit ramp of velocity, and so there is also a velocity error term $c_1 t$, shown by the dashed curve ③, which is equal to the velocity t multiplied by the velocity error coefficient c_1. Curve ③ is added to the solid composite transient term to

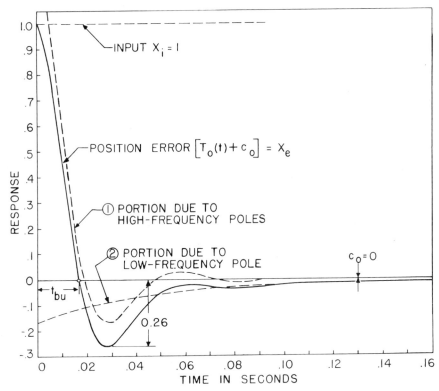

Figure 14.3-3   Position transient for the error response. Reprinted with permission from Ref [14.1]© 1955 AIEE (now IEEE).

obtain the total error for the step of acceleration, shown as curve ④. The dashed curve ② is the component due to the low-frequency real pole. The small difference between curve ② and the solid curve is the component for the high-frequency pole pair. This shows that the acceleration transient is almost entirely characterized by the low-frequency real pole.

Rate-of-Acceleration Transient. The solid curve of Fig 14.3-6 is the composite transient term for a step of rate-of-acceleration, designated $(c_3 + T_3[t])$. This composite transient has a final value equal to the rate-of-acceleration error coefficient c_3. The dashed curve ② is the component for the low-frequency real pole, and the very small difference between curve ② and the solid curve is the component for the high-frequency complex pole pair. Note that the values of this plot are negative.

The total error for a step of rate-of-acceleration is shown in Fig 14.3-7. Curve ① is the rate-of-acceleration composite transient obtained from Fig 14.3-6. The unit step of rate-of-acceleration produces a ramp of acceleration

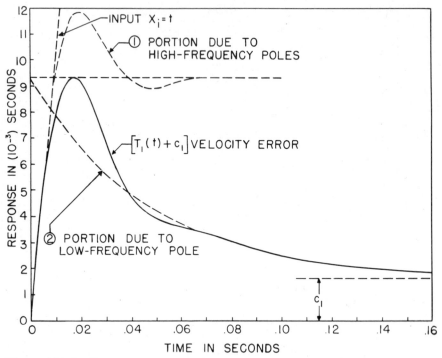

Figure 14.3-4 Composite velocity transient for the error response. Reprinted with permission from Ref [14.1]© 1955 AIEE (now IEEE).

equal to t, and a parabolic velocity equal to $t^2/2$. Curve ② is the acceleration error $c_2 t$, and curve ③ is the velocity error term $c_1 t^2/2$. Curve ④ is the sum of the acceleration error ② and the rate-of-acceleration error ①. This shows that the rate-of-acceleration error term has approximately the effect of a time delay applied to the acceleration error ②. The velocity error ③ is added to curve ④ to obtain the total error response, shown as curve ⑤. The input is shown as curve ⑥.

Impulse and Doublet Response. The feedback-signal responses to a unit impulse and a unit doublet are shown in Figs 14.3-8 and -9. These are the transient terms $T'_{-1}[t]$ and $T'_{-2}[t]$ for unit steps of the first and second integrals of the input. The initial values of these are the initial-value coefficients. As shown in Eqs 14.3-25, -26, the initial value a_1 of the impulse response is zero, and the initial value a_2 of the doublet response is 1.85×10^4 sec^{-2}. The very small dashed curves labeled ② are the components of these responses due to the low-frequency real pole. Thus, the impulse and doublet responses are characterized primarily by the high-frequency complex pole pair.

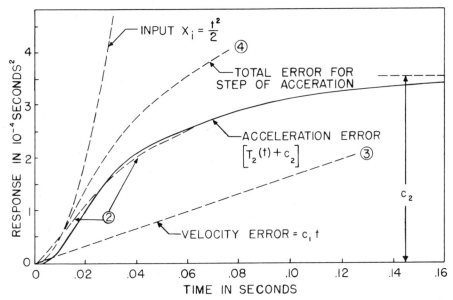

Figure 14.3-5 Composite acceleration transient for the error response. Reprinted with permission from Ref [14.1]© 1955 AIEE (now IEEE).

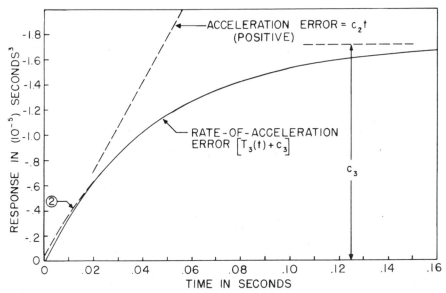

Figure 14.3-6 Composite rate-of-acceleration transient for the error response. Reprinted with permission from Ref [14.1]© 1955 AIEE (now IEEE).

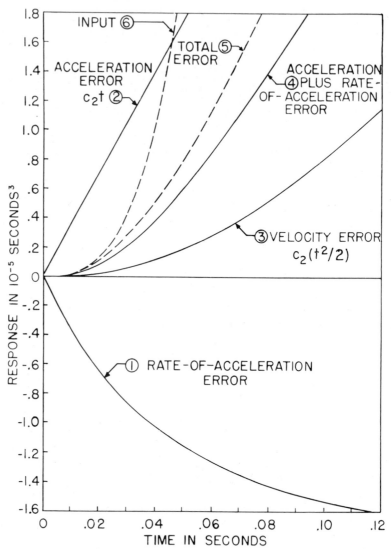

Figure 14.3-7   Components of response for a step of rate-of-acceleration.

14.3.3   Calculation of Response to Arbitrary Input

Curve ① in Fig 14.3-10 shows the arbitrary input position signal considered for the example. The system is initially at rest at an angle of $0°$. At time $t = 0$, the angular position jumps discontinuously to $-1.5°$, and then follows the smooth curve ①. This input is broken down as follows.

First the step at $t = 0$ is subtracted from the rest of the curve and considered separately. Then the smooth portion of this position curve is

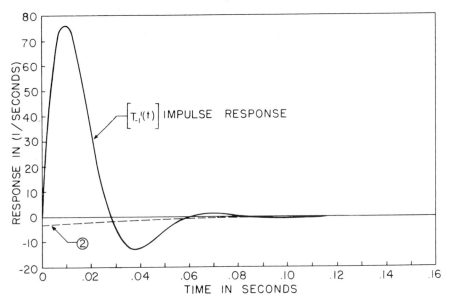

Figure 14.3-8 Impulse transient. Reprinted with permission from Ref [14.1]© 1955 AIEE (now IEEE).

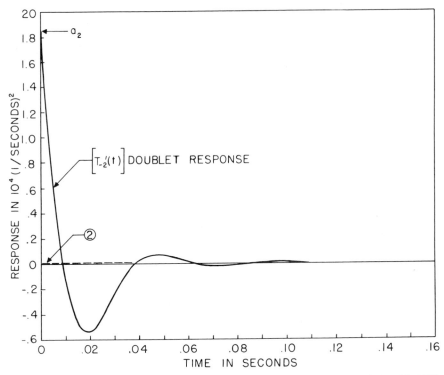

Figure 14.3-9 Doublet transient. Reprinted with permission form Ref [14.1]© 1955 AIEE (now IEEE).

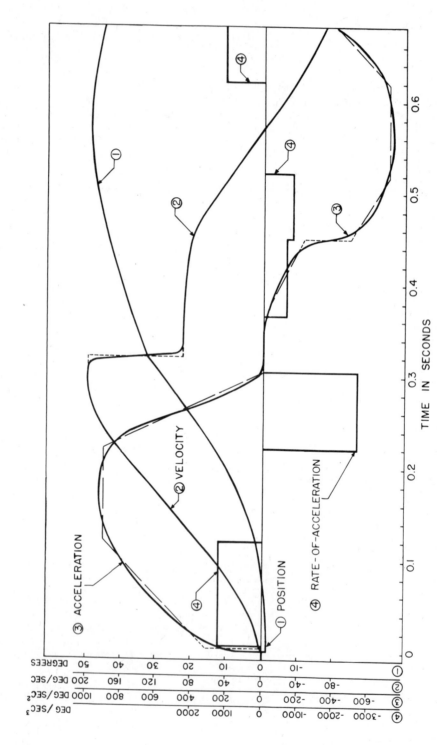

Figure 14.3-10 Illustrative arbitrary input and its derivatives. Reprinted with permission from Ref [14.1]© 1955 AIEE (now IEEE).

differentiated to yield the velocity curve ②. At time $t = 0.32$ sec, the velocity curve has a steep drop. Since the fall time of this drop is about 0.008 sec, which is much less than the rise time of the velocity transient, the drop has approximately the effect of a step of velocity. Hence, the drop is approximated as a step.

The smooth portion of the approximated velocity curve is differentiated to yield curve ③, the acceleration. At times $t = 0$ and $t = 0.45$ sec, the acceleration curve rises significantly faster then the acceleration transient, and so is approximated by steps at these points. The rest of the acceleration curve is approximated by straight-line segments, as shown by the long broken lines. When the broken-line approximation is differentiated, it yields the rate-of-acceleration plot, curve ④, which is a series of steps. Thus, except for the portions neglected in the various approximations, the input is broken down into a series of steps of various derivatives of the input.

The composite transient plots in Figs 14.3-3 to -6 are used to construct the response of the loop to the arbitrary input of Fig 14.3-10. In Fig 14.3-11, the approximated curves of velocity, acceleration, and rate of acceleration are multiplied by the corresponding error coefficients to yield curves ①′, ②′, and ③′. The velocity curve includes a single step approximation at $t = 0.32$ sec; the acceleration curve includes step approximations at $t = 0.03$ sec and $t = 0.45$ sec; and the rate-of-acceleration curve is entirely composed of step approximations. At each of the discontinuities of these curves, composite transient terms are added to produce the resultant curves of velocity error, acceleration error, and rate-of-acceleration error labeled ①, ②, ③. The individual transient terms are shown as dashed curves; the complete error components are shown as solid curves; and curves ①′, ②′, and ③′ are shown as dot-dashed curves in the regions where they do not correspond to the actual error curves ①, ②, ③.

The largest transient in Fig 14.3-11 occurs at 0.32 sec and is caused by the approximate step of velocity at that point. The transient is obtained by multiplying the curve of $(T_1[t] + c_1)$ in Fig 14.3-4 by the step of velocity, which is -108 deg/sec. When the transient is added to curve ①′ (which is $c_1 Dx_i[t]$), the resultant curve ① does not have a discontinuity. The reason is that the discontinuity in curve ①′ is c_1 times the step of velocity, which is the final value of the transient.

The acceleration error curve has two transients, at 0.03 sec and at 0.45 sec. Adding the two transients to the discontinuous curve ②′ gives the continuous acceleration error curve ②. The rate-of-acceleration error is entirely composed of steps, there being eight in all. For each of these steps there is a transient which adds to curve ③′ to form the continuous rate-of-acceleration curve ③.

Curves ①, ②, and ③, the three error components in Fig 14.3-11, are added together to form curve ① in Fig 14.3-12, which is the total error

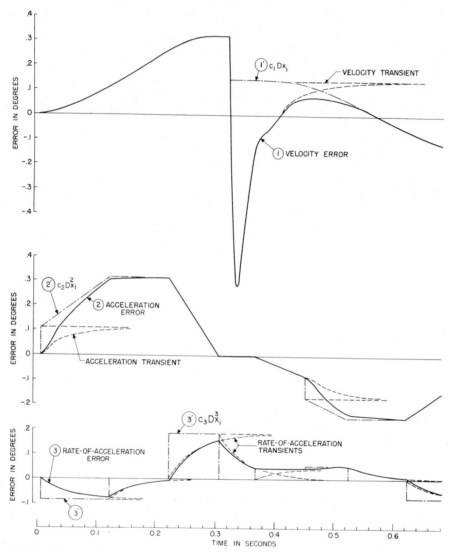

Figure 14.3-11 Components of the error response to the arbitrary input. Reprinted with permission from Ref [14.1]© 1955 AIEE (now IEEE).

caused by the derivatives of the input. Curve ② in Fig 14.3-12 is the error component caused by the $-1.5°$ step of position that occurs at $t = 0$. Curve ② is obtained by multiplying the position transient term in Fig 14.3-3 by the value of the step, $-0.15°$. The total error response, curve ③, is obtained by adding curve ② to curve ①.

Although discontinuities did not occur in the error-response components for the derivatives, the step of position produces a discontinuity in the position

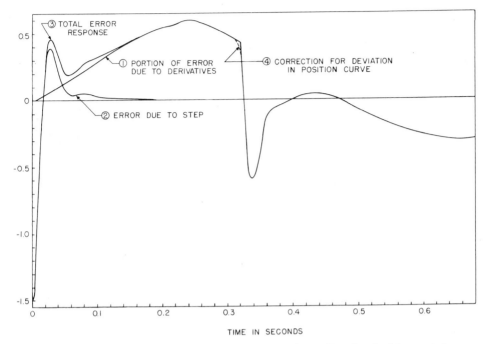

Figure 14.3-12   Total error response to the arbitrary input. Reprinted with permission from Ref [14.1]© 1955 AIEE (now IEEE).

component, curve ②, and in the total error response, curve ③. If the feedback response were plotted instead of the error, there would be no discontinuity.

14.4   DETERMINATION OF ACCURACY OF CALCULATED RESPONSE

14.4.1   Remainder due to nth-Derivative Deviation

The calculation of the response to an arbitrary input requires that approximations be made on the input. The effect of these approximations on the accuracy of the calculated response can be determined by calculating an upper bound to the inaccuracy caused by the approximation. If the approximation is performed properly, this upper bound can be kept quite small with respect to the total error response, and hence the per-unit accuracy of the calculated response is guaranteed to be high.

An expression for the upper bound to the inaccuracy of an error-coefficient expansion was first derived by Bower [14.2]. However, the upper bound calculated herein is based on an integral derived by Arthurs and Martin [14.3], because their integral yields a lower value for the upper bound.

The response to a deviation neglected in the nth derivative can be calculated as follows. The transform of the feedback signal can be expressed as

$$X_b = G_{ib} X_i = \frac{G_{ib}}{s^n} s^n X_i \qquad (14.4\text{-}1)$$

where $s^n X_i$ is the Laplace transform of the nth derivative of the input:

$$s^n X_i = \mathscr{L}\left[D^n x_i[t] \right] \qquad (14.4\text{-}2)$$

By Eqs 14.2-5, -6, G_{ib}/s^n can be expanded as

$$\frac{G_{ib}}{s^n} = \frac{c_0'}{s^n} + \frac{c_1'}{s^{n-1}} + \cdots + \frac{c_{n-1}'}{s} + T_{n-1}'[s] \qquad (14.4\text{-}3)$$

Combining Eqs 14.4-1, -2, -3 gives

$$X_b = \left(\frac{c_0'}{s^n} + \frac{c_1'}{s^{n-1}} + \cdots + \frac{c_{n-1}'}{s} \right)(s^n X_i) + T_{n-1}'[s] \mathscr{L}\left[D^n x_i \right]$$

$$= c_0' X_i + c_1' s X_i + \cdots + c_{n-1}' s^{n-1} X_i + T_{n-1}'[s] \mathscr{L}\left[D^n x_i \right] \qquad (14.4\text{-}4)$$

Taking the inverse transform of this gives

$$x_b = c_0' x_i + c_1' D x_i + \cdots + c_{n-1} D^{n-1} x_i + \mathscr{R}_n[t] \qquad (14.4\text{-}5)$$

The last term, $\mathscr{R}_n[t]$, is the real convolution of $T_{n-1}'[t]$ and $D^n x_i[t]$, and can be expressed as

$$\mathscr{R}_n[t] = \int_0^\infty d\tau\, T_{n-1}[\tau] D^n x_i[t - \tau] \qquad (14.4\text{-}6)$$

This is obtained by applying the convolution integral described in Chapter 11, Section 11.2.1 (Eq 11.2-5). Equation 14.4-6 was first derived by Arthurs and Martin [14.3], and represents the exact value of the remainder of an error-coefficient expansion. Equation 14.4-5 shows that the effect on the calculated response of neglecting a deviation of the nth derivative is given by the remainder $\mathscr{R}_n[t]$ for that deviation. The steady-state coefficient terms in this expansion are automatically included in the calculated response when the procedure described previously is followed. Hence, it is only the term $\mathscr{R}_n[t]$ that is disregarded when the nth-derivative deviation is neglected.

14.4.2 The Remainder Coefficient

An exact calculation of $\mathcal{R}_n[t]$ for a given deviation would be very difficult, and is not necessary because $\mathcal{R}_n[t]$ should be small with respect to the total response $x_e[t]$ when the approximations are performed properly. An upper bound on the magnitude of $\mathcal{R}_n[t]$ is a measure of the accuracy of the calculated response. By Eq 14.4-6 the magnitude of the remainder is given by

$$|\mathcal{R}_n[t]| = \left| \int_0^\infty d\tau\, T_{n-1}[\tau] D^n x_i[t - \tau] \right| \tag{14.4-7}$$

The magnitude of the integral of a function can be no greater than the integral of the magnitude of the function. Hence this magnitude is bounded by

$$|\mathcal{R}_n[t]| \le \int_0^\infty d\tau |T_{n-1}[\tau]| |D^n x_i[t - \tau]| \tag{14.4-8}$$

The maximum value of the magnitude of the nth-derivative deviation is designated as $\text{Max}|D^n x_i|$, and so the following holds for all values of t and τ:

$$|D^n x_i[t - \tau]| \le \text{Max}|D^n x_i| \tag{14.4-9}$$

Thus, the right-hand side of Eq 14.4-8 is bounded by

$$\int_0^\infty d\tau |T_{n-1}[\tau]| |D^n x_i[t - \tau]| \le \int_0^\infty d\tau |T_{n-1}[\tau]| \text{Max}|D^n x_i| \tag{14.4-10}$$

By Eq 14.4-8, the magnitude of the remainder, $|\mathcal{R}_n[t]|$, is bounded by the right-hand integral in Eq 14.4-10. Since $\text{Max}|D^n x_i|$ is a constant, it can be taken outside this integral to give the following upper bound for the magnitude of the remainder:

$$|\mathcal{R}_n[t]| \le \text{Max}|D^n x_i| \int_0^\infty d\tau |T_{n-1}[\tau]| \tag{14.4-11}$$

The integral in Eq 14.4-11 is defined as the nth-derivative remainder coefficient, which is designated r_n:

$$r_n = \int_0^\infty d\tau |T_{n-1}[\tau]| \tag{14.4-12}$$

The upper bound given in Eq 14.4-11 can therefore be expressed as

$$|\mathcal{R}_n[t]| \le r_n \text{Max}|D^n x_i| \tag{14.4-13}$$

This shows that the upper bound for the magnitude of the error caused by neglecting a deviation at the nth derivative is equal to the nth-derivative remainder coefficient r_n multiplied by the maximum value of the magnitude of the deviation.

The remainder coefficient r_n for the nth derivative can be readily derived from a plot of the transient term of the nth derivative. Equation 14.2-29 gives the following expression for the nth-derivative transient term:

$$T_n[t] - T_n[0] = \int_0^t T_{n-1}[t]\, dt \tag{14.4-14}$$

Comparing this with Eq 14.4-12 shows that $(T_n[t] - T_n[0])$ is the integral of $T_{n-1}[t]$, while r_n is the integral of the magnitude of $T_{n-1}[t]$, evaluated at time $t = \infty$. Hence the remainder coefficient can be calculated as shown in Fig 14.4-1. In diagram a, the solid curve Ⓐ is a plot of the $(n-1)$th derivative transient term $T_{n-1}[t]$, while the dashed curve Ⓑ is a plot of the magnitude of this, which is $|T_{n-1}[t]|$. In diagram b, the solid curve Ⓒ is the integral of the solid curve Ⓐ of diagram a, and so by Eq 14.4-14 is a plot of $(T_n[t] - T_n[0])$. Curve Ⓒ can be formed by plotting the nth derivative transient term $T_n[t]$, and shifting the curve vertically until it is zero at time $t = 0$.

The dashed curve Ⓓ in diagram b is a plot of the integral of the dashed curve Ⓑ in diagram a, which is $|T_{n-1}[t]|$. Hence, by Eq 14.4-12, the final value of curve Ⓓ is the nth-derivative remainder coefficient r_n, as shown. Curve Ⓓ can be readily constructed from the solid curve Ⓒ as follows.

In the regions where curve Ⓒ has a positive slope, curve Ⓓ has the same shape as curve Ⓒ, and in the regions where curve Ⓒ has a negative slope, curve Ⓓ has the shape of the mirror image of curve Ⓒ reflected about the horizontal axis. Thus to form curve Ⓓ, the segments of curve Ⓒ with positive slope, and the mirror images of the segments with negative slope, are shifted vertically to form a curve that starts at zero and progresses as a monotonically increasing curve.

14.4.3 Error Bound for mth-Integral Deviation

An equivalent bound can be calculated as follows for the inaccuracy caused by neglecting a deviation of the mth integral. The transform of the feedback signal can be expressed as

$$X_b = G_{ib} X_i = s^m G_{ib} \frac{X_i}{s^m} \tag{14.4-15}$$

where X_i/s^m is the Laplace transform of the mth integral of the input:

$$\frac{X_i}{s^m} = \mathscr{L}\left[D^{-m} x_i\right] \tag{14.4-16}$$

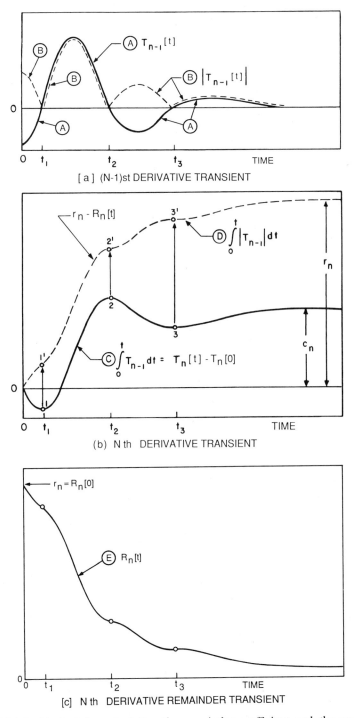

Figure 14.4-1 Method for calculating the remainder coefficient and the remainder transient: (*a*) transient term for $(n-1)$st derivative; (*b*) transient term for nth derivative; (*c*) remainder transient for nth derivative. Reprinted with permission from Ref [14.1]© 1955 AIEE (now IEEE).

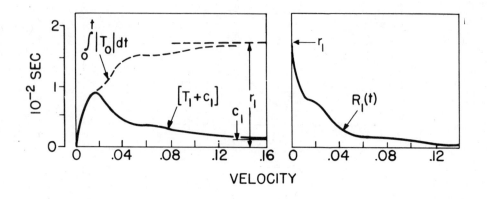

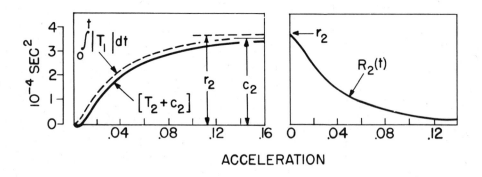

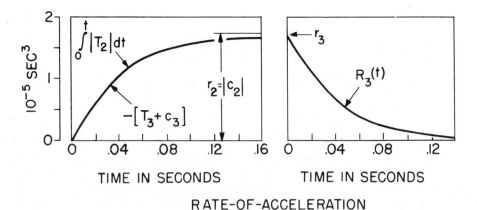

Figure 14.4-2 Calculation of remainder coefficients for illustrative system. Reprinted with permission from Ref [14.1]© 1955 AIEE (now IEEE).

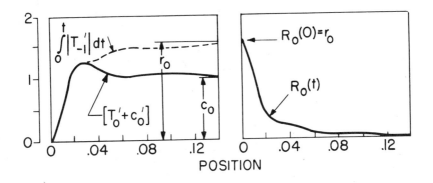

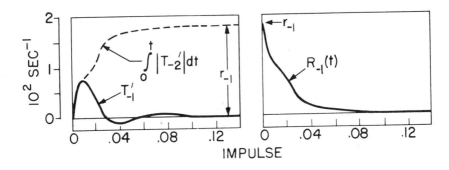

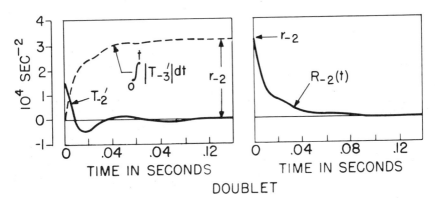

Figure 14.4.2 Continued.

By Eqs 14.2-35, -36, $s^m G_{ib}$ can be expanded as

$$s^m G_{ib} = a_0 s^m + a_1 s^{m-1} + \cdots + a_m + T'_{-(m+1)}[s] \qquad (14.4\text{-}17)$$

Substituting this into Eq 14.4-15 gives

$$X_b = \left(a_0 s^m + a_1 s^{m-1} + \cdots + a_m \right) \frac{X_i}{s^m} + T'_{-(m+1)}[s] \frac{X_i}{s^m} \qquad (14.4\text{-}18)$$

Replace X_i/s in the last term with Eq 14.4-16, and simplify:

$$X_b = a_0 X_i + a_1 \frac{X_i}{s} + \cdots + a_m \frac{X_i}{s^m} + T'_{-(m+1)}[s] \mathscr{L}[D^{-m} x_i] \qquad (14.4\text{-}19)$$

The inverse transform of this is

$$x_b = a_0 x_i + a_1 D^{-1} x_i + \cdots + a_m D^{-m} x_i + \mathscr{R}_{-m}[t] \qquad (14.4\text{-}20)$$

where $\mathscr{R}_{-m}[t]$ is the convolution of $T'_{-(m+1)}[t]$ and $D^{-m} x_i$, which is

$$\mathscr{R}_{-m}[t] = \int_0^\infty d\tau \, T'_{-(m+1)}[t] D^{-m} x_i \qquad (14.4\text{-}21)$$

This is the same as the expression for $\mathscr{R}_n[t]$ given in Eq. 14.4-6 if n is replaced by $-m$. The magnitude of this is bounded by the mth integral remainder coefficient r_{-m} multiplied by $\text{Max}|D^{-m} x_i|$.

14.4.4 Calculation of Accuracy for Example

Figure 14.4-2 shows the constructions for calculating the remainder coefficients for the illustrative feedback loop. These are calculated from the transient terms given previously for steps of position, velocity, acceleration, and rate of acceleration, and for steps of the first and second integral (impulse response and doublet response). These yield the remainder coefficients r_n given in Table 14.4-1 for values of n from -2 to $+3$.

Note that the remainder coefficient r_n is very much greater than the magnitude of the error coefficient c_n for steps of position and velocity. However, for a step of acceleration and all higher derivatives, r_n is approximately equal to $|c_n|$. This illustrates the general principle that in an adequately damped feedback loop, the magnitude of the error coefficient becomes roughly equal to the remainder coefficient at a certain derivative, and often approaches even closer at a higher derivative, but can never become greater.

This first derivative at which $|c_n|$ is approximately equal to r_n is very significant, because the final straight-line approximation of the signal should be made at this derivative. When this is done, the per-unit inaccuracy in the

TABLE 14.4-1   Remainder Coefficients for Illustrative Feedback Loop

Response	Coefficient	Value
Doublet	r_{-2}	$3.10 \times 10^4 \ \text{sec}^{-2}$
Impulse	r_{-1}	$182 \ \text{sec}^{-1}$
Position	r_0	1.55
Velocity	r_1	$1.707 \times 10^{-2} \ \text{sec}$
Acceleration	r_2	$3.57 \times 10^{-4} \ \text{sec}^2$
Rate of Acceleration	r_3	$1.715 \times 10^{-5} \ \text{sec}^3$

calculated response is no greater (essentially) than the per-unit inaccuracy of the approximation.

For example, in the illustrative system, the final straight-line approximations were made on the acceleration curve. There is an acceleration component in the computed error response equal to c_2 multiplied by the approximate acceleration curve, and an acceleration component of inaccuracy equal to r_2 multiplied by the acceleration deviation. Since $|c_2|$ is approximately equal to r_2, if the acceleration curve is approximated within 5%, the resultant inaccuracy in the calculated response can be no greater than 5% of the acceleration component of error. Since r_n can never be less than $|c_n|$, nothing is gained by performing the approximation at a higher derivative.

Plots of the deviations associated with the error-coefficient expansion of the input are shown in Fig 14.4-3. Let us use the remainder-coefficient values of Table 14.4-1 to calculate the error bounds corresponding to these deviations. Figure 14.4-3c gives a plot of the acceleration deviation corresponding to the straight-line approximation of the input acceleration. The maximum value of the magnitude of the deviation is 55 deg/sec^2. Using the value of r_2 in Table 14.4-1 gives the following upper bound to the inaccuracy caused by this approximation:

$$r_2 \text{Max}|D^2 x_i| = \left(3.57 \times 10^{-4} \ \text{sec}^2\right)\left(55 \ \text{deg/sec}^2\right) = 0.0196° \quad (14.4\text{-}22)$$

The acceleration-error contribution of curve ② in Fig 14.3-11 was obtained by multiplying c_2 by the approximated acceleration curve, not the exact curve. The reason for this is explained by the remainder expansion of Eq 14.4-5. The expansion shows that the remainder coefficient bounds the inaccuracy from neglecting a deviation only when that deviation is not included in the response. Thus, the inaccuracy bound of Eq 14.4-6 holds only if the approximated acceleration curve is used to calculate the acceleration component of error.

Step approximations of a derivative should be considered to be the integrated effect of line-segment approximations of the lower-order derivative. For example, the step approximation of the velocity curve in Fig 14.3-10 at 0.32 sec should be considered to be a straight-line approximation of the position curve. In Fig 14.4-4, the solid curve shows the actual input position,

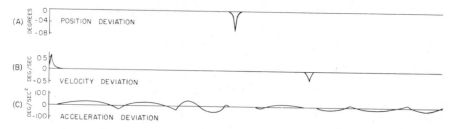

Figure 14.4-3 Deviation components for approximations of the arbitrary input. Reprinted with permission from Ref [14.1]© 1955 AIEE (now IEEE).

and the dashed straight lines show the line-segment approximation, which yields the step approximation of velocity. The deviation of the position curve corresponding to this approximation is plotted in Fig 14.4-3a. Similarly, the step approximations of acceleration at 0 and 0.45 sec are considered to be line-segment approximations of velocity, and the deviations corresponding to these approximations are shown in Fig 14.4-3b.

The peak magnitude of the velocity deviation of Fig 14.4-3b is 0.57 deg/sec, and so the corresponding inaccuracy bound for this deviation is

$$r_1 \text{Max}|Dx_i| = \left(1.707 \times 10^{-2} \text{ sec}^2\right)(0.57 \text{ deg/sec}) = 0.0097° \quad (14.4\text{-}23)$$

The peak magnitude of the position deviation in Fig 14.4-3a is 0.072°, and so the corresponding inaccuracy bound is

$$r_0 \text{Max}|x_i| = 1.55(0.072°) = 0.122° \quad (14.4\text{-}24)$$

Although it is convenient to consider a step approximation to be a line-segment approximation of the lower-order derivative, it is not necessary to perform the line-segment approximation directly. Consider for example Fig 14.4-5, which shows a portion of the $(n-1)$th derivative (diagram a), which is approximated by straight lines, and the nth derivative (diagram b), with the corresponding step approximation. The step approximation can be performed directly on the nth derivative by making the areas between the approximate curve and the exact curve equal, as shown in diagram b. This area A is equal to the maximum deviation of the $(n-1)$th derivative from the line-segment approximation. Hence the inaccuracy bound is equal to the area A, obtained from the nth derivative, multiplied by the remainder coefficient for the $(n-1)$th derivative.

14.4.5 The Remainder Transient

A basic limitation on the use of the remainder coefficient is that the upper bound to the inaccuracy calculated from it applies for all values of time. This can be remedied by generalizing the remainder-coefficient concept to obtain

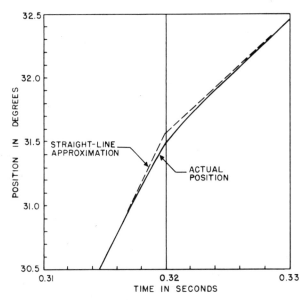

Figure 14.4-4 Line-segment approximation of position, which corresponds to the step approximation of velocity. Reprinted with permission from Ref [14.1]© 1955 AIEE (now IEEE).

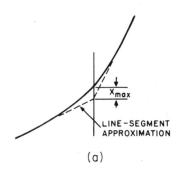

(a)

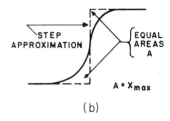

(b)

Figure 14.4-5 Comparison of a step approximation of a derivative with the equivalent line-segment approximation of the lower-order derivative: (a) line-segment approximation of $(n-1)$st derivative; (b) step approximation of nth derivative. Reprinted with permission from Ref [14.1]© 1955 AIEE (now IEEE).

the remainder transient, which provides a quantitative measure of the period of time over which the inaccuracy applies.

As shown in Fig 14.4-1b, the dashed curve ⒟ is designated ($r_n - R_n[t]$), where $R_n[t]$ is defined as the remainder transient. The plot of the remainder transient $R_n[t]$ is shown as curve ⒠ of diagram c, and is found by inverting curve ⒟ and shifting it vertically until its final value is zero. Since curve ⒟ is equal to the integral of $|T_{n-1}[t]|$, this gives

$$r_n - R_n[t] = \int_0^t |T_{n-1}[t]| dt \qquad (14.4\text{-}25)$$

Solve for $R_n[t]$, and set r_n equal to the expression of Eq 14.4-5:

$$R_n[t] = r_n - \int_0^t |T_{n-1}[t]| dt$$

$$= \int_0^\infty |T_{n-1}[t]| dt - \int_0^t |T_{n-1}[t]| dt = \int_t^\infty |T_{n-1}[t]| dt \quad (14.4\text{-}26)$$

It can be seen from this equation, as well as from the plot of curve ⒠ in Fig 14.4-1c, that the initial value of the remainder transient (at $t = 0$) is equal to the remainder coefficient.

The remainder transient can be applied by assuming that the nth derivative is approximated as shown in Fig 14.4-6a. The magnitude of the deviation is assumed to be no greater than $\text{Max}|D^n x_i|$ for values of time less than t_1, and zero thereafter:

$$|D^n x_i[t]| \leq \text{Max}|D^n x_i| \qquad \text{for} \quad t \leq t_1 \qquad (14.4\text{-}27)$$

$$D_n x_i[t] = 0 \qquad \text{for} \quad t > t_1 \qquad (14.4\text{-}28)$$

(Note that $\text{Max}|D^n x_i|$ is represented by N in Fig 14.4-6.) The general expression for the remainder associated with any approximation is given by Eq 14.4-6 as

$$\mathscr{R}_n[t] = \int_0^\infty d\tau \, T_{n-1}[\tau] D^n x_i[t - \tau] \qquad (14.4\text{-}29)$$

As shown in Fig 14.4-6a, τ represents a time variable measured in the direction of negative time from the present time t. At values of τ less than $\tau_1 = (t - t_1)$, the nth-derivative deviation $D^n x_i$ is zero. Hence the integral of

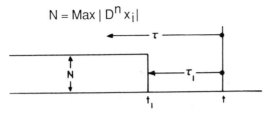

$$N = \text{Max} \, | \, D^n x_i |$$

(A) A DEVIATION BOUND FOR nth DERIVATIVE

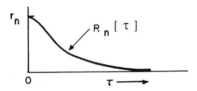

(B) A TYPICAL REMAINDER TRANSIENT

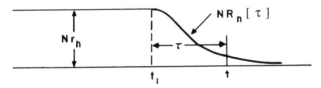

(C) UPPER BOUND TO INACCURACY FOR (A)

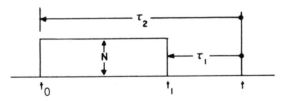

(D) A DEVIATION BOUND FOR nth DERIVATIVE

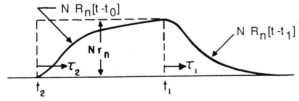

(E) UPPER BOUND TO INACCURACY FOR (D)

Figure 14.4-6 Application of remainder transient. Reprinted with permission from Ref [14.1]© 1955 AIEE (now IEEE).

Eq 14.4-29 should be split into two regions as follows:

$$\mathscr{R}_n[t] = \int_0^{\tau_1} d\tau \, T_{n-1}[\tau] D^n x_i[t - \tau] + \int_{\tau_1}^{\infty} d\tau \, T_{n-1}[\tau] D^n x_i[t - \tau] \quad (14.4\text{-}30)$$

The first of the two integrals is zero, because the nth-derivative deviation $D^n x_i$ is zero for that range of τ. In the second integral, the magnitude of the deviation is no greater than $\text{Max}|D^n x_i|$. Therefore, the upper bound on the magnitude of the remainder at time t is

$$|\mathscr{R}_n[t]| \leq \text{Max}|D^n x_i| \left| \int_{\tau_1}^{\infty} d\tau \, |T_{n-1}[\tau]| \right| \quad (14.4\text{-}31)$$

Since $\tau_1 = t - t_1$, the integral is equal to $R_n[t - t_1]$. Hence

$$|\mathscr{R}_n[t]| \leq \text{Max}|D^n x_i| R_n[t - t_1] \quad (14.4\text{-}32)$$

In Fig 14.4-6, diagram b is a plot of a typical remainder transient $R_n[t]$. Diagram c is the upper bound to the inaccuracy, obtained by applying this remainder transient to the deviation bound of diagram a. Prior to time t_1, the remainder bound is equal to $\text{Max}|D^n x_i| r_n$. After the deviation ends at time t_1, the remainder bound decays along a curve proportional to the remainder transient, which is equal to $\text{Max}|D^n x_i| R_n[t - t_1]$.

Let us now assume the deviation bound of diagram d of Fig 14.4-6, which is zero before time t_0 and after time t_1, and is limited by $\text{Max}|D^n x_i|$ between t_0 and t_1. By extending the preceding analysis, it can be shown that the bound on the remainder is as given in diagram e. This remainder bound rises after time t_0 along a curve equal to $\text{Max}|D^n x_i|$ times $(r_n - R_n[t - t_0])$, and reaches a final value of $\text{Max}|D^n x_i| r_n$. After time t_1, the bound decays from this value along the curve equal to $\text{Max}|D^n x_i|$ times $R_n[t - t_1]$.

The remainder transients for the illustrative feedback loop are plotted in Fig 14.4-2 for all of the transient terms. The time for a remainder transient to settle can be regarded as the "settling time" for a step of that particular derivative or integral of the input.

14.4.6   Lower Error Bound for Unipolar Deviation

In many cases the nth-derivative deviation is either completely positive or completely negative (i.e., it is unipolar), at least in a local region. For such cases, a lower error bound can be used. This lower bound also applies when the deviation is positive in one region and negative in another, provided that the two regions are separated by more than the settling time for the remainder transient of that derivative.

To develop this lower bound, let us review the use of the remainder expression, which was given in Eqs 14.4-6 and -21. The application of this expression is illustrated in Fig 14.4-7, which shows the waveforms for calculating the remainder $\mathscr{R}_n[t]$ at a specific time $t = t_1$. Diagram a is the assumed

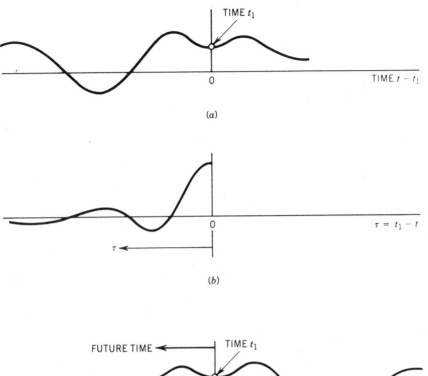

(a)

(b)

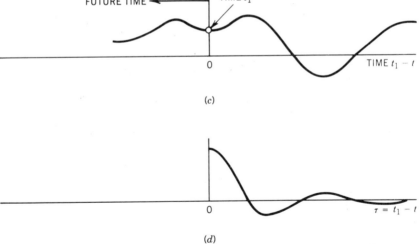

(c)

(d)

Figure 14.4-7 Waveforms associated with calculation of error caused by nth-derivative deviation of signal: (a) deviation for the nth derivative, $D^n x_i[t - t_1]$; (b) transient term $T_{n-1}[t]$ for $(n - 1)$st derivative, projected backward in time from time t_1; (c) deviation for nth derivative, projected backward in time from time t_1; (d) transient term for $(n - 1)$st derivative, projected forward in time.

plot of the nth-derivative deviation $D^n x_i[t]$, with the axis located at time t_1. Diagram b is the $(n-1)$th derivative transient term $T_{n-1}[\tau]$, which is expressed as a function of the parameter $\tau = t - t_1$, and so is projected backward in time relative to time t_1. The value of the remainder at time t_1, which is designated $\mathcal{R}[t_1]$, is obtained by multiplying the plots of diagrams a and b, and finding the area under the resultant product curve (i.e., integrating the product curve).

This process can also be illustrated as shown in diagrams c and d of Fig 14.4-7. In diagram c the nth derivative of the input, $D^n x_i$, is projected backward in time relative to time t_1, and in diagram d the $(n-1)$th-derivative transient term $T_{n-1}[\tau]$ is projected forward in time. The time variable is τ, which is equal to $(t_1 - t)$. Curves c and d are multiplied together, and the product is integrated to obtain $\mathcal{R}_n[t_1]$.

Figure 14.4-8 shows the steps for calculating the upper bound to the remainder for a bipolar (positive and negative) deviation. Diagram a is the $(n-1)$th derivative transient term. Diagrams b and c are the assumed worst-case plots of the nth-derivative deviation $D^n x_i$, which are projected backward in time relative to time t_1. The waveform in diagram b generates the most positive value of the remainder $\mathcal{R}_n[t_1]$, while diagram c generates the most negative value of this remainder. The magnitude of the remainder $\mathcal{R}_n[t]$ is the same for both. In diagram b, the nth-derivative deviation has the maximum positive value, $+\mathrm{Max}|D^n x_i|$, when $T_{n-1}[t]$ is positive, and the maximum negative value, $-\mathrm{Max}|D^n x_i|$, when $T_{n-1}[t]$ is negative. The plot of diagram c is the negative of that of diagram b.

The remainder coefficient is derived from the nth-derivative transient term $T_n[\tau]$, which is plotted in diagram d of Fig 14.4-8. Point 1 on the plot is the initial value $T_n[0]$. Points 2, 3, 4, and 5 are points of zero slope of $T_n[t]$, which occur when $T_{n-1}[t]$ in diagram a goes through zero. The differences of the values of $T_n[\tau]$ between adjacent zero-slope points are designated ΔT_n. In accordance with Fig 14.4-1, the remainder coefficient r_n is equal to

$$r_n = \overset{(+)}{\sum}|\Delta T_n| + \overset{(-)}{\sum}|\Delta T_n| \qquad (14.4\text{-}33)$$

The first summation, labeled $(+)$, is the summation of the magnitude of ΔT_n in the regions of positive slope of $T_n[\tau]$ (regions 1–2, 3–4, 5–∞). The second summation, labeled $(-)$, is the summation of the magnitude of ΔT_n in the regions of negative slope of $T_n[\tau]$ (regions 2–3, 4–5). The maximum positive and negative bounds on the remainder $\mathcal{R}_n[t]$, which result from the worst-case deviation plots of diagrams b and c, are as follows:

$$\mathcal{R}_n[t] \le \left\{ \overset{(+)}{\sum}|\Delta T_n| + \overset{(-)}{\sum}|\Delta T_n| \right\} \mathrm{Max}|D^n x_i| = r_n \mathrm{Max}|D^n x_i| \qquad (14.4\text{-}34)$$

$$\mathcal{R}_n[t] \ge - \left\{ \overset{(+)}{\sum}|\Delta T_n| + \overset{(-)}{\sum}|\Delta T_n| \right\} \mathrm{Max}|D^n X_i| = -r_n \mathrm{Max}|D^n X_i| \qquad (14.4\text{-}35)$$

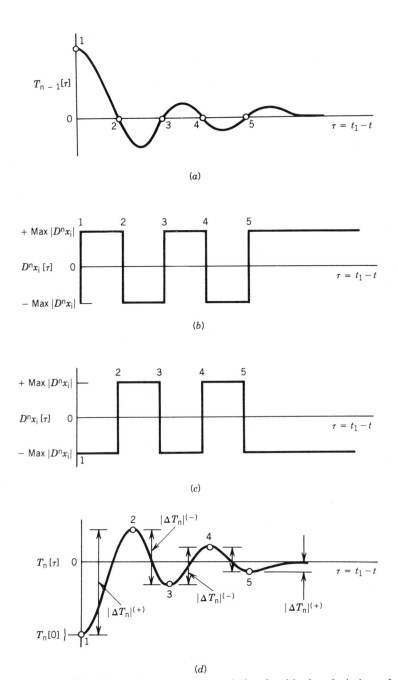

Figure 14.4-8 Calculation of worst-case remainder for bipolar deviation of nth derivative: (a) transient term for $(n-1)$st derivative; (b) worst-case nth-derivative deviation for positive remainder; (c) worst-case nth-derivative deviation for negative remainder; (d) transient term for nth derivative.

Since the initial value of $T_n[t]$ is $T_n[0]$ and the final value is zero, the following holds:

$$T_n[0] + \overset{(+)}{\sum} |\Delta T_n| - \overset{(-)}{\sum} |\Delta T_n| = 0 \qquad (14.4\text{-}36)$$

Combining Eqs 14.4-33, -36 gives

$$\overset{(+)}{\sum} |\Delta T_n| = \tfrac{1}{2}(r_n - T_n[0]) \qquad (14.4\text{-}37)$$

$$\overset{(-)}{\sum} |\Delta T_n| = \tfrac{1}{2}(r_n + T_n[0]) \qquad (14.4\text{-}38)$$

Now, let us consider the case where the deviation $D^n x_i$ is unipolar. For simplicity, it is assumed that $D^n x_i[t]$ is either positive or zero, but the same result holds when it is negative or zero. The resultant worst-case assumptions are shown in Fig 14.4-9. Diagram a is the $(n - 1)$th-derivative transient term, and diagram b is the nth-derivative transient term. Diagram c is the nth-derivative deviation that results in the maximum positive remainder $R_n[t]$, while diagram d is the nth-derivative deviation that results in the maximum negative remainder.

The deviation of diagram c is equal to $\text{Max}|D^n x_i|$ when the slope of $T_n[\tau]$ is positive, and is zero when the slope is negative. Hence it results in the following positive bound on the remainder:

$$\mathscr{R}_n[t] \le \left\{ \overset{(+)}{\sum} |\Delta T_n| \right\} \text{Max}|D^n x_i| \qquad (14.4\text{-}39)$$

The deviation of diagram d is equal to $\text{Max}|D^n x_i|$ when the slope of $T_n[\tau]$ is negative, and is zero when that slope is positive. Hence, it results in the following negative bound on the remainder:

$$\mathscr{R}_n[t] \ge - \left\{ \overset{(-)}{\sum} |\Delta T_n| \right\} \text{Max}|D^n x_i| \qquad (14.4\text{-}40)$$

Substituting Eqs 14.4-37, -38 into Eqs 14.4-39, -40 gives the following for the remainder bounds when the nth-derivative deviation is unipolar:

$$\mathscr{R}_n[t] \le \tfrac{1}{2}(r_n - T_n[0])\text{Max}|D^n x_i| \qquad (14.4\text{-}41)$$

$$\mathscr{R}_n[t] \ge - \tfrac{1}{2}(r_n + T_n[0])\text{Max}|D^n x_i| \qquad (14.4\text{-}42)$$

These can be replaced with the following single bound on the absolute value of

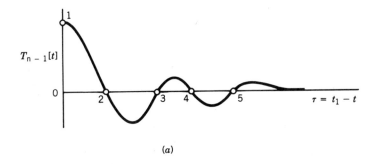

(a)

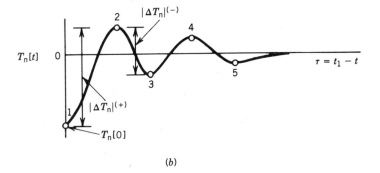

(b)

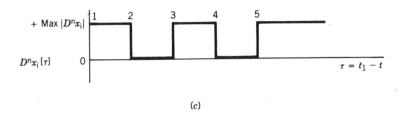

(c)

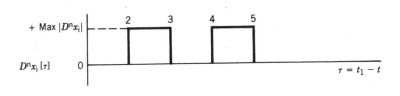

(d)

Figure 14.4-9 Calculation of worst-case remainder for unipolar deviation of nth derivative: (a) transient term for $(n-1)$st derivative; (b) transient term for nth derivative; (c) worst-case nth-derivative deviation for positive remainder; (d) worst-case nth-derivative deviation for negative remainder.

the remainder:

$$|\mathcal{R}_n[t]| \le \tfrac{1}{2}\big(r_n + |T_n[0]|\big)\mathrm{Max}|D^n x_i| \qquad (14.4\text{-}43)$$

Let us consider separately the bounds for a deviation of a derivative and an integral. For a derivative, the initial value of the nth-derivative transient term is equal to the negative of the steady-state coefficient c_n:

$$T_n[0] = -c_n \qquad (14.4\text{-}44)$$

Hence, the remainder bound of Eq 14.4-43 is

$$|\mathcal{R}_n[t]| \le \tfrac{1}{2}\big(r_n + |c_n|\big)\mathrm{Max}|D^n x_i| \qquad (14.4\text{-}45)$$

For an integral of the input, the initial value of the mth-integral transient term is equal to the mth-integral initial-value coefficient a_m:

$$T'_{-m}[0] = a_m \qquad (14.4\text{-}46)$$

Hence, the remainder bound of Eq 14.4-43 is

$$|\mathcal{R}_{-m}[t]| \le \tfrac{1}{2}\big(r_{-m} + |a_m|\big)\mathrm{Max}|D^{-m} x_i| \qquad (14.4\text{-}47)$$

As shown in the expansion of Eq 14.4-5, when a deviation for a derivative is neglected, the error-coefficient term for that deviation is not included in the expansion. However, Eq 14.4-20 shows that when a deviation for an integral is neglected, the initial-value term for that integral is still included in the response. Often it is desirable that this mth-integral initial-value term be ignored. Hence, a bound on the following combined expression is needed:

$$\{\mathcal{R}_{-m}[t] + a_m D^{-m} x_i[t]\} \qquad (14.4\text{-}48)$$

Assuming that $D^{-m} x_i[t]$ is positive, the maximum positive bound on this expression is (from Eq 14.4-41)

$$\mathcal{R}_{-m}[t] + a_m D^{-m} x_i[t] \le \tfrac{1}{2}\big(r_{-m} - a_m\big)\mathrm{Max}|D^{-m} x_i| + a_m \mathrm{Max}|D^{-m} x_i| \qquad (14.4\text{-}49)$$

where $T_n[0]$ in Eq 14.4-41 was replaced by a_m. This simplifies to

$$\mathcal{R}_{-m}[t] + a_m D^{-m} x_i[t] \le \tfrac{1}{2}\big(r_{-m} + a_m\big)\mathrm{Max}|D^{-m} x_i| \qquad (14.4\text{-}50)$$

From Eq 14.4-32, the maximum negative limit on the expression of Eq 14.4-48 is

$$\mathcal{R}_{-m}[t] + a_m D^{-m} x_i[t] \ge -\tfrac{1}{2}\big(r_{-m} + a_m\big)\mathrm{Max}|D^{-m} x_i| + a_m \mathrm{Max}|D^{-m} x_i| \qquad (14.4\text{-}51)$$

This simplifies to

$$\mathcal{R}_{-m}[t] + a_m D^{-m} x_i[t] \geq -\tfrac{1}{2}(r_{-m} - a_m)\text{Max}|D^{-m}x_i| \quad (14.4\text{-}52)$$

Equations 14.4-50, -52 can be replaced by the following bound on the magnitude of the expression of Eq 14.4-48:

$$\left|\mathcal{R}_{-m}[t] + a_m D^{-m} x_i[t]\right| \leq \tfrac{1}{2}(r_{-m} + |a_m|)\text{Max}|D^{-m}x_i| \quad (14.4\text{-}53)$$

This is the same as the bound on the magnitude of the remainder term alone $\mathcal{R}_{-m}[t]$ that was given in Eq 14.4-40. Thus, when the deviation of the mth integral is unipolar, the error bound is the same whether or not the mth-integral initial-value term is included in the calculated response.

The preceding can be summarized as follows. If the deviation for a derivative or integral is unipolar (either positive or negative), a reduced bound for the remainder can be obtained by replacing the remainder coefficient r_n (or r_{-m}) by $\tfrac{1}{2}(r_n + |c_n|)$ for an nth-derivative deviation, or by $\tfrac{1}{2}(r_{-m} + |a_m|)$ for an mth-integral deviation. The resultant bound for an integral holds regardless of whether the mth-integral initial-value term is included in the calculated response. In applying this principle, two deviations of opposite sign can still be regarded as unipolar if they are separated in time by an interval greater than the settling time of the remainder transient of that derivative or integral.

14.5 APPROXIMATION OF HIGH-FREQUENCY COMPONENTS OF THE INPUT

14.5.1 Approximating a Fast Rise with a Step

Chapter 5 stated the principle that when a portion of an input signal rises faster than the step response of a feedback loop, that portion should be approximated by a step. The following analysis will show that the error of this approximation is proportional to the rise time of the input, and is approximately equal to 20% of the amplitude of the step, when the rise time of the input is equal to the rise time of the step response.

Diagram a of Fig 14.5-1 shows an ideal gradual step input which rises linearly from zero to 1.0 in a time which is designated the buildup time T_{bu}. Since the rise time T_R is the time to rise from 10% to 90%, for this waveform it is equal to $0.8T_{bu}$.

The gradual step in diagram a of Fig 14.5-1 is approximated by a sharp step indicated by the dashed plot. As shown in diagram b, this is equivalent to a straight-line approximation of the integral of the input. The maximum deviation for the integral of diagram b, designated $\text{Max}|D^{-1}x_i|$, is equal to one

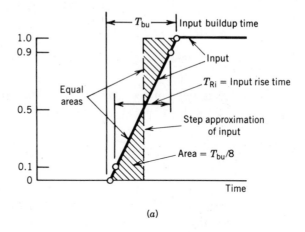

(a)

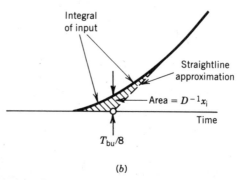

(b)

Figure 14.5-1 Deviation of first integral of input, corresponding to step approximation of a gradual input step: (a) step approximation of a gradual input step; (b) corresponding straight-line approximation of the integral of the input.

of the two equal cross-hatched areas of diagram *a*, which is $T_{bu}/8$:

$$\text{Max}|D^{-1}x_i| = T_{bu}/8 \qquad (14.5\text{-}1)$$

In Fig 14.5-2, the solid curve shows a typical feedback-signal response to a unit step. Points *A*, *B* are the instants of 10% and 90% response, and so the time between these points is the rise time T_R. The average slope of the step response between points *A* and *B*, which is the slope of the dashed line, is $0.8/T_R$. However, the maximum slope is greater than this. A reasonable estimate of the maximum slope is $1/T_R$:

$$\text{Maximum slope of unit-step response} \doteq 1/T_R \qquad (14.5\text{-}2)$$

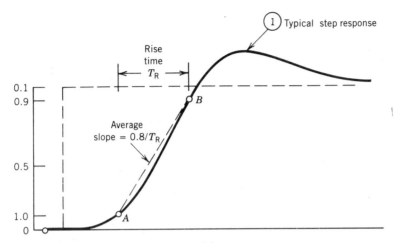

Figure 14.5-2 Typical feedback-signal step response, showing average slope during rise of response.

The derivative of the step response is the impulse response. Hence, Eq 14.5-2 shows that the maximum value of the impulse response is approximately equal to $1/T_R$:

$$\text{Max}|T'_{-1}[t]| \doteq 1/T_R \tag{14.5-3}$$

Diagram a in Fig 14.5-3 shows an impulse response which has an initial value of zero. If the system has good stability, the first overshoot of the step response is generally no greater than 25% of the maximum value. If we ignore subsequent overshoots and assume that the first overshoot is 25%, the remainder coefficient for the impulse response is approximately

$$r_{-1} \doteq (1 + 1 + 0.25 + 0.25)\text{Max}|T'_{-1}[t]|$$
$$\doteq 2.5\,\text{Max}|T'_{-1}[t]| \doteq 2.5/T_R \tag{14.5-4}$$

The approximation of Eq 14.5-3 has been substituted for $\text{Max}|T'_{-1}[t]|$. Diagram b shows an impulse response which has a large initial value. For this case, the sum of the remainder coefficient plus the magnitude of the initial value coefficient is approximately

$$|a_1| + r_{-1} \doteq (1 + 1 + 0.25 + 0.25)\text{Max}|T'_{-1}[t]|$$
$$\doteq 2.5\,\text{Max}|T'_{-1}[t]| \doteq 2.5/T_R \tag{14.5-5}$$

This is the same as the limit of Eq 14.5-4. By Eqs 14.4-53, 14.5-5, the following is the bound on the remainder associated with the step approximation of the

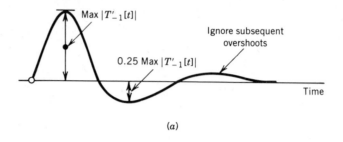

(a)

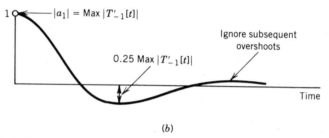

(b)

Figure 14.5-3   Typical impulse responses, illustrating approximation of the impulse-response remainder term $(r_{-1} + |a_1|)$: (a) with zero initial value; (b) with large initial value.

input of Fig 14.5-1:

$$|\text{error}| \le \tfrac{1}{2}(r_{-1} + |a_1|)\text{Max}|D^{-1}x_i| = \frac{1}{2}\frac{2.5}{T_r}\frac{T_{bu}}{8}$$

$$\le 0.156\frac{T_{bu}}{T_R} = 0.20\frac{T_{Ri}}{T_R} \tag{14.5-6}$$

where T_{Ri} is the rise time of the input waveform, which is equal to $0.8T_{bu}$.

   Now let us consider the more realistic gradual step input waveform of Fig 14.5-4. It rises from a local minimum (or inflection point) at point 1 to a local maximum (or inflection point) at point 2, with a rise time T_{Ri}, specified as the time between the 10% and 90% points. For convenience, the amplitude difference between points 1 and 2 is normalized to unity. An equivalent buildup time is defined equal to $1.25T_{Ri}$, by extending the line between points 3, 4 to points 5, 6 at the levels of the local minimum and maximum. The approximate step is drawn so that the two cross-hatched areas are equal. Each of these areas is somewhat greater than $T_{bu}/8$, but does not exceed $T_{bu}/8$ by a

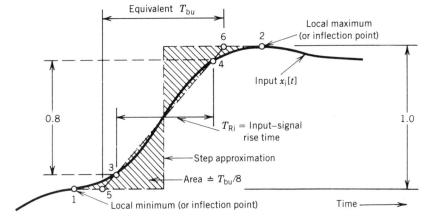

Figure 14.5-4 Step approximation of a rapidly rising portion of a realistic input curve.

large factor. Hence, the preceding analysis still applies approximately to this waveform. By Eq 14.5-6, the bound on the remainder for the step approximation is approximately

$$|error| \le 0.20 \frac{T_{Ri}}{T_R} \tag{14.5-7}$$

Let us compare these approximations with results for the example studied in Section 14.3. From the step response (Fig 14.3-3), it can be shown that the rise time T_R is 0.01155 sec. By Eq 14.5-3, the approximation for the maximum value of the impulse response is

$$\text{Max}|T'_{-1}[t]| \doteq \frac{1}{T_R} = \frac{1}{0.01155 \text{ sec}} = 86.6 \text{ sec}^{-1} \tag{14.5-8}$$

As shown in Fig 14.3-8, the actual maximum value of the impulse response is 76 sec^{-1}. By Eq 14.5-4, the approximate value of the remainder coefficient for the impulse response is

$$r_{-1} \doteq \frac{2.5}{T_R} = \frac{2.5}{0.01155 \text{ sec}} = 216 \text{ sec}^{-1} \tag{14.5-9}$$

As shown in Table 14.4-1, the actual value of the impulse-response remainder coefficient r_{-1} is 182 sec^{-1}.

Figure 14.5-5 was obtained from the author's 1955 paper [14.1]. The figure compares the exact and approximate responses of the feedback-loop example with the ideal gradual-step input waveform of Fig 14.5-1. The buildup time T_{bu} of the gradual step is set equal to the time for the step response to rise

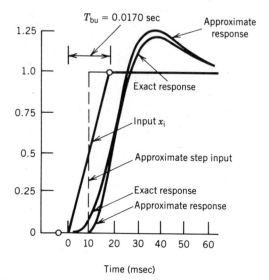

Figure 14.5-5 Comparison of approximate and exact responses of the illustrative feedback loop to a gradual step with $T_{bu} = 0.0170$ sec.

from zero to 100%, which is 17.0 msec $= 0.0170$ sec. Hence the rise time of the input waveform is

$$T_{Ri} = 0.8T_{bu} = 0.8(0.0170 \text{ sec}) = 0.0136 \text{ sec} \qquad (14.5\text{-}10)$$

By Eq 14.5-7, the approximate error bound for the step approximation is

$$|\text{error}| \le 0.20 \frac{T_{Ri}}{T_R} = 0.20 \frac{0.0136}{0.01155} = 0.24 \qquad (14.5\text{-}11)$$

The actual error bound is

$$|\text{error}| \le \tfrac{1}{2}(r_{-1} + |a_1|)D^{-1}x_i$$

$$\le \tfrac{1}{2}(182 \text{ sec}^{-1})\frac{T_{bu}}{8} = 0.19 \qquad (14.5\text{-}12)$$

where T_{bu} was set equal to 0.0170 sec. The actual maximum error is 0.11. The paper [14.1] also considered an input with a buildup time of half this value (0.0085 sec). The actual peak error for the step response for that case is 0.023. The error bound as given by Eq 14.5-12 would be 0.085.

14.5.2 Approximating a Pulse with an Impulse

As was stated earlier in Chapter 5, if an input rises and falls, like a pulse, in a time interval that is shorter than the step-response rise time, it should be approximated as an impulse. The following analysis shows that when the pulse width is half the rise time of the step response, the error of this approximation

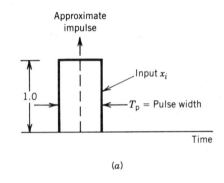

(a)

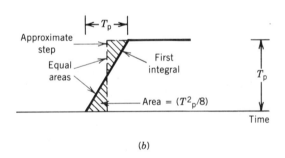

(b)

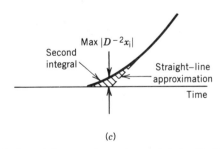

(c)

Figure 14.5-6 Calculation of the error bound associated with the approximation of a pulse with an impulse: (a) input pulse $x_i[t]$; (b) first integral of input pulse; (c) second integral of input pulse.

is about 10% of the amplitude of the pulse. For narrower pulses, the error is proportional to the square of the pulse width.

This concept is analyzed in Fig 14.5-6. Diagram a is an ideal square input pulse of unit amplitude and pulse width T_p. Diagram b is the integral of this, which is approximated as a step, so that the cross-hatched areas are equal. Each of these areas is equal to $T_p^2/8$. Diagram c is the second integral of the input. The step approximation of the first integral of the input is equivalent to a straight-line approximation of the second integral. The maximum deviation for the second integral is equal to one of the equal cross-hatched areas of diagram b, and so is

$$\text{Max}|D^{-2}x_i| = T_p^2/8 \tag{14.5-13}$$

It is convenient to normalize this as follows in terms of the rise time T_R of the step response:

$$\text{Max}|D^{-2}x_i| = \tfrac{1}{8}T_R^2\left(T_p/T_R\right)^2 \tag{14.5-14}$$

By Eq 14.4-53, the error bound for the approximation is

$$|\text{error}| \le \tfrac{1}{2}\left(r_{-2} + |a_2|\right)\text{Max}|D^{-2}x_i|$$
$$\le \tfrac{1}{16}\left(r_{-2} + |a_2|\right)T_R^2\left(T_p/T_R\right)^2 \tag{14.5-15}$$

For the example of Section 14.3, the rise time T_R is 0.01155 sec. As shown in Eq 14.3-26 and Table 14.4-1, for the doublet response the initial-value coefficient a_2 is 1.885×10^4 sec^{-2}, and the remainder coefficient r_{-2} is 3.10×10^4 sec^{-2}. Hence the error bound of Eq 14.5-15 is equal to

$$|\text{error}| \le \tfrac{1}{16}(3.10 + 1.85)(10^4 \text{ sec}^{-2})(0.01155 \text{ sec})^2\left(\frac{T_p}{T_R}\right)^2$$
$$\le 0.413\left(T_p/T_R\right)^2 \tag{14.5-16}$$

If $T_p = T_R/2$, the error bound is 0.10. Thus, if the pulse width T_p is less than $\tfrac{1}{2}$ of the step-response rise time T_R, the error associated with approximating the pulse with an impulse is no greater than 10% of the pulse amplitude.

Chapter 15

A Servo for Integrated-Circuit Photolithography

This chapter describes a unique control system that was developed by the author. This was the basis for the GCA wafer-stepping photolithography equipment for fabricating integrated circuits, which exposes the image directly on the silicon wafer. The control system positions an optical stage to a precision of $\frac{1}{15}$ of a wavelength of light over a 6-in. travel. This servo (along with the wafer-stepping approach) provided greatly improved registration accuracy, which allowed the line widths of integrated circuits to be reduced from 5 μm (micrometres, microns) to less than 1 μm. This equipment has resulted in a tremendous increase in the density of integrated circuitry since 1977.

15.1 THE GCA PHOTOREPEATER® FOR WAFER-STEPPING FABRICATION OF INTEGRATED CIRCUITS

This chapter describes the servo system of a step-and-repeat camera (Photorepeater®*) for fabricating integrated circuits, which the author designed in 1974–1976 while working at GCA Corporation. The servo system positions an optical stage in two dimensions to a precision of ± 0.040 μm, which is about $\frac{1}{15}$ of a wavelength of visible light. This control system has had a revolutionary effect on integrated circuitry. It was a key element in the development of wafer-stepping photolithography equipment, which allowed the line widths of integrated circuits to be reduced from about 5 μm to less than 1 μm, thereby increasing the circuit density by more than $25:1$. (The author was not involved directly in the development of the wafer-stepping process itself, which occurred after he left GCA.)

Step-and-repeat cameras for making integrated circuits built prior to 1976 generally operate in the following manner. A photographic mask, containing

---

*"Photorepeater" is a registered trademark of GCA Corporation.

the image of one or more integrated circuits for one processing stage, is projected onto a photographic glass plate, typically with a reduction of 10 : 1. The plate is mounted on a photographic stage, which is accurately stepped, one field at a time, so that the image is repeated many times across the plate. After the photographic emulsion on the plate is exposed and developed, the plate is accurately registered on a silicon wafer, and its image is contact-printed onto the wafer. This procedure is repeated with several masks to perform the various stages of the photolithography process. When the processing is completed, the wafer is cut into pieces to make the individual integrated circuits.

The contact-printing process eventually wears the image on the glass plate, and the plate must be replaced. Nevertheless, one glass plate can expose a large number of wafers.

The extremely high precision of the photographic-stage control system designed by the author allowed a new approach to photolithography: the wafer-stepping process. In this process, the image is projected directly onto the silicon wafer, instead of onto a glass plate. Fiduciary marks or details on the first image are sensed optically and used to align subsequent images. The stage is precisely positioned so that each image is accurately registered with the earlier images. Since image registration is performed locally, rather than across the whole wafer, much more accurate registration can be achieved. The increase in registration accuracy has allowed line widths of circuitry elements to be reduced from about 5 μm to less than 1 μm.

The wafer-stepping process resulted in a tremendous increase in the demand for the new GCA step-and-repeat cameras, because the process requires one of these instruments for each assembly station. With the earlier process, one step-and-repeat camera can serve about 50 assembly stations.

A description of this control system is useful, not only because of the practical importance of the process itself, but also because the same control principles can be applied to other systems where ultrahigh positional accuracy is required. One of the requirements for achieving ultrahigh positional accuracy is an ultra-accurate sensor for measuring position and velocity. This was provided by the Hewlett-Packard 5501A laser interferometer, which is discussed in Section 15.2. The stage-control system for this GCA step-and-repeat camera is described in Section 15.3.

15.2 THE HEWLETT-PACKARD 5501A LASER INTERFEROMETER

The Hewlett-Packard 5501A laser interferometer [15.1] is based on a principle called the Zeeman effect. A laser generates two optical waves, which are linearly cross-polarized relative to one another, and which are separated in frequency by about 1.8 MHz. Very accurate distance measurements can be achieved by comparing these optical waves.

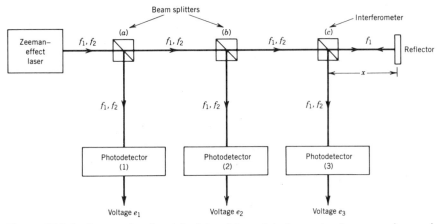

Figure 15.2-1 System showing principle of laser interferometer for measuring position.

A helium–neon laser is used with an output optical power of 1 mW (milliwatt). It operates at a wavelength λ and frequency f of

$$\lambda = 0.632\,999\,37\,\mu\text{m} \tag{15.2-1}$$

$$f = \frac{c}{\lambda} = \frac{2.997\,925 \times 10^8 \text{ m/sec}}{0.632\,999\,37 \times 10^{-6} \text{ m}}$$

$$= 4.73606 \times 10^{14} \text{ Hz} = 4.73606 \times 10^8 \text{ MHz} \tag{15.2-2}$$

The laser light is visible, and has a red color. The optical frequency f is 2.6×10^8 times greater than the 1.8-MHz frequency difference between the two cross-polarized waves.

Figure 15.2-1 illustrates the use of this laser for interferometer distance measurement. The laser transmits two optical frequencies f_1, f_2. Part of this beam is diverted by a beam splitter (a) to a photodetector (1). The output voltage from the photodetector (1) is proportional to the power in the optical wave, and so is proportional to the square of the optical signal. If filtering action is neglected, the signal e_1 from the photodetector (1) is ideally expressed as

$$\text{Ideal}[e_1] = (\cos[\omega_1 t] + \cos[\omega_2 t])^2 \tag{15.2-3}$$

The parameters ω_1, ω_2 are equal to $2\pi f_1, 2\pi f_2$ respectively, where f_1, f_2 are the two optical frequencies separated by 1.8 MHz. Calculating the square in Eq

15.2-3 gives

$$\text{Ideal}[e_1] = \cos^2[\omega_1 t] + \cos^2[\omega_2 t] + 2\cos[\omega_1 t]\cos[\omega_2 t]$$

$$= \frac{1 + \cos[2\omega_1 t]}{2} + \frac{1 + \cos[2\omega_2 t]}{2}$$

$$+ \cos[(\omega_1 + \omega_2)t] + \cos[(\omega_1 - \omega_2)t] \qquad (15.2\text{-}4)$$

The double-optical-frequency components at the frequencies $2\omega_1$, $2\omega_2$, and $(\omega_1 + \omega_2)$ are at such high frequency they are eliminated in the detection process. Hence, the actual photodetector signal reduces to

$$e_1 = 1 + \cos[(\omega_1 - \omega_2)t] \qquad (15.2\text{-}5)$$

The first term is a constant DC term, which is used to measure the signal strength. The interferometer position measurement is derived from the second, AC term, which has a frequency of 1.8 MHz:

$$e_{1(AC)} = \cos[(\omega_1 - \omega_2)t] \qquad (15.2\text{-}6)$$

In Fig 15.2-1, the laser beam is intercepted by a second beam splitter (b), and part of the beam is fed to a second photodetector (2). The optical waves at the surface of the photodetector can be expressed as $\cos[\omega_1 t - \Psi_1]$ and $\cos[\omega_2 t - \Psi_2]$, where the phase shifts Ψ_1, Ψ_2 are equal to

$$\Psi_1 = 2\pi \frac{y}{\lambda_1} \qquad (15.2\text{-}7)$$

$$\Psi_2 = 2\pi \frac{y}{\lambda_2} \qquad (15.2\text{-}8)$$

The parameter y is the additional optical path length to the second photode-tector, relative to the first photodetector. Since the wavelengths λ_1, λ_2 for the two frequencies are essentially equal (to within one part in 2.6×10^8), the phase shifts Ψ_1, Ψ_2 are essentially equal. Therefore, the AC signal at the output of photodetector (2) is

$$e_{2(AC)} = \cos[(\omega_1 - \omega_2)t - (\Psi_1 - \Psi_2)] = \cos[(\omega_1 - \omega_2)t] \quad (15.2\text{-}9)$$

This is the same as that in Eq 15.2-6. Thus, the AC signal at the output of a photodetector is independent of the optical path length from the laser.

The laser beam is fed to a third optical element (c), called an interferome-ter. This device uses the polarization difference between the two waves to separate the f_1-wave from the f_2-wave, and feeds the f_1-wave along an additional path x to the reflector. It is reflected back to the interferometer

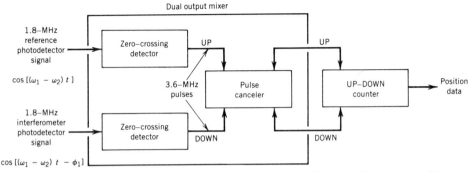

Figure 15.2-2 Block diagram of basic circuitry for decoding interferometer position data.

and combined with the f_2-wave. The combined f_1, f_2 waves are fed to the photodetector (3). The AC output signal from detector (3) is

$$e_{3(AC)} = \cos\left[(\omega_2 - \omega_1)t - \phi_1\right] \qquad (15.2\text{-}10)$$

where ϕ_1 is the additional phase shift given to the f_1-signal, which is equal to

$$\phi_1 = \frac{2\pi}{\lambda}2x \qquad (15.2\text{-}11)$$

The variable x is the distance of the reflector from the interferometer. Since the f_1 optical wave travels this distance twice, the variable x is multiplied by 2.

Thus the distance x can be measured by comparing the phase of the 1.8-MHz signal from photodetector (3) at the output of the interferometer with the 1.8-MHz signal from one of the other photodetectors, (1) or (2), which acts as a reference.

Figure 15.2-2 shows how the phase information of the interferometer signal is converted to digital position data. The 1.8-MHz reference and interferometer signals from the photodetectors are fed to zero-crossing detectors, which generate two pulses per cycle of the sinusoid. The resultant 3.6-MHz pulse trains are fed to a pulse-canceler circuit. If there is an excess of UP or DOWN counts, the excess is counted in an UP–DOWN pulse counter. Whenever the phase ϕ_1 of the interferometer signal increases by 180°, an UP pulse is applied to the counter; and whenever ϕ_1 decreases by 180°, a DOWN pulse is applied.

Because of the factor of 2 in Eq 15.2-11, a phase change of 180° occurs whenever x changes by $\frac{1}{4}$ wavelength. It is convenient to consider displacements in microinches. Since one metre is equal to 39.37 inches, one micrometre is equal to 39.37 microinches. Thus, $\frac{1}{4}$ wavelength in micrometres is

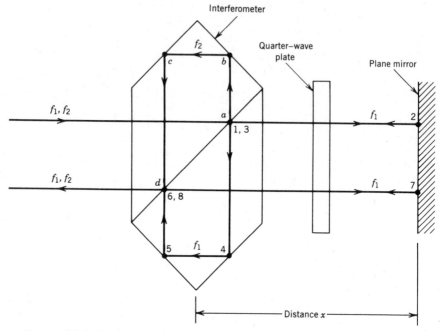

Figure 15.2-3   Paths of laser optical frequencies for plane-mirror interferometer.

(from Eq 15.2-1):

$$\Delta x = \frac{\lambda}{4} = \left(\tfrac{1}{4}\right)(0.6330\ \mu\text{m})(39.37\ \text{in./m}) = 6.2\ \mu\text{in}. \quad (15.2\text{-}12)$$

Thus an extra pulse is counted for every 6.2-μin. motion of the reflector.

The HP 5501A laser interferometer system may use either a retroflector or a plane-mirror reflector to reflect the f_1 beam. When a plane-mirror reflector is used, a double reflection occurs off the mirror, which doubles the resolution of the interferometer. The resultant optical path is shown in Fig 15.2-3.

In Fig 15.2-3, the reference frequency f_2 travels along path a–b–c–d. Since the f_1 component is cross-polarized, it does not reflect at a. It separates from the f_2 component at point 1, which corresponds to point a, and passes through the quarter-wave plate to point 2 on the mirror. The f_1 component is reflected back through the quarter-wave plate to point 3, at the same location as point 1. The two passes through the quarter-wave plate rotate the polarization of the f_1 component by 90°. Hence the return f_1 component reflects at point 3 along path 3–4–5–6. At point 6 the f_1 component is reflected through the quarter-wave plate to point 7 on the mirror, then back through the quarter-wave plate to point 8. Since the two passes through the quarter-wave plate provide another 90° polarization shift, the reflected f_1 component passes

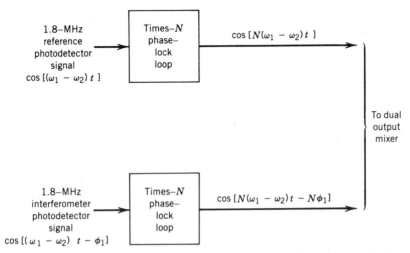

Figure 15.2-4 Block diagram of phase-lock loop circuitry for increasing resolution.

through the mirror and combines with the f_2 component at point 8, which corresponds to point d.

The phase lag imparted to the f_1 component relative to the f_2 component is

$$\phi_1 = \frac{2\pi}{\lambda} 4x \qquad (15.2\text{-}13)$$

where x is the distance of the mirror from the midpoint of the interferometer.

The double reflection off the mirror results in a factor of 4 in Eq 15.2-13. Hence, the basic resolution of the system with the plane-mirror interferometer is $\frac{1}{2}$ of that given in Eq 15.2-12, and so is 3.1 μin. rather than 6.2 μin.

Greater resolution can be achieved by using phase-lock loops in the manner illustrated in Fig 15.2-4. The reference signal and the interferometer signal (which are at 1.8 MHz) are fed to phase-lock loops, which generate phase-coherent frequencies that are N times the input frequency (i.e., sinusoids at 1.8N MHz). This may be achieved by a single phase-lock loop of frequency ratio N, or by cascading loops having a lower frequency ratio. The phase-lock loops increase the phase shift by the factor N. The output signals from the phase-lock loops are fed to the dual output mixer shown in Fig 15.2-2, in place of the photodetector signals. Thus, the resolution is increased by the factor N of the phase-lock loop.

Hewlett-Packard provides phase-lock-loop electronics to extend the resolution by a factor of 16:1. Therefore, with a plane-mirror interferometer, the resolution can be decreased by a factor of 16 from 3.1 to 0.2 μin.

15.3   THE SERVO FOR THE GCA STEP-AND-REPEAT CAMERA

15.3.1   Early Step-and-Repeat Cameras

Step-and-repeat cameras that were built in the early 1970s for making integrated circuits typically operate in the following manner. The position of the photographic stage is sensed by linear optical encoders, using analog interpolation to measure a small fraction of an encoder line width. The stage is driven by motor-tachometer servos that are similar to the servo discussed in Chapter 7.

As was shown in Chapter 7, the accuracy of a conventional motor-tachometer servo driving an optical stage is limited by static friction. The example computed a peak static error of 0.2×10^{-3} in. However, this is optimistic, because it ignores the effect of structural dynamics. Probably an accuracy of $\pm 0.5 \times 10^{-3}$ in. is more realistic, which is equivalent to ± 12.5 μm.

This is the accuracy obtainable with conventional positional control. The early step-and-repeat camera achieved much greater accuracy by taking advantage of the principle that friction is nearly constant when the direction of the velocity is constant.

The photographic stage is positioned as follows. The stage is driven to the desired location in a programmed manner. As the desired position is approached, the velocity is decreased linearly. Finally the motor is declutched, and the stage is brought to rest by friction. Since the friction is nearly constant, the stopping trajectory is highly repeatable. The accuracy of positioning is very much better than the ± 12.5 μm achievable by conventional positional control: it is about ± 1.0 μm. Note that this approach cannot allow position corrections to be made after the stage is positioned, a capability that is required in wafer-stepping lithography equipment.

15.3.2   Principles for Eliminating Friction

Much greater positioning accuracy became possible with the availability of the Hewlett-Packard 5501A laser interferometer, described in Section 15.2, which can provide extremely accurate position and velocity data. For a step-and-repeat camera to take advantage of the potential accuracy of this laser interferometer, it must have considerably reduced friction.

There are three basic principles for eliminating friction in a positional control system:

1. Hydraulic bearings.
2. Air bearings.
3. Flexure (spring) mounts.

Hydraulic bearings are not applicable to step-and-repeat cameras, because the hydraulic fluid would contaminate the optical process.

Air bearings have been applied with some success to step-and-repeat cameras. However, an air bearing is inherently a rather soft spring. In order for it to be reasonably stiff, the size of the air-bearing pad must be quite large, and the air gap must be very small—much less than 0.001 in. With such a small air gap, accurate alignment and contamination from dust particles can be serious problems.

The GCA step-and-repeat camera uses the third approach. The fine position stage is supported on flexure (or spring) mounts. The basic principle of flexure mounts is well known. Although the force from a flexure can be quite significant, unlike static friction, this force varies smoothly, and so its effect can be compensated for by an integral network. Hence, a stage mounted on flexures can be positioned to extremely high accuracy.

Because the motion allowable with flexure supports is limited, a two-stage servo is needed when flexures are used. A coarse stage is positioned by conventional means to within about 0.001 in. of the desired location. The coarse stage carries the fine stage, which is mounted on flexures. The fine servo drives the fine stage relative to the coarse stage, to achieve the required ultrahigh accuracy.

15.3.3 Control Configuration of GCA Step-and-Repeat Camera Stage

Figure 15.3-1 shows the control-system configuration of the flexure-mounted photographic stage of the GCA step-and-repeat camera, for a single axis. The position of the coarse stage is controlled by the motor-tachometer servo drive. The fine stage is flexure-mounted onto the coarse stage. The linear actuator (which has no friction) positions the fine stage relative to the coarse stage. The linear variable differential transformer (LVDT) measures the relative displace-

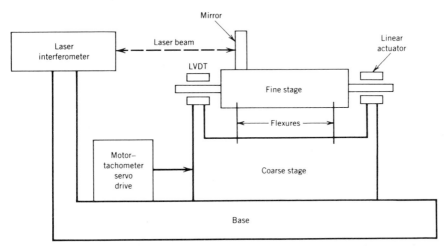

Figure 15.3-1 Basic elements of photographic-stage control system for one axis.

ment between the fine and coarse stage. The laser interferometer measures the absolute position of the fine stage by means of a laser beam, which is reflected off the mirror mounted on the fine stage.

The laser interferometer provides position and velocity data for absolute control of the fine stage. The LVDT measures the relative displacement between the coarse and fine stages. The combination of the laser-interferometer data and the LVDT signal provides position information for control of the coarse stage.

Each axis of the coarse stage is positioned by a tachometer servo, which operates according to the principles discussed in Chapter 7, except that position feedback is derived from a combination of the signals from the laser and the LVDT. When the motor-tachometer drive mechanism positions the coarse stage to approximately the desired position, the motor is declutched, and the coarse stage is brought to rest by friction. Accurate positioning is performed by driving the fine stage relative to the coarse stage by means of the

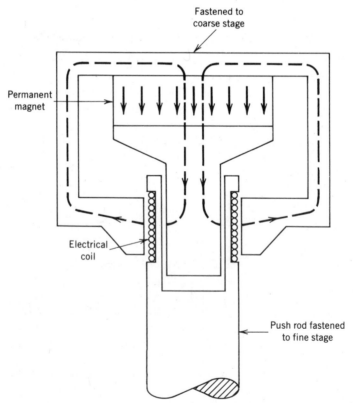

Figure 15.3-2   Principle of frictionless voice-coil-type electromagnet actuator for driving fine stage.

linear actuator. The fine-servo control system derives position and velocity feedback signals from the laser interferometer.

It is essential that the linear actuator and the LVDT have no friction. If appropriately mounted, the LVDT device (which was described in Chapter 8) is frictionless. A frictionless actuator can be achieved by applying the principles of the voice-coil drive in a loudspeaker, in the manner shown in Fig 15.3-2.

The actuator of Fig 15.3-2 consists of a permanent magnet mounted on the coarse stage, and a coil-driven push rod that is coupled to the fine stage. Current flowing through the coil on the push rod applies a force to the fine stage, which is proportional to the coil current.

This actuator mechanism differs from a loudspeaker voice coil in that the push-rod-and-coil mechanism must be very stiff, to provide a high mechanical resonant frequency in the drive. Rather large air gaps can result from the necessary support structure and to allow for motion between the fine and coarse stages. Fortunately, these problems can be counteracted by using a samarium-cobalt permanent magnet, which has an extremely strong magnetic field.

Note that the actuator mechanism that positions the flexure-supported fine stage provides only small motion, whereas an actuator that positions a stage supported on air bearings must provide large motion. It is much more difficult to achieve a frictionless actuator mechanism when the displacement is large.

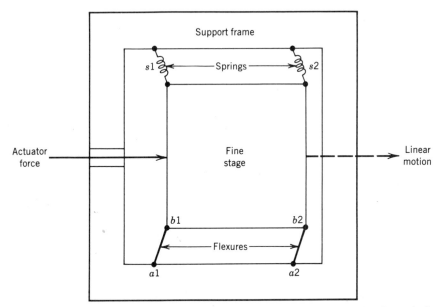

Figure 15.3-3 Basic principle of flexure support of fine stage for motion in a single dimension.

Consequently, the accuracy of a step-and-repeat camera stage supported on air bearings may be degraded by friction in the actuator mechanism.

15.3.4   Flexure Support of Fine Stage

Figure 15.3-3 shows the basic principle for flexure support of the fine stage for motion in a single dimension. The fine stage is supported within the support frame by means of the two flexible supports $a1-b1$, $a2-b2$, and by the springs $S1$, $S2$. The springs keep the flexible supports in tension. The two flexible supports form a parallelogram ($a1-b1-b2-a2$), which keeps the lines $a1-a2$ and $b1-b2$ parallel. Thus, the fine stage can only move in a single dimension, without rotation, within the support frame.

Figure 15.3-4 gives a more practical mounting scheme, which is equivalent to that of Fig 15.3-3. The fine stage is supported by two straps $a1-b1-c1-d1$ and $a2-b2-c2-d2$. One end of each strap ($a1$, $a2$) is connected to the support frame, and the other end ($d1$, $d2$) is connected to one of the blocks ($B1$, $B2$). The blocks $B1$, $B2$ are coupled to the frame by means of the strap springs $S1$, $S2$, which keep the support straps $a1-b1-c1-d1$ and $a2-b2-c2-d2$ in tension.

The flexure support mechanism in Fig 15.3-4 keeps the surface of the fine stage in a horizontal plane (parallel to the surface of the paper). The stage is allowed to move (without rotation) only in a single dimension.

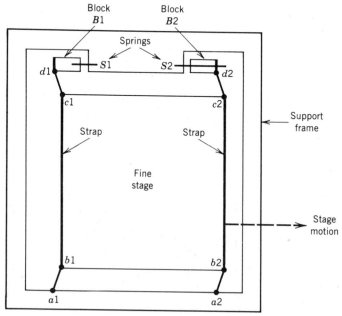

Figure 15.3-4   Practical design of flexure support of fine stage for motion in a single dimension.

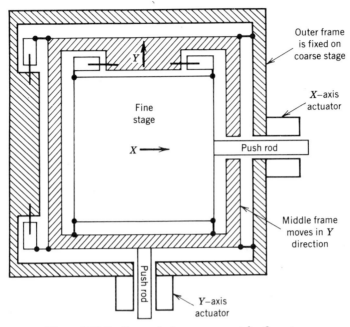

Figure 15.3-5   Two-axis flexure support for fine stage.

Figure 15.3-5 shows the principle of a two-axis flexure mount for support-ing the fine stage on the coarse stage. The fine stage is fastened by means of flexures to the middle frame. The flexures allow the fine stage to move relative to the middle frame in the X-direction. The middle frame is fastened by means of flexures to the outer frame (attached to the coarse stage). These flexures allow the middle frame to move in the Y-direction, relative to the coarse stage. The fine stage moves in the Y-direction along with the middle frame. The X and Y actuators drive the fine stage relative to the coarse stage in the X and Y directions.

The two-axis flexure mount illustrated in Fig 15.3-5 shows the well-known principle of supporting an instrument stage by means of flexures, to allow two-axis motion. The actual flexure support used in the GCA step-and-repeat camera applies this principle, but differs in mechanical details.

15.3.5   Summary of Operation of Photographic-Stage Servo

The coarse stage of the GCA step-and-repeat camera is driven in the X and Y directions by tachometer-feedback DC electric servos, which are similar to the servo discussed in Chapter 7. Each servo has a current feedback loop, a tachometer velocity loop, and a position loop. The error signal for the position loop is derived from (1) the digital position command, (2) the digital position feedback signal obtained from the laser interferometer, and (3) the analog signal from the LVDT, which measures the displacement between the fine and coarse stages. These servos drive the coarse stage to approximately the desired position, to an accuracy of the order of 0.001 in. Then the motor is declutched, and the coarse stage is held fixed by friction.

The fine stage is driven in the X and Y directions by means of electromagnetic voice-coil-type actuators. A current feedback loop provides precise control of the actuator current, which is proportional to the force exerted by the actuator on the fine stage.

The position of the fine stage is measured to a precision of 1.5 μin. by the Hewlett-Packard 5501A laser interferometer. A flat-mirror interferometer is used, which has a basic resolution of 3 μin. A phase-lock loop increases this resolution by a factor of 2 : 1, to achieve a final position resolution of 1.5 μin., which is equivalent to 0.040 μm.

Velocity feedback information is achieved by using phase-lock loops to provide laser position data of very fine resolution. These fine laser-interferometer pulses are counted over a fixed time interval to obtain a velocity signal. For example, phase-lock-loop electronics supplied by Hewlett-Packard can provide a 16 : 1 increase in resolution. For a flat-mirror laser interferometer, which has a basic resolution of 3 μin., the resolution with a 16 : 1 phase-lock-loop ratio is 0.2 μin. If these 0.2-μin. pulses are counted over a period of 1 millisecond (msec), the resultant velocity resolution is

$$\Delta V = (0.2\ \mu in)/(1\ msec) = 2 \times 10^{-4}\ in./sec \qquad (15.3\text{-}1)$$

Nominal values for the primary dynamic parameters of the fine position servo are as follows:

Velocity-loop gain crossover frequency:

$$\omega_{cv} = 250\ rad/sec\ (40\ Hz) \qquad (15.3\text{-}2)$$

Position-loop gain crossover frequency:

$$\omega_{cp} = 100\ rad/sec\ (16\ Hz) \qquad (15.3\text{-}3)$$

Break frequency of position-loop integral network:

$$\omega_{ip} = 25\ rad/sec\ (4\ Hz) \qquad (15.3\text{-}4)$$

For the position-loop gain crossover frequency ω_{cp} of Eq 15.3-3, the velocity error ΔV of Eq 15.3-1 would produce the following error in position:

$$X_e = \frac{\Delta V}{\omega_{cp}} = \frac{2 \times 10^{-4} \text{ in./sec}}{100 \text{ sec}^{-1}}$$

$$= 2 \times 10^{-6} \text{ in.} = 2 \ \mu\text{in.} \qquad (15.3\text{-}5)$$

On the other hand, the servo system averages the returns from many velocity samples, and so the resultant position error is appreciably smaller than this.

The resolution achievable from the laser velocity signal is critically dependent on noise in the electronic circuitry that processes the laser-interferometer signal. A strong factor in the extremely fine precision achieved by the GCA equipment was the excellent low-noise design of the phase-lock loops and other circuits for processing the laser signal. This circuitry was designed at GCA by Edward Radziewicz.

In order for this servo system to step quickly, it includes a great many nonlinear controls, which allow the system to switch rapidly from coarse to fine control, with minimum transients. The position of the fine stage relative to the coarse stage is controlled during the coarse step in order to minimize settling time. In tests performed by the author, the system could step 0.1 in. and be within an error of $\pm 6.2 \ \mu$in. ($\pm 0.16 \ \mu$m) in 0.5 sec. The final position was held to a precision of $\pm 1.5 \ \mu$in.

15.3.6   A Structural-Resonance Problem Encountered during Development

As in the design of any high-performance servo system, structural dynamics was an important issue. However, a particularly acute structural-resonance problem was experienced in the development of this equipment, which serves as a good illustration of the basic principle: In servo development, structural dynamics should be treated with a great deal of respect.

The first prototype step-and-repeat camera had been constructed, and I started to optimize the feedback loops of the fine X and Y control systems. Dynamic parameters were soon achieved that were reasonably close to the desired values. The next day, I attempted to repeat the tests, and the servo started to oscillate at 40 Hz. I struggled with this problem for several days, but nothing I did could stop this erratic oscillation, which would strangely appear and disappear.

Finally, I received assistance from a well-seasoned engineer, with tremendous engineering insight. After watching me struggle with the problem for a while, Paul Dippolito quietly stated, "George, you have a rotary oscillation mode."

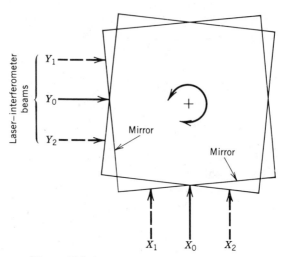

Figure 15.3-6 Rotary oscillation of fine stage.

Once he explained the problem, it was obvious what was happening. The fine stage was oscillating in a rotary fashion about the vertical axis, as shown in Fig 15.3-6. When the coarse stage was positioned in the center of the 6-in. range of motion, the laser beams would reflect off the mirrors along lines X_0 and Y_0. For these beam positions, the rotary oscillation did not cause any changes of optical path length, and so did not couple into the X and Y control loops. However, when the coarse stage was positioned off center, the light beams were in locations such as X_1, X_2, Y_1, Y_2. For these locations, the rotation caused a change in the measured distance. The rotary motion would couple into the position and velocity loops, causing the fine servo to oscillate.

Thus, when the tests were performed with the coarse stage near the center of the field, the servos would work. However, when the coarse stage was moved sufficiently off from the center of the field, the rotary mode was excited, and the servo would oscillate.

Several of us began to examine the flexure support of the fine stage to determine the cause of the rotary oscillation. Remember, from the discussion in Section 15.3.4, that the flexures are supposed to keep the fine stage from rotating, and to allow only linear motion in the X and Y directions. However, the collection of straps and blocks in the flexure-support mechanism made the problem seem very complicated.

While I was missing the point, Paul Dippolito again came forth with the answer, along with mechanical measurements he had taken to prove his conclusion. The middle frame in Fig 15.3-5 was distorting from the shape of a rectangle to a parallelogram, because it was weak in bending at its corners. This distortion of the middle stage allowed the fine stage to rotate. The outer frame could not distort, because it was mounted firmly onto the coarse stage.

Tests were performed on this rotary oscillation mode by recording the decay of the oscillation after the excitation was removed. The oscillation decayed by 1.8% from one cycle to the next, which indicated that the logarithmic decrement δ was 1.8%. In accordance with the equations in Chapter 10, Section 10.4.2, the corresponding value of resonance amplification factor Q was 170, and the damping ratio ζ was 0.003.

This rotary oscillation mode, with its extremely high Q, was weakly coupled to the control loops of the fine stage. Consequently, the mode could not be observed in an open-loop frequency-response test. However, when the loops were closed, enough energy could couple into this high-Q mode to cause oscillation.

The rotary-oscillation problem was corrected by fastening a plate across the under side of the middle frame. This plate, which passed under the fine stage, greatly increased the stiffness of the middle frame, and so prevented the parallelogram distortion. With this increase in stiffness, the 40-Hz oscillation was eliminated.

As pointed out in the Preface, I became acutely aware of structural resonance in the early 1950s, when performing frequency-response tests on an aircraft fire-control system. Over the years, I had encountered structural-dynamics problems again and again. Yet, here in my tests on this control system, I was floored by the mysterious oscillation. Instinctively, I felt it was due to structural resonance, but the precise cause eluded me.

Difficulties like this can be frustrating. However, the struggle to overcome them is what makes feedback control such an exciting area in which to work.

Appendix D

Equations Relating True and Pseudo Complex Frequencies

This appendix derives the equations used in Chapter 9, Section 9.3, which express σ and ω, the real and imaginary parts of the complex frequency s, as functions of α and W, the real and imaginary parts of the sampled-data complex pseudo frequency p. The following variables are defined:

$$s = \sigma + j\omega \qquad \text{(D-1)}$$

$$\bar{z} = e^{-sT} \qquad \text{(D-2)}$$

$$U = \frac{1 - \bar{z}}{1 + \bar{z}} \qquad \text{(D-3)}$$

Substituting Eqs D-1, D-2 into Eq D-3 gives

$$U = \frac{1 - e^{-(\sigma + j\omega)T}}{1 + e^{-(\sigma + j\omega)T}} \qquad \text{(D-4)}$$

Rationalize this expression by multiplying the numerator and denominator by $(1 + e^{-(\sigma - j\omega)T})$:

$$U = \frac{(1 - e^{-(\sigma + j\omega)T})(1 + e^{-(\sigma - j\omega)T})}{(1 + e^{-(\sigma + j\omega)T})(1 + e^{-(\sigma - j\omega)T})}$$

$$= \frac{1 - e^{-2\sigma T} + e^{-\sigma T}(e^{j\omega T} - e^{-j\omega T})}{1 + e^{-2\sigma T} + e^{-\sigma T}(e^{j\omega T} + e^{-j\omega T})} \qquad \text{(D-5)}$$

Replace the complex exponentials by the following trigonometric expressions:

$$(e^{j\omega T} - e^{-j\omega T}) = j2\sin[\omega T] \qquad \text{(D-6)}$$

$$(e^{j\omega T} + e^{-j\omega T}) = 2\cos[\omega T] \qquad \text{(D-7)}$$

This gives for U

$$U = \frac{1 - e^{-2\sigma T} + j2e^{-\sigma T}\sin[\omega T]}{1 + e^{-2\sigma T} + 2e^{-\sigma T}\cos[\omega T]}$$

$$= \frac{(e^{\sigma T} - e^{-\sigma T}) + j2\sin[\omega T]}{(e^{\sigma T} + e^{-\sigma T}) + 2\cos[\omega T]} \tag{D-8}$$

This becomes

$$U = \frac{\sinh[\sigma T] + j\sin[\omega T]}{\cosh[\sigma T] + \cos[\omega T]} \tag{D-9}$$

The pseudo complex-frequency p is defined as follows:

$$U = p\frac{T}{2} \tag{D-10}$$

The real and imaginary parts of p are defined as α and W, respectively. Thus,

$$p = \alpha + jW \tag{D-11}$$

$$U = p\frac{T}{2} = \frac{\alpha T}{2} + j\frac{WT}{2} \tag{D-12}$$

Comparing Eqs D-9 and D-12 gives

$$\alpha\frac{T}{2} = \frac{\sinh[\sigma T]}{\cosh[\sigma T] + \cos[\omega T]} \tag{D-13}$$

$$W\frac{T}{2} = \frac{\sin[\omega T]}{\cosh[\sigma T] + \cos[\omega T]} \tag{D-14}$$

These equations can be expressed as

$$\cosh[\sigma T] + \cos[\omega T] = \frac{2}{\alpha T}\sinh[\sigma T] \tag{D-15}$$

$$\cosh[\sigma T] + \cos[\omega T] = \frac{2}{WT}\sin[\omega T] \tag{D-16}$$

Since the left-hand sides of the two equations are equal, the right-hand sides must be equal. Hence,

$$W\sinh[\sigma T] = \alpha\sin[\omega T] \tag{D-17}$$

Solve Eq D-15 for $\cos[\omega T]$ and Eq D-16 for $\cosh[\sigma T]$:

$$\cos[\omega T] = \frac{2}{\alpha T}\sinh[\sigma T] - \cosh[\sigma T] \qquad \text{(D-18)}$$

$$\cosh[\sigma T] = \frac{2}{WT}\sin[\omega T] - \cos[\omega T] \qquad \text{(D-19)}$$

Square both sides of Eqs D-18, -19:

$$\cos^2[\omega T] = \left(\frac{2}{\alpha T}\right)^2 \sinh^2[\sigma T] + \cosh^2[\sigma T]$$
$$-2\left(\frac{2}{\alpha T}\right)\sinh[\sigma T]\cosh[\sigma T] \qquad \text{(D-20)}$$

$$\cosh^2[\sigma T] = \left(\frac{2}{WT}\right)^2 \sin^2[\omega T] + \cos^2[\omega T]$$
$$-2\left(\frac{2}{WT}\right)\sin[\omega T]\cos[\omega T] \qquad \text{(D-21)}$$

Substitute into these the following identities:

$$\cos^2[\omega T] = 1 - \sin^2[\omega T] \qquad \text{(D-22)}$$
$$\cosh^2[\sigma T] = 1 + \sinh^2[\sigma T] \qquad \text{(D-23)}$$

Equations D-20, -21 become

$$1 - \sin^2[\omega T] = \left(\frac{2}{\alpha T}\right)^2 \sinh^2[\sigma T] + \left(1 + \sinh^2[\sigma T]\right)$$
$$-2\left(\frac{2}{\alpha T}\right)\sinh[\sigma T]\cosh[\sigma T] \qquad \text{(D-24)}$$

$$1 + \sinh^2[\sigma T] = \left(\frac{2}{WT}\right)^2 \sin^2[\omega T] + \left(1 - \sin^2[\omega T]\right)$$
$$-2\left(\frac{2}{WT}\right)\sin[\omega T]\cos[\omega T] \qquad \text{(D-25)}$$

Equation D-17 yields the following:

$$\sin[\omega T] = \frac{W}{\alpha}\sinh[\sigma T] \qquad \text{(D-26)}$$

$$\sinh[\sigma T] = \frac{\alpha}{W}\sin[\omega T] \qquad \text{(D-27)}$$

Substitute Eq D-26 into the left side of Eq D-24, and Eq D-27 into the left side of Eq D-25. Subtract unity from both sides. This gives

$$-\left(\frac{W}{\alpha}\right)^2 \sinh^2[\sigma T] = \left(\frac{2}{\alpha T}\right)^2 \sinh^2[\sigma T] + \sinh^2[\sigma T]$$

$$-\left(\frac{4}{\alpha T}\right)\sinh[\sigma T]\cosh[\sigma T] \qquad \text{(D-28)}$$

$$\left(\frac{\alpha}{W}\right)^2 \sin^2[\omega T] = \left(\frac{2}{WT}\right)^2 \sin^2[\omega T] - \sin^2[\omega T]$$

$$-\left(\frac{4}{WT}\right)\sin[\omega T]\cos[\omega T] \qquad \text{(D-29)}$$

Combining similar terms in Eqs D-28, -29 gives

$$\frac{4}{\alpha T}\sinh[\sigma T]\cosh[\sigma T] = \left[\left(\frac{2}{\alpha T}\right)^2 + 1 - \left(\frac{W}{\alpha}\right)^2\right]\sinh^2[\sigma T] \quad \text{(D-30)}$$

$$\frac{4}{WT}\sin[\omega T]\cos[\omega T] = \left[\left(\frac{2}{WT}\right)^2 - 1 - \left(\frac{\alpha}{W}\right)^2\right]\sin^2[\omega T] \quad \text{(D-31)}$$

Dividing the right-hand sides of the equations by the left-hand sides gives

$$1 = \left(\frac{1}{\alpha T} + \frac{\alpha T}{4} - \frac{W^2 T}{4\alpha}\right)\tanh[\sigma T]$$

$$= \frac{1}{\alpha T}\left(1 + \left(\frac{\alpha T}{2}\right)^2 + \left(\frac{WT}{2}\right)^2\right)\tanh[\sigma T] \qquad \text{(D-32)}$$

$$1 = \left(\frac{1}{WT} - \frac{WT}{4} - \frac{\alpha^2 T}{4W}\right)\tan[\omega T]$$

$$= \frac{1}{WT}\left(1 - \left(\frac{WT}{2}\right)^2 - \left(\frac{\alpha T}{2}\right)^2\right)\tan[\omega T] \qquad \text{(D-33)}$$

Solve Eq D-32 for tanh[σT] and Eq D-33 for tan[ωT]. This gives

$$\tanh[\sigma T] = \frac{\alpha T}{1 + \left[(\alpha T/2)^2 + (WT/2)^2\right]} \qquad \text{(D-34)}$$

$$\tan[\omega T] = \frac{WT}{1 - \left[(\alpha T/2)^2 + (WT/2)^2\right]} \qquad \text{(D-35)}$$

The magnitude of the complex pseudo frequency p is equal to

$$|p| = \sqrt{\alpha^2 + W^2} \qquad \text{(D-36)}$$

Hence, Eqs D-34, -35 can be expressed as follows in terms of $|p|$:

$$\tanh[\sigma T] = \frac{\alpha T}{1 + (|p|T/2)^2} \qquad \text{(D-37)}$$

$$\tan[\omega T] = \frac{WT}{1 - (|p|T/2)^2} \qquad \text{(D-38)}$$

Appendix E

Integration of Runge-Kutta Polynomial

Equation 9.4-31 of Section 9.4 gave the following expression, used in the Runge–Kutta integration routine, for the integral of $\dot{y}[t]$ over the sampling period T:

$$\Delta y = \frac{T}{6}\left(\dot{y}_{(1)} + 4\dot{y}_{(2)} + \dot{y}_{(3)}\right) \tag{E-1}$$

This is the best estimate of the integral of $\dot{y}[t]$ over the time interval T, assuming one is given the values of $\dot{y}[t]$ only at the beginning, middle, and end of that time interval, which are designated $\dot{y}_{(1)}, \dot{y}_{(2)}, \dot{y}_{(3)}$, respectively. This expression will now be derived.

Three values of $\dot{y}[t]$ are specified at time t_1, t_2, t_3, where the two time intervals, $t_3 - t_2$ and $t_2 - t_1$, are equal. The best estimate of the continuous signal $\dot{y}[t]$ corresponding to these three values is obtained by fitting to them a polynomial of the following form:

$$\dot{y}[t] = a + bt + ct^2 \tag{E-2}$$

To simplify the fitting of this curve, the zero-time axis is chosen such that time t is zero at time t_2. Since the time interval between t_1 and t_3 is the sample period T, then $t_2 = 0$, $t_1 = -T/2$, and $t_3 = +T/2$. The values of $\dot{y}[t]$ at times t_1, t_2, and t_3 are therefore related as follows to the coefficients a, b, c:

$$\dot{y}_{(1)} = a + bt_1 + ct_1^2 = a + b(-T/2) + c(-T/2)^2 \tag{E-3}$$

$$\dot{y}_{(2)} = a + bt_2 + ct_2^2 = a \tag{E-4}$$

$$\dot{y}_{(3)} = a + bt_3 + ct_3^2 = a + b(T/2) + c(T/2)^2 \tag{E-5}$$

Solving these for the polynomial coefficients a, b, c gives

$$a = \dot{y}_{(2)} \tag{E-6}$$

$$b = \frac{1}{T}\left(\dot{y}_{(3)} - \dot{y}_{(1)}\right) \tag{E-7}$$

$$c = \frac{2}{T^2}\left(\dot{y}_{(3)} + \dot{y}_{(1)} - 2\dot{y}_{(2)}\right) \tag{E-8}$$

The integral of the polynomial expression for $\dot{y}[t]$ given in Eq E-2, over the interval from t_1 to t_3, is

$$\begin{aligned}
\Delta y &= \int_{-T/2}^{+T/2}\left(a + bt + ct^2\right) dt \\
&= \left\{a(T/2) + (b/2)(T/2)^2 + (c/3)(T/2)^3\right\} \\
&\quad - \left\{a(-T/2) + (b/2)(-T/2)^2 + (c/3)(-T/2)^3\right\} \\
&= aT + (c/12)T^3 \tag{E-9}
\end{aligned}$$

Substituting Eqs E-6, -8 into this gives the expression of Eq E-1, which was to be proved.

Appendix F

Summary of Matrix Formulas

F.1 SOME BASIC MATRIX DEFINITIONS

The elements of a matrix are labeled with a double subscript notation of the form a_{ij}. The first subscript i is the row of the matrix, and the second subscript j is the column. Thus, a 3-by-3 matrix has the form

$$\underset{\sim}{A} = \begin{bmatrix} a_{11} & a_{12} & a_{13} \\ a_{21} & a_{22} & a_{23} \\ a_{31} & a_{32} & a_{33} \end{bmatrix} \tag{F-1}$$

In this book, a matrix is designated by underlining its symbol with a tilde ($\sim$). If the number of rows equals the number of columns, the matrix is called *square*. The *diagonal* is defined only for a square matrix, and for Eq F-1 consists of the elements a_{11}, a_{22}, a_{33}. The *trace* of a square matrix, designated tr[$\underset{\sim}{A}$], is the sum of the diagonal elements, which for this matrix is

$$\mathrm{tr}\big[\underset{\sim}{A}\big] = \sum_{i=1}^{n} a_{ii} = a_{11} + a_{22} + a_{33} \tag{F-2}$$

The following are special types of square (n by n) matrices:

Diagonal matrix. A diagonal matrix, designated D or D_n, is a square matrix in which all of the off-diagonal elements are zero.

Unit matrix. A unit matrix, designated I or I_n, is a diagonal matrix in which all of the diagonal elements are unity.

Triangular matrix. An upper triangular matrix is a square matrix in which all of the upper off-diagonal elements are zero. A lower triangular matrix is a square matrix in which all of the lower off-diagonal elements are zero.

For a square 3-by-3 matrix, an upper triangular matrix, a diagonal matrix, and a unit matrix have the following forms:

Upper triangular matrix:

$$T_3 = \begin{bmatrix} a_{11} & 0 & 0 \\ a_{21} & a_{22} & 0 \\ a_{31} & a_{32} & a_{33} \end{bmatrix} \tag{F-3}$$

Diagonal matrix:

$$D_3 = \begin{bmatrix} a_{11} & 0 & 0 \\ 0 & a_{22} & 0 \\ 0 & 0 & a_{33} \end{bmatrix} \tag{F-4}$$

Unit matrix:

$$I_3 = \begin{bmatrix} 1 & 0 & 0 \\ 0 & 1 & 0 \\ 0 & 0 & 1 \end{bmatrix} \tag{F-5}$$

A *zero matrix* is a matrix in which all elements are zero.

F.2 MATRIX ADDITION

Matrix addition can be performed only on matrices having the same number of rows and the same number of columns. It is performed by adding the corresponding elements of the two matrices. For example:

$$\begin{bmatrix} a_{11} & a_{12} \\ a_{21} & a_{22} \\ a_{31} & a_{32} \end{bmatrix} + \begin{bmatrix} b_{11} & b_{12} \\ b_{21} & b_{22} \\ b_{31} & b_{32} \end{bmatrix} = \begin{bmatrix} a_{11} + b_{11} & a_{12} + b_{12} \\ a_{21} + b_{21} & a_{22} + b_{22} \\ a_{31} + b_{31} & a_{32} + b_{32} \end{bmatrix} \tag{F-6}$$

Matrix subtraction is defined in a similar manner.

F.3 MATRIX MULTIPLICATION

When a matrix is multiplied by a scalar, all elements of the matrix are multiplied by the scalar. For example,

$$\begin{bmatrix} a_{11} & a_{12} \\ a_{21} & a_{22} \end{bmatrix} \times K = \begin{bmatrix} a_{11}K & a_{12}K \\ a_{21}K & a_{22}K \end{bmatrix} \tag{F-7}$$

Two matrices can be multiplied together only if the number of columns of the first matrix is equal to the number of rows of the second matrix. One or both of these matrices can consist of a single column or row (i.e., it can be a vector). Consider the following matrix product:

$$\underset{\sim}{C} = \underset{\sim}{A}\underset{\sim}{B} \tag{F-8}$$

The elements of the product matrix $\underset{\sim}{C}$ are given by the formula

$$c_{ij} = \sum_{k=1}^{n} a_{ik} b_{kj} \tag{F-9}$$

where n is the number of columns of the matrix $\underset{\sim}{A}$, which is equal to the number of rows of the matrix $\underset{\sim}{B}$. Consider for example, the following matrix multiplication:

$$\underset{\sim}{C} = \begin{bmatrix} a_{11} & a_{12} \\ a_{21} & a_{22} \\ a_{31} & a_{32} \\ a_{41} & a_{42} \end{bmatrix} \begin{bmatrix} b_{11} & b_{12} & b_{13} \\ b_{21} & b_{22} & b_{23} \end{bmatrix} \tag{F-10}$$

The product matrix $\underset{\sim}{C}$ is

$$\underset{\sim}{C} = \begin{bmatrix} a_{11}b_{11} + a_{12}b_{21} & a_{11}b_{12} + a_{12}b_{22} & a_{11}b_{13} + a_{12}b_{23} \\ a_{21}b_{11} + a_{22}b_{21} & a_{21}b_{12} + a_{22}b_{22} & a_{21}b_{13} + a_{22}b_{23} \\ a_{31}b_{11} + a_{32}b_{21} & a_{31}b_{12} + a_{32}b_{22} & a_{31}b_{13} + a_{32}b_{23} \\ a_{41}b_{11} + a_{42}b_{21} & a_{41}b_{12} + a_{42}b_{22} & a_{41}b_{13} + a_{42}b_{23} \end{bmatrix} \tag{F-11}$$

The matrix product $\underset{\sim}{A}\underset{\sim}{B}$ is generally not equal to $\underset{\sim}{B}\underset{\sim}{A}$. Matrix division is not defined.

F.4   TRANSPOSE OF A MATRIX

The transpose of a matrix $\underset{\sim}{A}$ is designated $\underset{\sim}{A}^T$, and is obtained by interchanging the rows and columns of the matrix $\underset{\sim}{A}$. For example, the transpose of the 3-by-3 matrix $\underset{\sim}{A}$ of Eq F-1 is

$$\underset{\sim}{A}^T = \begin{bmatrix} a_{11} & a_{21} & a_{31} \\ a_{12} & a_{22} & a_{32} \\ a_{13} & a_{23} & a_{33} \end{bmatrix} \tag{F-12}$$

For a square matrix, the transpose can be formed by rotating the matrix about its diagonal. Consider also the following 3-by-2 matrix:

$$\underset{\sim}{F} = \begin{bmatrix} f_{11} & f_{21} \\ f_{12} & f_{22} \\ f_{13} & f_{23} \end{bmatrix} \tag{F-13}$$

The transpose of this matrix is

$$\underset{\sim}{F}^{T} = \begin{bmatrix} f_{11} & f_{12} & f_{13} \\ f_{21} & f_{22} & f_{23} \end{bmatrix} \tag{F-14}$$

The square of any matrix F^2 is obtained by multiplying the matrix by its transpose:

$$\underset{\sim}{F}^{2} = \underset{\sim}{F}\underset{\sim}{F}^{T} \tag{F-15}$$

An important matrix equation involving the transpose is

$$(\underset{\sim}{A}\underset{\sim}{B})^{T} = \underset{\sim}{B}^{T}\underset{\sim}{A}^{T} \tag{F-16}$$

The *complex conjugate transpose* (or Hermitian conjugate) of a matrix $\underset{\sim}{A}$ is designated $\underset{\sim}{A}^{\dagger}$. Its elements are the complex conjugates of the elements of the matrix $\underset{\sim}{A}^{T}$, which is the transpose of the matrix $\underset{\sim}{A}$. An important relation concerning this matrix is

$$(\underset{\sim}{A}\underset{\sim}{B})^{\dagger} = \underset{\sim}{B}^{\dagger}\underset{\sim}{A}^{\dagger} \tag{F-17}$$

The following are definitions for special types of matrices:

$\underset{\sim}{A}$ is symmetric if	$\underset{\sim}{A} = \underset{\sim}{A}^{T}$	(F-18)
$\underset{\sim}{A}$ is skew-symmetric if	$\underset{\sim}{A} = -\underset{\sim}{A}^{T}$	(F-19)
$\underset{\sim}{A}$ is Hermitian if	$\underset{\sim}{A} = \underset{\sim}{A}^{\dagger}$	(F-20)
$\underset{\sim}{A}$ is skew-Hermitian if	$\underset{\sim}{A} = -\underset{\sim}{A}^{\dagger}$	(F-21)
$\underset{\sim}{A}$ is normal if	$\underset{\sim}{A}^{\dagger}\underset{\sim}{A} = \underset{\sim}{A}\underset{\sim}{A}^{\dagger}$	(F-22)
$\underset{\sim}{A}$ is unitary if	$\underset{\sim}{A}^{\dagger}\underset{\sim}{A} = \underset{\sim}{I}$	(F-23)
$\underset{\sim}{A}$ is orthogonal if	$\underset{\sim}{A}^{\dagger}\underset{\sim}{A} = \underset{\sim}{D}$	(F-24)

F.5 DETERMINANT OF A MATRIX

The determinant is defined only for a square matrix A and is designated $\det[A]$. The determinant of an n-by-n matrix is a sum of $n!$ terms, each of which is the product of n elements, multiplied by ± 1. Consider the following 2-by-2 matrix:

$$A = \begin{bmatrix} a_{11} & a_{12} \\ a_{21} & a_{22} \end{bmatrix} \qquad \text{(F-25)}$$

The determinant of this matrix is

$$\det[A] = a_{11}a_{22} - a_{21}a_{12} \qquad \text{(F-26)}$$

The determinants for higher-order matrices can be computed by the method of cofactors, which reduces a determinant to those of matrices of lower order. Let us apply this to the following 3-by-3 matrix A, which was given in Eq F-1:

$$A = \begin{bmatrix} - & a_{11} & - & a_{12} & - & a_{13} & - \\ & & & | & & & \\ & a_{21} & & a_{22} & & a_{23} & \\ & & & | & & & \\ & a_{31} & & a_{32} & & a_{33} & \end{bmatrix} \qquad \text{(F-27)}$$

For each element of a matrix there is a submatrix called a *minor*, which is obtained by deleting the row and column of the element. The *cofactor* of an element is the determinant of its minor multiplied by $(-1)^{i+j}$, where i and j are the column and row indices of the element. Thus, the minor of the element a_{12} in the matrix A of Eq F-27 is

$$\text{Minor}[a_{12}] = A_{12} = \begin{bmatrix} a_{21} & a_{23} \\ a_{31} & a_{33} \end{bmatrix} \qquad \text{(F-28)}$$

This minor is formed by deleting the elements indicated by the lines in the matrix of Eq F-27, which are the elements in the row and column of element a_{12}. The cofactor of element a_{12} is the determinant of this minor matrix of Eq F-28 multiplied by $(-1)^{1+2}$, which is

$$\text{Cof}[a_{12}] = (-1)^{1+2}\det\begin{bmatrix} a_{21} & a_{23} \\ a_{31} & a_{33} \end{bmatrix}$$

$$= (-1)^{3}(a_{21}a_{33} - a_{31}a_{23}) = a_{31}a_{23} - a_{21}a_{33} \qquad \text{(F-29)}$$

The determinant of the matrix A is the sum, over any row or column, of each element multiplied by its cofactor. Let us apply this rule to the elements

a_{11}, a_{12}, a_{13} of the first row:

$$\det[A] = a_{11}\text{Cof}[a_{11}] + a_{12}\text{Cof}[a_{12}] + a_{13}\text{Cof}[a_{13}]$$

$$= a_{11}(-1)^{1+1}\det\begin{bmatrix} a_{22} & a_{23} \\ a_{32} & a_{33} \end{bmatrix} + a_{12}(-1)^{1+2}\det\begin{bmatrix} a_{21} & a_{23} \\ a_{31} & a_{33} \end{bmatrix}$$

$$+ a_{11}(-1)^{1+3}\det\begin{bmatrix} a_{21} & a_{22} \\ a_{31} & a_{32} \end{bmatrix} \qquad \text{(F-30)}$$

Solving for the determinants of the minors gives

$$\det[A] = a_{11}(a_{22}a_{33} - a_{32}a_{23}) - a_{12}(a_{21}a_{33} - a_{31}a_{23})$$

$$+ a_{11}(a_{21}a_{32} - a_{31}a_{22}) \qquad \text{(F-31)}$$

F.6 ADJOINT OF A MATRIX

The adjoint, designated adj[A], is defined for a square matrix. The adjoint of a square matrix A is the transpose of a related matrix B having elements that are the cofactors of the elements of matrix A. Thus

$$\text{adj}[A] = B^T \qquad \text{(F-32)}$$

where each element b_{ij} of the matrix B is the cofactor of the corresponding element a_{ij} of matrix A:

$$b_{ij} = \text{Cof}[a_{ij}] = (-1)^{i+j}\det[A_{ij}] \qquad \text{(F-33)}$$

The matrix A_{ij} is the minor matrix of the element a_{ij}, obtained by deleting the row and column of the element a_{ij}. Let us find the adjoint matrix for the 3-by-3 matrix A given in Eq F-1. The corresponding matrix B is defined as

$$B = \begin{bmatrix} b_{11} & b_{12} & b_{13} \\ b_{21} & b_{22} & b_{23} \\ b_{31} & b_{32} & b_{33} \end{bmatrix} \qquad \text{(F-34)}$$

Consider for example the element b_{31} of this matrix. This is equal to

$$b_{31} = \text{Cof}[a_{31}] = (-1)^{3+1}\det\begin{bmatrix} a_{12} & a_{13} \\ a_{22} & a_{23} \end{bmatrix}$$

$$= (-1)^4(a_{12}a_{23} - a_{22}a_{13}) = a_{12}a_{23} - a_{22}a_{13} \qquad \text{(F-35)}$$

By repeating this procedure, all nine elements of the matrix B can be calculated. By Eq F-32, the adjoint of A is expressed as follows in terms of these nine elements:

$$\text{adj}[A] = B^T = \begin{bmatrix} b_{11} & b_{21} & b_{31} \\ b_{12} & b_{22} & b_{32} \\ b_{13} & b_{23} & b_{33} \end{bmatrix} \tag{F-36}$$

Let us calculate the adjoint of the 2-by-2 matrix A given in Eq F-25. The corresponding matrix B has the form

$$B = \begin{bmatrix} b_{11} & b_{12} \\ b_{21} & b_{22} \end{bmatrix} \tag{F-37}$$

The elements of this matrix B are

$$b_{11} = \text{Cof}[a_{11}] = (-1)^{1+1}\det[a_{22}] = a_{22} \tag{F-38}$$

$$b_{12} = \text{Cof}[a_{12}] = (-1)^{1+2}\det[a_{21}] = -a_{21} \tag{F-39}$$

$$b_{21} = \text{Cof}[a_{21}] = (-1)^{2+1}\det[a_{12}] = a_{12} \tag{F-40}$$

$$b_{22} = \text{Cof}[a_{22}] = (-1)^{2+2}\det[a_{11}] = -a_{11} \tag{F-41}$$

Since A is a 2-by-2 matrix, the minor of any element is a scalar, and the determinant of the minor is equal to that scalar. Substituting Eqs F-38 to -41 into Eq F-37 gives the matrix B:

$$B = \begin{bmatrix} a_{22} & -a_{21} \\ -a_{12} & a_{11} \end{bmatrix} \tag{F-42}$$

Applying Eq F-32 gives the following for the adjoint of A:

$$\text{adj}[A] = B^T = \begin{bmatrix} a_{22} & -a_{12} \\ -a_{21} & a_{11} \end{bmatrix} \tag{F-43}$$

This is the transpose of the matrix B given in Eq F-42.

F.7 INVERSE OF A MATRIX

The inverse of a matrix A is designated A^{-1}, and is defined by

$$AA^{-1} = I \tag{F-44}$$

where I is an identity (unit) matrix, which was defined in Section F.1. If the matrix $\underset{\sim}{A}$ is square and its determinant is not zero, the inverse of the matrix $\underset{\sim}{A}$ is given as follows by Cramer's rule:

$$\underset{\sim}{A}^{-1} = \frac{\text{adj}[\underset{\sim}{A}]}{\text{det}[\underset{\sim}{A}]} \qquad\qquad \text{(F-45)}$$

Let us apply Eq F-45 to calculate the inverse of the 2-by-2 matrix $\underset{\sim}{A}$ of Eq F-25. The determinant of this matrix was given in Eq F-26, and the adjoint was given in Eq F-43. Substituting Eqs F-26, -43 into Eq F-45 yields the inverse matrix:

$$\underset{\sim}{A}^{-1} = \frac{\text{adj}[\underset{\sim}{A}]}{\text{det}[\underset{\sim}{A}]}$$

$$= \frac{1}{a_{11}a_{22} - a_{21}a_{12}} \begin{bmatrix} a_{22} & -a_{12} \\ -a_{21} & a_{11} \end{bmatrix} \qquad\qquad \text{(F-46)}$$

Appendix G

Values for Noise-Bandwidth Integrals

Frequency-response calculations of noise bandwidth and mean square error can be greatly simplified by using tables of integrals of the general form

$$I_n = \frac{1}{2\pi j} \int_{-j\infty}^{+j\infty} H[s]H[-s]\, ds \tag{G-1}$$

The transfer function $H[s]$ is defined as follows as a ratio of two polynomials in s, where the order of the numerator is less than that of the denominator:

$$H[s] = \frac{c_{n-1}s^{n-1} + \cdots + c_1 s + c_0}{d_n s^n + d_{n-1}s^{n-1} + \cdots + d_1 s + d_0} \tag{G-2}$$

The parameter n is the order of s in the denominator of $H[s]$. This appendix gives equations for the integral I_n for values of n from 1 to 6, as functions of the coefficients c, d of the numerator and denominator of $H[s]$. These were obtained from Appendix E of Newton, Gould, and Kaiser [11.5]. That reference shows how these integrals are derived, and gives expressions for the integrals for values of n from 1 to 10. The integrals I_n, for $n = 1$ to 6, are as follows:

$$I_1 = \frac{c_0^2}{2d_0 d_1} \tag{G-3}$$

$$I_2 = \frac{c_1^2 d_0 + c_0^2 d_2}{2d_0 d_1 d_2} \tag{G-4}$$

$$I_3 = \frac{1}{2D_3}\left[c_2^2 d_0 d_1 + \left(c_1^2 - 2c_0 c_2 \right) d_0 d_3 + c_0^2 d_2 d_3 \right] \tag{G-5}$$

$$D_3 = d_0 d_3 (d_1 d_2 - d_0 d_3) \tag{G-6}$$

$$I_4 = \frac{1}{2D_4}\left[c_3^2 m_{40} + \left(c_2^2 - 2c_1c_3\right)m_{41} + \left(c_1^2 - 2c_0c_2\right)m_{42} + c_0^2 m_{43}\right]$$

(G-7)

$$m_{40} = d_0 d_1 d_2 - d_0^2 d_3 \tag{G-8}$$

$$m_{41} = d_0 d_1 d_4 \tag{G-9}$$

$$m_{42} = d_0 d_3 d_4 \tag{G-10}$$

$$m_{43} = d_2 d_3 d_4 - d_1 d_4^2 \tag{G-11}$$

$$D_4 = d_0 d_4\left(d_1 d_2 d_3 - d_0 d_3^2 - d_1^2 d_4\right) \tag{G-12}$$

$$I_5 = \frac{1}{2D_5}\left[c_4^2 m_{50} + \left(c_3^2 - 2c_2c_4\right)m_{51}\right.$$

$$\left. + \left(c_2^2 - 2c_1c_3 + 2c_0c_4\right)m_{52} + \left(c_1^2 - 2c_0c_2\right)m_{53} + c_0^2 m_{54}\right] \tag{G-13}$$

$$m_{51} = d_1 d_2 - d_0 d_3 \tag{G-14}$$

$$m_{52} = d_1 d_4 - d_0 d_5 \tag{G-15}$$

$$m_{53} = \frac{d_2 m_{52} - d_4 m_{51}}{d_0} \tag{G-16}$$

$$m_{54} = \frac{d_2 m_{53} - d_4 m_{52}}{d_0} \tag{G-17}$$

$$m_{50} = \frac{d_3 m_{51} - d_1 m_{52}}{d_5} \tag{G-18}$$

$$D_5 = d_0\left(d_1 m_{54} - d_3 m_{53} + d_5 m_{52}\right) \tag{G-19}$$

$$I_6 = \frac{1}{2D_6}\left[c_5^2 m_{60} + \left(c_4^2 - 2c_3c_5\right)m_{61} + \left(c_3^2 - 2c_2c_4 + 2c_1c_5\right)m_{62}\right.$$

$$\left. + \left(c_2^2 - 2c_1c_3 + 2c_0c_4\right)m_{63} + \left(c_1^2 - 2c_0c_2\right)m_{64} + c_0^2 m_{65}\right] \tag{G-20}$$

$$m_{61} = d_0 d_3^2 + d_1^2 d_4 - d_0 d_1 d_5 - d_1 d_2 d_3 \tag{G-21}$$

$$m_{62} = d_0 d_3 d_5 + d_1^2 d_6 - d_1 d_2 d_5 \tag{G-22}$$

$$m_{63} = d_0 d_5^2 + d_1 d_3 d_6 - d_1 d_4 d_5 \tag{G-23}$$

$$m_{64} = \frac{d_2 m_{63} - d_4 m_{62} + d_6 m_{61}}{d_0} \tag{G-24}$$

$$m_{65} = \frac{d_2 m_{64} - d_4 m_{63} + d_6 m_{62}}{d_0} \tag{G-25}$$

$$m_{60} = \frac{d_4 m_{61} - d_2 m_{62} + d_0 m_{63}}{d_6} \tag{G-26}$$

$$D_6 = d_0\left(d_1 m_{65} - d_3 m_{64} + d_5 m_{63}\right) \tag{G-27}$$

Appendix H

Structural Resonant Frequencies of Servo-Controlled Antennas

Figure H-1 shows the values of servo structural resonant frequency of 160 servo-controlled parabolic antennas, expressed as a function of antenna diameter. This figure was summarized in Chapter 10, Fig. 10.1-1. These data have been compiled over the past 25 years by Denny D. Pidhayny. (Additions would be appreciated. Please mail to Denny D. Pidhayny, Aerospace Corp., P.O. Box 92957, Los Angeles, CA 90009.)

The names and manufacturers of the antennas that correspond to the numbers indicated on Fig H-1 are as follows:

1. Hughes Aircraft Co., GAR-1A

2. Radiation Inc., 35 GHz horn, mini-pedestal

3. Teledyne-Ryan, 622

4. TRW, 3-axis K_a-band antenna

5. Hughes Aircraft Co., 94-GHz Scanner (spin-stabilized)

6. MIT/EE, AN/APG-27

7. Radiation Inc., CONDOR (tuned)

8. Hughes Aircraft Co., AN/APG-37

9. Hughes Aircraft Co., AN/APG-40

10. Cubic Corp., QRC-501

11. Westinghouse Electric, AN/APG-66 and -68 (F-16 aircraft)

12. RCA/D.V., Lunar Excursion Module torquer

13. Harris Corp., NESP (Naval 20/44 GHz antenna)

14. Westinghouse Electric Corp., AN/APQ-120

15. Datron Systems Inc., NESP (Naval 20/44 GHz antenna)

16. Westinghouse Electric Corp., AN/APQ-72 and AN/APQ-109

17. Westinghouse Electric Corp., AN/AWG-10

18. Bell Aerospace Corp., (helicopter antenna)

19. Hughes Aircraft Co., AN/APG-63 (F-15 aircraft)

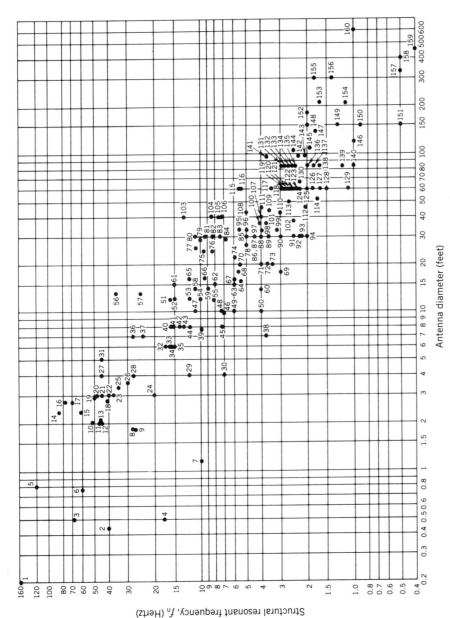

Figure H-1 Servo structural-resonant frequencies of servo-driven parabolic antennas, plotted as a function of antenna diameter (compiled by Pidhayny).

595

20. Bell Aircraft Corp., AN/ASC-22
21. Bell Aircraft Corp., AN/ASC-28
22. Hughes Aircraft Co.
23. Hughes Aircraft Co., AN/AWG-9
24. Cubic Corp., AN/APQ-131
25. RCA, SHF Antenna
26. Hughes Aircraft Corp., AN/ASG-18 (hydraulic)
27. General Dynamics, 3-axis torquer (graphite)
28. Collins, AN/WSC-6
29. Datron Systems Inc., AN/WCS-2
30. OE-81C/WSC-1
31. Westinghouse Electric Corp., AN/APQ-81
32. NASA 35 GHz
33. Harris Corp., shipboard
34. Cubic Corp., LAMPS
35. SCR-584 Anti-Aircraft Antenna (WW II)
36. Naval Research Lab., DMAR
37. Datron Systems Inc., (torquer)
38. MET 8000
39. RCA, AN/TPQ-27
40. Tasker Systems Division
41. Datron Systems Inc., (torquer)
42. MK-73 TAR-TAR
43. Datron Systems Inc.
44. SCR-784 Anti-Aircraft Antenna
45. Scientific Atlanta, DL
46. Harris Corp., GAIN-1
47. WECO Titan guidance
48. Radiation Inc., TRANSTERM
49. Harris Corp., XY-1
50. Scientific Atlanta, AN/UPD-4
51. Sperry, AN/FPQ-10
52. RCA, AN/MPS-36
53. RCA, AN/FPS-16 (hydraulic)
54. RCA, CAPRI
55. DYNASOAR

56. Westinghouse Electric Corp., AN/APD-7
57. Scientific Atlanta, 3893 Torquer
58. General Electric, ATLAS (torquer)
59. Prelort (SGLS)
60. Satcom
61. Datron Systems Inc., 8902
62. University of Texas, 75 GHz (X–Y)
63. Aerospace Corp./Rohr, 90 GHz (hydraulic)
64. Scientific Atlanta, TDRSS
65. AN/FPS-16V
66. RCA, AN/MPS-25, T-AGM-8
67. Radiation Inc., SHIPS
68. MK-1
69. Cubic Corp., AGUVA
70. Harris Corp., GAIN-2
71. Datron Systems Inc., LANDSAT
72. Scientific Atlanta, THAS
73. Datron Systems Inc., LANDSAT
74. Radiation Inc., SHIPS
75. Canoga/Tasker Systems Div., SDA
76. Canoga, PS2W7
77. AM/FSS-7
78. Harris Corp., GAIN-3
79. Datron Systems Inc., DMFT
80. Antenna Systems Inc., (hydraulic)
81. Sperry, ARS
82. ANT LAB, AN/GKR-7
83. Collins, AN/MSC-44 (hydraulic, without trailer)
84. MIPIR AN/TPQ-18
85. AN/FPQ-6 and AN/TPQ-18
86. Harris Corp., (kingpost)
87. TRIAX (3-axis torquer)
88. Rohr/Collins, X–Y mount (NASA)
89. Scientific Atlanta, LANDSAT (3-axis)
90. Datron Systems Inc., LANDSAT
91. TEL-3

92. Reeves, TRIAX
93. TAA-3
94. Radiation Inc., MGS
95. Canoga/Tasker, TTS-6A
96. Scientific Atlanta, OSCS-111, Camp Parks
97. Scientific Atlanta, GPS/GA (4 systems)
98. Datron Systems Inc.
99. Datron Systems Inc., ARTS
100. Harris Corp., W and T1
101. Rohr, Kitt Peak (45 GHz)
102. Harris Corp., MT
103. AMF-MARS
104. Rohr
105. Sperry, ARS
106. Sperry, ARS
107. ALCOR, (Marshall Islands)
108. Rohr
109. Harris Corp, W and T2
110. Datron Systems Inc.
111. Philco, SGLS
112. UK TCS, Harris Corp.
113. Harris Corp., W and T3
114. Datron Systems Inc., DLT
115. AMRAD, AN/FPS-62 (RAMPART)
116. Naval Research Laboratories
117. Radiation Inc., (X–Y) (HOT-LINE, Rohr Comsat)
118. Radiation Inc., SMS
119. Radiation Inc., UK-TCS
120. Radiation Inc., TT and C (3)
121. Harris Corp., TDRSS Ku-band
122. Ford Aerospace, AN/FSC-78 and -79
123. GTE Sylvania, ADVENT
124. Harris Corp., W and T4
125(a). JPL/Philco, S-band (polar, hydraulic)
125(b). MIT Lincoln Lab., Camp Parks, X-band
126. Westinghouse Electric Corp., STELLAR
127. Stanford University

128. Philco, T. and D. (3-axis hydraulic)
129. TLM-18
130. FACC, No. 1
131. Rohr, (X–Y)
132. RCA, ANFPS-49
133. P-4
134. Philco, LAAS
135. TV
136. Stonehouse, Millstone, Tradex
137. Rohr
138. Westinghouse Electric Corp., RTMS
139. Jet Propulsion Laboratories
140. Rohr, TAA-2
141. Rohr, Comsat
142. Philco, Comsat
143. Collins, (hydraulic)
144. Harris Corp., CST
145. JPL/TIW-Systems, X-band (3 systems)
146. Haystack
147. National Science Foundation
148. Radiation Inc., ALTAIR, Marshall Islands
149. Sugar Grove
150. Naval Research Laboratories, SRI
151. BAY HOUSE
152. Bell Telephone Laboratories, TELSTAR horn
153. Goldstone: Rohr (1), Collins (3)
154. CSIRO
155. WT NRT
156. Sugar Grove (estimated)
157. Max Planck Institute (Germany)
158. Stanford (estimated)
159. NEROC (estimated)
160. Naval Radio Research Station, Sugar Grove, W. Va. (estimated)

This information was supplied by the following individuals:

Fritz M. Amtsberg	J. H. Hamel
John Barber	J. R. Horsch
John G. Barnes	R. W. Larkins
David K. Barton	John Lozier
Charles Bass	T. Lund
Cary Bedford	A. R. Mace
George Bendis	Walter R. Manzke
Robert Berquist	John Marlowe
George Biernson	George Newton, Jr.
Guy R. Blatt	C. D. Phillips
R. N. Bracewell	Denny D. Pidhayny
C. O. Burns	Arthur Shaw
C. James Butts	Hugh Smith
Theodore A. Cameron	J. H. Staehlin
Jerry Cantrell	Charles P. Swanson
Leopold Cantafio	James W. Titus
A. A. Clark	John Todd
R. J. Conlon	Robert Tryon
Arthur Connolly	Robert Wallace
Dunley Cottler	John Ward
Charles Fassnacht	Herb Weiss
Robert Finefrock	Pierce Wetter

Appendix I

Gear-Train Dissipation During Resonance

During resonance, the structural compliance is placed under stress over a quarter cycle, as energy is transferred from kinetic energy stored in the inertia to stress energy stored in the compliance; and over the next quarter cycle, the energy is transferred back from stress to momentum. Energy is dissipated as the compliance is placed under stress, and an equal amount of energy is dissipated as the stress energy is released. (This assumes that the resonance is forced, so that the amplitude of the oscillation is constant.) This appendix calculates the energy that is dissipated in a gear train during a quarter cycle of the resonance, as the compliance is placed under stress.

Let us consider the lowest structural resonance, where the motor velocity experiences a resonant null. At this frequency, the motor moves very little, and so the motor is assumed to be stationary in the following analysis.

Figure I-1a shows the model of the gear train. The gear ratio of the gear train is normalized to unity, just as was done in Section 10.3. The torque T_L is the maximum torque exerted by the load on the gear train, and Θ_L is the resultant angular displacement of the load during a quarter cycle. The torque T_m is the resultant torque exerted by the gear train on the motor shaft, which is equal to the reaction torque exerted by the motor shaft on the gear train. The angular displacement Θ_m of the motor shaft is assumed to be zero.

The gear-train efficiency is designated η. The torque T_m applied by the gear train to the motor shaft is equal to the torque T_L applied by the load to the other end of the gear train, multiplied by the gear efficiency η:

$$T_m = \eta T_L \tag{I-1}$$

Hence, the friction torque lost in the gear train is

$$T_f = T_L - T_m = (1 - \eta)T_L \tag{I-2}$$

The model shown in Fig. I-1a assumes that the friction torque is separated into three equal torques, designated ΔT, which are applied at separate points

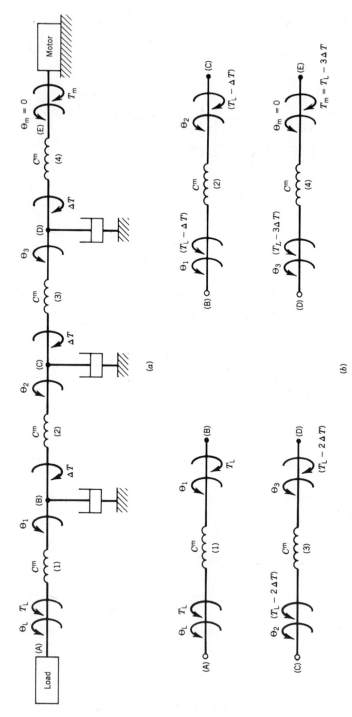

Figure I-1 Mechanical model for gear train with distributed compliance and friction.

(B), (C), (D) along the gear train. By Eq I-2, these separate friction torques are equal to

$$\Delta T = \frac{T_f}{3} = \frac{(1 - \eta)T_L}{3} \tag{I-3}$$

The model assumes that the compliance of the gear train is separated into four equal compliances in series, which are designated $(1), (2), (3), (4)$. The friction torques ΔT are applied at the intersections between these compliances.

During a quarter cycle in which the compliance of a gear train is placed under stress, the actual time variation of load torque T_L, or that of the corresponding angular deflection of the gear train, need not be considered. Only the maximum torques and maximum angular deflections are important, because these determine the values of stored and dissipated energy during the quarter cycle.

In diagram b of Fig I-1, the four compliances are isolated, to show the torques exerted at both ends of each compliance and the corresponding angular displacements of the ends. The compliance of each element $(1), (2), (3), (4)$ is designated C^m. The angular compression of an individual compliance element, which is designated $\Delta\Theta$, is related as follows to the torque T applied at each end of the compliance:

$$\Delta\Theta = C^m T \tag{I-4}$$

Relating Eq I-4 to the four compliance elements in diagram b gives

Element (1): $\qquad\qquad \Theta_L - \Theta_1 = C^m T_L \qquad\qquad$ (I-5)

Element (2): $\qquad\qquad \Theta_1 - \Theta_2 = C^m(T_L - \Delta T) \qquad\qquad$ (I-6)

Element (3): $\qquad\qquad \Theta_2 - \Theta_3 = C^m(T_L - 2\,\Delta T) \qquad\qquad$ (I-7)

Element (4): $\qquad\qquad \Theta_3 = C^m(T_L - 3\,\Delta T) \qquad\qquad$ (I-8)

Substituting Eqs I-1 to -3 into Eqs I-5 to -8 gives

$$\Theta_3 = C^m(\eta)T_L \tag{I-9}$$

$$\Theta_2 = \frac{C^m(1 + 5\eta)T_L}{3} \tag{I-10}$$

$$\Theta_1 = C^m(1 + 2\eta)T_L \tag{I-11}$$

$$\Theta_L = 2C^m(1 + \eta)T_L \tag{I-12}$$

The energy imparted by the load inertia to the gear train during the quarter cycle is

$$U_{in} = T_L\Theta_L \tag{I-13}$$

The energy dissipated by the gear train friction is

$$U_d = \Theta_1 \, \Delta T + \Theta_2 \, \Delta T + \Theta_3 \, \Delta T = (\Theta_1 + \Theta_2 + \Theta_3) \, \Delta T \qquad (I\text{-}14)$$

The energy stored in the gear-train compliance during the quarter cycle (U_{st}) is equal to the input energy U_{in} minus the dissipated energy U_d, which is

$$U_{st} = U_{in} - U_d = T_L \Theta_L - (\Theta_1 + \Theta_2 + \Theta_3) \, \Delta T \qquad (I\text{-}15)$$

Substituting Eqs I-3 and I-9 to -12 into Eqs I-14, -15 gives the following for the dissipated and stored energy during the quarter cycle:

$$U_d = C^m T_L^2 \frac{(1 - \eta)(4 + 14\eta)}{9} \qquad (I\text{-}16)$$

$$U_{st} = C^m T_L^2 \frac{14 + 8\eta + 14\eta^2}{9} \qquad (I\text{-}17)$$

Dividing Eq I-16 by Eq I-17 gives the following for the ratio of dissipated to stored energy during the quarter cycle:

$$\frac{U_d}{U_{st}} = \frac{(1 - \eta)(4 + 14\eta)}{14 + 8\eta + 14\eta^2} \qquad (I\text{-}18)$$

Appendix J

Computation of Steady-State and Initial-Value Coefficients

The long-division method for calculating the steady-state and initial-value coefficients will now be described by applying it to the following feedback transfer function:

$$G_{ib} = \frac{5 + 5.5s}{(s + 1)(s + 5)} = \frac{5 + 5.5s}{5 + 6s + s^2} \tag{J-1}$$

The steady-state coefficients are obtained by expressing the numerator and denominator polynomials in ascending powers of s, and dividing numerator by denominator in a long-division process as follows:

$$
\begin{array}{r}
1 - 0.1s - 0.08s^2 + 0.116s^3 \\
5 + 6s + s^2 \overline{)\, 5 + 5.5s} \\
5 + 6.0s + 1.0s^2 \\
\hline
-0.5s - 1.0s^2 \\
-0.5s - 0.6s^2 - 0.10s^3 \\
\hline
-0.4s^2 + 0.10s^3 \\
-0.4s^2 - 0.48s^3 - 0.08s^4 \\
\hline
0.58s^3 + 0.08s^4
\end{array}
$$

$$\tag{J-2}$$

Thus, G_{ib} can be expanded in the series

$$G_{ib} = 1 - 0.1s - 0.08s^2 + 0.116s^3 + \cdots$$

$$= c'_0 + c'_1 s + c'_2 s^2 + c'_3 s^3 + \cdots \tag{J-3}$$

The steady-state coefficients c'_n for the feedback signal are shown in the first column of Table J-1. When the terms of the polynomials are reversed, so they are expressed in descending powers of s, the long division process yields the

TABLE J-1 Feedback-Signal Steady-State and Initial-Value Coefficients for G_{ib} of Eq. J-1

Steady State	Initial value
$c'_0 = 1$	$a_0 = 0$
$c'_1 = -0.1$ sec	$a_1 = 5.5$ sec^{-1}
$c'_2 = -0.08$ sec^2	$a_2 = -28$ sec^{-2}
$c'_3 = 0.116$ sec^3	$a_3 = 140.5$ sec^{-3}

initial-value coefficients, as follows:

$$
s^2 + 6s + 5 \;\overline{)\;\begin{array}{l} 5.5/s - 28/s^2 + 140.5/s^3 \\ \hline 5.5s \quad\quad +5 \\ 5.5s \quad +33 \quad +27.5/s \\ \hline \quad\quad -28 \quad -27.5/s \\ \quad\quad -28 \quad -168/s - 140/s^2 \\ \hline \quad\quad\quad\quad 140.5/s + 140/s^2 \end{array}}
$$

$$\text{(J-4)}$$

Hence G_{ib} can be expanded as

$$
G_{ib} = \frac{5.5}{s} - \frac{28}{s^2} + \frac{140.5}{s^3} + \cdots
$$

$$
= a_0 + \frac{a_1}{2} + \frac{a_2}{s^2} + \frac{a_3}{s^3} + \cdots \tag{J-5}
$$

The initial-value coefficients are shown in the second column of Table J-1.

References to Volume Two

CHAPTER 1

1.1 A. C. Hall, *The Analysis and Synthesis of Linear Servomechanisms*, Technology Press, Cambridge, Mass., 1943.

1.2 H. M. James, N. B. Nichols, and R. S. Phillips, *Theory of Servomechanisms*, McGraw-Hill, 1947.

1.3 G. S. Brown and D. P. Campbell, *Principles of Servomechanisms*, Wiley, 1948.

1.4 H. Chestnut and R. W. Mayer, *Servomechanism and Regulating System Design*, Vol. 1, Wiley, 1950.

1.5 J-C Gille, M. J. Pelegrin, and P. Decaulne, *Feedback Control Systems*, *Analysis and Synthesis*, McGraw-Hill, 1959.

1.6 J. L. Bower and P. M. Schultheiss, *Introduction to the Design of Servomechanisms*, Wiley, 1958.

1.7 S. J. Mason, "Feedback Theory: Some Properties of Signal-Flow Graphs", *Proc. IRE*, vol. 41, 1953, pp 1144–1156.

1.8 S. J. Mason, "Feedback Theory: Further Properties of Signal-Flow Graphs", *Proc. IRE*, vol. 44, 1956, pp 920–926.

1.9 S. J. Mason and H. J. Zimmerman, *Electronic Circuits, Signals, and Systems*, Wiley, 1960.

1.10 H. Harris, Jr., M. J. Kirby, and E. F. Von Arx, "Servomechanisms Transient Performance from Decibel–Log Frequency Response", *AIEE Transactions*, vol. 70, pt II, 1951, pp 1452–1459.

1.11 G. A. Biernson, "Quick Methods for Evaluating the Closed-Loop Poles of Feedback Control Systems", *AIEE Transactions*, vol. 72, pt II, pp 53–70, May 1953 (with discussions by N. L. Kusters and W. J. M. Moore, and by Maurice J. Kirby and D. C. Beaumariage).

1.12 Robert E. Graham, "Linear Servo Theory", *Bell. Syst. Technical Journal*, Oct. 1946, p. 616.

1.13 W. R. Evans, *Control System Dynamics*, McGraw-Hill, 1954.

1.14 G. A. Biernson, *Optimal Radar Tracking Systems* (in print).

1.15   G. A. Biernson, "A Feedback Control Model of Human Vision", *IEEE Proceedings*, vol. 54, No. 6, pp 858–872, June 1966.

CHAPTER 8

8.1   John E. Gibson and Franz B. Tuteur, *Control System Components*, McGraw-Hill, 1958.

8.2   Sidney A. Davis and Byron K. Ledgerwood, *Electromechanical Components for Servomechanisms*, McGraw-Hill, 1961.

8.3   A. Sobczyk, "Stabilization of Carrier-Frequency Servomechanisms", *Journal of Franklin Institute*, July, Aug., Sept. 1948.

8.4   G. Bjornson (Biernson); "Network Synthesis by Graphical Methods for AC Servomechanisms", *AIEE Transactions*, vol. 70, pt II, 1951.

CHAPTER 9

9.1   J. R. Ragazzini and G. F. Franklin, *Sampled Data Control Systems*, McGraw-Hill, 1958.

9.2   A. V. Oppenheim, A. S. Willsky, and I. T. Young, *Signals and Systems*, Prentice-Hall, 1983 (p. 527, Fig 8.15).

9.3   G. W. Johnson, D. P. Lindorff, and G. A. Nordling, "Extension of Continuous Data Systems Design Techniques to Sample-Data Control Systems", *Trans. AIEE*, vol. 74, pt II, pp 252–263, Sept. 1955.

9.4   D. P. Lindorff, *Theory of Sample Data Control Systems*, Wiley, New York, 1965.

9.5   K. P. Phillips, "Current-source Inverter for AC Motor Drives", *IEEE Trans. Industry Applications*, vol. IA-8, Nov./Dec. 1972, pp 679–683.

9.6   E. D. Rainville and P. E. Bedient, *Elementary Differential Equations*, Macmillan, Fifth Ed., 1974 (pp 389–392).

9.7   W. K. Linvill, "Sample-Data Control Systems Studied through Comparison of Sampling with Amplitude Modulation", *Trans. AIEE*, vol. 70, pt II, 1951, pp 1779–1788.

9.8   J. R. Ragazzini and L. H. Zadeh, "The Analysis of Sampled-Data Systems", *Trans. AIEE*, vol. 71, pt II, 1952, pp 225–234.

9.9   H. Hurewicz, "Filters and Servo Systems with Pulsed Data", in Ref 1.2, Chapter 5.

9.10   Julius T. Tou, *Digital and Sample-Data Control Systems*, McGraw-Hill, 1959.

CHAPTER 10

10.1   B. J. Lazan, "Energy Dissipation Mechanisms in Structures with Particular Reference to Materials Damping", in *Structural Damping*, Jerome E. Ruzicka, editor, ASME, 1959.

10.2   Colin J. Smithells, *Metals Reference Book*, Fifth Ed., Butterworths, London and New York 1976 (pp 980, 981, Table 7).

10.3   G. A. Biernson and R. E. Pike, "A Study of Structural Limitations on Servo Performance", Final Report on Contract DA28-043-AMC-00023(E), 1 April 1964 to 1 March 1965 (at Applied Research Laboratory, Sylvania Electronic

Systems, Waltham, Mass), Defense Documentation Center, Attn: TSIA, Cameron Station, Alexandria, Va.

10.4 "Advent Communication Satellite Ground-Station Antenna System, Final Report, vol. III, Compliance Report", Sylvania Electronic Systems, Waltham, Mass.; submitted to U.S. Army Signal Supply Agency, Ft. Monmouth, N.J., 1 Aug. 1962.

CHAPTER 11

11.1 William Feller, *An Introduction to Probability Theory and its Applications*, vol. 1, Wiley, second Ed., 1957 (p 166).

11.2 Al Ryan and Tim Scranton, "DC Amplifier Noise Revisited", *Analog Dialog*, vol. 18, No. 1, 1984 (Analog Devices, Norwood, Mass.).

11.3 J. G. Truxal, *Automatic Feedback Control System Synthesis*, McGraw-Hill, 1955.

11.4 R. S. Phillips, "Statistical Properties of Time-Varying Data", and "RMS Error Criterion in Servomechanisms Design", in Ref [1.2] (Chapters 6, 7).

11.5 G. C. Newton, L. A. Gould, and J. F. Kaiser, *Analytical Design of Linear Feedback Controls*, Wiley, 1957.

11.6 G. Biernson, "Fundamental Equations for the Application of Statistical Techniques to Feedback Control Systems", *IRE Trans. Automatic Control*, vol. PGAC-2, 1957, pp 56–78.

11.7 James G. Titus, "Wind-Induced Torques Measured on a Large Antenna", U.S. Naval Research Laboratory Report 5549, 27 Dec. 1960.

11.8 Robert S. Briggs, Jr., "Control System for a Large Steerable Tracking Antenna", MSEE Thesis, University of Texas, Jan. 1967.

CHAPTER 12

12.1 G. A. Biernson, "How the Bandwidth of a Servo Affects its Saturated Response", *IRE Trans. Automatic Control*, vol. AC-4, 1958, pp 3–14.

12.2 P. Travers, "Motor Saturation in Servomechanisms", MSEE Thesis, Mass. Inst. of Tech., Sept. 1948.

12.3 W. Gibbs, *Collected Works*, Longmans, Green, New York, 1928.

12.4 John E. Gibson, *Nonlinear Automatic Control*, McGraw-Hill, 1963.

12.5 K. Ogata, *Modern Control Engineering*, Prentice-Hall, 1970.

12.6 John E. Gibson and Franz B. Tuteur, *Control System Components*, McGraw-Hill, 1958.

12.7 R. Kochenburger, "Frequency Response Method for Analysis of a Relay Servomechanism", *Trans. AIEE*, vol. 69, 1950, pp 270–283.

12.8 Eugene M. Grabbe, Simon Ramo, and Dean E. Wooldridge, *Handbook of Automatic Computation and Control*, vol. 1, "Control Fundamentals", Wiley, 1958.

12.9 W. M. Gaines, "Nonlinear Systems", in Ref [12.8] (Chapter 25).

12.10 R. Sridhar, "A General Method for Deriving the Describing Functions for a Certain Class of Nonlinearities", *IRE Trans. Automatic Control*, vol. AC-5, 1060, pp 135–141.

12.11 E. C. Johnson, "Sinusoidal Analysis of Feedback Control Systems Containing Nonlinear Elements", *Trans. AIEE*, pt II, vol. 71, 1952, pp 169–181.

12.12 N. B. Nichols, "Backlash in a Velocity Lag Servomechanism", *Trans. AIEE*, pt II, vol. 73, 1954, pp 462–467.

12.13 D. E. Gray, Editor, *American Institute of Physics Handbook*, Second Ed., McGraw-Hill, 1963.

CHAPTER 13

13.1 W. E. Sollecito and S. G. Reque, "Stability", in Ref [12.8] (Chapter 21).

13.2 B. C. Kuo, *Analysis and Synthesis of Sampled-Data Control Systems*, Prentice-Hall, 1963.

13.3 N. L. Kusters and W. J. M. Moore, "A Generalization of the Frequency Response Method for Study of Feedback Control Systems", in *Automatic and Manual Control*, by A. Tustin, Butterworth, London, 1952.

13.4 G. Biernson, "A General Technique for Approximating Transient Response from Frequency Response Asymptotes", *AIEE Transactions*, vol. 75, pt II, 1956, pp 245–273.

CHAPTER 14

14.1 G. Biernson, "A Simple Method for Calculating the Time Response of a System to an Arbitrary Input", *AIEE Transactions*, vol. 74, pt II, 1955, pp 227–245.

14.2 J. L. Bower, "A Note on the Error Coefficients of a Servomechanism", *Journal of Applied Physics*, vol. 21, July 1950, p 723.

14.3 E. Arthurs and L. H. Martin, "A Closed Expansion of the Convolution Integral (A Generalization of Servomechanisms Error Coefficients)", *Journal of Applied Physics*, vol. 26, Jan. 1955, pp 58–60.

CHAPTER 15

15.1 "Hewlett-Packard Application Note 197-2", Hewlett-Packard, 1501 Page Mill Road, Palo Alto, CA 94304.

References to Volume One

CHAPTER 1

1.1   A. C. Hall, *The Analysis and Synthesis of Linear Servomechanisms*, Technology Press, Cambridge, Mass., 1943.

1.2   H. M. James, N. B. Nichols, and R. S. Phillips, *Theory of Servomechanisms*, McGraw-Hill, 1947.

1.3   G. S. Brown and D. P. Campbell, *Principles of Servomechanisms*, Wiley, 1948.

1.4   H. Chestnut and R. W. Mayer, *Servomechanism and Regulating System Design*, Vol. 1, Wiley, 1950.

1.5   J-C Gille, M. J. Pelegrin, and P. Decaulne, *Feedback Control Systems*, *Analysis and Synthesis*, McGraw-Hill, 1959.

1.6   J. L. Bower and P. M. Schultheiss, *Introduction to the Design of Servomechanisms*, Wiley, 1958.

1.7   S. J. Mason, "Feedback Theory: Some Properties of Signal-Flow Graphs", *Proc. IRE*, vol. 41, 1953, pp 1144–1156.

1.8   S. J. Mason, "Feedback Theory: Further Properties of Signal-Flow Graphs", *Proc. IRE*, vol. 44, 1956, pp 920–926.

1.9   S. J. Mason and H. J. Zimmerman, *Electronic Circuits, Signals, and Systems*, Wiley, 1960.

1.10   H. Harris, Jr., M. J. Kirby, and E. F. Von Arx, "Servomechanisms Transient Performance from Decibel–Log Frequency Response", *AIEE Transactions*, vol. 70, pt II, 1951, pp 1452–1459.

1.11   G. A. Biernson, "Quick Methods for Evaluating the Closed-Loop Poles of Feedback Control Systems", *AIEE Transactions*, vol. 72, pt II, pp 53–70, May 1953 (with discussions by N. L. Kusters and W. J. M. Moore, and by Maurice J. Kirby and D. C. Beaumariage).

1.12   Robert E. Graham, "Linear Servo Theory", *Bell. Syst. Technical Journal*, Oct. 1946, p. 616.

1.13   W. R. Evans, *Control System Dynamics*, McGraw-Hill, 1954.

1.14   G. A. Biernson, *Optimal Radar Tracking Systems* (in print).

1.15   G. A. Biernson, "A Feedback Control Model of Human Vision", *IEEE Proceedings*, vol. 54, No. 6, pp 858–872, June 1966.

CHAPTER 2

2.1   M. F. Gardner and J. C. Barnes, *Transients in Linear Systems*, Wiley, 1942.

2.2   Robert B. Angus, *Advanced Circuits*, 1983, Bowen Publishing Co., P.O. Box 270, Bedford, MA 01730-0270.

2.3   "American Standard Terminology for Automatic Control", American Standards Association Document ASA C85.1-1963, American Society of Mechanical Engineers, New York, 1963.

2.4   E. I. Green, "The Decilog, a Unit for Logarithmic Measurement", *Electrical Engineering*, vol. 73, No. 7, pp 597–599, July 1954.

CHAPTER 3

3.1   H. W. Bode, *Network Analysis and Feedback Amplifier Design*, Van Nostrand, Princeton, N.J., 1945.

3.2   P. Travers, "A Note on the Design of Conditionally Stable Feedback Systems", *AIEE Transactions*, vol. 70, pp 626–630, 1951.

CHAPTER 5

5.1   D. Graham and R. C. Lathrop, "The Synthesis of Optimum Response: Criteria and Standard Forms", *AIEE Transactions*, pt II, vol. 72, 1953, pp 273–288.

CHAPTER 6

6.1   G. A. Biernson, "Relation between Structural Compliance and Allowable Friction in a Servomechanism", *IEEE Trans. Automatic Control*, vol. AC-10, Jan. 1969, pp 59–66.

Problems

The following problems are numbered in terms of the section or chapter of the book to which they apply.

CHAPTER 8

8.2-1. A bandpass filter operating at a carrier frequency of 400 Hz has the following equivalent signal-frequency transfer function:

$$H = \frac{1}{1 + s/\omega_f}$$

The break frequency is $\omega_f/2\pi = 25$ Hz. Determine the parameters for implementing this filter with the *RLC* circuit of Fig P8.2-1. The inductor L has an inductance of 0.10 H and a Q of 40 at 400 Hz. The resistor R_p is the equivalent parallel circuit resistance corresponding to this Q, which is

$$R_p = Q\omega_r L$$

where $\omega_r/2\pi = 400$ Hz. Find the capacitance C and resistance R_1. What is the gain E_o/E_i at the carrier frequency?

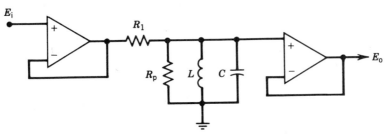

Figure P8.2-1   Circuit of RLC bandpass filter.

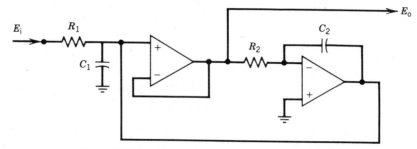

Figure P8.2-2 Circuit of feedback RC bandpass filter.

8.2-2. Assume that the bandpass filter in Problem 8.2-1 is implemented by the feedback *RC* circuit of Fig P8.2-2. The capacitance values are $C_1 = 1$ μF, $C_2 = 0.01$ μF. Find the values for resistors R_1 and R_2. What is the gain E_o/E_i at the carrier frequency.

8.2-3. A bandpass filter operating at a carrier frequency of 400 Hz has the following equivalent signal-frequency transfer function:

$$H = \frac{1}{1 + 2\zeta(s/\omega_n) + (s/\omega_n)^2}$$

where $\omega_n/2\pi = 50$ Hz and $\zeta = 0.707 = 1\sqrt{2}$. Assume that two circuits of the form of Fig P8.2-2 are cascaded to implement this transfer function, where $C_1 = 1$ μF, $C_2 = 0.01$ μF. Find the values for resistors R_1 and R_2 for both circuits. What is the total gain E_o/E_i for the two circuits at the carrier frequency?

8.5-1. As a simplified radar-system model, assume that the IF amplifier of a radar receiver operates at a carrier frequency of 60 MHz and contains the simple bandpass filter stage shown in Fig P8.5-1. The filter input signal is the current source I_i, and the output signal is the voltage E_o. The circuit parameters are $L = 1.0$ μH, $C = 7.036$ pF, and $R = 2.5$ kΩ. This is a bandpass filter tuned to 60 MHz. The input to the IF amplifier is a 60-MHz pulse with a rectangular pulse modulation, of pulse width 0.1 μsec. Plot the envelope of the 60-MHz wave at the output of the filter.

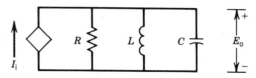

Figure P8.5-1 Circuit of tuned RLC IF amplifier stage.

8.5-2. Assume a 60-MHz amplifier, as described in Problem 8.5-1, which uses a two-stage stagger-tuned filter, where each stage has the form of Fig 8.5-1. The equivalent signal-frequency transfer function of this band-pass filter has the same form shown in Problem 8.2-3, where $\omega_n/2\pi = 6.37$ MHz, $\zeta = 0.5$. Find the values of R and C for the two stages, assuming $L = 1.0\ \mu$H. Plot the envelope of the 60-MHz IF signal at the output of the filter, assuming the input signal is the same as for Problem 8.5-1.

CHAPTER 9

9.2-1. Calculate the sampled-data algorithm to simulate the following lead-network transfer function:

$$\frac{Y}{X} = \frac{1 + s/\omega_L}{1 + s/\alpha\omega_L}$$

To simplify the computations, define the factor $2/\omega_L T$ as B. Give the numerical values of the algorithm constants for $T = 0.0001$ sec, $\alpha = 10$, $\omega_L = 200$ sec^{-1}.

9.2-2. Assume that the stage positioning servo of Fig 9.2-1 in Chapter 9 has the following parameters, expressed in sec^{-1}:

$$\omega_e = 800, \quad \omega_{cc} = 500, \quad \omega_{cv} = 270, \quad \omega_{cp} = 80, \quad \omega_{ip} = 16$$

Note that $\tau_e = 1/\omega_e$. Use serial simulation to simulate the response to a unit step of command position x_k. Plot the controlled position, velocity, and acceleration. Ignore the effect of back EMF of the motor. Use a sample period T of 0.5×10^{-3} sec, and plot for 200 ms.

9.2-3. Using the servo of Problem 9.2-2, plot the error response x_e to the following normalized ramp input: $x_k = \omega_{cp}t$. The velocity ω_{cp} of the input ramp is chosen so that the peak ramp error is approximately unity. Perform this simulation for the following two values of the integral-network break frequency: $\omega_{ip} = 16$ sec^{-1}, 10 sec^{-1}.

9.2-4. Using the approximation technique presented in Chapter 5, Section 5.3, plot approximate ramp error responses for the two cases considered in Problem 9.2-3. Plot to the same scales as the plots obtained in Problem 9.2-3, so that the approximate results can be compared with the exact results obtained from the simulation.

9.2-5. For the servo described in Problem 9.2-2, set the following limits on maximum velocity and maximum acceleration:

$$\text{Max}[V] = 3 \text{ in/sec}, \quad \text{Max}[A] = 80 \text{ in./sec}^2$$

Also limit the integral-network correction signal so it does not exceed

Max[V]/β. Simulate the feedback response of the position loop to step inputs of 0.01 inch and 0.20 inch, for $\beta = 4$. Repeat the response to the 0.20-inch step, with $\beta = 1$. Plot position; also velocity and acceleration for 0.20 inch step ($\beta = 4$)

9.4-1. For the ELF transmitter circuit of Fig 9.4-2 in Chapter 9, approximate the source current I_s with the following waveform: $I_s = I_{s0}\sin[\omega_o t]$. The waveform starts at time $t = 0$; I_s is zero prior to that point. Assume the frequency $f_o = \omega_o/2\pi = 100$ Hz, $I_{s0} = 100$ A, $R_a = 10$ Ω. The antenna is tuned to the frequency $\omega_o = 2\pi f_o$, and has a Q of 5 at this frequency. The inductor has a Q of 20 at the frequency f_o. Hence the circuit parameters are $L_a = 79.5$ mH, $C_a = 31.8$ μf, $L = 15.9$ mH, $R = 0.5$ Ω, $C_1 = C_2 = 159$ μf. Simulate the circuit, and plot the antenna current for 4 cycles of I_s (to 40 msec). The sample period is $T = 0.1$ msec. Plot points every 0.4 msec. This response approximates the startup transient of the transmitter.

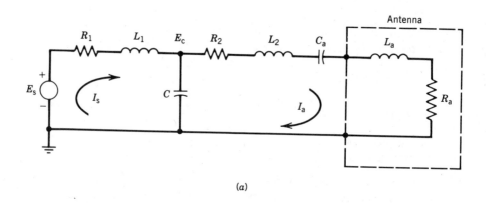

(a)

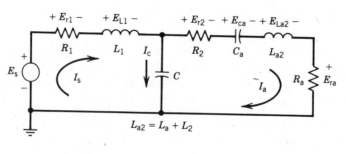

(b)

Figure P9.4-1   ELF transmitter circuit with voltage-source inverter, used for serial simulation: (a) complete circuit; (b) simplified form used in simulation.

9.4-2. Another possible circuit for driving the ELF antenna described in Section 9.4.1 is shown in diagram a of Fig P9.4-1. The circuit uses a thyristor voltage-source inverter, which generates a voltage waveform equivalent to the current waveform I_s of Fig 9.4-1. (The voltage-source thyristor inverter is commonly used at megawatt power levels in unin-terruptable AC power supplies to protect critical facilities, such as computers and broadcast stations, from loss of prime AC power. The 60-Hz prime power is rectified and stored in batteries. A voltage-source inverter converts the DC battery power back to AC power, which drives the critical equipment.) The values of L_1, L_2 in Fig P9.4-1 are the same as L in Problem 9.4-1; R_1, R_2 are the same as R; C is the same as C_1 or C_2; and the values for R_a, L_a, and C_a are the same. To develop the signal-flow diagram of the circuit, it is convenient to combine the series-connected inductors L_a and L_2 into a single inductor L_{a2} equal to $(L_a + L_2)$, as shown in diagram b. Do the following:

(a) Draw a signal-flow diagram for the circuit.

(b) Write the state equations for the circuit.

(c) Express these state equations in matrix format.

(d) Convert the signal-flow diagram for the circuit to the form for serial simulation.

(e) Perform the serial simulation described in Problem 9.4-1, using this circuit, in response to the signal $E_s = E_{s0}\sin[\omega_o t]$ from the inverter, for $E_{s0} = 1000$ V.

CHAPTER 10

10.6-1. Use serial simulation to simulate the stage positioning servo with structural resonance described in Section 10.6-2. Perform the simula-tion tests (a),(b),(c),(d) described in Section 10.6.3.3. Assume the parameters given in Table 10.6-1, with the following changes: $Q = 10$, $\omega_{n2} = 1400$ sec^{-1}. Plot the step response (c) of the current loop to 25 msec, plotting every point. Plot the step response (b) of the velocity loop to 100 msec, plotting every point to $t = 10$ msec, and every other point thereafter. Plot the other transients (a),(d) to 200 msec, using every fourth point.

CHAPTER 11

11.2-1. Find the noise bandwidth in hertz of the following transfer function:

$$H = \frac{1 + s/20}{(1 + s/10)(1 + s/100)}$$

11.3-1. A radar operating at a frequency of 6 GHz has a parabolic steerable antenna with an antenna efficiency of 55%. The radar is tracking a target having a radar cross section of 2 m^2 at a range of 80 km. The transmitter generates 300-kW peak power with rectangular pulses 0.07 μsec in duration. Assume that the total waveguide circuit loss (transmition pluse reception) is 2.5 dB. The noise temperature at the input of the receiver (including noise from the antenna and the atmosphere) is 300°C. The matching loss of the receiver is 0.7 dB. Find the following:

(a) The wavelength.

(b) The peak antenna gain, expressed in decibels and as a power ratio.

(c) The approximate half-power beamwidth. (Assume this is the exact value in the following.)

(d) The beamwidth between -1.5-dB points.

(e) The range resolution of the radar.

In steps (f) to (k), assume that the target is located at the peak of the antenna beam, and find:

(f) The peak power received from the target at the input to the receiver.

(g) The energy received from one pulse at that point.

(h) The noise power density referenced to the input of the receiver.

(i) The signal/noise power ratio for an ideal matched receiver.

(j) The actual signal/noise power ratio at the output of the receiver.

(k) The output signal from the receiver is fed through a peak detector. Find the (approximate) noncoherent detection loss of the detector, and the effective detected signal/noise ratio.

(l) Repeat steps (j) and (k), assuming that the target is 0.5° from the center of the beam.

11.3-2. A radar transmits rectangular pulses 0.15 μsec wide, with a pulse-repetition frequency of 600 Hz. The signal/noise ratio at the output of a radar receiver (prior to the envelope detector) is 2.0 (or 3 dB). The matching loss of the receiver is 1.1 dB. Find the following:

(a) The approximate loss in the noncoherent detector.

(b) The signal/noise ratio E_s/p_n of an ideal matched receiver.

(c) The ratio E_s/p_{nd} of signal energy to detected noise power density.

(d) Assuming the range-gate expression of Eq 11.3-19 in Chapter 11, find the RMS range-gate timing error per pulse, and the RMS range error per pulse.

(e) The noise bandwidth of the IF amplifier, and the gate width τ_g corresponding to the range-gate expression used in part (d).

(f) The noise bandwidth of the range-tracking loop in hertz, assuming that the range-tracking loop has the following loop transfer function:

$$G = \frac{\omega_{cr}(1 + \omega_{ir}/s)}{s}$$

where $\omega_{cr} = 80 \text{ sec}^{-1}$, $\omega_{ir} = 40 \text{ sec}^{-1}$.

(g) The effective integration time of the range-tracking loop.

(h) The effective number of pulses integrated by the range-tracking loop.

(i) The RMS noise tracking error of the range-tracking loop.

(j) The peak (3-sigma) range-tracking error due to noise.

(k) The peak error at lock-on due to target velocity, for a maximum target velocity of 400 m/sec. Compare this with the linear region of the range gate.

(l) The maximum tracking error due to target motion after the lock-on transient has settled, assuming a maximum target acceleration of $4g$ (39.2 m/sec^2).

11.3-3. A monopulse tracking radar transmits 0.12-μsec rectangular pulses at a pulse-repetition frequency of 500 Hz. The signal/noise ratio at the output of the sum-channel receiver, prior to the detector, is 2.5 (or 4 dB). The matching loss of the receiver is 0.9 dB. The normalized monopulse slope k_m is 1.8, and the antenna beam has a 3-dB half-power beamwidth of 1.5°. The monopulse detector has a range-gate width τ_{ga} of 0.2 μsec. Find the following:

(a) The signal/noise ratio E_s/p_n of an ideal receiver.

(b) The RMS angular monopulse error due to noise for a single pulse.

(c) The noise bandwidth of the angle-tracking loop in hertz, and the effective integration time, assuming that the loop transfer function of the angle-tracking loop is approximated by the expression of Eq 11.3-52 in Chapter 11, where $\omega_i = 3$ sec^{-1}, $\omega_c = 9$ sec^{-1}, and $\omega_f = 27$ sec^{-1}.

(d) The effective number of pulses integrated by the angle-tracking loop.

(e) The RMS angle-tracking error per axis due to noise, expressed in milliradians.

(f) The peak (3-sigma) two-axis angle-tracking error due to noise, expressed in milliradians.

(g) The maximum angle-tracking error due to target motion, expressed in milliradians, for a target following a straight-line course at a velocity of 400 m/sec, reaching a minimum range of 500 m.

CHAPTER 12

12.2-1. Use serial simulation to simulate the feedback-control system in Fig. P12.2-1. Blocks (A), (B), and (C) are saturation elements, which limit the maximum absolute values of the outputs from those elements, using the saturation-limit computations described in Chapter 9, Section 9.2.4. Limit the acceleration α to $\pm \alpha_{Max}$; limit Ω_k to $\pm \Omega_{Max}$; and limit Ω_{cor} to $\pm \Omega_{Max}/\beta$. Assume the parameters $\omega_e = 200$ sec^{-1}, $\omega_{cv} = 100$ sec^{-1}, $\omega_{cp} = 30$ sec^{-1}, $\alpha_{Max} = 3$ rad/sec. Simulate responses for a step of command angle Θ_k equal to 0.1 rad, for the following conditions:

(a) $\Omega_{Max} = 0.2$ rad/sec, $\omega_{ip} = 0$.

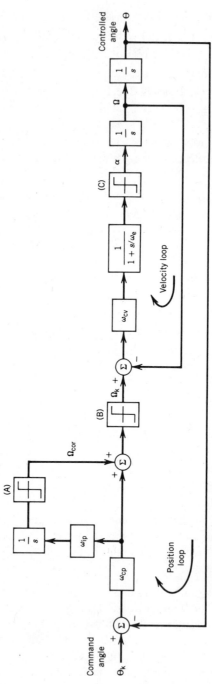

Figure P12.2-1 Simplified mathematical diagram of servo with velocity feedback and integral-network compensation, with saturation limits on acceleration α, command velocity, Ω_k, and velocity correction signal Ω_{cor} stored on integral network.

(b) $\Omega_{Max} = 0.4$ rad/sec, $\omega_{ip} = 0$.

(c) $\Omega_{Max} = 0.6$ rad/sec, $\omega_{ip} = 0$.

(d) $\Omega_{Max} = 0.4$ rad/sec, $\omega_{ip} = 10$ sec^{-1}, $\beta = 4$.

(e) $\Omega_{Max} = 0.4$ rad/sec, $\omega_{ip} = 10$ sec^{-1}, $\beta = 2$.

Compare the responses with the general equations developed in Section 12.2.

12.4-1. Consider a vacuum-chamber pressure-control system using the same simple valve and shaped valve described in Section 12.4. The system parameters are

$$V = 10 \text{ liter} \quad \text{(volume of vacuum chamber)}$$

$$Q_i = 1.0 \text{ torr-liter/sec} \quad \text{(input flow rate)}$$

$$P_{min} = 0.040 \text{ torr} \quad \text{(minimum operating pressure)}$$

$$P_{Max} = 2.5 \text{ torr} \quad \text{(maximum operating pressure)}$$

$$E_{Max} = 10 \text{ V} \quad \text{(saturated output from amplifier)}$$

$$\delta\Theta_m = 9° \quad \text{(step size for stepper motor)}$$

$$\dot{M}_{Max} = 600 \text{ pulse/sec} \quad \text{(maximum stepping rate of motor)}$$

$$N = 120 \quad \text{(gear ratio)}$$

$$K_p = 2.5 \quad \text{V/torr} \quad \text{(sensitivity of pressure transducer)}$$

$$\omega_f = 80 \text{ sec}^{-1} \quad \text{(break frequency of lowpass filter)}$$

$$\alpha_L = 15 \quad \text{(attenuation factor of lead network)}$$

Set the lead-network break frequency ω_L equal to the asymptotic gain crossover frequency (ω_{cc}) of the vacuum-chamber loop, at minimum operating pressure (P_{min}). For control systems using each valve type, set the amplifier gain K_a so that, at maximum operating pressure (P_{Max}), the pressure loop has optimum response in accordance with the criteria of Chapter 3, Section 3.4. Find:

(1) The break frequency ω_L of the lead network.

(2) The value of the gain crossover frequency of the pressure loop at P_{Max}.

(3) The value of Max$|G_{ib}|$ at the pressure P_{Max}.

(4) The values of amplifier gain K_a for control systems using the two throttle valves.

(5) The resolution of the valve angle $\delta\theta_v$ that corresponds to the stepper-motor resolution.

Consider step changes of pressure between the following pressures:

(a) 2.50 to 2.49 torr.

(b) 2.49 to 2.50 torr.

(c) 0.04 to 0.05 torr.

(d) 0.05 to 0.04 torr.

For the simple valve, find the steady-state value of M corresponding to 2.50 torr. Set this as an initial condition, along with the value of the pressure, and run the simulation until steady state is achieved at 2.50 torr. Then perform step (a) followed by step (b). Repeat this for 0.04 torr, to perform steps (c) and (d). Then repeat the complete procedure for the shaped valve. Plot the following variables: pressure P, valve flow Q_o, amplifier voltage E_a, and ideal linear-amplifier voltage E_o. A comparison of E_o and E_a indicates the degree of saturation.

Author Index

This index applies to both volumes. The pages following (1) refer to Volume 1, and those following (2) refer to Volume 2.

Subject Index

This index applies to both volumes. The pages following (1) refer to Volume 1, and those following (2) refer to Volume 2. Pages 1–27 of Volume 1 are in Chapter 1, Introduction, and so are also in Volume 2.